BEE PRODUCTS AND THEIR APPLICATIONS IN THE FOOD AND PHARMACEUTICAL INDUSTRIES

BEE PRODUCTS AND THEIR APPLICATIONS IN THE FOOD AND PHARMACEUTICAL INDUSTRIES

Edited by

DILEK BOYACIOGLU

Retired Professor, Department of Food Engineering,
Istanbul Technical University, Istanbul, Turkey

ELSEVIER

ACADEMIC PRESS

An imprint of Elsevier

Academic Press is an imprint of Elsevier
125 London Wall, London EC2Y 5AS, United Kingdom
525 B Street, Suite 1650, San Diego, CA 92101, United States
50 Hampshire Street, 5th Floor, Cambridge, MA 02139, United States
The Boulevard, Langford Lane, Kidlington, Oxford OX5 1GB, United Kingdom

ISBN: 978-0-323-85400-9

For information on all Academic Press publications
visit our website at https://www.elsevier.com/books-and-journals

Publisher: Nikki P. Levy
Acquisitions Editor: Nina Bandeira
Editorial Project Manager: Andrea Dulberger
Production Project Manager: Bharatwaj Varatharajan
Cover Designer: Mark Rogers

Typeset by STRAIVE, India

Dedication

This book is dedicated to the beekeepers who try hard to save bees and nature, scientists who carry out scientific research to elucidate the power of the products of these phenomenal animals for human health, and honest manufacturers who know the value of these precious products and try to retain their natural properties.

Contents

Contributors

Saboor Ahmad
Institute of Apicultural Research/Key Laboratory of Pollinating Insect Biology, Ministry of Agriculture, Chinese Academy of Agricultural Sciences, Beijing, China

Lin Ang
KM Science Research Division, Korea Institute of Oriental Medicine, Daejeon, Korea

Ofelia Anjos
Polythecnic Institute of Castelo Branco, Castelo Branco; Forest research centre, School of Agriculture (ISA, Instituto Superior de Agronomia), University of Lisbon, Lisbon; Center for Plant Biotechnology of Beira interior, Castelo Branco, Portugal

Elena Azzini
CREA—Research Centre for Food and Nutrition, Via Ardeatina, Rome, Italy

Isabella Letícia Esteves Barros
Postgraduate Program in Health Sciences, Health Sciences Center; Medical Mycology Laboratory, Department of Clinical Analysis and Biomedicine, State University of Maringá, Maringá, PR, Brazil

Dilek Boyacioglu
Freelance Consultant for the Food Industry, Lee's Summit, MO, United States; Scientific Bio Solutions, Inc., Istanbul, Turkey

Marcos Luciano Bruschi
Laboratory of Research and Development of Drug Delivery Systems, Department of Pharmacy, State University of Maringá, Maringá, PR, Brazil

Maria G. Campos
Observatory of Drug-Herb Interactions/Faculty of Pharmacy, University of Coimbra, Heath Sciences Campus; Coimbra Chemistry Centre (CQC, FCT Unit 313) (FCTUC), University of Coimbra, Coimbra, Portugal

Esra Capanoglu
Faculty of Chemical and Metallurgical Engineering, Department of Food Engineering, Istanbul Technical University, Maslak, Istanbul, Turkey

Jakeline Luiz Corrêa
Postgraduate Program in Health Sciences, Health Sciences Center; Medical Mycology Laboratory, Department of Clinical Analysis and Biomedicine, State University of Maringá, Maringá, PR, Brazil

Carlos A. Fuenmayor
Universidad Nacional de Colombia, Sede Bogotá, Instituto de Ciencia y Tecnología de Alimentos, Bogotá, Colombia

Siew Hua Gan
School of Pharmacy, Monash University Malaysia, Jalan Lagoon Selatan, Bandar Sunway, Selangor, Malaysia

Róbert Gáspár
Albert Szent-Györgyi School of Medicine, Department of Pharmacology and Pharmacotherapy, University of Szeged, Szeged, Hungary

Büşra Gültekin Subaşı
Faculty of Chemical and Metallurgical Engineering, Department of Food Engineering, Istanbul Technical University, Maslak, Istanbul; Hafik Kamer Ornek Vocational School, Sivas Cumhuriyet University, Sivas, Turkey

Sibel Karakaya
Faculty of Engineering, Department of Food Engineering, Ege University, Izmir, Turkey

Sevgi Kolayli
Faculty of Science, Department of Chemistry, Karadeniz Technical University, Trabzon, Turkey

Aleksandar Ž. Kostić
Faculty of Agriculture, Chair of Chemistry and Biochemistry, University of Belgrade, Belgrade, Serbia

Anna Kurek-Górecka
Faculty of Pharmaceutical Sciences in Sosnowiec, Department of Community Pharmacy, Medical University of Silesia in Katowice, Sosnowiec, Poland

Hye Won Lee
KM Convergence Research Division, Korea Institute of Oriental Medicine, Daejeon, Korea

Myeong Soo Lee
KM Science Research Division, Korea Institute of Oriental Medicine, Daejeon, Korea

Kashif Maroof
School of Pharmacy, Monash University Malaysia, Jalan Lagoon Selatan, Bandar Sunway, Selangor, Malaysia

Danijel D. Milinčić
Faculty of Agriculture, Chair of Chemistry and Biochemistry, University of Belgrade, Belgrade, Serbia

Mahaneem Mohamed
Department of Physiology; Unit of Integrative Medicine, School of Medical Sciences, Universiti Sains Malaysia, Kubang Kerian, Kelantan, Malaysia

Karsten Münstedt
Department of Gynecology and Obstetrics, Ortenau Klinikum Offenburg-Kehl, Offenburg, Germany

Mamdouh Ibrahim Nassar
Faculty of Science, Cairo University, Cairo, Egypt

Melyssa Negri
Postgraduate Program in Health Sciences, Health Sciences Center; Medical Mycology Laboratory, Department of Clinical Analysis and Biomedicine, State University of Maringá, Maringá, PR, Brazil

Victor Udo Nna
Department of Physiology, Faculty of Basic Medical Sciences, College of Medical Sciences, University of Calabar, Calabar, Cross River State, Nigeria

Paweł Olczyk
Faculty of Pharmaceutical Sciences in Sosnowiec, Department of Community Pharmacy, Medical University of Silesia in Katowice, Sosnowiec, Poland

Zaidatul Akmal Othman
Department of Physiology, School of Medical Sciences, Universiti Sains Malaysia, Kubang Kerian, Kelantan; Unit of Physiology, Faculty of Medicine, Universiti Sultan Zainal Abidin, Kuala Terengganu, Malaysia

Tugba Ozdal
Faculty of Engineering and Natural Sciences, Department of Food Engineering, Istanbul Okan University, Tuzla, Istanbul, Turkey

Beste Özsezen
Faculty of Medicine, Department of Pediatric Pulmonology, Hacettepe University, Ankara, Turkey

Mirjana B. Pešić
Faculty of Agriculture, Chair of Chemistry and Biochemistry, University of Belgrade, Belgrade, Serbia

Huseyin Sahin
Espiye Vocational School, Giresun University, Espiye, Giresun, Turkey

Asli Elif Tanugur Samanci
Scientific Bio Solutions, Inc., Istanbul, Turkey

Taylan Samanci
Scientific Bio Solutions, Inc., Istanbul, Turkey

Adrienn B. Seres
Faculty of Pharmacy, Department of Pharmacodynamics and Biopharmacy, University of Szeged, Szeged, Hungary

Eunhye Song
Global Cooperation Center, Korea Institute of Oriental Medicine, Daejeon, Korea

Joseph Bagi Suleiman
Department of Science Laboratory Technology, Akanu Ibiam Federal Polytechnic, Unwana, Afikpo, Ebonyi State, Nigeria

Terezinha Inez Estivalet Svidzinski
Postgraduate Program in Health Sciences, Health Sciences Center; Medical Mycology Laboratory, Department of Clinical Analysis and Biomedicine, State University of Maringá, Maringá, PR, Brazil

Živoslav Lj. Tešić
Faculty of Chemistry, Chair of Analytical Chemistry, University of Belgrade, Belgrade, Serbia

Flávia Franco Veiga
Postgraduate Program in Health Sciences, Health Sciences Center; Medical Mycology Laboratory, Department of Clinical Analysis and Biomedicine, State University of Maringá, Maringá, PR, Brazil

Zaida Zakaria
Department of Physiology, School of Medical Sciences, Universiti Sains Malaysia, Kubang Kerian, Kelantan, Malaysia

Carlos M. Zuluaga-Dominguez
Universidad Nacional de Colombia, Sede Bogotá, Facultad de Ciencias Agrarias, Departamento de Desarrollo Rural y Agroalimentario, Bogotá, Colombia

Preface

Humans have used bee products for their healthy attributes as medicinal foods for many ages. These products have always attracted scientists who wished to understand their properties in more detail and producers who produced mass amounts and incorporated them into their products. Recently, there has been increasing attention from different industry sectors and scientists based on technological and scientific developments. The food industry is interested in utilizing the functional properties of these products for developing new food products with improved features based on technological developments in the last decade. On the other hand, some bee products, such as royal jelly, propolis, and bee venom, attracted the attention of the food supplement sector due to their biological actions proved by scientific studies. Current research showing their action mechanisms as therapeutics and many applications to protect, manage, and cure various diseases attracted the attention of scientists and R&D specialists in relevant sectors such as pharmacology and medicine. Moreover, bee products have also attracted the attention of the cosmetics industry as they can be used in various applications such as hair products, toothpastes, sunscreen creams, lip balms, and face/hand/body moisturizing creams.

This book evaluates recent scientific research based on the medicinal and pharmacological functions of bee products on human health. The main bee products discussed in this book are raw honey, propolis, royal jelly, bee pollen (bee bread), bee venom, and beeswax. Although honey is now widely known for its health benefits based on many scientific articles, this book intends to fill the gap for a comprehensive approach to include other bee products that have the potential to be used in complementary medicine and drug discovery research.

The chapters focus on the therapeutic value and action mechanisms of various bee products with respect to their bioactive compounds to protect and cure various diseases based on solid scientific evidence. The first three chapters evaluate the overall health benefits of bee products, followed by chapters focusing on a particular health benefit of each product, aiming to provide more compact and detailed information about each activity for a specific interest. The last two chapters present a general approach to evaluate the benefits and risks associated with the consumption of bee products and also present the outlook based on the trending research using computational tools for drug discovery to develop therapeutic agents from bee products, advanced analytical technology in enhancing the knowledge about their components and developing international standards, and factors such as consumer demands, climate change, international trade, harmonization of standards, and safety.

Some additional comments about this book are appropriate in this context. The content preparation, invitations to scientists to contribute to the book, and follow-up communication were somewhat disrupted because of the COVID-19 pandemic. Under these circumstances that were beyond the control of the editor, it was necessary to reduce the number of chapters and find new authors when most of the chapters had been submitted. Therefore, the completion of this book was considerably delayed. However, ironically, as the science world is trying to cure COVID-19 patients worldwide, bee products, particularly propolis, have attracted the attention of many scientists and raised expectations against SARS CoV-2 in drug delivery research, confirming the correct timing of this book.

This book was made possible by the time spent and intense efforts taken by the authors under challenging conditions. The editor owes an immense debt of gratitude to the authors representing 14 different countries who dedicated their valuable time researching these precious bee products and for sharing their knowledge despite the unexpected difficulties experienced during the pandemic. The editor thanks all the staff at Elsevier for their diligent work in helping to transform the manuscript into a book form.

CHAPTER 1

Bee products—An overview of their pharmacological properties and medicinal applications

Karsten Münstedt
Department of Gynecology and Obstetrics, Ortenau Klinikum Offenburg-Kehl, Offenburg, Germany

Contents

Introduction

Honey, propolis, and other crude drugs have been used since prehistoric times for the preparation of substances from natural sources. There is definite evidence that honeybees were exploited in Neolithic times (Roffet-Salque et al., 2015), although it is not clear whether bee products were primarily considered to be food or remedy. But the medical use of beeswax as a dental filling was confirmed on a Neolithic human tooth (Bernardini et al., 2012). Reports on the medical uses of other bee products date back

Bee Products and Their Applications in the Food and Pharmaceutical Industries
https://doi.org/10.1016/B978-0-323-85400-9.00005-8

thousands of years to ancient Egypt, Greece, and China. Healing properties and nutritional benefits are mentioned in many religious texts, including the Veda, Bible, and Qur'an.

Early pharmacology focused on herbalism and natural substances, mainly plant extracts but also honey and other bee products as crude drugs. The problem with crude drugs is that they are adulterated with substances other than the active ingredients. To understand how therapeutic drugs and poisons produced their effects, the first pharmacology department was set up by Rudolf Buchheim in 1847 (Schmiedeberg, 1912). To do this, pharmacologists separated inert or toxic substances from active ingredients, which showed biological effects, and tried to elucidate the cellular and organismal function in relation to these chemicals. Some of these substances may be developed further to make medicinal drugs with one or more active substances. This principle was widely applied in conventional medicine, which uses highly standardized drugs with clearly defined substances with many known interactions. The development of pharmacology and conventional medicine achievements have significantly contributed to the continually improving life expectancy. The development of life expectancy between 1871 and 2019 is presented in Fig. 1.1. Apart from much better surgical techniques, anticoagulants, antibiotics, vaccines, and cancer drugs can be considered key elements for this development and must be attributed to modern pharmacology and conventional medicine.

Complementary and alternative medicine

Despite the advances in modern medicine, many people have become dissatisfied with conventional medicine and have turned to complementary and alternative medicine (CAM). Although this is neither the time nor place to delve into the whys and wherefores behind the change, it should be mentioned that the terms "alternative" and "complementary" are often used interchangeably, even though they refer to different concepts:

- "Complementary" use means a nonmainstream practice used **together with** conventional medicine, while
- "Alternative" use means the use of a nonmainstream practice **in place of** conventional medicine (National Center for Complementary and Integrative Health, 2018).

There is growing evidence for several methods from this field. Contrary to what is often communicated by many protagonists from the field of CAM, conventional medicine is a dynamic field that integrates new

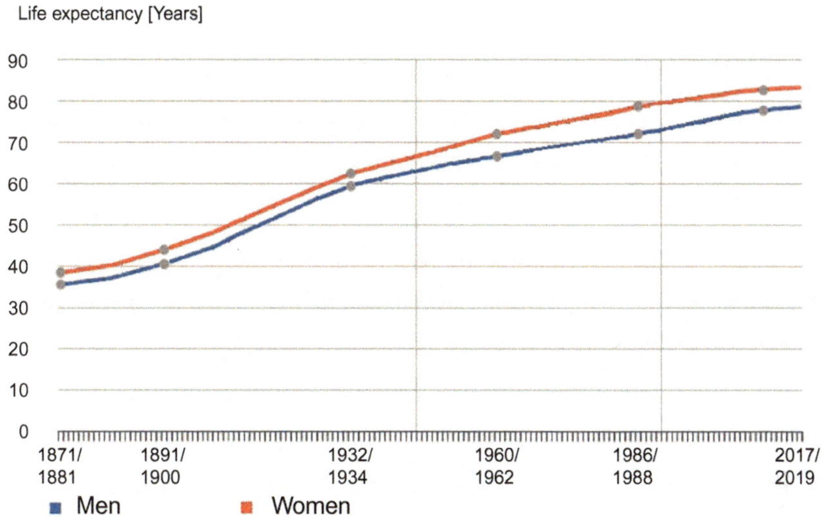

Fig. 1.1 Development of life expectancy in Germany between 1871 and 2019. *(Graphic modified from Federal Statistical Office, 2019. Deaths, Life Expectancy. Available from: https://www.destatis.de/EN/Themes/Society-Environment/Population/Deaths-Life-Expectancy/_node.html. [10 April 2021].)*

methods once they have proven their value but also abandons methods when superior treatments have been found or where severe adverse effects have been discovered.

There are few comparisons between CAM and conventional medicine, but there are some in the field of oncology. The use of alternative medicine (therapies replacing conventional medicine) resulted in poorer survival (Johnson et al., 2018a). But complementary medicine (therapies in addition to conventional medicine) was shown to be a problem too if patients using complementary medicine refused surgery, chemotherapy, radiotherapy, and/or hormone therapy. In such cases, the 5-year overall survival was poorer (Johnson et al., 2018b).

Since alternative and complementary practitioners, in general, spend more time on their patients, their dissatisfaction increases with conventional medicine. The fact that conventional physicians cannot spend a lot of time on their patients is due to the enormous economic pressure under which they operate. As a result, the use of methods from CAM has increased. Apitherapy should be considered as part of the field of complementary and alternative medicine. However, apitherapy must be approached differently because there two differing fields:

(1) Holistic apitherapy—here recommendations for treatments with bee products are based on personal opinions, eventually personal experiences.

(2) Rational apitherapy—it evaluates bee products for the potential use of bee products according to evidence-based medicine concepts.

Very often, there is no differentiation between the two forms of apitherapy. However, some authors refer to rational apitherapy as apimedicine to achieve a better distinction between the two. In this respect, rational apitherapy or apimedicine also must prove its value. Here, investigations considering pharmacological principles and empirical methods could be an option to prove the value of apitherapy.

Bee product pharmacology

Many products are derived from the beehive. They include classical bee products such as honey, propolis, royal jelly, pollen, bee venom, and beeswax. More recently, various "new" bee products like bee podmore, apilarnil, or beehive air have been introduced. Other types of bee medicine have a mysterious character like "sleep on a beehive" and are only recommended by holistic apitherapists. In this case, bees are believed to sense bad vibrations in the human body and produce vibrations supposed to antagonize them.

Bee products may be grouped into six classes:

(1) Natural products collected by bees and, largely/generally, unmodified by bees: pollen, perga

(2) Natural products collected by bees and modified by bees: honey, propolis

(3) Products produced by bees: royal jelly, bee venom, beeswax

(4) Products produced by humans from bees: podmore, apilarnil

(5) Esoteric "bee products": sleep on a beehive

(6) Others: beehive air

There has been an intensive investigation, although it has mainly been concentrated on the classical bee products. Based on this research, several pharmacological properties have been identified. These are summarized in Table 1.1.

There are numerous publications on the various components of bee products from all over the world. The publications show enormous differences between them, revealing that honey from one part of the world is significantly different from honey from elsewhere. The problems become even more pronounced when the honey from different bee species is compared. Honey from the Western honeybee (*Apis mellifera*) differs significantly from

Table 1.1 Biological and pharmacological properties of bee products as tested in cell and animal experiments.

Product	Biological and pharmacological properties
Honey	Antibacterial, antifungal, antiviral, antioxidative, probiotic, antiinflammatory, anticarcinogenic
Pollen	Antibacterial, antifungal, antioxidative, immunomodulating, radioprotective, against anemia, atherosclerosis, diarrhea, and osteoporosis
Royal jelly	Antibacterial, antifungal, antiviral, antioxidative, bio stimulating, immunomodulating, radioprotective, anticarcinogenic, antifatigue, antistress, antihypoxia, antitumor, antiinflammatory, antihypertensive, antihypotensive, vasodilative, antiatherosclerosis, cardio-protective, tranquilizing, protects the central nervous system, antiosteoporosis
Propolis	Antibacterial, antifungal, antiviral, antioxidative, antiparasitic, immunomodulating, antiinflammatory, anticarcinogenic, local anesthetic, pain soothing, spasmolytic, hepatodetoxiquant (*hepatic detoxifier*) and hepatoprotecting, anticoagulant, improving blood circulation
Bee venom	Antibacterial, antiinflammatory, immunoactivating, immunosuppressive, analgesic, radioprotective, anticarcinogenic, accelerates heartbeat, increases blood circulation, lowers blood pressure, improves hemoglobin synthesis, anticoagulant, lowers cholesterol levels, membrane effects on blood cells, influences immuno-active blood cells and hormone levels, antiarrhythmic, heart stimulating therapeutic effects, improvement in hypertonia and atherosclerosis

Modified from Münstedt, K., Bogdanov, S., 2009. Bee products and their potential use in modern medicine. J. ApiProduct ApiMedical Sci. 1(3), 57–63. https://doi.org/10.3896/IBRA.4.01.3.01.

honey from stingless bee species, which again differ from each other. Thus, with respect to pharmacology, bee products create many problems. They can be summarized as;

- Many substances in the products have yet to be identified (especially in propolis)
- Some substances in bee products, honey, or propolis may have adverse effects
- There is insufficient knowledge about most of the components of bee products, even on the ones already identified
- Differences in the composition of the raw products because of
  - ➢ seasonal variations (different flowering periods of different plants)

- ➢ regional divergences (different plants in different regions, temperature, and humidity may have a substantial impact on plant metabolite production)
- ➢ differences between bee species (conventional honeybees vs stingless honeybees)

- Un- or nonstandardized extraction methods in case of propolis extracts (aqueous extracts vs ethanolic extracts, the concentration of alcohol)
- Unstandardized production methods in case of royal jelly (potential influence of feeding)
- Unstandardized processing methods (e.g., honey—heating, filtration)
- Limited regulations regarding standardization and characterization procedures

As a consequence, little is known about the important processes in pharmacology which are pharmacokinetics (influence of the organism on the given substance), pharmacodynamics (influence of the given substance on the organism), and pharmacogenetics (influence of the genetic variation between patients on pharmacological effects) concerning the bee products. With respect to choosing the right dose and the dosing schedule, the information on these aspects is particularly important. Otherwise, overdosage and potentially adverse effects are likely to occur. These concerns underline the importance of pharmacology and the need to understand the working mechanisms of pharmacologically active multicomponent mixtures.

The main problem with using bee products in medicine is that they must be safe, effective, and of appropriate quality. Because bee products are multicomponent mixtures, the standardization thereof is more complicated. Apart from the problems mentioned above, regulations on medicinal products vary from country to country. For all these reasons, it is difficult for consumers to discriminate between high- and low-quality products. Due to their complex multicomponent nature, the production process of any bee product designated to be used for medical purposes must be strictly monitored, starting with the correct authentication and proof of being untainted by any adulterants and/or contaminants.

If the production includes extraction, the extraction solvent and the conditions need to be optimized to enrich the bioactive constituents in the extract. Appropriate quality assessment measures must be put in place at each production stage. Apart from the traditional quality parameters (water content, hydroxymethylfurfural (HMF), inverted sugars, pollen, and color), the pharmacological quality of bee products can be determined based on marker constituents or analytical fingerprints using a wide range of methods ranging from macroscopic, microscopic and DNA-based authentication methods to spectroscopic methods as well as chromatographic and hyphenated methods which include HPLC (high-performance liquid chromatography), GC-MS (gas chromatography-mass spectrometry), and LC-MS (liquid chromatography-mass spectrometry) (Pferschy-Wenzig and Bauer, 2015).

Many aspects need to be considered when performing pharmacological and toxicological studies. To guarantee valid results, it is essential to consider the ADME-Tox (Absorption, Distribution, Metabolism, Excretion, and Toxicity) characteristics of the bioactive constituents and administration route. Determining the marker constituents which are believed to be involved in the working mechanisms is of great importance. For example, methylglyoxal content is important for the antibacterial action of manuka honey, which consequently determines the quality of the honey (Johnston et al., 2018). Accordingly, the range of essential substances required for their medical uses must be known.

Food fraud is another big problem. Honey is one of the six most common animal products subjected to food fraud (Hong et al., 2017). Economic reasons for this prevail. For example, the price value of different types of honey varies more than 100-fold (Aliaño-González et al., 2019). This relates especially to manuka honey (Beitlich et al., 2016). Honey adulteration is likely to interfere with possible health benefits. A review on the topic even concludes that there might be resultant toxic effects (Fakhlaei et al., 2020).

Another problem is the reporting quality of the clinical trials on the use of varying bee products for humans. It is often meager or poor that impairs the pooling of clinical data for systematic reviews and meta-analyses. All these facts explain why bee products have lost their importance over time. Bee products only play marginal roles in modern Westernized medicine. However, people in developing countries and followers of holistic approaches still use bee products on a larger scale.

However, a good example of the application of pharmacological principles to bee products is sweet bee venom. Sweet bee venom resembles purified bee venom and is mainly composed of melittin, which is considered less allergenic because the main allergenic substances like phospholipase A2 have been removed. As shown, this reduces the problem significantly but not completely (Jo and Roh, 2015).

In summary, it seems almost impossible to assure consistent product quality over time. Marker constituents that are believed to be involved in the working mechanisms are of great importance.

Ethical considerations with the use of bee products

In today's medicine, numerous ways to treat diseases have been developed. In contrast to the imputations of some protagonists of alternative medicine, conventional medicine is a dynamic concept that integrates new methods once they have proven efficacy and weeds out methods with a lower efficacy or those resulting in a poor relationship between efficacy and side effects. This approach is important because it must be understood that the patient deserves the best available treatment and care. To achieve this, modern medical guidelines based on examining current evidence within the paradigm of evidence-based medicine have been issued. Some guidelines may have problems. Because of this, they have been classified in Germany. Here, the S3-guideline has the highest value because it is certified, indicating that the guideline went through all systematic development elements (logic, decision and outcome analysis, evaluation of scientific studies' clinical relevance, and regular reviews) (AWMF, n.d.).

This concept clearly differs from the concepts of protagonists of CAM. Rather than selecting the best possible approach among the plethora of offerings, protagonists of alternative medicine want to treat every disease with one treatment concept, i.e., a homeopath wants to treat exclusively with homeopathy and an apitherapist with apitherapy. However, homeopathic or apitherapeutic treatments are not likely always the best option when considering the variety of offerings from conventional medicine, phytotherapy, and other concepts from complementary and alternative medicine. Therefore, responsible counseling of patients must be done on a rational basis and clearly separated from wishful thinking. Those criteria must be well-thought-out when considering a medical treatment:

Efficacy: It is understood that medical treatment must have proven efficacy. The evidence can be rated based on the Levels of Evidence from the Oxford Centre for Evidence-Based Medicine (Phillips et al., 2009).

Safety: Absence or low prevalence of undesired effects is essential. If a method is associated with substantial side effects, it should be investigated and determined if they can be antagonized employing supportive care.

These two dimensions and the suggested resulting consequences for a physician were published by Cohen and Eisenberg (2002). However, the situation is not complete. Two further dimensions need to be considered which were not mentioned:

Cost: Money always matters. If there are two methods with comparable efficacy and safety, costs should be considered, and the less expensive method should be preferred.

Availability: Various methods from complementary and alternative medicine are closely related to selected practitioners. They are often not available on the open market. Furthermore, many methods depend on practitioners who are not always available for the patients. Methods that allow the patient to use them whenever required should be preferred. For example, acupressure can be easily taught to patients and can be applied whenever needed. In contrast, acupuncture which was not found to be superior is generally applied by practitioners.

Therefore, Fig. 1.2 combines all four dimensions with the respective recommendations regarding use or nonuse. All considerations regarding the use of treatment should thus be four-dimensional. Bee products could be convincing concerning costs and availability, especially when original hive products are used. Furthermore, other methods should also be considered. If there are other therapeutic possibilities, they should be acknowledged too. If their profile regarding efficacy, safety, costs, and availability is superior, the choice for an apitherapeutic method should be rethought.

Therefore, if bee products should be used in modern medicine, their use is only justified if

- bee products were found to be superior to other methods
- bee products had healing properties which established methods did not have

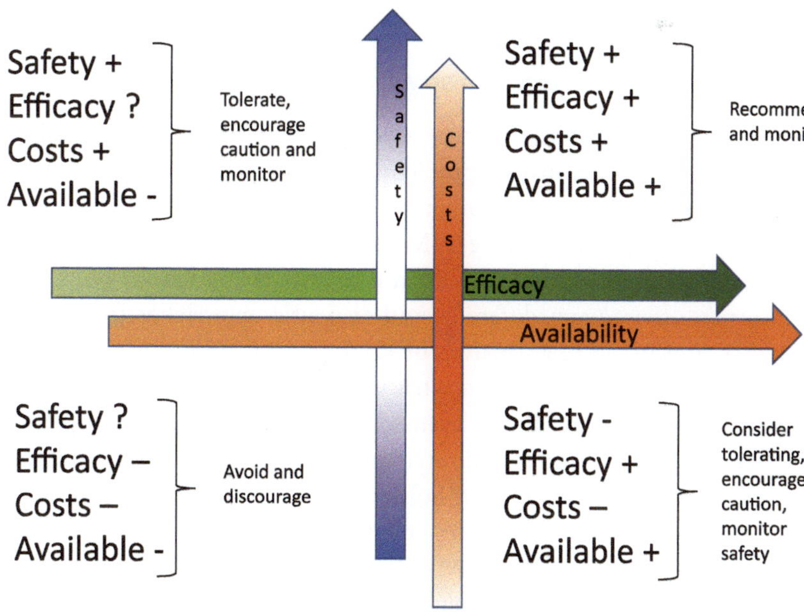

Fig. 1.2 Four-dimensional justification chart for recommending bee products to treat diseases.

- bee products had similar healing properties but fewer side effects than other methods
- the healing properties of the bee product were universal, irrespective of the variances in the composition of the product or if the quality could be assured by marker constituents
- products comparable to bee products could not be produced
- the working principle was not only associated with one substance but with the interactions of various substances present in the bee product.

Notably, there are many high-priced apitherapy preparations that cannot be considered cost-effective at all. Prices also do not reflect effectiveness. The question is whether the price of manuka honey is justified. It has been shown that Slovakian honeydew honey has exceptional antibacterial activity against the multidrug-resistant *S. maltophilia* isolates and was more efficient than manuka honey (UMF 15+) (Majtan et al., 2011).

All aspects mentioned above should be considered before a treatment decision is made. Obviously, the concepts of many apitherapists, who believe that all diseases can and should be treated with bee products exclusively or in combination with other methods from complementary and alternative medicine, do not reflect this rational mode of operation. As mentioned before, this type of medicine is considered holistic apitherapy. The background of treatment decisions in holistic apitherapy remains to be elucidated.

Problems of holistic apitherapy

Several analyses of holistic apitherapy have recently been published. Analyses on dysmenorrhea, menopausal problems, seasonal allergic rhinitis, cancer prevention and treatment, and benign prostatic hyperplasia have shown that apitherapy is not a uniform concept (Münstedt, 2018; Münstedt and Männle, 2020a,b,c,d). Holistic apitherapy, which is mainly promoted via books and the internet, offers many different apitherapeutic approaches for a single problem. For example, in an analysis of 129 books on apitherapy, there were 29 different recommendations on what was thought to be best for seasonal allergic rhinitis treatment. This is even more remarkable since the topic was only mentioned in 50 books (Münstedt and Männle, 2020d). Comparisons between the recommendations of holistic apitherapists and the findings of clinical studies show that only a few recommendations are comprehensive (Münstedt, 2018; Münstedt and Männle, 2020a,b,c,d). It shows that most apitherapy books have very particular concepts that seem to be based on very personal views, possibly experience too, rather than clinical evidence. As a result, holistic apitherapy can only be attributed to low credibility.

As mentioned before, many apitherapists believe that all diseases can and should be treated with bee products (Gupta and Stangaciu, 2014; Zhu and Wongsiri, 2011). This concept was recently examined by our group in the field of gynecology (unpublished data). It shows that gynecological topics are not well represented in the apitherapeutic literature and, although the analysis focused only on the 20 most relevant topics of gynecology, the coverage of these topics in apitherapeutic books is extremely low (average 1.6 topics, maximum 6). Even if all the available information from apitherapeutic books could be gathered, it would not be sufficient because important topics like endometriosis, ovarian cysts, pelvic pain, and vulvodynia are among the most important topics, which are not covered.

Another common misconception in the field of holistic apitherapy is the deduction of a specific effect based on the chemical proof of the presence of a specific component in a bee product. For example, the proof of minerals and vitamins in honey causes holistic apitherapists to suggest honey as an excellent source of minerals and vitamins in deficiency cases. However, such a recommendation is not justified because, in most cases, the quantities needed for efficient supplementation would be in the range of kilograms, which is far beyond the quantity any patient can consume.

Bias in interpretation of scientific data

Another critical point is the soundness of apitherapeutic publications. An analysis on meta-analyses on honey and oral mucositis due to radiotherapy in the combination of with chemotherapy or without found that the quality of these meta-analyses was low. Especially the more recent meta-analyses included trials in which radiotherapy was not part of the medical intervention or where substances other than pure honey were used (Münstedt and Männle, 2020e). Such meta-analyses cannot be considered reliable. It can be assumed that these problems are not detected by common readers because even the reviewers did not detect them. However, such flaws will prevent experts from accepting the concept into clinical guidelines.

Another quite common problem is the overinterpretation of results. A recent systematic review and meta-analysis on the effectiveness of honey for symptomatic relief in upper respiratory tract infections analyzed the improvement of the symptomatic relief with honey (Abuelgasim et al.,

2020). Surprisingly, the review concluded, "*It (honey) provides a widely available and cheap alternative to antibiotics. Honey could help efforts to slow the spread of antimicrobial resistance, but further high quality, placebo-controlled trials are needed.*" If and how far honey could replace antibiotics in upper respiratory tract infections has not yet been investigated. Therefore, it is not scientifically sound to suggest such a conclusion. It also seems too early to speculate on honey's therapeutic and preventive properties and its bioactive compounds in treating cancer. As already mentioned, honey is a reasonable treatment option for oral mucositis, but only preclinical studies support its therapeutic and preventive use. It is likely that some readers may overinterpret titles like "*Anatolian honey is not only sweet but can also protect from breast cancer: …*" (Seyhan et al., 2017) or passages like "*Sidr honey can be used as an antimicrobial agent but can be used as an anticancer agent with care as it stimulated cell growth of some lines (e.g., Hala) and inhibited another (e.g., HepG2)*" (Ghramh et al., 2020). In this respect, authors of apitherapy books and apitherapy papers tend not to differentiate between in vitro and in vivo evidence.

Acceptance of bee products as medicine

On being confronted with a disease and then having a treatment option with honey, one would think that many people may believe that this would be a really sweet option. However, one trial has shown that about one-third of all patients refuse to consume about 30 g of honey (Rajan et al., 2002). Another trial showed similar problems regarding pollen (Münstedt et al., 2015). A survey among patients consulting general physicians and gynecologists showed that other bee products, especially live bee stings, apilarnil, or the inhalation of beehive air, did not appeal to most of the patients at all (Münstedt et al., 2019a). Other reasons for refusing to use honey therapeutically are related to the ways bees are kept and the perception of honey regarding price, health, and handling (Münstedt et al., 2020). It is also known that vegans will refuse the use of bee products because they believe that humans exploit bees. It must be realized that the number of vegetarians and vegans has increased worldwide over the past decades (Sebastiani et al., 2019). Thus, it appears important to know about patients' willingness to tolerate apitherapeutic treatments.

In summary, the taste of some bee products (pollen, royal jelly), the perceived danger of others (allergy to bee venom), the pain associated with the application (bee sting therapy, apipuncture), the quantity of a bee product to be consumed, the sweetness as well as the vegan orientation make it challenging to promote bee products as medicine for everybody which clearly impairs their usefulness.

Possibilities for uses of bee products in medicine

As mentioned before, the advice to use bee products for medical purposes must be based on a rational process. Accordingly, the suggestions here are based on the Levels of Evidence from the Oxford Centre for Evidence-Based Medicine (Phillips et al., 2009). Therefore, the following recommendations are made on using this base and rely on systematic reviews and meta-analyses.

Treatment of various types of wounds

Wound infections

A localized tissue defect invaded by pathogenic organisms triggers the body's immune response, causes inflammation and tissue damage, and slows down the healing process. Symptoms can include increasing pain, swelling, and redness. Some infections, if left untreated, may become more severe and require medical intervention.

Honey has shown activity against antibiotic-resistant bacteria. Several mechanisms of action have been identified in honey: (1) high sugar concentration, (2) hydrogen peroxide, (3) low pH, (4) methylglyoxal, and (5) an antimicrobial peptide bee product called defensin-1. The antibacterial activity of honey is extraordinarily complex due to the involvement of multiple compounds (Kwakman and Zaat, 2012). This multitude of working mechanisms also allows the treatment of multiresistant bacteria (Broznić et al., 2018).

A recent systematic review concluded that honey in acute and chronic wounds provided rapid epithelization and wound contraction in wound healing, had an antiinflammatory and debridement effect, decreased the pain, ensured infection control, shortened the time of wound healing, and was cost-effective (Yilmaz and Aygin, 2020). However, it should be noted that based on in vitro data, it seems that buckwheat, manuka, linden, and forest honey seem to be more suitable for the treatment of wound infections than other types of honey.

Furthermore, it was shown that the treatment of wounds with honey leads to a reduction of the scar width. Thus, honey can improve the healing process of the surgical wound and its final esthetic results (Goharshenasan et al., 2016).

Tonsillectomy

A systematic review and meta-analysis of randomized controlled trials on the role of honey after a tonsillectomy analyzed eight trials that met the inclusion criteria (Lal et al., 2017). Compared to control, honey significantly decreased postoperative pain, consumption of painkillers, and the number of times the patient awakened at night due to pain. It also improved the healing of tonsillar fossa after tonsillectomy with no relevant adverse effects. Unfortunately, the quality of most trials was low (Lal et al., 2017). Two more recent studies on honey confirmed the positive effects on postoperative pain after tonsillectomy (Hatami et al., 2018; Geißler et al., 2020).

Diabetic foot ulcers

There is also a systematic review and meta-analysis on the effectiveness of honey dressing in the treatment of diabetic foot ulcers (Wang et al., 2019). Based on 11 trials, the authors concluded that honey dressing effectively shortened the wound debridement time, wound healing time, and bacterial clearance time. Honey dressings increased the wound healing rate and bacterial clearance rates (Wang et al., 2019).

Radiotherapy and/or chemotherapy-induced oral mucositis

Cancer treatment is associated with numerous adverse effects, including oral mucositis. It occurs during radiotherapy for head and neck cancer. Since radiotherapy is often combined with chemotherapy, the problem of oral mucositis is more pronounced. Patients undergoing chemotherapy suffer from oral mucositis too.

Oral mucositis is currently one of the most intensively researched areas regarding the use of bee products. Many systematic reviews and meta-analyses have been published on the topic; however, as shown in critical analysis on the meta-analyses, the quality of said analyses was exceptionally low (Münstedt and Männle, 2020e). It was found that recent meta-analyses included trials in which substances other than pure honey were used or where radiotherapy was not part of the medical intervention (Münstedt and Männle, 2020e). Generally, though, honey was found to be effective. A comprehensive systematic review outlined that the type of honey may explain the controversial results previously published in this area. Studies using manuka honey failed to show a benefit for honey, while trials using conventional kinds of honey presented evidence for its usefulness (Münstedt et al., 2019b). Perhaps an understanding of these differences between honey kinds led to the recently issued positive recommendation for conventional honey for oral mucositis by the highly respected Mucositis Study Group of the Multinational Association of Supportive Care in Cancer and International Society of Oral Oncology (Yarom et al., 2020).

Interestingly, the use of honey for chemotherapy-associated mucositis was only investigated in pediatric oncology patients. A review found four papers, which provide Grade C evidence that honey is effective as a preventative and therapeutic measure for oral mucositis (Friend et al., 2018). A more recent trial confirmed that honey mouthwash effectively prevents and reduces the severity of mucositis and weight loss. It can be recommended for patients undergoing chemotherapy (Khanjani Pour-Fard-Pachekenari et al., 2019).

There is also some evidence that propolis is another exciting option for both radiotherapy and chemotherapy-induced oral mucositis, although the evidence is not comparable to honey (Münstedt and Männle, 2019).

Burns

A systematic review with meta-analysis analyzed six randomized controlled trials on antibacterial properties and healing times of honey and silver in burns. The analysis showed that honey was more efficacious for wound healing than silver. The number of days needed for the wounds to heal was less with honey, and honey turned out to have better antibacterial qualities than silver (Lindberg et al., 2015; Aziz and Hassan, 2017).

Malignant fumigating wounds

A systematic review on malignant fumigating wounds concluded that odor and exudates were significantly decreased by the use of honey and silver dressings (Tsichlakidou et al., 2019).

Herpes virus-associated skin lesions (HSV-1, HSV-2)

A systematic review comprising three trials on honey and six trials on propolis found evidence that both these two bee products can be considered alternatives to acyclovir. The evidence from these trials suggests that propolis is a good option in treating herpetic skin lesions, especially those related to HSV-1. Honey could be an interesting alternative for herpetic lesions in the oral cavity (Münstedt, 2019; Abdel-Naby Awad and Hamad, 2018). Lately, another trial has been published. It confirmed that the lesions healed faster (complete encrustation or epithelization), there were less typical herpes symptoms (pain, burning and itching, tension and swelling), and no safety problems (Jautová et al., 2019).

Honey for acute coughs in children

Three systematic reviews have analyzed the current evidence regarding honey and acute coughs. Interestingly, all the studies that have been done have researched the effect on children (Pecoraro et al., 2020; Oduwole et al., 2018; Nitsche and Carreño, 2016). All the reviewers concluded that honey relieves cough symptoms to a greater extent than no treatment or a placebo and even diphenhydramine, respectively. There does not seem to be a big difference between honey and dextromethorphan. The cough duration is better with honey if compared to a placebo or salbutamol. However, all reviews concluded that there is a need for more high-quality, large randomized controlled trials confirming the effectiveness and practical application of honey.

Oral health—Dental plaque and gingival inflammation

A systematic review of 9 clinical trials, comprising 333 subjects, showed that propolis mouthwashes showed excellent efficacy on plaque and gingivitis. However, the authors of the systematic review reported that methodological limitations along with small sample sizes in some of the included studies weakened the strength of the evidence. Thus, they recommended further well-designed clinical trials with large sample sizes and adequate follow-up periods (Halboub et al., 2020).

Poststroke shoulder pain

Bee venom could be an interesting option for poststroke shoulder pain. Several randomized clinical trials have found significant beneficial effects on motor function and pain relief. Again, the quantity and quality of the evidence are not yet sufficient to draw definitive conclusions on this topic (Kim and Han, 2020).

Musculoskeletal pain

Bee venom acupuncture involving injecting diluted bee venom into acupoints could be effective in musculoskeletal pain management. A meta-analysis produced suggestive evidence for its usefulness for arthritis, pain, and rheumatoid diseases. However, future trials with larger patient samples for more extended treatment periods and appropriate controls are needed to promote this method in general (Lee et al., 2008).

Bee products must be superior to other approaches

Considering the ethical aspects mentioned before, the use of bee products can only be justified if it surpasses the quality of other approaches. There have not been many investigations that have critically reviewed the usefulness of various treatment approaches from both conventional and complementary and alternative medicine so far. In a review of apitherapy books, only one of the recommendations of 1 author out of 70 books matched the current evidence, which shows that royal jelly and honey can improve primary dysmenorrhea (Münstedt, 2018). But with all due respect to the evidence, royal jelly and honey cannot be the best options. (1) Local heat applications, (2) exercise, (3) aromatherapy with lavender oil or lavender oil massage, (4) vitamin B1, (5) omega-3 fatty acids, (6) acupressure, (7) ginger, and (8) chaste berry must be considered better options if it comes to simplicity, safety, costs, and availability. Honey and royal jelly may be considered the ninth-best when considering CAM methods (Münstedt, 2018). However, the evidence does not always play a role. When considering patients' decisions, they would opt first for pain relievers and then contraceptive pills from conventional medicine, then further down the line, homeopathy, order therapy, local heat applications, and diet from CAM (Münstedt and Riepen, 2019). Honey and royal jelly were among the least appealing methods (Münstedt and Riepen, 2019). This example shows that no aspect qualifies honey and royal jelly as attractive therapeutic options for primary dysmenorrhea. In this respect, it is also not reasonable to suggest bee products unless new studies show that bee products deserve a better evaluation.

From the scientific point of view, bee products can be promoted in five situations indications

(1) Honey for radiotherapy and/or chemotherapy-induced oral mucositis
(2) Honey for wound infections
(3) Honey for burns
(4) Propolis for herpes virus-associated skin lesions
(5) Honey for acute coughs in children

Of course, there are several other potential fields where bee products could be beneficial. Areas where honey or other bee products could be helpful to include:

- the prevention of xerostomia with honey,
- improving sleep quality through milk with honey,

- treating skin diseases with honey,
- treating wounds with propolis,
- treating tooth and gum problems with propolis and honey,
- and treating chronic sinus infections with honey,

to name just the most interesting ones (Hbibi et al., 2020; Kuo et al., 2018). In these areas, further studies need to be carried out to underline the potential importance of such treatment with bee products. In order to present the potential of bee products, the scientific principles should be taken into account so that the efficiency and safety of the method can be explained against potential critics. Although apitherapy maybe thus of minor importance against the background of the wide range of complementary and alternative medicine in the industrialized nations, it could become more critical when considering the background of scarce resources in developing countries. Perhaps due to the lack of other treatment options, apitherapy could gain more importance.

Bringing bee products back into medicine—Discussion and concluding remarks

The requirement applies that patients deserve the best possible treatment. From the rational point of view, conventional medicine seems to be most appropriate. However, a patient's desire for natural treatment options must be respected. Disrespect for conventional medicine is often not justified. Conventional medicine is often perceived to be chemically orientated, which is not necessarily true. Many active substances are derived from natural products but only produced by chemical syntheses since production from natural resources would be too expensive or a threat to nature. For example, the production of paclitaxel, a cancer drug, would require the bark of many Pacific yew trees (*Taxus brevifolia*) to meet the demand of only a few oncological patients. Thus, the substance was produced from precursors in the leaves and twigs of the tree and now through cell culture. Many of conventional medicine's treatments are based on natural substances.

It must be noted that the desire for complementary medicine often hides the patient's wishes to be more involved in the therapy options and to get more understandable information. Doctors should be able to take these wishes more into account (Bonacchi et al., 2015). It will be a future task to give physicians the possibility (and the time) to care for and advise patients as to their needs which cannot be done in the time frames currently provided. However, it must be realized that even considering the justified criticism of conventional medicine, alternative medicine or pseudo-medicine, in particular, is hardly ever a real alternative. In the cases of illnesses with a psychosomatic background, one may achieve something positive for the sick with the help of the placebo effect in one case or another. In more severe diseases, for example, cancer, delays in the use of effective therapy can result in situations that rule out healing or even make them significantly less likely. Those concerns have been proven several times (Johnson et al., 2018a,b).

If apitherapy wants to gain a real status in medicine, a radical reorientation appears to be necessary. Apitherapy should refrain from propagating therapy concepts that are ineffective or harmful (e.g., ear candles with beeswax). Instead, the areas mentioned above should be researched further, and apitherapeutic concepts should only be propagated if proven helpful. In contrast to the other methods of complementary and alternative medicine, apitherapy has rationally comprehensible treatment options. Therefore, it is significantly better than homeopathy, where the mode of action is essentially based on the placebo effect.

Reviewing the data mentioned above shows that bee products can only be recommended for a few indications based on our current knowledge. These include treating various types of wounds, herpes virus-associated skin lesions, and acute coughs in children. Eventually, further research will result in the expansion of this list in the future.

References

Abdel-Naby Awad, O.G., Hamad, A.H., 2018. Honey can help in herpes simplex gingivostomatitis in children: prospective randomized double-blind placebo controlled clinical trial. Am. J. Otolaryngol. 39 (6), 759–763. https://doi.org/10.1016/j.amjoto.2018.09.007.

Abuelgasim, H., Albury, C., Lee, J., 2020. Effectiveness of honey for symptomatic relief in upper respiratory tract infections: a systematic review and meta-analysis. BMJ Evid. Based Med. 26 (2), 57–64. https://doi.org/10.1136/bmjebm-2020-111336.

Aliaño-González, M.J., Ferreiro-González, M., Espada-Bellido, E., Palma, M., Barbero, G.F., 2019. A screening method based on headspace-ion mobility spectrometry to identify adulterated honey. Sensors (Basel) 19 (7), 1621. https://doi.org/10.3390/s19071621.

AWMF, n.d. Ararbeitsgemeinschaft der Wissenschaftliche Medizinische Fach Societies, AWMF Rules Guidelines: Classification According to Systematics. Available from: https://www.awmf.org/leitlinien/awmf-regelwerk/ll-entwicklung/awmf-regelwerk-01-planung-und-organisation/po-stufenklassifikation.html. [8 April 2021].

Aziz, Z., Hassan, B.A.R., 2017. The effects of honey compared to silver sulfadiazine for the treatment of burns: a systematic review of randomized controlled trials. Burns 43 (1), 50–57. https://doi.org/10.1016/j.burns.2016.07.004.

Beitlich, N., Lübken, T., Kaiser, M., Ispiryan, L., Speer, K., 2016. Fluorescent pteridine derivatives as new markers for the characterization of genuine monofloral New Zealand manuka (Leptospermum scoparium) honey. J. Agric. Food Chem. 64 (46), 8886–8891. https://doi.org/10.1021/acs.jafc.6b03984.

Bernardini, F., Tuniz, C., Coppa, A., Mancini, L., Dreossi, D., Eichert, D., Turco, G., Biasotto, M., Terrasi, F., De Cesare, N., Hua, Q., Levchenko, V., 2012. Beeswax as dental filling on a neolithic human tooth. PLoS One 7 (9). https://doi.org/10.1371/journal.pone.0044904, e44904.

Bonacchi, A., Toccafondi, A., Mambrini, A., Cantore, M., Muraca, M.G., Focardi, F., Lippi, D., Miccinesi, G., 2015. Complementary needs behind complementary therapies in cancer patients. Psychooncology 24, 1124–1130. https://doi.org/10.1002/pon.3773.

Broznić, D., Ratkaj, I., Malenica Staver, M., Kraljević Pavelić, S., Žurga, P., Bubalo, D., Gobin, I., 2018. Evaluation of the antioxidant capacity, antimicrobial and antiproliferative potential of fir (Abies alba Mill.) honeydew honey collected from Gorski kotar (Croatia). Food Technol. Biotechnol. 56 (4), 533–545. https://doi.org/10.17113/ftb.56.04.18.5666.

Cohen, M.H., Eisenberg, D.M., 2002. Potential physician malpractice liability associated with complementary and integrative medical therapies. Ann. Intern. Med. 136 (8), 596–603. https://doi.org/10.7326/0003-4819-136-8-200204160-00009.

Fakhlaei, R., Selamat, J., Khatib, A., Razis, A.F.A., Sukor, R., Ahmad, S., Babadi, A.A., 2020. The toxic impact of honey adulteration: a review. Foods 9 (11), 1538. https://doi.org/10.3390/foods9111538.

Friend, A., Rubagumya, F., Cartledge, P., 2018. Global Health Journal Club: is honey effective as a treatment for chemotherapy-induced mucositis in paediatric oncology patients? J. Trop. Pediatr. 64 (2), 162–168. https://doi.org/10.1093/tropej/fmx092.

Geißler, K., Schulze, M., Inhestern, J., Meißner, W., Guntinas-Lichius, O., 2020. The effect of adjuvant oral application of honey in the management of postoperative pain after tonsillectomy in adults: a pilot study. PLoS One 15 (2). https://doi.org/10.1371/journal.pone.0228481, e0228481.

Ghramh, H.A., Ibrahim, E.H., Kilany, M., 2020. Study of anticancer, antimicrobial, immunomodulatory, and silver nanoparticles production by Sidr honey from three different sources. Food Sci. Nutr. 8 (1), 445–455. https://doi.org/10.1002/fsn3.1328.

Goharshenasan, P., Amini, S., Atria, A., Abtahi, H., Khorasani, G., 2016. Topical application of honey on surgical wounds: a randomized clinical trial. Forsch. Komplementmed. 23 (1), 12–15. https://doi.org/10.1159/000441994.

Gupta, R.K., Stangaciu, S., 2014. Apitherapy: holistic healing through the honeybee and bee products in countries with poor healthcare system. In: Gupta, R.K., Reybroeck, P.R., van Veen, J., Gupta, A. (Eds.), Beekeeping for Poverty Alleviation and Livelihood Security. Springer, pp. 413–446, https://doi.org/10.1007/978-94-017-9199-1_15. 2014.

Halboub, E., Al-Maweri, S.A., Al-Wesabi, M., Al-Kamel, A., Shamala, A., Al-Sharani, A., Koppolu, P., 2020. Efficacy of propolis-based mouthwashes on dental plaque and gingival inflammation: a systematic review. BMC Oral Health 20 (1), 198. https://doi.org/10.1186/s12903-020-01185-5.

Hatami, M., Mirjalili, M., Ayatollahi, V., Vaziribozorg, S., Zand, V., 2018. Comparing the efficacy of peritonsillar injection of tramadol with honey in controlling post-tonsillectomy pain in adults. J. Craniomaxillofac. Surg. 29 (4), e384–e387. https://doi.org/10.1097/SCS.0000000000004393.

Hbibi, A., Sikkou, K., Khedid, K., El Hamzaoui, S., Bouziane, A., Benazza, D., 2020. Antimicrobial activity of honey in periodontal disease: a systematic review. J. Antimicrob. Chemother. 75 (4), 807–826. https://doi.org/10.1093/jac/dkz527.

Hong, E., Lee, S.Y., Jeong, J.Y., Park, J.M., Kim, B.H., Kwon, K., Chun, H.S., 2017. Modern analytical methods for the detection of food fraud and adulteration by food category. J. Sci. Food Agric. 97 (12), 3877–3896. https://doi.org/10.1002/jsfa.8364.

Jautová, J., Zelenková, H., Drotarová, K., Nejdková, A., Grünwaldová, B., Hladiková, M., 2019. Lip creams with propolis special extract GH 2002 0.5% versus aciclovir 5.0% for herpes labialis (vesicular stage): randomized, controlled double-blind study. Wien. Med. Wochenschr. 169 (7–8), 193–201. https://doi.org/10.1007/s10354-018-0667-6.

Jo, N., Roh, J., 2015. Systemic immediate hypersensitive reactions after treatment with sweet bee venom: a case report. J. Pharmacopuncture 18 (4), 59–62. https://doi.org/10.3831/KPI.2015.18.039.

Johnson, S.B., Park, H.S., Gross, C.P., Yu, J.B., 2018a. Use of alternative medicine for cancer and its impact on survival. J. Natl. Cancer Inst. 110 (1), 121–124. https://doi.org/10.1093/jnci/djx145.

Johnson, S.B., Park, H.S., Gross, C.P., Yu, J.B., 2018b. Complementary medicine, refusal of conventional cancer therapy, and survival among patients with curable cancers. JAMA Oncol. 4 (10), 1375–1381. https://doi.org/10.1001/jamaoncol.2018.2487.

Johnston, M., McBride, M., Dahiya, D., Owusu-Apenten, R., Nigam, P.S., 2018. Antibacterial activity of Manuka honey and its components: an overview. AIMS Microbiol. 4 (4), 655–664. https://doi.org/10.3934/microbiol.2018.4.655.

Khanjani Pour-Fard-Pachekenari, A., Rahmani, A., Ghahramanian, A., Asghari Jafarabadi, M., Onyeka, T.C., Davoodi, A., 2019. The effect of an oral care protocol and honey mouthwash on mucositis in acute myeloid leukemia patients undergoing chemotherapy: a single-blind clinical trial. Clin. Oral Investig. 23 (4), 1811–1821. https://doi.org/10.1007/s00784-018-2621-9.

Kim, M., Han, C.H., 2020. Pharmacopuncture for stroke survivors: a systematic review of randomized controlled trials in South Korea. Complement. Ther. Clin. Pract. 40. https://doi.org/10.1016/j.ctcp.2020.101179, 101179.

Kuo, C.C., Wang, R.H., Wang, H.H., Li, C.H., 2018. Meta-analysis of randomized controlled trials of the efficacy of propolis mouthwash in cancer therapy-induced oral mucositis. Support. Care Cancer 26, 4001–4009. https://doi.org/10.1007/s00520-018-4344-5.

Kwakman, P.H., Zaat, S.A., 2012. Antibacterial components of honey. IUBMB Life 64 (1), 48–55. https://doi.org/10.1002/iub.578.

Lal, A., Chohan, K., Chohan, A., Chakravarti, A., 2017. Role of honey after tonsillectomy: a systematic review and meta-analysis of randomised controlled trials. Clin. Otolaryngol. 42 (3), 651–660. https://doi.org/10.1111/coa.12792.

Lee, M.S., Pittler, M.H., Shin, B.C., Kong, J.C., Ernst, E., 2008. Bee venom acupuncture for musculoskeletal pain: a review. J. Pain 9 (4), 289–297. https://doi.org/10.1016/j.jpain.2007.11.012.

Lindberg, T., Andersson, O., Palm, M., Fagerström, C., 2015. A systematic review and meta-analysis of dressings used for wound healing: the efficiency of honey compared to silver on burns. Contemp. Nurse 51 (2–3), 121–134. https://doi.org/10.1080/10376178.2016.1171727.

Majtan, J., Majtanova, L., Bohova, J., Majtan, V., 2011. Honeydew honey as a potent antibacterial agent in eradication of multi-drug resistant Stenotrophomonas maltophilia isolates from cancer patients. Phytother. Res. 25 (4), 584–587. https://doi.org/10.1002/ptr.3304.

Münstedt, K., 2018. Meaningfulness of apitherapeutic approaches using the example of primary dysmenorrhea. J. Apitherapy 3, 9–16. https://doi.org/10.5455/ja.20170917090943.

Münstedt, K., 2019. Bee products and the treatment of blister-like lesions around the mouth, skin and genitalia caused by herpes viruses—a systematic review. Complement. Ther. Med. 43, 81–84. https://doi.org/10.1016/j.ctim.2019.01.014.

Münstedt, K., Männle, H., 2019. Using bee products for the prevention and treatment of oral mucositis induced by cancer treatment. Molecules 24 (17), 3023. https://doi.org/10.3390/molecules24173023.

Münstedt, K., Männle, H., 2020a. Apitherapy for menopausal problems. Arch. Gynecol. Obstet. 302 (6), 1495–1502. https://doi.org/10.1007/s00404-020-05692-2.

Münstedt, K., Männle, H., 2020b. Bee products and their role in cancer prevention and treatment. Complement. Ther. Med. 51. https://doi.org/10.1016/j.ctim.2020.102390, 102390.

Münstedt, K., Männle, H., 2020c. Benign prostatic hyperplasia—is apitherapy a reasonable therapeutic option? Urol. Nurs. 40 (5), 239–243. https://doi.org/10.7257/1053-816X.2020.40.5.

Münstedt, K., Männle, H., 2020d. Seasonal allergic rhinitis and the role of apitherapy. Allergolog. immunopathol. (Madr) 48 (6), 582–586. https://doi.org/10.1016/j.aller.2020.03.015.

Münstedt, K., Männle, H., 2020e. What is wrong with the meta-analyses on honey and oral mucositis due to cancer therapies? Complement. Ther. Med. 49. https://doi.org/10.1016/j.ctim.2019.102286, 102286.

Münstedt, K., Riepen, T., 2019. Patients' decisions regarding the treatment of primary dysmenorrhoea. Complement. Ther. Med. 45, 1–6. https://doi.org/10.1016/j.ctim.2019.05.016.

Münstedt, K., Voss, B., Kullmer, U., Schneider, U., Hübner, J., 2015. Bee pollen and honey for the alleviation of hot flushes and other menopausal symptoms in breast cancer patients. Mol. Clin. Oncol. 3 (4), 869–874. https://doi.org/10.3892/mco.2015.559.

Münstedt, K., Funk, D., Riepen, T., Berkes, E., Hübner, J., 2019a. Acceptance of apitherapeutic methods in patients consulting general physicians or gynaecologists. Complement. Ther. Clin. Pract. 35, 154–157. https://doi.org/10.1016/j.ctcp.2019.02.005.

Münstedt, K., Momm, F., Hübner, J., 2019b. Honey in the management of side effects of radiotherapy- or radio/chemotherapy-induced oral mucositis. A systematic review. Complement. Ther. Clin. Pract. 34, 145–152. https://doi.org/10.1016/j.ctcp.2018.11.016.

Münstedt, K., Männle, H., Riepen, T., 2020. Survey of reasons why women utilize honey therapeutically, and reasons for not utilizing honey. Heliyon 6 (10). https://doi.org/10.1016/j.heliyon.2020.e05231, e05231.

National Center for Complementary and Integrative Health, 2018. Complementary, Alternative, or Integrative Health: What's in a Name? US Department of Health and Human Services. Available from https://www.nccih.nih.gov/health/complementary-alternative-or-integrative-health-whats-in-a-name. 10 April 2021.

Nitsche, M.P., Carreño, M., 2016. Is honey an effective treatment for acute cough in children? Medwave 16 (Suppl 2). https://doi.org/10.5867/medwave.2016.6454, e6454.

Oduwole, O., Udoh, E.E., Oyo-Ita, A., Meremikwu, M.M., 2018. Honey for acute cough in children. Cochrane Database Syst. Rev. 4 (4), CD007094. https://doi.org/10.1002/14651858.CD007094.pub5.

Pecoraro, L., Flore, A.I., Dalle Carbonare, L., Piacentini, G., Pietrobelli, A., 2020. Honey and children: only a grandma's panacea or a real useful tool? Int. J. Food Sci. Nutr., 1–8. https://doi.org/10.1080/09637486.2020.1811958.

Pferschy-Wenzig, E.M., Bauer, R., 2015. The relevance of pharmacognosy in pharmacological research on herbal medicinal products. Epilepsy Behav. 52, 344–362. https://doi.org/10.1016/j.yebeh.2015.05.037.

Phillips, B., Ball, C., Sackett, D., Badenoch, D., Straus, S., Haynes, B., Dawes, M., 2009. Oxford Centre for Evidence-Based Medicine-Levels of Evidence. Available from http://www.cebm.ox.ac.uk/resources/levels-of-evidence/oxford-centre-for-evidence-based-medicine-levels-of-evidence-march-2009. 10 April 2021.

Rajan, T.V., Tennen, H., Lindquist, R.L., Cohen, L., Clive, J., 2002. Effect of ingestion of honey on symptoms of rhinoconjunctivitis. Ann. Allergy Asthma Immunol. 88 (2), 198–203. https://doi.org/10.1016/S1081-1206(10)61996-5.

Roffet-Salque, M., Regert, M., Evershed, R.P., Outram, A.K., Cramp, L.J., Decavallas, O., Dunne, J., Gerbault, P., Mileto, S., Mirabaud, S., Pääkkönen, M., Smyth, J., Šoberl, L., Whelton, H.L., Alday-Ruiz, A., Asplund, H., Bartkowiak, M., Bayer-Niemeier, E., Belhouchet, L., Bernardini, F., Budja, M., Cooney, G., Cubas, M., Danaher, E.M., Diniz, M., Domboróczki, L., Fabbri, C., González-Urquijo, J.E., Guilaine, J., Hachi, S., Hartwell,

B.N., Hofmann, D., Hohle, I., Ibáñez, J.J., Karul, N., Kherbouche, F., Kiely, J., Kotsakis, K., Lueth, F., Mallory, J.P., Manen, C., Marciniak, A., Maurice-Chabard, B., McGonigle, M.A., Mulazzani, S., Özdoğan, M., Perić, O.S., Perić, S.R., Petrasch, J., Pétrequin, A.M., Pétrequin, P., Poensgen, U., Pollard, C.J., Poplin, F., Radi, G., Stadler, P., Stäuble, H., Tasić, N., Urem-Kotsou, D., Vuković, J.B., Walsh, F., Whittle, A., Wolfram, S., Zapata-Peña, L., Zoughlami, J., 2015. Widespread exploitation of the honeybee by early Neolithic farmers. Nature 527 (7577), 226–230. https://doi.org/10.1038/nature15757.

Schmiedeberg, O., 1912. Rudolf Buchheim, sein Leben und seine Bedeutung für die Begründung der wissenschaftlichen Arzneimittellehre und Pharmakologie. Arch. Exp. Pathol. Pharmakol. 67, 1–54.

Sebastiani, G., Herranz Barbero, A., Borrás-Novell, C., Alsina Casanova, M., Aldecoa-Bilbao, V., Andreu-Fernández, V., Pascual Tutusaus, M., Ferrero Martínez, S., Gómez Roig, M.D., García-Algar, O., 2019. The effects of vegetarian and vegan diet during pregnancy on the health of mothers and offspring. Nutrients 11 (3), 557. https://doi.org/10.3390/nu11030557.

Seyhan, M.F., Yılmaz, E., Timirci-Kahraman, Ö., Saygılı, N., Kısakesen, H.İ., Eronat, A.P., Ceviz, A.B., Bilgiç Gazioğlu, S., Yılmaz-Aydoğan, H., Öztürk, O., 2017. Anatolian honey is not only sweet but can also protect from breast cancer: elixir for women from artemis to present. IUBMB Life 69 (9), 677–688. https://doi.org/10.1002/iub.1652.

Tsichlakidou, A., Govina, O., Vasilopoulos, G., Kavga, A., Vastardi, M., Kalemikerakis, I., 2019. Intervention for symptom management in patients with malignant fungating wounds—a systematic review. J. BUON 24 (3), 1301–1308.

Wang, C., Guo, M., Zhang, N., Wang, G., 2019. Effectiveness of honey dressing in the treatment of diabetic foot ulcers: a systematic review and meta-analysis. Complement. Ther. Clin. Pract. 34, 123–131. https://doi.org/10.1016/j.ctcp.2018.09.004.

Yarom, N., Hovan, A., Bossi, P., Ariyawardana, A., Jensen, S.B., Gobbo, M., Saca-Hazboun, H., Kandwal, A., Majorana, A., Ottaviani, G., Pentenero, M., Nasr, N.M., Rouleau, T., Lucas, A.S., Treister, N.S., Zur, E., Ranna, V., Vaddi, A., Barasch, A., Lalla, R.V., KKF, C., Elad, S., Mucositis Study Group of the Multinational Association of Supportive Care in Cancer/International Society of Oral Oncology (MASCC/ISOO), 2020. Systematic review of natural and miscellaneous agents, for the management of oral mucositis in cancer patients and clinical practice guidelines—part 2: honey, herbal compounds, saliva stimulants, probiotics, and miscellaneous agents. Support. Care Cancer 28 (5), 2457–2472. https://doi.org/10.1007/s00520-019-05256-4.

Yilmaz, A.C., Aygin, D., 2020. Honey dressing in wound treatment: a systematic review. Complement. Ther. Med. 51. https://doi.org/10.1016/j.ctim.2020.102388, 102388.

Zhu, F., Wongsiri, S., 2011. A brief introduction to apitherapy health care. J. Thai Tradit. Altern. Med. 6 (3), 303–312.

CHAPTER 2

Bee products and skin therapy

Anna Kurek-Górecka and Paweł Olczyk
Faculty of Pharmaceutical Sciences in Sosnowiec, Department of Community Pharmacy, Medical University of Silesia in Katowice, Sosnowiec, Poland

Contents

Introduction

Bee products such as honey, propolis, bee pollen, royal jelly, beeswax, and bee venom constitute important pharmaceutical and cosmetic components. Each bee product is characterized by the content of the active substance,

Bee Products and Their Applications in the Food and Pharmaceutical Industries
https://doi.org/10.1016/B978-0-323-85400-9.00016-2

which differentiates one bee product from another, and causes that each of them is worth using for a different skin problem. Flavonoids and phenolic acids play a crucial role in the influence of those products on the skin. Honey, propolis, and pollen are used to heal burn wounds. Moreover, bee venom called apitoxin contains active peptides and amines used in the wound healing process. Research on wound dressing containing honey, propolis, or bee venom presents the potential of these products for wound healing therapy. The advantages of pharmaceuticals and cosmetics based on bee products are high effectiveness with minimal side effects. Therefore, bee products may become a new strategy in skin therapy.

This chapter discusses the pharmacological potential of various bee products for skin therapy in each section. Their bioactive contents and the action of mechanisms responsible for these effects are explained through various activities such as antioxidant, antimicrobial, immunomodulatory, and antiinflammatory.

Honey

Honey is a bee-derived, supersaturated sugar solution. The honey components which have a beneficial effect on the skin include sugars, mainly fructose and glucose, amino acids, organic acids, vitamins, minerals, proteins, and phytocompounds such as phenols and methylglyoxal. Honey is particularly suitable for cosmetic products and dressings for different wounds, including burn wounds. Moreover, many clinical trials have revealed positive effects of honey for acne, herpes, skin rashes, contact dermatitis, diaper dermatitis, psoriasis, and hemorrhoids (Burlando and Cornara, 2013; Julianti et al., 2017; Kurek-Górecka et al., 2020). In addition, honey has been included in the official pharmacopeias of various countries, e.g., European pharmacopeia, US Pharmacopeia. Floral sources allow honey categorization in terms of its biological properties.

Types of honey

Among different types of honey, there are those derived from the tree genus *Leptospermum* from New Zealand and Australia. The first type of honey derived from the Manuka tree *Leptospermum scoparium* is known as manuka bush and derived from the Manuka tree Leptospermum polygalifoliumis, known as a jelly bush, respectively. These two kinds of honey exhibit a high antibiotic activity. Antibiotic activities vary depending on the type of honey and its components. Several factors include high sugar concentration, low pH, hydrogen peroxide releasing from glucose oxidase, methylglyoxal, and antimicrobial peptide bee defensin polyphenols (McLoone et al., 2016). Ninety percent honey solution in water shows a high antifungal effect against *Malassezia,* making them effective in treating seborrheic dermatitis and dandruff (Al-Waili, 2001; Gupta et al., 2004). However, the active constituents like methylglyoxal in Manuka honey or polyphenols and biological properties like antimicrobial of specific honey types play a crucial role in determining the use of honey in the design and synthesis of selected drug forms and cosmetics ointments, shampoos, conditioners, creams, and lotions. The acacia and manuka honey attract attention in the field mentioned above.

Acacia honey type, produced from black locust *Robinia pseudoacacia,* is commonly used in cosmetics due to the higher concentration of fructose than glucose. Type of acacia honey is classified as slow-crystallizing. Its average F/G ratio is equal 1.42 (Balos, 2020). The F/G ratio is known as a typical feature of honey types. Fructose is responsible for the hygroscopic properties of honey. If the ratio F/G is less than 1.11, the honey crystallizes quickly. Thanks to these proportions, the risk of crystallizing sugar is reduced in cosmetic products (Burlando and Cornara, 2013; Vranic et al., 2019). The cosmetic properties of honey allow its implementation to cosmetic products as an emollient, humectant, and moisturizing component.

The role of honey constituents

Derivatives of honey may appear in the composition of cosmetics, e.g., creams, hair conditioners. Among the derivatives of honey there are Mel Extract, Hydrogenated Honey, and Hydroxypropyltrimonium Honey. Mel Extract exhibits moisturizing properties. Hydrogenated Honey plays a role as a humectant in cosmetics (skin-conditioning agents-humectant; skin-conditioning agents-miscellaneous), while Hydroxypropyltrimonium honey has antistatic properties. Due to hydrating and regenerative effects, honey is used in skin care products like hair conditioners (Belsito et al., 2019). A high content of fructose and glucose and the possibility of forming hydrogen bridges with water provide a moisturizing effect. Moreover, among the moisturizing honeycomb components, the following have been enumerated: amino acids (proline, arginine, glutamic acid alanine, lysine, glycine, leucine), and organic acids (gluconic acid, lactic acid, citric acid, formic acid, acetic acid).

The presence of carbohydrates and fruit acids is responsible for the regenerative effects. Due to osmosis and microcirculation caused by metabolic processes are stimulated, and the dermal tissue is nourished and oxygenated in a better way. Apart from that, honey also improves skin elasticity, revitalizing its color and smoothing out wrinkles. Phytocompounds like phenols play a role as antioxidants and are responsible for antiaging action. Honey-containing cosmetics keep the skin juvenile and protect against wrinkles formation. Fruit acids like citric, lactic, acetic, and malic acids exfoliate dead skin cells (Bouhlali et al., 2019). Therefore, honey possesses a keratolytic effect, which is associated with improving elasticity and depigmenting properties. Due to the exfoliation of the superficial epidermal papillae, a signal is created to stimulate divisions in the basal layer. Besides, cytokines stimulate the production of collagen and extracellular matrix in the dermis. By decreasing the adhesion of keratinocytes in the epidermis, honey reduces excessive pigmentation (Marion, 2000). Honey can be used as a peeling agent as a natural sugar mixture.

Furthermore, honey fatty acids like myristic, myristoleic, palmitic, palmitoleic, stearic, and arachidonic acids, and mineral substances like sodium, potassium, calcium, iron, copper, zinc relieve xerosis (Kapoulas et al., 1977). Thus, honey soothes skin irritation, chapped lips, rough, cracked hands, and frostbites. Due to the content of flavonoids, honey can also play an important role in sun protection by preventing skin irritation. Additionally, carbohydrates, essential oils, and bio elements influence the suitability of honey as a component of balms and bath products due to its toning, relaxing, and conditioning properties. In cosmetics, honey is used in lower concentrations in the range of 0.5%–5%. A higher concentration of honey of about 10%–15% is used in ointments (Krell, 1996).

Therapeutic properties of honey

The therapeutic properties of honey are connected with its antimicrobial, antifungal, immunomodulatory effect, and physiological mechanism of action. Due to its properties, honey promotes angiogenesis, granulation, epithelialization and influences molecular markers connected with tissue repair. Thanks to the antiseptic effect, antimicrobial activities, and ability to promote wound healing, honey plays an important role in treating acute and chronic wounds. The antimicrobial mechanism of honey is based on the action of the following factors: releasing of hydrogen peroxide by the enzyme glucose oxidase, the presence of phenolic acids, mainly caffeic acid, ferulic acid as well as *p*-coumaric acid, flavonoids, especially quercetin, and the presence of lysozyme (Bogdanov, 1997). The antimicrobial mechanism of honey action is associated with peroxide-mediated and nonperoxide-mediated action. The first is connected to the action of hydrogen peroxide, and the second is based on low pH and the presence of methylglyoxal in Manuka honey. In addition, Manuka honey exhibits the ability to impair bacterial cell cycle progression.

Additionally, among antibacterial factors in some honey were found bee-defensin-1 and melanoidins (Kwakman et al., 2010; Mavric et al., 2008; Rufián-Henares and Morales, 2007). The mechano-physical mechanism of antimicrobial honey action is connected with hampering the production of biofilms due to its hygroscopic properties. Due to its physicochemical properties, it creates a barrier between the wound bed and external factors. Honey protects against superinfection with microorganisms. Due to its high osmolarity, it leads to the lysis of bacterial cells. Moreover, honey exhibits antimicrobial effects against *Escherichia coli, Pseudomonas aeruginosa, Staphylococcus aureus, Acinetobacter, Stenotrophomonas*, and *Enterococcus* (Blaser et al., 2007; George and Cutting, 2007; Simka, 2006; Tan et al., 2009).

Overall, it is known that floral sources determine the power of the antibacterial activity. It has been found that European honey displays weaker antibacterial effects than manuka honey which contains a larger amount of methylglyoxal. The antibiotic activity of manuka honey is estimated by Unique Manuka Factor (UMF) and methylglyoxal (MGO) markers. The antimicrobial activity of manuka honey is also evident in undiluted honey, and it is not abolished by catalase, which differentiates manuka honey from other types of honey (Hołderna-Kędzia et al., 2008). Another type of honey with antibacterial activity is Revamil from the Netherlands. The antibiotic factor in Revamil is the peptide defensin-1. Bee defensin-1 permeabilizes bacteria and inhibits their RNA, DNA, and protein synthesis (Mandal and Mandal, 2011; Sojka et al., 2016). However, in other varieties of honey the phenolics are also responsible for antibacterial activities. Honey exhibits the antifungal effect by reducing *Malassezia* development on the skin surface. Honey is particularly suitable for alleviating the symptoms of labial and genital herpes. The antimicrobial activity of honey is the reason for using a combination of beeswax and olive oil in treating pityriasis versicolor, tinea cruris, tinea corporis, and tinea faciei.

Besides, this mixture is effectively used in diaper dermatitis, psoriasis, and hemorrhoids. The components of honey responsible for antifungal properties are colon pinocembrin and lysozyme (Al-Waili, 2004, 2005a; Kędzia and Hołderna-Kędzia, 2017). In addition to its antimicrobial and antifungal properties, honey exhibits an immunomodulatory effect in wound healing shown by Tonks et al. (2003). The mentioned researchers provided detailed information that manuka honey increases the levels of tumor necrosis factor-α (TNF-α), interleukin-1β (IL-β), interleukin-6 (IL-6), and prostaglandin E_2 (PGE_2). Thus, on the one hand, honey extends the inflammation stage of wound healing. Still, on the other hand, further immunomodulation promotes the antimicrobial effect of honey, decreasing the biofilm formation and bacterial cell cycle progression.

Honey participates in the immunologic cascade, including tumor necrosis factor-β (TNF-β), metalloproteinase-9 (MMP-9) during proliferative and remodeling phases. The increased levels of MMP-9 and TGF-β are associated with increasing epithelialization and the next phase—wound remodeling (Lee et al., 2011). The study by Subrahmanyam (1998) and Subrahmanyam et al. (2001) has shed more light on the role of honey during the wound healing process among burnt patients. The effect of honey dressing was compared with silver sulfadiazine in two groups of 25 patients. In the case of wounds treated with honey, 84% showed satisfactory epithelialization by the 7th day among 84% of patients, while in wounds treated silver sulfadiazine among 72% patients. Moreover, on the 7th day, the reparative activity of honey dressing was observed in 80% of patients with minimal inflammation. With silver sulfadiazine, lower reparative activity was obtained, and the sustained inflammatory reaction was noted even on epithelialization.

Many studies indicate that honey reduces inflammation, relieves pain, and decreases healing time. Honey exhibits higher efficiency than silver sulfadiazine or other dressings in burn treatment (Wijesinghe et al., 2009). The use of honey for surgical and traumatic wounds leads to accelerated wound healing. It may be an attractive option in surgical wound care and was confirmed in a randomized clinical trial in a University hospital in Tehran. The impact of honey dressing for healing postsurgical wounds among 52 patients after different surgical procedures as follow mammoplasty, abdominoplasty, breast prosthesis, mastopexy, and face incision was documented by Goharshenasan et al. (2016). The obtained results demonstrated improvement of the healing process and final esthetic result by using honey dressing. Therefore, the width of the scars was smaller while using a honey dressing compared to a conventional dressing.

Additionally, applying honey dressings may lead to a lower frequency of complications like marked erythema, dehiscence, and infection among women after plastic surgeries. Honey plays a vital role as the apitherapeutic agent in topical wound treatment, displaying bacteria-killing, ability to bacterial biofilm penetration, lowering wound pH, reducing pain and inflammation, promoting fibroblast migration, and keratinocyte and wound closure and promoting collagen type I and III depositions. Therefore, honey is incorporated into the biomaterial leading to template regeneration of tissues. Honey may be used for electrospun templates, cryogels, or hydrogels. Due to the application of the honey to the biomaterial or incorporate it into the biomaterial is delivered appropriate concentrations of honey and achieved these optimal effects. Thanks to the controlled release of honey, cytotoxic effects are avoided, and the prolonged presence of the honey in the wounds is ensured (Minden-Birkenmaier and Bowlin, 2018). Actually, honey is used in wound treatment as a polyacrylamide polymer dressing, hydroactive, hydrogel-coated, polyurethane wound dressing, alginate dressing, and a wound care gel combined with lanolin, polyethylene glycol, vitamins A and E, and sterile 100% Manuka honey (Lee et al., 2011). The concentrations of honey in the dressing are in the range of 80%–95%, and in the case of sterile manuka, honey may be dispensed from a tube.

The main challenge of topical honey application is plant pollen allergy, so honey should be filtrated to remove pollen grains. Moreover, honey applied on wounds should be sterilized by gamma radiation. Other problems connected with the honey using are honey cytotoxicity of high concentrations, e.g., honey treatment with methylglyoxal levels above 1.8 mg/mL caused cilia denudation and squamous metaplasia (Paramasivan et al., 2014; Simka, 2006) (Table 2.1).

Table 2.1 Honey components and their impact on skin.

Components of honey	Impact on skin	References
Methylglyoxal, bee defensin-1, melanoidins, phenolic acids (i.e., caffeic acid, ferulic acid, p-coumaric acid), quercetin, Lysozyme	Antimicrobial effect	McLoone et al. (2016) Bogdanov (1997)
Pinocembrin, lysozyme	Antifungal effect	Al-Waili (2001, 2004), Ali et al. (2015), Gupta et al. (2004)
Carbohydrates, amino acids, organic acids	Moisturizing effect	Burlando and Cornara (2013)
Carbohydrates, fruit acids	Regenerative effect	Kurek-Górecka et al. (2020)
Polyphenols	Antiaging effect	Kurek-Górecka et al. (2020)
Fruit acids, carbohydrates	Exfoliating and depigmenting effect	Marion (2000)

Propolis

Propolis is a complex, resinous substance collected by honeybees from various plant sources. It consists of diverse natural products. The classification of propolis is based on its geographical location, color, and agricultural characteristics. It is also classified according to the flora characteristic for the origin of propolis. According to plant sources, propolis has been divided into the seven following groups: *Populus, Eucalyptus, Baccharis, Clusia, Macaranga, Betula,* and Mediterranean propolis. Taken together all classifications, great impact on propolis composition and its influence on skin is the local flora (Abu-Mellal et al., 2012; Hernández et al., 2005; Huang et al., 2007; Pereira et al., 2003; Popova et al., 2010; Wang et al., 2016; Zhang et al., 2016a). Among the propolis components that have a beneficial effect on the skin are phenolic acids, flavonoids, terpene compounds, alcohols, minerals, vitamins, and enzymes. Propolis is known as one of the richest sources of phenolic acids and flavonoids. The latter ones belong to the most important biologically active compounds (Garedew et al., 2004).

Propolis is particularly useful as an antiseptic, antioxidant, antiinflammatory, and anesthetic, photoprotective agent suitable as a constituent of cosmetic products and commonly used in treatments for dermal pathologies. It is a valuable component of dressing for wounds and burns. In the 17th century, propolis was included as an official medicine to British pharmacopeias (da Silva et al., 2020). Propolis is used as aqueous or ethanol extract; however, the latter is most frequently cited (Kurek-Górecka et al., 2012).

Propolis against photodamage of skin

An aqueous propolis extract is used in antifungal cosmetics, while propolis dissolved in fats is used to produce lipsticks. Moreover, ethanol extract of propolis and its main polyphenolic components are included in the cosmeceuticals class as active ingredients of commercial sunscreen formulations due to their protective and preventive properties confirmed in many studies (Bolfa et al., 2013; Fonseca et al., 2011; Gregoris et al., 2011). Propolis is considered a good ingredient of sun blockers (creams, lotions, sticks, and lipsticks). It is used as a natural filter and an antioxidative, antiinflammatory, and regenerative agent. Propolis absorbs UV light due to the presence of caffeic acid, coumaric acid, and ferulic acid (Saewan and Jimtaisong, 2015). It was shown that propolis protects skin from UVB damage, reduces cutaneous inflammation, immunosuppression, and lipid peroxidation induced by UV exposure. Romanian propolis exhibits strong photoprotective effects against UVB. Propolis may be used to prevent sunburns (Bolfa et al., 2013; Cole et al., 2010; Saric and Sivamani, 2016). Therefore, propolis smoothes out wrinkles and has antiaging properties protecting by photodamage.

Due to high antioxidant activity, propolis is used in commercial formulations for sun protection. Phenolic acids and flavonoids are responsible for the high antioxidant activity of propolis and its antiaging effect. Moreover, nanoemulsions based on aqueous propolis extract and lycopene extract improve skin's ability to defend itself from the sunlight (Butnariu and Giuchici, 2011; Sawicka and Borawska, 2013).

Antimicrobial properties of propolis against infections

Due to its antimicrobial properties, propolis is used to treat viral, fungal, and bacterial infections. Propolis exhibits antiviral activity in cold sores caused by the herpes simplex virus. Flavonoids such as galangin, kaempferol, and quercetin, play a considerable role in antiviral action of propolis (Amoros et al., 1992). Both aqueous and ethanol extracts of propolis demonstrate high antiviral properties against herpes simplex type 2 (HSV-2) (Nolkemper et al., 2010).

In addition, propolis seems to be helpful against fungal infections due to its phenolic constituents like pinocembrin, pinobanksin, caffeic acid, and terpenes (Peng et al., 2012; Sawicka and Borawska, 2013). Several reports have indicated antifungal properties of propolis, especially against *Candida*. According to Boisard et al. (2015), poplar-derived propolis against *Candida* is correlated with flavonoid content. Due to its antimycotic activity, propolis is an active substance in shampoos used in dandruff treatments.

The results of conducted studies highlight the importance of propolis ethanol extract applied on the skin to limit *Cutibacterium* acnes. Besides antifungal activity, propolis exhibits antimicrobial activity against Gram-positive and Gram-negative bacteria, although it is considered more effective against Gram-positive bacteria. Propolis extract has curative effects in purulent skin infections, intertrigo, and thrush. Moreover, propolis demonstrates high effectiveness during acne therapy. The clinical and bacteriological significant efficacy of topical 10% ethanol propolis extract in treating facial acne vulgaris among 20 patients has been reported (Ali et al., 2015).

The antimicrobial activity of ethanol propolis extract has been correlated with the concentration of flavonoids and phenolic acids possessing antibacterial, antioxidant, and antiinflammatory properties (Ali et al., 2015). Mazzarello et al. (2018) confirmed that cream containing 20% propolis, 3% tea tree oil, and 10% *aloe vera* leaf juice effectively cleared papules and pustules, especially in mild and moderate acne. Moreover, this cream was more effective in reducing acne compared to 3% erythromycin cream. Topical propolis application is effective and promising for acne vulgaris treatment, so it can potentially be a cosmetics for the skin with acne and a pharmaceutical.

Wound healing properties of propolis

Propolis possesses not only antimicrobial and antiinflammatory effects but is also helpful for skin reepithelization and cicatrization, thus demonstrating promising actions in wound healing. Furthermore, propolis contains chrysin which is responsible for the analgesic effect of ethanol propolis extract. Polish propolis ethanol extract turned out to be highly effective in treating skin burns (Olczyk et al., 2014). Studies by Olczyk et al. (2013) have shown that propolis is more effective during burn wounds than the treatment of traditional damage such as SSD (silver sulfadiazine). Propolis used as an apitherapeutic agent to treat burn wounds in pigs increased fibroblasts' proliferation, activation, and growth. It was also shown that propolis stimulates glycosaminoglycan accumulation which is needed for granulation, tissue growth, and wound closure.

Moreover, propolis stimulates collagen type I and III accumulation in bed wound. Accumulation of collagen type I in a matrix of a skin injury accelerates the repair process as collagen type I is indispensable for the keratinocyte migration and reepithelization. Additionally, the accumulation of collagen type III creates a scaffold necessary for cell migration and regulates collagen I fibrillogenesis. Topically applied propolis ointment in the case of burn wounds contributes to reepithelization. Similarly, Han et al. (2005) compared the effect of 50% propolis cream and silver sulfadiazine during thermal injuries in an animal model. They found that propolis cream was effective in healing burns more than SSD. Therefore, it can be stated that propolis may be used in treating burns in all stages of the healing process leading to the beneficial effects of the management applied.

Other works have shown that topically applied propolis constitutes an apitherapeutic agent which may bring satisfactory effect in diabetic wounds treatment. Al-Waili et al. (2015) reported that propolis decreased persistent inflammation in diabetic wounds by normalizing neutrophil and neutrophil elastase counts. Neutrophils cleanse the wound of bacteria and dead cells and release elastase, facilitating the penetration of small blood vessels into the plug that leads to the start of the next stage. Therefore, propolis exhibits antiinflammatory action during wound healing. The primary role in the antiinflammatory effect of propolis is caffeic acid phenyl ester. It inhibits MMP-9, leading to decreased collagenolysis and proinflammatory proteinase, which are increased in diabetic foot ulcers. Therefore, propolis therapy was found to increase the accumulation of collagen type I and decrease matrix metalloproteinase activity and inflammation.

It is noteworthy that during diabetic wound healing in mice, propolis accelerated wound healing through angiogenesis stimulation. It was found that a crucial role in the mentioned process is playing by genistein (Tie et al., 2013). Other works have shown a positive effect of propolis on the healing process in different animal models, including those with burned and diabetic wounds. The diabetic mouse model propolis accelerated wound closure due to the mentioned bee product's influence on transforming growth factor promotion (Hozzein et al., 2015). It should be noted that Afkhamizadeh et al. (2018) confirmed the effectiveness of 5% propolis ointment applied in clinical trials among diabetic patients. Propolis ointment after 4 weeks of topical application diminished the ulceration area and sped up the wound healing process. What's important, it has been recently shown that propolis may be used as an active compound in a biodegradable wound dressing. Propolis may be released continuously throughout the healing process (Stojko et al., 2020). The electron paramagnetic resonance studies of blood from the organisms with burned wounds treated with electrospun nonwovens dressing releasing propolis have indicated its positive effect on burns regeneration. Apitherapeutic dressing exhibits the multifaceted influence of propolis on the prooxidative/antioxidative balance status determining to a large extent the repair of the tissue damage (Olczyk et al., 2020). Due to regenerative, repairing, and protecting properties, propolis is mainly used in the concentration of 5%–20% to produce antibedsores preparations since it firms the dermal tissue and protects it against pathogenic bacteria (Sawicka and Borawska, 2013) (Table 2.2).

Bee pollen

Bee pollen is a mixture of flower pollen derived from different plant species agglutinated by nectar and enzymes secreted by salivary bee glands. The chemical composition of bee pollen is diverse and depends on its geographical origin, ecological habitat, and seasons. It contains proteins, amino acids, sugars, lipids, vitamins, minerals, as well as polyphenols. Bee pollen's therapeutic and cosmetic value is expected due to the appropriate amount of flavonoids and phenolic acids. Flavonoids such as kaempferol, quercetin, and

Table 2.2 Propolis components and their impact on skin.

Components of propolis	Impact on skin	References
Caffeic acid, coumaric acid, ferulic acid	Photo protecting effect against UV radiation	Saewan and Jimtaisong (2015), Saric and Sivamani (2016), Cole et al. (2010)
Phenolic acids and flavonoids	Antiaging effect	Butnariu and Giuchici (2011)
Galangin, kaempferol, quercetin	Antiviral activity	Amoros et al. (1992), Nolkemper et al. (2010)
Pinocembrin, pinobanksin, caffeic acid and terpenes	Antifungal effect	Peng et al. (2012), Sawicka and Borawska (2013), Boisard et al. (2015)
Flavonoids, phenolic acids	Antibacterial effect, including the antiacne effect	Ali et al. (2015)
Caffeic acid phenyl ester	Antiinflammatory effect	Al-Waili et al. (2015)
Polyphenols	Antioxidant activity	Kurek-Górecka et al. (2020)

isorhamnetin are present commonly in pollens from different geographical regions. In the bee pollen from *Brassica campestris* L. quercetin, naringenin, kaempferol, isorhamnetin, rutin dominated, while in Croatian bee pollen from *Cystus incanus* L., galangin, kaempferol, chrysin, and pinocembrin were found most often (Šarić et al., 2009; Zhang et al., 2016b). Among bee pollen phenolic acids and their esters, the following compounds were identified: *p*-hydroxybenzoic acid, gallic acid, syringic acid, vanillic acid, protocatechuic acid, p-coumaric acid, ferulic acids, and caffeic acid. These compounds determine the antioxidant activity of bee pollen (Kocot et al., 2018). Bee pollen extracts exhibit high antioxidant activity, which depends on the kind and origin of plants. The mentioned bee pollen capacity changes in the following order: the highest one is found in pollen from *Brassica napus,* the pollen from *Papaver somniferum* and the smallest pollen from *Helianthus annuus* (Fatrcová-Šramková et al., 2013).

Polyphenols constitute important and valuable constituents of bee pollen and play a crucial role in bee pollen's antiinflammatory and antibacterial action. Moreover, due to flavonoids content, bee pollen exhibits antifungal, antiviral, and analgesic effects. Additionally, bee pollen turned out to be highly effective in skin regeneration (Javed, 2021). Bee pollen is used in ethanol, aqueous, glycerin, propylene glycols, lipid extracts, and lyophilized form.

Bee pollen and skin products

The cosmetic properties of bee pollen allow to include it as an essential and active component that may be used to prevent hyperpigmentation, skin aging including photoaging, and dry skin (Denisow and Denisow-Pietrzyk, 2016). Melanogenesis is a multistage process. During this process, activated by ultraviolet radiation, tyrosine is converted into dopachinone under the influence of tyrosinase and then into one of the melanin forms (Kilian-Pięta and Hoppe, 2019). Ethanol extract of bee pollen exhibits tyrosinase inhibitory activity. Furthermore, bee pollen is an active component of cosmetics that reduces hyperpigmentation. Kim et al. showed that the best ability to inhibit tyrosinase activity in 57.9% had bee pollen obtained in extraction with 75% ethanol mixture with ethyl acetate in methanol at a temperature of 10°C (Kim et al., 2015). Other studies confirmed strong tyrosinase inhibitory properties of extracts from different monofloral bee pollen. According to Zhang et al. (2015), 75% ethanol turned out to be an effective solvent for bee pollen to inhibit tyrosinase. Extracts of bee pollen from apricot, camellia, and sunflower showed high antimelanogenesis properties. Besides, nano-size bee pollen using wet-grinding technology had good results in reducing the melanin index, demonstrating its effect on melanin content of the skin (Pyeon et al., 2017).

Moreover, many studies have shown a correlation between the whitening activities of bee pollen and phenolic content. Polyphenols are responsible for preventing UV radiation and reducing pigmentation spots' appearance (Hu and Kitts, 2005; Zhang et al., 2015).

Due to the potent antioxidant activity of polyphenols, bee pollen has a beneficial effect on skin wrinkles and photoaging. Additionally, bee pollen contains vitamins A, C, E, which whitens melasma, delay skin aging, and decrease the number of wrinkles. Also, amino acids of pollen prevent the accumulation of melanin in melanosomes and reduce photoaging. Bee pollen amino acids, such as methionine, lysine, threonine, histidine, leucine, isoleucine, valine, phenylalanine, tryptophan, show moisturizing effects because they are important components of the natural moisturizing factor (NMF). Besides amino acids, bee pollen carbohydrates play an important role in preventing skin dryness (Xi et al., 2018).

A cosmetic ingredient for hair products

Bee pollen as a cosmetic ingredient is also used to produce shampoos and conditioners. It possesses sebo-balancing activity and normalizes sebaceous glands' activity due to the presence of zinc, methionine, and phospholipids.

Furthermore, sulfur-containing amino acids, mainly cysteine, present in bee pollen strengthen the hair shaft. Bee pollen is a vital ingredient in antidandruff shampoos because it reduces fungal growth and inhibits itching of the scalp. Furthermore, it also exhibits moisturizing, conditioning, and regenerating properties that positively impact scalp skin. Bee pollen could be mixed with ethyl esters of essential unsaturated fatty acids omega-3 and omega-6 acids and used to care for atopic skin, sensitive skin, and the skin that tends to get scares. In cosmetics, different extracts of bee pollen are used at 0.5%–5% (Basista and Sodzawiczny, 2011).

Burn healing properties

Therapeutic properties of bee pollen are connected with its antimicrobial, antifungal, antiinflammatory, anesthetic, and immunostimulating properties. This active bee agent also facilitates the granulation process during burn healing (Denisow and Denisow-Pietrzyk, 2016; Komosińska-Vassev et al., 2015; Olczyk et al., 2016). Pollen ethanol extract exhibits a strong antimicrobial action against *Staphylococcus aureus, Escherichia coli, Klebsiella pneumoniae, Pseudomonas aeruginosa,* and antifungal activity against *Candida albicans.* The antibacterial effect of bee pollen was observed by Olczyk et al. (2016) during the application of bee pollen ointment based on 5% multifloral bee pollen ethanolic extract and 95% of petroleum jelly in topical burns inflicted on polish domestic pigs. The flavonoid and phenolic acids content of pollen are responsible for its antibacterial and antifungal effects. Polyphenols form complexes with bacterial cell walls, leading to the disruption of cell wall integrity, ion channels, and electron flow inhibition in the electron transport chain. Moreover, pinocembrin, galangin, and caffeic acid phenethyl ester inhibit bacterial RNA-polymerase. In consequence of RNA polymerase inhibition, the genes necessary for the functioning of the cell are not transcribed, which results in the death of the bacteria (Rzepecka-Stojko et al., 2015; Skarżyńska et al., 2020; Takaisi-Kikuni and Schilcher, 1994). The mechanism of the antifungal action of bee pollen polyphenols is based on the degradation of the fungal cells' cytoplasm membrane, which leads to loss of potassium ions followed by cell autolysis. The antiinflammatory action of bee pollen is related to the quercetin content. Quercetin plays a crucial role in inhibiting the arachidonic acid metabolism by decreasing proinflammatory prostaglandins and inhibiting cyclooxygenase and lipoxygenase activities. Due to the antiinflammatory activity of bee pollen, the healing period is shortened, and the duration and intensity of ailments are reduced. The antiinflammatory properties of bee pollen are related to kaempferol, which inhibits hyaluronidase and elastase activities and suppresses inflammatory responses (Denisow and Denisow-Pietrzyk, 2016). Bee pollen as an apitherapeutic agent also possesses anesthetic and immune-modulating properties. The last accelerates the epithelialization stage during the wound healing process. The clinical and histopathological results of the study conducted by Olczyk et al. (2016) demonstrated the beneficial effect of the bee pollen ointment on the burn wound healing process. Bee pollen ointment turns out to reduce the time of burn wound treatment providing faster reepithelization and wound closure. Bee pollen may be considered an apitherapeutic agent for skin regeneration. It promotes the proliferation of endothelial and mesenchymal cells and the deposition of collagen matrix in tissue granulation of the burn wounds (Javed, 2021) (Table 2.3).

Royal jelly

Royal jelly is a glandular secretion produced by young worker bees between 5 and 14 days of their life. It is an acid colloid whose pH is ranged from 3.6 to 4.2 (Fratini et al., 2016a). It is composed mainly of water, proteins, lipids, sugar, vitamins, and mineral salt. The composition of royal jelly depends on geographical origin, season, regional feeding conditions, storage conditions, and genetic variability between Africanized honeybee colonies (Baitala et al., 2010; Boselli et al., 2003; Li et al., 2012; Liu et al., 2008).

Table 2.3 Bee pollen components and their impact on skin.

Components of bee pollen	Impact on skin	References
Polyphenols, vitamins (i.e., A, C, E)	Depigmenting effect	Zhang et al. (2015), Pyeon et al. (2017)
Amino acids, carbohydrates	Moisturizing effect	Xi et al. (2018)
Pinocembrin, galangin, and caffeic acid phenethyl ester (CAPE)	Antimicrobial action	Olczyk et al. (2016), Rzepecka-Stojko et al. (2015)
Polyphenols	Antifungal effect	Denisow and Denisow-Pietrzyk (2016), Kurek-Górecka et al. (2020)
Quercetin, kaempferol	Antiinflammatory	Denisow and Denisow-Pietrzyk (2016)
Polyphenols	Antioxidant action	Kurek-Górecka et al. (2020), Rzepecka-Stojko et al. (2015)

Different composition of royal jelly has an impact on its properties especially antibacterial ones. Citrus and camphor royal jelly from Egypt demonstrate more effective bactericidal and bacteriostatic action than royal jelly from China (Moselhy et al., 2013). Based on its particular composition involving several proteins and lipids, it is successfully used in cosmetics when dealing with skin problems. Moreover, it has been subjected to investigations on its pharmacological properties and application in medicine. Among the royal jelly components which have beneficial effects on the skin, there are the following ones: many proteins, i.e., eight Major Royal Jelly Proteins (MRJP1-8), Apismin, Royalisin, Royalactina, Jelleines, glucoside oxidase, Apolipophorin III, Defensin amino acids, fatty acids, e.g., 10-hydroxy-2-decenoic acid (10-HDA), vitamins especially from B group, sugars mainly consisting of fructose and glucose.

Moreover, royal jelly contains a small amount of the following flavonoids: hesperetin, isosakuranetin, naringenin, acacetin, apigenin, pinocembrin, chrysin, luteolin, isorhamnetin, kaempferol, coumestrol, formononetin, and genistein (Ahmad et al., 2020; Fratini et al., 2016a; López-Gutiérrez et al., 2014).

Royal jelly in skin care

Royal jelly is beneficial in skin care due to its skin whitening, tonic, moisturizing, and antiaging effects. Furthermore, due to the content of 10-hydroxy-2-decenoic acid, royal jelly reduces melanogenesis, inhibiting tyrosinase, TRP-1 (tyrosinase-related protein 1), and TRP-2 (tyrosinase-related protein 2) and microphthalmia-associated transcription factor (MITF) expression. The results of conducted studies by Peng et al. (2017) indicate that 10-HDA decreases the level of MITF, which plays a crucial role in synthesizing tyrosinase, TRP-1, and TRP-2. The inhibition of the protein and gene expression of tyrosinase, TRP-1, and TRP-2 consequences in the whitening effect of 10-HDA and downregulate MITF gene expression. Tyrosinase controls the first step of melanogenesis. During this step, L-tyrosine is hydroxylated to L-3,4-dihydroxyphenylalanine (L-DOPA), and then L-DOPA is converted to dopaquinone. TRP-2 takes part in the production of 5,6-dihydroxyindole-carboxylic acid. The last one is a substrate for TRP-1. TRP-1 converts 5,6-dihydroxyindole-carboxylic acid to indole-5,6-quinone carboxylic acid. Creams containing 10-HDA isolated from royal jelly in concentrations 0.5%, 1%, and 2% were found to have a similar depigmenting action to 1% kojic acid cream.

Besides the skin-whitening effect, royal jelly exerts a highly moisturizing action. It provides hydration of the stratum corneum by retaining water in it. As a result, the skin is more elastic and better moisturized. After applying a 1% royal jelly cream, the degree of hydration of the stratum corneum was higher than in the base cream without royal jelly. Application properties (absorption, stickiness, and greasiness) of creams with lyophilized royal jelly were satisfactory. However, lyophilized royal jelly causes a significant reduction in dynamic viscosity compared to cream without royal jelly. Therefore, the development of a moisturizing cream recipe containing royal jelly requires the addition of consistency-forming agents. After applying the cream with royal jelly, the degree of hydration of the stratum corneum is much higher than in a base cream. The highest values and the lowest decrease moisture at the same time at samples with 1% of the royal jelly was observed (Bocho-Janiszewska et al., 2013).

Moreover, royal jelly shows an antiaging effect due to the content of 10-hydroxy-trans-2-decenoic acid. It induces the fibroblast to produce a transforming growth factor (TGF-β) which stimulates fibroblasts to the synthesis of collagen. Consequently, royal jelly increases collagen production and enhances the deposition of this fibrillar protein in the dermis (Koya-Miyata et al., 2004). In cosmetics, royal jelly can be found in a lyophilized form which does not influence the emulsion stability. Besides, in the mentioned form, royal jelly is more stable, susceptible to light, heat, and oxidation under air. What is more, cosmetics with high amounts of royal jelly are well absorbed and do not leave a greasy film. Royal jelly at 20% concentration has strong antimicrobial activity within skin tissue, however in creams is used mainly in the range from 0.5% to 5%, respectively (Bocho-Janiszewska et al., 2013; Kurek-Górecka et al., 2020).

Therapeutic effects of royal jelly

It is worthy of note that besides cosmetic properties, royal jelly exhibits therapeutic effects. These effects are connected with a multitude of its pharmacological properties, i.e., antioxidant, antiinflammatory, and antibacterial ones (Kurek-Górecka et al., 2020; Pavel et al., 2011). Royal jelly is one of the bee products accelerating the wound healing process by considering antioxidant, antiinflammatory, and antibacterial activity attributes. Antioxidant action is due to the presence of phenolic compounds, peptides, amino acids, and fatty acids. Among phenolic compounds, flavonoids such as chrysin, pinocembrin, and pinostrobin chalcone are responsible for antioxidant activity. A positive correlation between total phenolic content in royal jelly and FRAP (fluorescence recovery after photobleaching) values have been found (Kolayli et al., 2016). Besides phenolic compounds, peptides, especially three dipeptides containing tyrosine residues at their C-termini (Lys-Tyr, Arg-Tyr, and Tyr-Tyr), demonstrate high hydrogen peroxide-scavenging activity. In addition, amino acids as proline and cysteine play an important role in antioxidant activity. The former possesses hydroxyl radical-scavenging activity, and the latter participates in the synthesis of an effective cellular antioxidant—glutathione (Nakajima et al., 2009; Silici et al., 2011). Moreover, hydroxy dicarboxylic fatty acids with 8–12 carbon atoms in the chain are responsible for the antioxidant activity of royal jelly, as well. Among carboxylic acids, there are 10-hydroxy-2-decanoic acid (10-HDA), 10-hydroxy-2-decenoic acid (10H2DA), and sebacic acid (SA) (Kocot et al., 2018). It was also shown that 10-hydroxy-2-decenoic acid found in royal jelly could alleviate oxidative stress. Thus, it may be concluded that royal jelly may serve as an effective free radical scavenger (Guo et al., 2009; Kolayli et al., 2016; Nabas et al., 2014).

The antioxidant properties of royal jelly significantly impact the wound healing process, as the accumulation of reactive oxygen species (ROS) increases cellular damage and impairs neovascularization. Impaired wound healing, especially among diabetic patients, results from the reduced activity of the antioxidant enzymes such as glutathione peroxidase (GSH Px), manganese superoxide dismutase (MnSOD), and catalase. It is generally known that a large ROS production and diminished antioxidant defense lead to oxidative stress and delay the wound healing process via reducing keratinocytes proliferation and migration and apoptosis stimulation (Deveci et al., 2005). Overall, the antioxidant activity of royal jelly components confirms this unique bee product's beneficial impact on wound healing. Thus, it should be emphasized that royal jelly seems to be a promising component of wound dressing due to its' antiinflammatory properties.

The antiinflammatory activity of royal jelly is associated with the ability of MRJP3 Major Royal Jelly Proteins-3 to inhibit the secretion of proinflammatory cytokines, such as TNF-α, IL-6, and IL-1. Moreover, royal jelly exhibits an antiinflammatory effect inhibiting capillary permeability (Kohno et al., 2004; Pavel et al., 2011). Fujii et al. (1990) confirmed the antiinflammatory action of royal jelly during healing wounds among diabetic rats. Royal jelly decreased exudation and collagen formation in granulation tissue.

It is worthy of note that royal jelly is an ideal ingredient of ointments that accelerate wound healing due to antimicrobial activity. The main components of royal jelly involved in its antibacterial action are Royalisins, Jelleines, especially Jelleine I, II, III, Apolipophorin III, and glucose oxidase. Furthermore, 10-hydroxy-2-decenoic acid is involved in the antibacterial activity of royal jelly (Šedivá et al., 2018; Sagona et al., 2015; Fratini et al., 2016a). The antimicrobial action of royal jelly peptides is based on four models, i.e., barrel-stave, carpet-like, toroidal pore, and aggregate channel. The first model is based on the formation of pores in the bacterial membrane created by antimicrobial peptides (AMPs). AMPs are polypeptides containing from 10 to 50 amino acids. They interact with the microbial surface and penetrate the cell membrane by bilayer phospholipids. AMPs can interact with the intracellular target and inhibit the synthesis of cell constituents. The second model is based on the interaction between antimicrobial peptides and the external surface of the bacterial membrane, especially with anionic phospholipids creating carpet which extends on the target membrane. This mechanism reduces the bacterial membrane lipid layer followed by membrane disruption leading to the collapse of the lipid structure. According to the third model, the formation of pores in the membrane is created in the double bacterial membrane layers. The lines of double phospholipid layers become a continuous structure with pores. In the fourth model, peptides bind to phospholipids bacterial membrane and are transported through the membrane without modification. Inside the bacterial cell, they interact with intracellular targets such as DNA, RNA, and proteins. Overall, the antibacterial royal jelly action is connected with the electrostatic interactions between antimicrobial peptides and the bacterial membrane, which determine the permeability of the bacterial structure in question (Fratini et al., 2016a; Li et al., 2012).

It should be noted that royal jelly shows a high antimicrobial action against Gram-positive bacteria. Moreover, royal jelly's ether-soluble fraction exhibits higher antibacterial properties than crude royal jelly (Eshraghi, 2005). The beneficial effect of royal jelly in the wound healing process was confirmed by Siavash et al. (2015). The efficacy of topical application of sterile 5% royal jelly on diabetic foot ulcers has been studied among patients in Iran. Among the eight treated ulcers, seven were cured, and in one case, an improvement was obtained by decreasing the ulcer's length, width, and depth. In another study, cream with honey, royal jelly, and olive oil-propolis extract were applied to inflicted skin damage in diabetic rats (Rashidi et al., 2016). Effects of topically applied cream containing a mixture of honey, royal jelly, and propolis extract in nondiabetic and diabetic rats were satisfactory. It accelerates wound closure in both groups, nondiabetic and diabetic, but it caused faster wound closure in nondiabetic animals than phenytoin ointment.

Additionally, the results of a study conducted by Bucekova et al. (2017) indicated that royal jelly promotes wound reepithelization. Keratinocytes are responsible for the elevated production of MMP-9 (matrix metalloproteinase-9) after incubation with a water extract of royal jelly. Royal jelly increased keratinocyte migration and wound closure rates. Defensin − 1 is a component of royal jelly responsible for stimulating MMP-9 production. Defensin-1 promotes reepithelization, and it is responsible for cutaneous wound closure by enhancing keratinocyte and MMP-9 secretion. Therefore, royal jelly may be considered a potential component of a wound dressing. The pharmaceutical properties of royal jelly allow using it as a valuable component of creams and ointments (Table 2.4).

Beeswax

Beeswax is produced by glands located in the bee abdomen. Wax is obtained from the honeycombs of bees containing fatty acids like 10-hydroxy-trans-2-decenoic acid, esters of fatty acids (mainly palmitic and

Table 2.4 Royal jelly components and their impact on skin.

Components of royal jelly	Impact on skin	References
Royalisins, jellines (especially I, II, III), apolipophorin III	Antibacterial effect	Moselhy et al. (2013), Šedivá et al. (2018), Fratini et al. (2016a), Li et al. (2012), Eshraghi (2005)
Amino acids, carbohydrates	Moisturizing effect	Bocho-Janiszewska et al. (2013)
10-Hydroxy-2-decenoic acid	Antiaging effect Depigmenting effect	Koya-Miyata et al. (2004) Peng et al. (2017)
Chrysin, pinocembrin, pinostrobin, proline, cysteine, 10-Hydroxy-2-decanoic acid (10-HDA), 10-hydroxy-2-decenoic acid (10H2DA), and sebacic acid (SA)	Antioxidant activity	Kolayli et al. (2016), Silici et al. (2011), Nakajima et al. (2009), Kocot et al. (2018), Nabas et al. (2014), Guo et al. (2009)
Major Royal Jelly Proteins-3 (MRJP3)	Antiinflammatory effect	Kohno et al. (2004), Pavel et al. (2011), Fujii et al. (1990)

cerotinic acids), and alcohols (triacontanol, octacosanol, hexacosanol, and tetracosanol). Moreover, it contains flavonoids such as chrysin, pinocembrin, pinobanksin, galangin, and tectochrysin (Fratini et al., 2016b; Tomás-Barberán et al., 1993). The beeswax is used as a component of cosmetics and medicines to treat skin diseases, especially with disturbed skin barrier function as follows psoriasis or atopic dermatitis. It constitutes a valuable ingredient used in cosmetology and pharmacy. Beeswax is a natural fatty product for cosmetic purposes and is reported to have high hydrophobicity and moisture resistance. It is used as a thickener and a humectant in the composition of cosmetic formulations.

Beeswax, as a component of cosmetics and medicines, is known as yellow wax (Cera Flava) or white wax (Cera alba), which is obtained by chemical whitening of yellow wax (using chlorine or potassium permanganate). They are used to produce creams, ointments, liniments, and they have been described in pharmacopeias, e.g., Polish Pharmacopeia, European Pharmacopeia.

Beeswax in cosmetics

Beeswax as an emulsifying agent of cosmetics provides elasticity, plasticity and increases skin adhesiveness. Due to its properties, it is a valuable ingredient in the manufacture of lipsticks, sticks, and creams (Kasparaviciene et al., 2016). The mechanical properties of beeswax are diverse depending on the type of wax. The wax from *Apis dorsata* L. bees is stiffer and has a higher yield than *Apis mellifera* L. and *Apis cerana* L. beeswaxes, which have medium strength and stiffness. Beeswax of the *Apis andreniformis* L. species is the least strong, stiff, and elastic. The location of nests of particular species of bees influences on mechanical properties of beeswax (Goik et al., 2016).

Beeswax seems to be a favorable candidate for the development of nanosystems in cosmetics. Souza et al. (2017) demonstrated that beeswax-based nanoparticles containing 5% beeswax, which was incorporated into a gel-cream formulation, decreased transepidermal water loss (TEWL) and increased stratum corneum water content (SCWC). Consequently, beeswax-based nanoparticles in topical application protect the skin from environmental factors and increase hydration.

Beeswax in skin care

Due to the impact of beeswax-based nanoparticles on improving the functioning of the skin barrier, beeswax can play an important role in skin care, especially in the case of dry skin and dermatoses such as atopic dermatitis or psoriasis (Souza et al., 2017). When one considers the impact of beeswax on patients with atopic dermatitis and psoriasis vulgaris, it is evident that beeswax is suitable for use in cosmetics and the treatment of these diseases. Al-Waili (2003) showed that a mixture containing honey, beeswax, and olive oil in ratio 1:1:1 helped treat the mentioned diseases in 21 patients with atopic dermatitis. The results showed a significant and faster improvement in pruritis, scaling, and oozing on the application side of the mixture. Based on the available literature, beeswax seems to be a potentially active compound of ointments using in atopic dermatitis or psoriasis.

Kubota et al. (1983) showed that a mixture with zinc oxidate-beeswax improves patients who suffer from eczema and psoriasis. The ointment contains 10% zinc oxide, soybean oil, and white wax. The clinical efficiency of the mixture has been demonstrated on six patients with chronic eczema and seven patients with psoriasis vulgaris.

Antimicrobial effects of beeswax

Moreover, due to antiseptic, antibacterial, and antifungal properties, beeswax is used in wound treatment. Thanks to the content of squalene, 10-hydroxy-trans-2-decenoic acid, and chrysin, it plays a protective role for the skin against pathogenic microorganisms. Among these compounds, 10-hydroxy-trans-2-decenoic acid exhibits a strong antibacterial effect. Beeswax is effective against Gram-positive and Gram-negative bacteria. It exhibits antimicrobial effects against *Pseudomonas aeruginosa, Streptococcus aureus, Streptococcus enterica,* and *Escherichia coli* (Kacániová et al., 2012). Besides the antibacterial effects, it also exhibits antifungal activity against *Candida albicans.* Therefore, it can be used to treat dermatosis associated with fungal infection as follows diaper dermatitis, pityriasis versicolor, and tinea cruris. Al-Waili (2005a) showed that beeswax combined with honey and olive oil in ratio 1:1:1 inhibited the growth of *Candida albicans* among eight boys and four girls, aged 3–18 months suffer from diaper dermatitis. The study results using honey, beeswax, and olive oil mixture (1:1:1, *v*/v) demonstrated that this mixture inhibits the growth of *Staphylococcus aureus* and *Candida albicans.* In addition, the beeswax combined with propolis in ratio 1:1 inhibits the growth of *Streptococcus aureus* and *Candida albicans* (Castaldo and Capasso, 2002).

Antioxidative and antiinflammatory activity of beeswax

Beeswax exhibits antioxidative and antiinflammatory activity. Mainly flavonoids are responsible for the antioxidant effect. They can inhibit the activity of enzymes producing reactive oxygen species (ROS). Also, they are responsible for chelating ions of metals involved in free radical creation and scavenging reactive oxygen species. In this way, flavonoids inhibit the cascade of reactions leading to the peroxidation of lipids (Kurek-Górecka et al., 2014). Also, the antiinflammatory activity of beeswax is associated with flavonoids, mainly chrysin. Chrysin takes part in inflammatory cytokines regulation. It inhibits the production of IFN-c and TNF-α by activating macrophages through the NF-kB pathway (Ciftci and Ozdemir, 2011; Ha et al., 2010). It also inhibits prostaglandin-E2, COX-2, and NF-kB (O'Leary et al., 2004). In addition to chrysin, D-002 consisting of an alcoholic mixture of high molecular mass, which occurs in the beeswax, is responsible for antiinflammatory action. It reduces leukotriene B4 and thromboxane B2 (Carbajal et al., 1998; Noa and Mas, 1998; Ravelo et al., 2011).

Due to antimicrobial, antioxidant, and antiinflammatory properties, beeswax is added to ointments, liniments, and creams used to treat various dermatoses, i.e., atopic dermatitis, psoriasis, diaper dermatitis, considered as active ingredients alone or in combination with honey or propolis (Table 2.5).

Bee venom

Bee venom (BV) is known as an apitoxin. Apitoxin consists of a complex mixture of peptides, enzymes, and nonpeptide components. Peptides are the main components of bee venom and to them belong apamin, adolapin,

Table 2.5 Beeswax components and their impact on skin.

Components of beeswax	Impact on skin	References
Squalene, 10-hydroxy-trans-2-decenoic acid, chrysin	Antibacterial effect	Kačániová et al. (2012)
Squalene, 10-hydroxy-trans-2-decenoic acid, chrysin	Antifungal effect	Al-Waili (2005a, 2005b)
Flavonoids	Antioxidant effect	Kurek-Górecka et al. (2020)
Chrysin, D-002	Antiinflammatory effect	Ha et al. (2010), Ciftci and Ozdemir (2011), O'Leary et al. (2004), Carbajal et al. (1998), Ravelo et al. (2011), Noa and Mas (1998)

melittin, secapin, procamine, tertiapin, and mast cell degranulating peptide (MCD-peptides) (El-Seedi et al., 2020; Frangieh et al., 2019). There are also enzymes; phospholipase A2, hyaluronidase, alfa-glucosidase, acid phosphatase, acid phosphomonoesterase, and lysophospholipase. Among components other than peptides, there are carbohydrates (mainly glucose and fructose), amino acids such as γ-aminobutyric acid and β-aminoisobutyric acid, and minerals. Among the latter, magnesium, calcium, and phosphor can be listed.

Bee venom is a transparent liquid drying up even at room temperature, odorless, bitter taste, and pH value ranges from 4.5 to 5.5 (Ali, 2012). It is used as a cosmetic ingredient as a light yellow powder obtained using a bee venom collector without harming the honey bee. Bee venom as a substance for cosmetics and medicines should be purified. The purification is based on dissolving bee venom in a solvent to form a bee venom solution. Afterward, an adsorbent is applied to the bee venom solution to form a mixture. After removing the adsorbent through filtration, the filtrate is lyophilized at −20°C or colder to produce pure bee venom as a powder for use as a cosmetic ingredient (Han et al., 2013a). It shows multipotential action, including antiaging, antiinflammatory, and antibacterial (Han et al., 2017; Kim et al., 2019).

Cosmetic applications of bee venom

Bee venom may be used to produce antiphotoaging and antiacne cosmetic products because of its biological properties. Due to its antiaging properties, it can be helpful in cosmetic applications. Bee venom used in the case of facial creams exhibits the ability to the reduction of wrinkles. This effect was confirmed in clinical trials conducted among Korean women. It was used as an antiwrinkle agent included in serum at a concentration of 0.006% at an amount of 4 mL twice a day for 12 weeks. It caused a decrease in total wrinkle area, total wrinkle count, and wrinkle depth and size. At 8 weeks using a serum with bee venom, Han et al. (2015) observed about 6.64% decrement of wrinkles in all patients, and at 12 weeks, about 11.83% decrement of wrinkles compared with the serum without bee venom. Total wrinkle area was decreased in a time-dependent manner after application of the bee-venom serum.

The highest reduction of wrinkle area was observed at the 12th week after BV enriched serum application, from $26.04 \pm 3.46\,mm^2$ to $24.44 \pm 2.69\,mm^2$. Moreover, the highest reduction of wrinkle depth and total wrinkles count was observed at the 12th week compared to the situation before serum with BV application. The critical role in wrinkles formation plays UVA and UVB exposure. Ultraviolet radiation stimulates interstitial collagenase (MMP-1) and stromelysin-1 (MMP-3) production in human dermal fibroblasts, leading to the photoaging process. MMP-1 and MMP-3 belong to a superfamily of zinc-dependent endopeptidases. They are responsible for the degradation of extracellular matrix (ECM) components, including collagen fibers, resulting in the loss of elasticity and degradation of biomechanical properties of the skin. Han et al. (2007, 2017) investigated the effect of honey bee venom on UVB irradiated human dermal fibroblasts. They found that bee venom exhibited photoprotective activity by reducing UV-induced MMP-1 and MMP-3 activity by 50%–80% and 50%–85%, respectively, compared to the control (Han et al., 2007).

Pharmacological value of bee venom

Topical administration of bee venom has also been effective in acne vulgaris treatment, as reported by Marques Pereira et al. (2020). Acne vulgaris is a dermatological disease characterized by increased sebum production, bacterial colonization, and inflammation. Bee venom seems to be a good antibacterial agent. It demonstrates an inhibitory effect on *Propionibacterium acnes*. The bacteriostatic and bactericidal effect of BV is associated with melittin. Melittin is responsible for the destruction of the bacterial cell membrane. It leads to a loss of membrane integrity. Therefore, melittin causes changes in the cell morphology, including cell distortion and loss of cytoplasm content.

It was reported that Gram-positive bacteria, including *Propionibacterium acnes,* were more susceptible to melittin, compared to Gram-negative bacteria, due to the nature of the organism's cell membrane. It should be noted that melittin can penetrate the peptideoglycan layer of the Gram-positive bacteria cell membrane better than the Gram-negative ones because the latter possess a lipopolysaccharide layer protecting the membrane against melittin (Lüderitz et al., 1982; Navarre and Schneewind, 1999). The crucial role in the antibacterial action of melittin depends on the presence of proline residue in the peptide chain. The absence of proline reduces the antimicrobial action of melittin, which was found in the analog peptide (Jamasbi et al., 2014). Besides melittin, another peptide, namely secapin, plays an important role in the antimicrobial actions of bee venom. It directly binds to the bacteria cell walls (Lee et al., 2016). During incubation of *Propionibacterium acnes* with bee venom at various concentrations from 0.005 to 1.0 μg/mL. Han et al. (2013b) showed that the mentioned bee product exhibited antimicrobial activity against *Propionibacterium acnes* at the concentration of 0.5 μg, reducing the number of *P. acnes* the colony-forming unit (CFU) by approximately 6 logs.

Moreover, it was reported that bee venom enriched cosmetics, while topically applying in selected individuals (all subjects must have had mild to moderate acne) at the concentration of 0.06 mg/mL, leading to significant adenosine triphosphate (ATP) level decrease being equal to 57.5%. The level of ATP was evaluated using the ATP swab test, which indicates the remaining *P. acnes* deposited on the skin. The measured level of ATP indicates the reduction of *Propionibacterium acnes* amount on the skin because bacteria release ATP during rupturing the cell wall (Han et al., 2013b). Therefore, bee venom seems to be a good component of cosmetics and medicines for the skin of patients suffering from acne vulgaris due to its antibacterial and antiinflammatory effects. The latter one is associated with bee venom melittin, adolapin. Melittin blocks the expression of inflammatory genes via NF-kB, which directly controls genes involved in immune reactions (Park et al., 2004). Adolapin inhibits prostaglandin synthesis, which takes part in the inflammatory process by inhibiting cyclooxygenase activity (Bellik, 2015; Dadar et al., 2019). Han et al. (2010) showed that bee venom decreases the production and release of Il-8 and TNF-α from keratinocytes and monocytes, stimulated by *Propionibacterium acnes* responsible for inflammatory acne.

The antiinflammatory effect of bee venom allows it to be used to treat atopic dermatitis, a chronic disease characterized by defects in epidermal barrier function and cutaneous inflammation. It was found that patients suffering from atopic dermatitis and using emollient with bee venom and silk protein twice a day for 4 weeks demonstrated lower eczema area severity index, i.e., decreased about 34.69%, lower transepidermal water loss equal to 6.97%, and lower visual analog scale value for pruritis about −1.5 compared to patients who applied emollient without bee venom (You et al., 2016). It is also noteworthy that the antiinflammatory effect of bee venom brings satisfactory results in patients with psoriasis, a chronic, inflammatory disease associated with overproduction of inflammatory cytokines and chemokines activating T-cells. A significant reduction in serum IL-1β level was observed in patients receiving bee venom intradermally twice a week for 3 months. The IL-1β serum concentration was reduced from 222 ± 4.13 before treatment to 113 ± 3.73 after treatment (Hegazi et al., 2013). Moreover, Kwon et al. (2002) indicated that bee venom might also exhibit antiinflammatory action by reducing IL-6 levels usually increased during the disease in question.

The biological activities of bee venom, such as antiinflammatory ones, allow use in wound treatment. Many studies have shown that bee venom treatment may be helpful during all the phases of the healing process, such as hemostasis, inflammation, proliferation, and remodeling. During hemostasis, bee venom upregulates the expression of transforming growth factor (TGF-β) and vascular endothelial growth factor (VEGF). TGF-β is responsible for recruiting monocytes and neutrophils and initiating an inflammatory response, while VEGF stimulates endothelial cells to the angiogenesis process. During the second phase of wound healing, i.e., inflammation, BV reduces proinflammatory cytokines such as IL-1, IL-6, TNF-α, thus limiting a prolonged inflammation. In the third stage of the healing process, i.e., proliferation, the therapy with BV increases the collagen type I formation. Due to stimulation of human epidermal keratinocytes proliferation and migration and reduction of the IL-8, TNF-β level, bee venom influences the reepithelialization phase. Furthermore, bee venom is responsible for better neovascularization and angiogenesis by decreasing the caspase-3, -8, and -9 activity, upregulation of TGF-β and VEGF expression. Angiogenesis is activated by TGF-β and VEGF and stimulates the repair of tissues. It is worthy of note that the impaired wound healing process, e.g., in diabetes mellitus, is connected with caspase cascades that take part in the induction of apoptosis and stimulated endothelial progenitors cells in tissues confirmed by Hozzein et al. (2018).

During the fourth phase of wound healing, a bee venom increases collagen type I content in the bed wound and decreases expression of activating transcription factor-3 (ATF-3) and inducible nitric oxide synthase (iNOS). Due to the regulated expression of ATF-3 and iNOS and their impact on oxidative stress, a bee venom improves cellular differentiation and wound remodeling. In this stage, the extracellular matrix is rebuilt, and the production of collagen type I increases. The scar results from fibroblast apoptosis and extracellular matrix degradation (Badr et al., 2016; Kurek-Górecka et al., 2021). Summing up, bee venom may be considered a valuable component of wound dressing found in studies on animal models. Amin and Abdel-Raheem (2014) demonstrated that bee venom-loaded dressing containing a 4% apitoxin brings beneficial action in the wound repair process. Besides, 4% of bee venom reduced carrageenan-induced rat paw edema at a similar degree as 1% diclofenac (Amin and Abdel-Raheem, 2014). Bee venom in one topical formulation reduced PGE2 in the serum of rats (Amin et al., 2008). Additionally, bee venom at a concentration of 6% with chitosan exhibited a similar antiinflammatory effect like indomethacin.

Table 2.6 Bee venom components and their impact on skin.

Components of bee venom	Impact on skin	References
Not reported	Antiaging effect	Han et al. (2015)
Not reported	Photo protecting effect	Han et al. (2007, 2017)
Melittin, secapin	Antibacterial effect	Jamasbi et al. (2014), Lee et al. (2016)
Melittin, adolapin	Antiinflammatory effect	Park et al. (2004), Bellik (2015), Han et al. (2010)

In the light of available information, it is pretty promising that bee venom-loaded wound dressing with polyvinyl alcohol-chitosan and chitosan may be a potential agent in a novel strategy of wound healing management (Table 2.6).

Conclusion

Tables 2.7 and 2.8. summarize the possibilities of using bee products as a component of cosmetics and medicines to treat skin diseases or skin problems and mode of actions in wound healing. It is evident that bee products are valuable

Table 2.7 Bee products as a component of cosmetics and medicines for the treatment of skin diseases or skin problems.

Skin disease or skin problem	Bee products	References
Acne vulgaris	Honey, propolis, bee venom	Julianti et al. (2017), Mazzarello et al. (2018), Han et al. (2010), McLoone et al. (2016)
Psoriasis	Beeswax, bee venom, honey, propolis	Al-Waili (2003), Kubota et al. (1983), Hegazi et al. (2013)
Atopic dermatitis	Beeswax, bee venom, honey, propolis	Souza et al. (2017), Al-Waili (2003), You et al. (2016)
Diaper dermatitis	Beeswax, honey, propolis	Al-Waili (2005a, 2005b), Kurek-Górecka et al. (2020)
Acne vulgaris	Bee venom	Navarre and Schneewind (1999), Han et al. (2013b)

Table 2.8 Effect of bee products on wound healing process.

Bee products	Impact on wound healing	Application on skin	References
Honey	Decreasing the biofilm, reducing inflammation formation, increasing epithelialization,	Polyacrylamide polymer dressing, hydroactive, hydrogel-coated, polyurethane wound dressing, alginate dressing, gel, sterile 100% honey	Lee et al. (2011), Wijesinghe et al. (2009), Minden-Birkenmaier and Bowlin (2018)
Propolis	Increasing proliferation, reepithelization and cicatrization	Ointment, cream, biodegradable nanofiber dressing	Olczyk et al. (2013, 2020)
Bee pollen	Increasing epithelialization	Ointment	Olczyk et al. (2016)
Royal jelly	Increasing reepithelization	Cream, ointment, wound dressing	Siavash et al. (2015), Bucekova et al. (2017)
Bee venom	Impact on hemostasis, limiting prolonged inflammation, increasing proliferation and remodeling.	Polyvinyl alcohol–chitosan hydrogel, chitosan blend films	Hozzein et al. (2015), Badr et al. (2016), Amin et al. (2008), Amin and Abdel-Raheem (2014), Kurek-Górecka et al. (2020)

ingredients for skin treatment and care. Many scientific experiments demonstrated the abilities of honey, propolis, royal jelly, and pollen to heal wounds and explained their numerous functional properties such as antibacterial, antiinflammatory, antioxidant, disinfectant, antifungal, and antiviral.

The bee products offer many advantages for pharmaceuticals and cosmetics based on their high effectiveness with minimal side effects. It is evident that bee products have great potential as natural alternative sources for skin therapy and cosmetic formulations.

References

Abu-Mellal, A., Koolaji, N., Duke, R.K., Tran, V.H., Duke, C.C., 2012. Prenylated cinnamate and stilbenes from Kangaroo Island propolis and their antioxidant activity. Phytochemistry 77, 251–259.

Afkhamizadeh, M., Aboutorabi, R., Ravari, H., Fathi Najafi, M., Ataei Azimi, S., Javadian Langaroodi, A., Yaghoubi, M.A., Sahebkar, A., 2018. Topical propolis improves wound healing in patients with diabetic foot ulcer: a randomized controlled trial. Nat. Prod. Res. 32 (17), 2096–2099.

Ahmad, S., Campos, M.G., Fratini, F., Altaye, S.Z., Li, J., 2020. New insights into the biological and pharmaceutical properties of royal jelly. Int. J. Mol. Sci. 21 (2), 382.

Ali, M.A.A.S.M., 2012. Studies on bee venom and its medical uses. Int. J. Adv. Res. Technol. 1 (2), 69–83.

Ali, B.M.M., Ghoname, N.F., Hodeib, A.A., Elbadawy, M.A., 2015. Significance of topical propolis in the treatment of facial acne vulgaris. Egypt. J. Dermatol. Venerol. 35 (1), 29–36.

Al-Waili, N.S., 2001. Therapeutic and prophylactic effects of crude honey on chronic seborrheic dermatitis and dandruff. Eur. J. Med. Res. 6 (7), 306–308.

Al-Waili, N.S., 2003. Topical application of natural honey, beeswax and olive oil mixture for atopic dermatitis or psoriasis: partially controlled, single-blinded study. Complement. Ther. Med. 11 (4), 226–234.

Al-Waili, N.S., 2004. An alternative treatment for pityriasis versicolor, tinea cruris, tinea corporis and tinea faciei with topical application of honey, olive oil and beeswax mixture: an open pilot study. Complement. Ther. Med. 12 (1), 45–47.

Al-Waili, N.S., 2005a. Clinical and mycological benefits of topical application of honey, olive oil and beeswax in diaper dermatitis. Clin. Microbiol. Infect. 11 (2), 160–163.

Al-Waili, N.S., 2005b. Mixture of honey, beeswax and olive oil inhibits growth of *Staphylococcus aureus* and *Candida albicans*. Arch. Med. Res. 36 (1), 10–13.

Al-Waili, N., Hozzein, W.N., Badr, G., Al-Ghamdi, A., Al-Waili, H., Al-Waili, T., Salom, K., 2015. Propolis and bee venom in diabetic wounds; a potential approach that warrants clinical investigation. Afr. J. Tradit. Complement. Altern. Med. 12 (6), 1–11.

Amin, M.A., Abdel-Raheem, I.T., 2014. Accelerated wound healing and anti-inflammatory effects of physically cross linked polyvinyl alcohol–chitosan hydrogel containing honey bee venom in diabetic rats. Arch. Pharm. Res. 37 (8), 1016–1031.

Amin, M.A., Abdel-Raheem, I.T., Madkor, H.R., 2008. Wound healing and anti-inflammatory activities of bee venom-chitosan blend films. J. Drug Deliv. Sci. Technol. 18 (6), 424–430.

Amoros, M., Simõs, C.M.O., Girre, L., Sauvager, F., Cormier, M., 1992. Synergistic effect of flavones and flavonols against herpes simplex virus type 1 in cell culture. Comparison with the antiviral activity of propolis. J. Nat. Prod. 55 (12), 1732–1740.

Badr, G., Hozzein, W.N., Badr, B.M., Al Ghamdi, A., Saad Eldien, H.M., Garraud, O., 2016. Bee venom accelerates wound healing in diabetic mice by suppressing activating transcription factor-3 (ATF-3) and inducible nitric oxide synthase (iNOS)-mediated oxidative stress and recruiting bone marrow-derived endothelial progenitor cells. J. Cell. Physiol. 231 (10), 2159–2171.

Baitala, T.V., Faquinello, P., de Toledo, V.D.A.A., Mangolin, C.A., Martins, E.N., Ruvolo-Takasusuki, M.C.C., 2010. Potential use of major royal jelly proteins (MRJPs) as molecular markers for royal jelly production in Africanized honeybee colonies. Apidologie 41 (2), 160–168.

Basista, K., Sodzawiczny, K., 2011. Pyłek kwiatowy – nowy surowiec naturalny, możliwości wykorzystania w lecznictwie i kosmetologii (Bee pollen – a new natural material, possibilities of use in medicine and cosmetology). Gazeta Farmaceutyczna 12, 30–32 (in Polish).

Balos, M.M., Popov, N.S., Radulovic, J.Z.P., Stojanov, I.M., Jaksic, S.M., 2020. Sugar profile of different floral origin honeys from Serbia. J. Apic. Res.

Bellik, Y., 2015. Bee venom: its potential use in alternative medicine. Anti-Infect. Agents 13 (1), 3–16.

Belsito, M.D., Klaassen, C.D., Liebler, D.C., Marks Jr., J.G., Shank, R.C., 2019. Safety Assessment of Honey Ingredients as Used in Cosmetics. The 2019 Cosmetic Ingredient Review Expert Panel. Available from: https://www.cir-safety.org/sites/default/files/honey092019slr.pdf. (Accessed 5 May 2021).

Blaser, G., Santos, K., Bode, U., Vetter, H., Simon, A., 2007. Effect of medical honey on wounds colonised or infected with MRSA. J. Wound Care 16 (8), 325–328.

Bocho-Janiszewska, A., Sikora, A., Rajewski, J., Łobodzin, P., 2013. Zastosowanie mleczka pszczelego w kremach nawilżających (Application of royal jelly in moisturizing creams). Pol. J. Cosmetol. 16 (4), 314–320 (in Polish).

Bogdanov, S., 1997. Nature and origin of the antibacterial substances in honey. LWT-Food Sci. Technol. 30 (7), 748–753.

Boisard, S., Le Ray, A.M., Landreau, A., Kempf, M., Cassisa, V., Flurin, C., Richomme, P., 2015. Antifungal and antibacterial metabolites from a French poplar type propolis. Evid. Based Complement. Alternat. Med. 2015. https://doi.org/10.1155/2015/319240.

Bolfa, P., Vidrighinescu, R., Petruta, A., Dezmirean, D., Stan, L., Vlase, L., Damian, G., Catoi, C., Filip, A., Clichici, S., 2013. Photoprotective effects of Romanian propolis on skin of mice exposed to UVB irradiation. Food Chem. Toxicol. 62, 329–342.

Boselli, E., Caboni, M.F., Sabatini, A.G., Marcazzan, G.L., Lercker, G., 2003. Determination and changes of free amino acids in royal jelly during storage. Apidologie 34 (2), 129–137.

Bouhlali, E.D.T., Bammou, M., Sellam, K., El Midaoui, A., Bourkhis, B., Ennassir, J., Alem, C., Filali-Zegzouti, Y., 2019. Physicochemical properties of eleven monofloral honey samples produced in Morocco. Arab J. Basic Appl. Sci. 26 (1), 476–487.

Bucekova, M., Sojka, M., Valachova, I., Martinotti, S., Ranzato, E., Szep, Z., Majtan, V., Klaudiny, J., Majtan, J., 2017. Bee-derived antibacterial peptide, defensin-1, promotes wound re-epithelialisation in vitro and in vivo. Sci. Rep. 7 (1), 1–13.

Burlando, B., Cornara, L., 2013. Honey in dermatology and skin care: a review. J. Cosmet. Dermatol. 12 (4), 306–313.

Butnariu, M.V., Giuchici, C.V., 2011. The use of some nanoemulsions based on aqueous propolis and lycopene extract in the skin's protective mechanisms against UVA radiation. J. Nanobiotechnol. 9 (1), 1–9.

Carbajal, D., Molina, V., Valdés, S., Arruzazabala, M.D.L., Mas, R., Magraner, J., 1998. Anti-inflammatory activity of D-002: an active product isolated from beeswax. Prostaglandins Leukot. Essent. Fat. Acids 59 (4), 235–238.

Castaldo, S., Capasso, F., 2002. Propolis, an old remedy used in modern medicine. Fitoterapia 73, 1–6.

Ciftci, O., Ozdemir, I., 2011. Protective effects of quercetin and chrysin against 2, 3, 7, 8-tetrachlorodibenzo-p-dioxin (TCDD) induced oxidative stress, body wasting and altered cytokine productions in rats. Immunopharmacol. Immunotoxicol. 33 (3), 504–508.

Cole, N., Sou, P.W., Ngo, A., Tsang, K.H., Severino, J.A., Arun, S.J., Duke, C.C., Reeve, V.E., 2010. Topical 'Sydney'propolis protects against UV-radiation-induced inflammation, lipid peroxidation and immune suppression in mouse skin. Int. Arch. Allergy Immunol. 15 (2), 87–97.

Dadar, M., Shahali, Y., Chakraborty, S., Prasad, M., Tahoori, F., Tiwari, R., Dhama, K., 2019. Antiinflammatory peptides: current knowledge and promising prospects. Inflamm. Res. 68 (2), 125–145.

Denisow, B., Denisow-Pietrzyk, M., 2016. Biological and therapeutic properties of bee pollen: a review. J. Sci. Food Agric. 96 (13), 4303–4309.

Deveci, M., Gilmont, R.R., Dunham, W.R., Mudge, B.P., Smith, D.J., Marcelo, C.L., 2005. Glutathione enhances fibroblast collagen contraction and protects keratinocytes from apoptosis in hyperglycaemic culture. Br. J. Dermatol. 152 (2), 217–224.

El-Seedi, H., El-Wahed, A., Yosri, N., Musharraf, S.G., Chen, L., Moustafa, M., Zou, X., Al-Mousawi, S., Guo, Z., Khatib, A., Khalifa, S., 2020. Antimicrobial properties of Apis mellifera's bee venom. Toxins 12 (7), 451.

Eshraghi, S., 2005. An evaluation of the potent inhibitory effects of royal jelly fractions against Streptomyces bacteria. Pak. J. Med. Sci. 21 (1), 63–68.

Fatrcová-Šramková, K., Nôžková, J., Kačániová, M., Máriássyová, M., Rovná, K., Stričík, M., 2013. Antioxidant and antimicrobial properties of monofloral bee pollen. J. Environ. Sci. Health B 48 (2), 133–138.

Fonseca, Y.M., Marquele-Oliveira, F., Vicentini, F.T., Furtado, N.A.J., Sousa, J.P.B., Lucisano-Valim, Y.M., Fonseca, M.J.V., 2011. Evaluation of the potential of Brazilian propolis against UV-induced oxidative stress. Evid. Based Complement. Alternat. Med. 2011. https://doi.org/10.1155/2011/863917.

Frangieh, J., Salma, Y., Haddad, K., Mattei, C., Legros, C., Fajloun, Z., El Obeid, D., 2019. First characterization of the venom from apis mellifera syriaca, a honeybee from the middle east region. Toxins 11 (4), 191.

Fratini, F., Cilia, G., Mancini, S., Felicioli, A., 2016a. Royal Jelly: an ancient remedy with remarkable antibacterial properties. Microbiol. Res. 192, 130–141.

Fratini, F., Cilia, G., Turchi, B., Felicioli, A., 2016b. Beeswax: a minireview of its antimicrobial activity and its application in medicine. Asian Pac. J. Trop. Med. 9 (9), 839–843.

Fujii, A., Kobayashi, S., Kuboyama, N., Furukawa, Y., Kaneko, Y., Ishihama, S., Yamamoyo, H., Tamura, T., 1990. Augmentation of wound healing by royal jelly (RJ) in streptozotocin-diabetic rats. Jpn. J. Pharmacol. 53 (3), 331–337.

Garedew, A., Schmolz, E., Lamprecht, I., 2004. Microbiological and calorimetric investigations on the antimicrobial actions of different propolis extracts: an in vitro approach. Thermochim. Acta 422 (1–2), 115–124.

George, N.M., Cutting, K.F., 2007. Antibacterial honey (Medihoney™): in-vitro activity against clinical isolates of MRSA, VRE, and other multiresistant gram-negative organisms including *Pseudomonas aeruginosa*. Wounds 19 (9), 231–236.

Goharshenasan, P., Amini, S., Atria, A., Abtahi, H., Khorasani, G., 2016. Topical application of honey on surgical wounds: a randomized clinical trial. Complement. Med. Res. 23 (1), 12–15.

Goik, U., Goik, T., Załęska, I., 2016. Właściwości wosku pszczelego i jego zastosowanie w kosmetyce i kosmetologii (Beeswax properties and its use in cosmetics and cosmetology). Kosmetol. Estetyczna 6, 617–626 (in Polish).

Gregoris, E., Fabris, S., Bertelle, M., Grassato, L., Stevanato, R., 2011. Propolis as potential cosmeceutical sunscreen agent for its combined photoprotective and antioxidant properties. Int. J. Pharm. 405 (1–2), 97–101.

Guo, H., Kouzuma, Y., Yonekura, M., 2009. Structures and properties of antioxidative peptides derived from royal jelly protein. Food Chem. 113 (1), 238–245.

Gupta, A.K., Nicol, K., Batra, R., 2004. Role of antifungal agents in the treatment of seborrheic dermatitis. Am. J. Clin. Dermatol. 5 (6), 417–422.

Ha, S.K., Moon, E., Kim, S.Y., 2010. Chrysin suppresses LPS-stimulated proinflammatory responses by blocking NF-κB and JNK activations in microglia cells. Neurosci. Lett. 485 (3), 143–147.

Han, M.C., Durmus, A.S., Karabulut, E., Yaman, I., 2005. Effects of Turkish propolis and silver sulfadiazine on burn wound healing in rats. Rev. Med. Vet. 156 (12), 624–627.

Han, S., Lee, K., Yeo, J., Kweon, H., Woo, S., Lee, M., Baek, H., Park, K., 2007. Inhibitory effect of bee venom against ultraviolet B induced MMP-11 and MMP-3 in human dermal fibroblasts. J. Apic. Res. 46 (2), 94–98.

Han, S., Lee, K., Yeo, J., Baek, H., Park, K., 2010. Antibacterial and anti-inflammatory effects of honeybee (*Apis mellifera*) venom against acne-inducing bacteria. J. Med. Plant Res. 4 (6), 459–464.

Han, S.M., Lee, K.G., Park, K.K., 2013a. Method for Purifying Bee Venom on Mass Scale. Patent WO2013103178. Available from: https://patentscope.wipo.int/search/en/detail.jsf;jsessionid=5B0D17B5B63DB832690DC720E78EB937.wapp1nB?docId=WO2013103178&tab=FULLTEXT. (Accessed 16 May 2021).

Han, S.M., Lee, K.G., Pak, S.C., 2013b. Effects of cosmetics containing purified honeybee (*Apis mellifera* L.) venom on acne vulgaris. J. Integr. Med. 11 (5), 320–326.

Han, S.M., Hong, I.P., Woo, S.O., Chun, S.N., Park, K.K., Nicholls, Y.M., Pak, S.C., 2015. The beneficial effects of honeybee-venom serum on facial wrinkles in humans. Clin. Interv. Aging 10, 1587. https://doi.org/10.2147/CIA.S84940.

Han, S.M., Hong, I.P., Woo, S.O., Kim, S.G., Jang, H.R., Park, K.K., 2017. Evaluation of the skin phototoxicity and photosensitivity of honeybee venom. J. Cosmet. Dermatol. 16 (4), 68–75.

Hegazi, A.G., Abd Raboh, F.A., Ramzy, N.E., Shaaban, D.M., Khader, D.Y., 2013. Bee venom and propolis as new treatment modality in patients with localized plaque psoriases. Int. Res. J. of Med. Med. Sci. 1 (1), 27–33.

Hernández, I.M., Fernandez, M.C., Cuesta-Rubio, O., Piccinelli, A.L., Rastrelli, L., 2005. Polyprenylated benzophenone derivatives from Cuban propolis. J. Nat. Prod. 68 (6), 931–934.

Hołderna-Kędzia, E., Ostrowski-Meissner, H., Kędzia, B., 2008. Ocena aktywności antybiotycznej nowozelandzkiego miodu manuka metodą rozcieńczeń seryjnych w podłożu płynnym (Estimation of antibiotic activity of New Zealand manuka honey by the method of serial dilutions in liquid medium). Postępy Fitoterapii 2, 70–75 (in Polish).

Hozzein, W.N., Badr, G., Al Ghamdi, A.A., Sayed, A., Al-Waili, N.S., Garraud, O., 2015. Topical application of propolis enhances cutaneous wound healing by promoting TGF-beta/Smad-mediated collagen production in a streptozotocin-induced type I diabetic mouse model. Cell. Physiol. Biochem. 37 (3), 940–954.

Hozzein, W.N., Badr, G., Badr, B.M., Allam, A., Al Ghamdi, A., Al-Wadaan, M.A., Al-Waili, N.S., 2018. Bee venom improves diabetic wound healing by protecting functional macrophages from apoptosis and enhancing Nrf2, Ang-1 and Tie-2 signaling. Mol. Immunol. 103, 322–335.

Hu, C., Kitts, D.D., 2005. Dandelion (*Taraxacum officinale*) flower extract suppresses both reactive oxygen species and nitric oxide and prevents lipid oxidation in vitro. Phytomedicine 12 (8), 588–597.

Huang, W.J., Huang, C.H., Wu, C.L., Lin, J.K., Chen, Y.W., Lin, C.L., Chuang, S.E., Huang, C.Y., Chen, C.N., 2007. Propolin G, a prenylflavanone, isolated from Taiwanese propolis, induces caspase-dependent apoptosis in brain cancer cells. J. Agric. Food Chem. 55 (18), 7366–7376.

Jamasbi, E., Batinovic, S., Sharples, R.A., Sani, M.A., Robins-Browne, R.M., Wade, J.D., Separovic, F., Hossain, M.A., 2014. Melittin peptides exhibit different activity on different cells and model membranes. Amino Acids 46 (12), 2759–2766.

Javed, A., 2021. Efficacy of apitherapy for skin regeneration. Am. Sci. Res. J. Eng. Technol. Sci. 75 (1), 80–85.

Julianti, E., Rajah, K.K., Fidrianny, I., 2017. Antibacterial activity of ethanolic extract of cinnamon bark, honey, and their combination effects against acne-causing bacteria. Sci. Pharm. 85 (2), 19. https://doi.org/10.3390/scipharm85020019.

Kacániová, M., Vuković, N., Chlebo, R., Haščík, P., Rovna, K., Cubon, J., Dżugan, M., Pasternakiewicz, A., 2012. The antimicrobial activity of honey, bee pollen loads and beeswax from Slovakia. Arch. Biol. Sci. 64 (3), 927–934.

Kapoulas, V.M., Mastronicolis, S.K., Galanos, D.S., 1977. Identification of the lipid components of honey. Z. Lebensm. Unters. Forsch. 163 (2), 96–99.

Kasparaviciene, G., Savickas, A., Kalveniene, Z., Velziene, S., Kubiliene, L., Bernatoniene, J., 2016. Evaluation of beeswax influence on physical properties of lipstick using instrumental and sensory methods. Evid. Based Complement. Alternat. Med. 2016. https://doi.org/10.1155/2016/3816460.

Kędzia, B., Hołderna-Kędzia, E., 2017. Współczesne poglądy na mechanizm przeciwdrobnoustrojowego działania miodu (Contemporary opinions on the mechanism of antimicrobial action of honey). Postępy Fitoterapii 4, 290–297 (in Polish).

Kilian-Pięta, E., Hoppe, M., 2019. Wpływ melanogenezy na powstawanie przebarwień (The effect of melanogenesis on the formation of hyperpigmentation). Kosmetol. Estetyczna 8 (4), 419–422 (in Polish).

Kim, S.B., Jo, Y.H., Liu, Q., Ahn, J.H., Hong, I.P., Han, S.M., Hwang, B.Y., Lee, M.K., 2015. Optimization of extraction condition of bee pollen using response surface methodology: correlation between anti-melanogenesis, antioxidant activity, and phenolic content. Molecules 20 (11), 19764–19774.

Kim, H., Park, S.Y., Lee, G., 2019. Potential therapeutic applications of bee venom on skin disease and its mechanisms: a literature review. Toxins 11 (7), 374.

Kocot, J., Kiełczykowska, M., Luchowska-Kocot, D., Kurzepa, J., Musik, I., 2018. Antioxidant potential of propolis, bee pollen, and royal jelly: possible medical application. Oxidative Med. Cell. Longev. 2018. https://doi.org/10.1155/2018/7074209.

Kohno, K., Okamoto, I., Sano, O., Arai, N., Iwaki, K., Ikeda, M., Kurimoto, M., 2004. Royal jelly inhibits the production of proinflammatory cytokines by activated macrophages. Biosci. Biotechnol. Biochem. 68 (1), 138–145.

Kolayli, S., Sahin, H., Can, Z., Yildiz, O., Malkoc, M., Asadov, A., 2016. A member of complementary medicinal food: anatolian royal jellies, their chemical compositions, and antioxidant properties. J. Evid. Based Complement. Alternat. Med. 21 (4), 43–48.

Komosińska-Vassev, K., Olczyk, P., Kaźmierczak, J., Mencner, L., Olczyk, K., 2015. Bee pollen: chemical composition and therapeutic application. Evid. Based Complement. Alternat. Med. 2015. https://doi.org/10.1155/2015/297425.

Koya-Miyata, S., Okamoto, I., Ushio, S., Iwaki, K., Ikeda, M., Kurimoto, M., 2004. Identification of a collagen production-promoting factor from an extract of royal jelly and its possible mechanism. Biosci. Biotechnol. Biochem. 68 (4), 767–773.

Krell, R., 1996. Value-Added Products from Beekeeping. FAO Agricultural Services Bulletin no 124. Food and Agriculture Organization of the United Nations, Rome, Italy.

Kubota, K., Kumakiri, M., Miura, Y., Hine, K., Kori, N., Saito, H., Miyazaki, K., Arita, T., 1983. Clinical studies on zinc oxide ointment replacing boric acid and zinc oxide ointment (JP8). Hokkaido J. Med. Sci. 58 (4), 400–405.

Kurek-Górecka, A., Komosinska-Vasse, V.K., Rzepecka-Stojko, A., Olczyk, P., 2021. Bee venom in wound healing. Molecules 26 (1), 148. https://doi.org/10.3390/molecules26010148.

Kurek-Górecka, A.M., Sobczak, A., Rzepecka-Stojko, A., Górecki, M.T., Wardas, M., Pawłowska-Góral, K., 2012. Antioxidant activity of ethanolic fractions of polish propolis. Z. Naturforsch. C 67 (11–12), 545–550.

Kurek-Górecka, A., Rzepecka-Stojko, A., Górecki, M., Stojko, J., Sosada, M., Świerczek-Zięba, G., 2014. Structure and antioxidant activity of polyphenols derived from propolis. Molecules 19 (1), 78–101.

Kurek-Górecka, A., Górecki, M., Rzepecka-Stojko, A., Balwierz, R., Stojko, J., 2020. Bee products in dermatology and skin care. Molecules 25 (3), 556. https://doi.org/10.3390/molecules25030556.

Kwakman, P.H., Velde, A.A.T., de Boer, L., Speijer, D., Christina Vandenbroucke-Grauls, M.J., Zaat, S.A., 2010. How honey kills bacteria. FASEB J. 24 (7), 2576–2582.

Kwon, Y.B., Lee, H.J., Han, H.J., Mar, W.C., Kang, S.K., Yoon, O.B., Beitz, A.J., Lee, J.H., 2002. The water-soluble fraction of bee venom produces antinociceptive and anti-inflammatory effects on rheumatoid arthritis in rats. Life Sci. 71 (2), 191–204.

Lee, D.S., Sinno, S., Khachemoune, A., 2011. Honey and wound healing. Am. J. Clin. Dermatol. 12 (3), 181–190.

Lee, K.S., Kim, B.Y., Yoon, H.J., Choi, Y.S., Jin, B.R., 2016. Secapin, a bee venom peptide, exhibits anti-fibrinolytic, anti-elastolytic, and anti-microbial activities. Dev. Comp. Immunol. 63, 27–35.

Li, Y., Xiang, Q., Zhang, Q., Huang, Y., Su, Z., 2012. Overview on the recent study of antimicrobial peptides: origins, functions, relative mechanisms and application. Peptides 37 (2), 207–215.

Liu, J.R., Yang, Y.C., Shi, L.S., Peng, C.C., 2008. Antioxidant properties of royal jelly associated with larval age and time of harvest. J. Agric. Food Chem. 56 (23), 11447–11452.

López-Gutiérrez, N., del Mar Aguilera-Luiz, M., Romero-González, R., Vidal, J.L.M., Frenich, A.G., 2014. Fast analysis of polyphenols in royal jelly products using automated TurboFlow™-liquid chromatography–orbitrap high resolution mass spectrometry. J. Chromatogr. B 973, 17–28.

Lüderitz, O., Freudenberg, M.A., Galanos, C., Lehmann, V., Rietschel, E.T., Shaw, D.H., 1982. Lipopolysaccharides of gram-negative bacteria. In: Current Topics in Membranes and Transport. 17. Academic Press, pp. 79–151.

Mandal, M.D., Mandal, S., 2011. Honey: its medicinal property and antibacterial activity. Asian Pac. J. Trop. Biomed. 1 (2), 154–160.

Marion, C., 2000. Use of Unfermented Honey as an Agent for Depigmenting and/or Lightening the Skin, Body Hair or Head Hair and/or for Treating Liver Spots. Patent CN1121850C. Available from: https://patents.google.com/patent/CN1121850C/en. (Accessed 26 April 2021).

Marques Pereira, A.F., Albano, M., Bergamo Alves, F.C., Murbach Teles Andrade, B.F., Furlanetto, A., Mores Rall, V.L., et al., 2020. Influence of apitoxin and melittin from Apis mellifera bee on *Staphylococcus aureus* strains. Microb. Pathog. 141 (2020). https://doi.org/10.1016/j.micpath.2020.104011, 104011.

Mavric, E., Wittmann, S., Barth, G., Henle, T., 2008. Identification and quantification of methylglyoxal as the dominant antibacterial constituent of Manuka (*Leptospermum scoparium*) honeys from New Zealand. Mol. Nutr. Food Res. 52 (4), 483–489.

Mazzarello, V., Donadu, M.G., Ferrari, M., Piga, G., Usai, D., Zanetti, S., Sotgiu, M.A., 2018. Treatment of acne with a combination of propolis, tea tree oil, and *Aloe vera* compared to erythromycin cream: two double-blind investigations. Clin. Pharmacol. 10, 175–181.

McLoone, P., Warnock, M., Fyfe, L., 2016. Honey: a realistic antimicrobial for disorders of the skin. J. Microbiol. Immunol. Infect. 49 (2), 161–167.

Minden-Birkenmaier, B.A., Bowlin, G.L., 2018. Honey-based templates in wound healing and tissue engineering. Bioengineering 5 (2), 46. https://doi.org/10.3390/bioengineering5020046.

Moselhy, W.A., Fawzy, A.M., Kamel, A.A., 2013. An evaluation of the potent antimicrobial effects and unsaponifiable matter analysis of the royal jelly. Life Sci. J. 10 (2), 290–296.

Nabas, Z., Haddadin, M.S., Haddadin, J., Nazer, I.K., 2014. Chemical composition of royal jelly and effects of synbiotic with two different locally isolated probiotic strains on antioxidant activities. Pol. J. Food Nutr. Sci. 64 (3), 171–180.

Nakajima, Y., Tsuruma, K., Shimazawa, M., Mishima, S., Hara, H., 2009. Comparison of bee products based on assays of antioxidant capacities. BMC Complement. Altern. Med. 9 (1), 1–9.

Navarre, W.W., Schneewind, O., 1999. Surface proteins of gram-positive bacteria and mechanisms of their targeting to the cell wall envelope. Microbiol. Mol. Biol. Rev. 63 (1), 174–229.

Noa, M., Mas, R., 1998. Effect of D-002 on the pre-ulcerative phase of Carrageenan-induced colonic ulceration in the Guinea-pig. J. Pharm. Pharmacol. 50 (5), 549–553.

Nolkemper, S., Reichling, J., Sensch, K.H., Schnitzler, P., 2010. Mechanism of herpes simplex virus type 2 suppression by propolis extracts. Phytomedicine 17 (2), 132–138.

O'Leary, K.A., de Pascual-Tereasa, S., Needs, P.W., Bao, Y.P., O'Brien, N.M., Williamson, G., 2004. Effect of flavonoids and vitamin E on cyclooxygenase-2 (COX-2) transcription. Mutat. Res. Fundam. Mol. Mech. Mutagen. 551 (1–2), 245–254.

Olczyk, P., Wisowski, G., Komosińska-Vassev, K., Stojko, J., Klimek, K., Olczyk, M., Koźma, E.M., 2013. Propolis modifies collagen types I and III accumulation in the matrix of burnt tissue. Evid. Based Complement. Alternat. Med. 2013. https://doi.org/10.1155/2013/423809.

Olczyk, P., Komosińska-Vassev, K., Winsz-Szczotka, K., Stojko, J., Klimek, K., Gajewski, K., Olczyk, K., 2014. Ocena aktywności wybranych enzymów macierzy pozakomórkowej w przebiegu gojenia doświadczalnych uszkodzeń termicznych (The evaluation of chosen extracellular matrix enzymes activity during regeneration of experimental thermal injuries). Leczen. Ran 11 (3), 97–101 (in Polish).

Olczyk, P., Koprowski, R., Kaźmierczak, J., Mencner, L., Wojtyczka, R., Stojko, J., Olczyk, K., Komosinska-Vassev, K., 2016. Bee pollen as a promising agent in the burn wounds treatment. Evid. Based Complement. Alternat. Med. 2016. https://doi.org/10.1155/2016/8473937.

Olczyk, P., Komosinska-Vassev, K., Krzyminiewski, R., Kasperczyk, J., Ramos, P., Dobosz, B., Batoryna, O., Stojko, J., Stojko, M., Ivanova, D., Olczyk, K., Pilawa, B., 2020. The estimation of blood paramagnetic center changes during burns management with biodegradable propolis-nanofiber dressing. Oxidative Med. Cell. Longev. 2020. https://doi.org/10.1155/2020/3675603.

Paramasivan, S., Drilling, A.J., Jardeleza, C., Jervis-Bardy, J., Vreugde, S., Wormald, P.J., 2014. Methylglyoxal-augmented manuka honey as a topical anti-*Staphylococcus aureus* biofilm agent: safety and efficacy in an in vivo model. Int. Forum Allergy Rhinol. 4 (3), 187–195.

Park, H.J., Lee, S.H., Son, D.J., Oh, K.W., Kim, K.H., Song, H.S., Kim, G.J., Oh, G.T., Yoon, D.Y., Hong, J.T., 2004. Antiarthritic effect of bee venom: inhibition of inflammation mediator generation by suppression of NF-κB through interaction with the p50 subunit. Arthritis Rheum. 50 (11), 3504–3515.

Pavel, C.I., Mărghitaş, L.A., Bobiş, O., Dezmirean, D.S., Şapcaliu, A., Radoi, I., Mădaş, M.N., 2011. Biological activities of royal jelly-review. Sci. Pap. Anim. Sci. Biotechnol. 44 (2), 108–118.

Peng, L., Yang, S., Cheng, Y.J., Chen, F., Pan, S., Fan, G., 2012. Antifungal activity and action mode of pinocembrin from propolis against *Penicillium italicum*. Food Sci. Biotechnol. 21 (6), 1533–1539.

Peng, C.C., Sun, H.T., Lin, I.P., Kuo, P.C., Li, J.C., 2017. The functional property of royal jelly 10-hydroxy-2-decenoic acid as a melanogenesis inhibitor. BMC Complement. Altern. Med. 17 (1), 1–8.

Pereira, A.S., de Miranda Pereira, A.F., Trugo, L.C., de Aquino Neto, F.R., 2003. Distribution of quinic acid derivatives and other phenolic compounds in Brazilian propolis. Z. Naturforsch. C 58 (7–8), 590–593.

Popova, M.P., Graikou, K., Chinou, I., Bankova, V.S., 2010. GC-MS profiling of diterpene compounds in Mediterranean propolis from Greece. J. Agric. Food Chem. 58 (5), 3167–3176.

Pyeon, H.I., Bak, J., Seok, J.I., So, S., Suh, H.J., Oh, M., Kim, S., Chung-Eun Yang, C.E., Chung, I.K., Choi, Y.S., 2017. Effects of nano-sized bee pollen as a new cosmetic ingredient. Asian J. Beauty Cosmetol. 15 (1), 1–9.

Rashidi, M.K., Mirazi, N., Hosseini, A., 2016. Effect of topical mixture of honey, royal jelly and olive oil-propolis extract on skin wound healing in diabetic rats. Wound Med. 12, 6–9.

Ravelo, Y., Molina, V., Carbajal, D., Fernández, L., Fernández, J.C., Arruzazabala, M.L., Más, R., 2011. Evaluation of anti-inflammatory and antinociceptive effects of D-002 (beeswax alcohols). J. Nat. Med. 65 (2), 330–335.

Rufián-Henares, J.A., Morales, F.J., 2007. Functional properties of melanoidins: in vitro antioxidant, antimicrobial and antihypertensive activities. Food Res. Int. 40 (8), 995–1002.

Rzepecka-Stojko, A., Stojko, J., Kurek-Górecka, A., Górecki, M., Kabała-Dzik, A., Kubina, R., Moździerz, A., Buszman, E., 2015. Polyphenols from bee pollen: structure, absorption, metabolism and biological activity. Molecules 20 (12), 21732–21749.
Saewan, N., Jimtaisong, A., 2015. Natural products as photoprotection. J. Cosmet. Dermatol. 14 (1), 47–63.
Sagona, S., Turchi, B., Fratini, F., Giusti, M., Torracca, B., Nuvoloni, R., Cerri, D., Felicioli, A., 2015. Preliminary evaluation of glucose oxidase and its products in vitro antimicrobial activities on Paenibacillus larvae ATCC9545 vegetative form. Bull. Insectol. 68 (2), 233–237.
Saric, S., Sivamani, R.K., 2016. Polyphenols and sunburn. Int. J. Mol. Sci. 17 (9), 1521. https://doi.org/10.3390/ijms17091521.
Šarić, A., Balog, T., Sobočanec, S., Kušić, B., Šverko, V., Rusak, G., Likic, S., Bubaloc, D., Pinto, B., Reali, D., Marotti, T., 2009. Antioxidant effects of flavonoid from Croatian *Cystus incanus* L. rich bee pollen. Food Chem. Toxicol. 47 (3), 547–554.
Sawicka, D., Borawska, M.H., 2013. Zastosowanie propolisu w chorobach skórnych (the use of propolis in skin diseases). Dermatol. Estetyczna 1 (84), 13–17 (in Polish).
Šedivá, M., Laho, M., Kohútová, L., Mojžišová, A., Majtán, J., Klaudiny, J., 2018. 10-HDA, a major fatty acid of royal jelly, exhibits pH dependent growth-inhibitory activity against different strains of Paenibacillus larvae. Molecules 23 (12), 3236. https://doi.org/10.3390/molecules23123236.
Siavash, M., Shokri, S., Haghighi, S., Shahtalebi, M.A., Farajzadehgan, Z., 2015. The efficacy of topical royal jelly on healing of diabetic foot ulcers: a double-blind placebo-controlled clinical trial. Int. Wound J. 12 (2), 137–142.
Silici, S., Ekmekcioglu, O., Kanbur, M., Deniz, K., 2011. The protective effect of royal jelly against cisplatin-induced renal oxidative stress in rats. World J. Urol. 29 (1), 127–132.
da Silva, M.V.D., de Moura Jr, N.G.D., Motoyama, A.B., Ferreira, V.M., 2020. A review of the potential therapeutic and cosmetic use of propolis in topical formulations. J. Appl. Pharm. Sci. 10 (1), 131–141. https://doi.org/10.7324/JAPS.2018.8801.
Simka, M., 2006. Zastosowanie miodu w leczeniu ran (The use of honey in the wound management). Leczen. Ran 3 (4), 127–129 (in Polish).
Skarżyńska, M., Zając, M., Wasyl, D., 2020. Antibiotics and bacteria: mechanisms of action and resistance strategies. Adv. Microbiol. 59 (1), 49–63.
Sojka, M., Valachova, I., Bucekova, M., Majtan, J., 2016. Antibiofilm efficacy of honey and bee-derived defensin-1 on multispecies wound biofilm. J. Med. Microbiol. 65 (4), 337–344.
Souza, C., de Freitas, L.A.P., Campos, P.M.B.G.M., 2017. Topical formulation containing beeswax-based nanoparticles improved in vivo skin barrier function. AAPS PharmSciTech 18 (7), 2505–2516.
Stojko, M., Włodarczyk, J., Sobota, M., Karpeta-Jarzabek, P., Pastusiak, M., Janeczek, H., Dobrzyński, P., Starczynowska, G., Orchel, A., Stojko, J., Batoryna, O., Olczyk, P., Komosinska-Vassev, K., Olczyk, K., Kasperczyk, J., 2020. Biodegradable electrospun nonwovens releasing propolis as a promising dressing material for burn wound treatment. Pharmaceutics 12 (9), 883. https://doi.org/10.3390/pharmaceutics12090883.
Subrahmanyam, M., 1998. A prospective randomised clinical and histological study of superficial burn wound healing with honey and silver sulfadiazine. Burns 24 (2), 157–161.
Subrahmanyam, M., Sahapure, A., Nagane, N., 2001. Effects of topical application of honey on burn wound healing. Ann. Burns Fire Disasters 15 (3), 143–145.
Takaisi-Kikuni, N.B., Schilcher, H., 1994. Electron microscopic and microcalorimetric investigations of the possible mechanism of the antibacterial action of a defined propolis provenance. Planta Med. 60 (3), 222–227.
Tan, H.T., Rahman, R.A., Gan, S.H., Halim, A.S., Asma'Hassan, S., Sulaiman, S.A., Kirnpal-Kaur, B.S., 2009. The antibacterial properties of Malaysian tualang honey against wound and enteric microorganisms in comparison to manuka honey. BMC Complement. Altern. Med. 9 (34). https://doi.org/10.1186/1472-6882-9-34.

Tie, L., An, Y., Han, J., Xiao, Y., Xiaokaiti, Y., Fan, S., Liu, S., Chen, A.F., Li, X., 2013. Genistein accelerates refractory wound healing by suppressing superoxide and FoxO1/iNOS pathway in type 1 diabetes. J. Nutr. Biochem. 24 (1), 88–96.

Tomás-Barberán, F.A., Ferreres, F., Tomás-Lorente, F., Ortiz, A., 1993. Flavonoids from Apis mellifera beeswax. Z. Naturforsch. C 48 (1–2), 68–72.

Tonks, A.J., Cooper, R.A., Jones, K.P., Blair, S., Parton, J., Tonks, A., 2003. Honey stimulates inflammatory cytokine production from monocytes. Cytokine 21 (5), 242–247.

Vranic, D., Petronijevic, R., Koricanac, V., Stojanovic, J.D., Lilic, S., Borovic, B., Lukic, M., 2019. Evaluation of Serbian black locust honey quality parameters as a contribution to confirmation of its botanical origin. IOP Conf. Ser.: Earth Environ. Sci. 333 (1). https://doi.org/10.1088/1755-1315/333/1/012113, 012113.

Wang, X., Sankarapandian, K., Cheng, Y., Woo, S.O., Kwon, H.W., Perumalsamy, H., Ahn, Y.J., 2016. Relationship between total phenolic contents and biological properties of propolis from 20 different regions in South Korea. BMC Complement. Altern. Med. 16 (1), 1–12.

Wijesinghe, M., Weatherall, M., Perrin, K., Beasley, R., 2009. Honey in the treatment of burns: a systematic review and meta-analysis of its efficacy. N. Z. Med. J. 22 (122), 47–60.

Xi, X., Li, J., Guo, S., Li, Y., Xu, F., Zheng, M., Cao, H., Cui, X., Guo, H., Han, C., 2018. The potential of using bee pollen in cosmetics: a review. J. Oleo Sci., ess18048. https://doi.org/10.5650/jos.ess18048.

You, C.E., Moon, S.H., Lee, K.H., Kim, K.H., Park, C.W., Seo, S.J., Cho, S.H., 2016. Effects of emollient containing bee venom on atopic dermatitis: a double-blinded, randomized, base-controlled, multicenter study of 136 patients. Ann. Dermatol. 28 (5), 593–599.

Zhang, H., Wang, X., Wang, K., Li, C., 2015. Antioxidant and tyrosinase inhibitory properties of aqueous ethanol extracts from monofloral bee pollen. J. Apic. Sci. 59 (1), 119–128.

Zhang, J., Shen, X., Wang, K., Cao, X., Zhang, C., Zheng, H., Hu, F., 2016a. Antioxidant activities and molecular mechanisms of the ethanol extracts of Baccharis propolis and eucalyptus propolis in RAW64.7 cells. Pharm. Biol. 54 (10), 2220–2235.

Zhang, Y., Yang, F., Jamali, M.A., Peng, Z., 2016b. Antioxidant enzyme activities and lipid oxidation in rape (*Brassica campestris* L.) bee pollen added to salami during processing. Molecules 21 (11), 1439.

CHAPTER 3

Bee products and diabetes mellitus

Kashif Maroof and Siew Hua Gan
School of Pharmacy, Monash University Malaysia, Jalan Lagoon Selatan, Bandar Sunway, Selangor, Malaysia

Contents

Abbreviations

AGE advanced glycation end products
ATF activating transcription factor
BUN blood urea nitrogen
CAT catalase
DM diabetes mellitus
FBG fasting blood glucose
FFA free fatty acids
GPx glutathione peroxidase
GSH glutathione
HbA1c glycated hemoglobin
HDL-C high-density lipoprotein cholesterol
HFD high fat diet
IL interleukin
iNOS inducible isoform of nitric oxide synthase
LDL-C low-density lipoprotein cholesterol
MCP-1 monocyte chemoattractant protein-1

Bee Products and Their Applications in the Food and Pharmaceutical Industries
https://doi.org/10.1016/B978-0-323-85400-9.00012-5

MDA malondialdehyde
MMP matrix metalloproteinases
mTOR mammalian target of rapamycin
NO nitric oxide
NOS nitric oxide synthase
OLETF Otsuka Long Evans Tokushima Fatty rat
Px peroxidase
RAGE receptor for advanced glycation end products
SOD superoxide dismutase
STZ streptozotocin
TG triglyceride
TGF transforming growth factor
TNF tumor necrosis factor
VEGF vascular endothelial growth factor
VLDL-C very low-density lipoprotein cholesterol

Introduction

According to the International Diabetes Federation, 463 million individuals globally were diagnosed with diabetes worldwide in 2019, with that number anticipated to climb to 700 million by 2045 (Smokovski, 2021). Diabetes mellitus (DM) is a metabolic condition characterized by hyperglycemia resulting from either a deficit in insulin production or its action. Poorly treated diabetes can harm various organs, including the eyes, kidneys, nerves, and cardiovascular (CVS) system. Based on the origin and clinical aspects of DM, there are three primary kinds. Type 1 DM (T1DM), type 2 DM (T2DM), and gestational diabetes mellitus (GDM) are three types of DMs (Sami et al., 2017).

T1DM, a chronic autoimmune illness defined by high blood glucose (BG) levels (hyperglycemia), is caused by insulin insufficiency due to the loss of pancreatic islet cells. Generally, the symptoms appear quickly, in a matter of weeks (Katsarou et al., 2017). On the other hand, T2DM, the most prevalent type, is caused by various variables, with the most important being lifestyle, though genetic factors can also play a role. In T2DM, there is insulin resistance and a relative insulin deficiency (Oguntibeju, 2019). In comparison, GDM refers to any level of glucose intolerance that develops only during pregnancy (Bartáková et al., 2021).

The pathophysiological changes of diabetes are illustrated by (1) β-cell dysfunction in the periphery (e.g., the liver, skeletal muscle and adipose tissue) and (2) reactive oxygen species (ROS) produced locally by the tissue or immune cells interacting with the insulin receptor (IR) and their downstream signaling pathways causing an inability to respond appropriately to insulin levels (Galicia-Garcia et al., 2020). All these variables contribute to insulin resistance and chronic inflammation, which wreak havoc on BG management, leading to micro-and macrovascular problems (Wang and Lo, 2018).

The progression of macro and microvascular problems in the brain, eyes, kidneys, heart, liver, and other critical organs is attributed to continuous hyperglycemia. The development of diabetic complications is influenced by many factors, including oxidative stress, inflammatory cytokines, transcription factors and enzymes, protein kinase-c (PKC) activation, intracellular sorbitol and tissue AGEs accumulation, the polyol pathway, as well as mitochondrial superoxide productions (Vanessa Fiorentino et al., 2013; Sruthi and Raghu, 2021). The imbalance between ROS generation and the antioxidant system's ability to neutralize these chemicals results in oxidative stress (Sharma et al., 2019). Increased oxidative stress is caused by an upsurge in ROS production and/or a reduction in ROS scavenging capabilities, resulting in tissue damage (van der Pol et al., 2019).

Antioxidants are compounds that can help reduce oxidative stress. They can be (1) endogenous (reduced glutathione and antioxidant enzymes including catalase, superoxide dismutase, glutathione peroxidase, and reductase) or (2) exogenous (antioxidant vitamins like vitamins A, C, and E) (Adwas et al., 2019). By scavenging or stimulating the breakdown of free radicals, both forms of antioxidants can help prevent the production of the free radicals (Neha et al., 2019). Furthermore, they can transform reactive metabolites into less reactive ones (Lourenço et al., 2019). More prospects for developing novel focused treatments for diabetes treatment, prognosis and ultimately, prevention exist as more insights into the molecular underpinnings of this illness and its damaging repercussions are gained (Fitipaldi et al., 2018).

Apiculture is the study and practice of using honeybee colony products including honey, bee bread, bee venom, bee pollen, propolis, and royal jelly to either prolong, sustain, or maintain health (Pasupuleti et al., 2017). Bee products have been increasingly popular in both traditional and modern medicines in recent years. Due to their efficacies and wide availability, much research is conducted to investigate their specific health advantages and pharmacological aspects, resulting in the development of nutraceuticals and functional foods (Paolino et al., 2021). Functional food is defined as food that can promote improved physiological or psychological health, compared to traditional remediated and nutritional food (Ahmed and Uddin, 2020). These impacts contribute to better health maintenance, well-being, and a lower risk of chronic illness (Kumar and Krishnan, 2017). While the concept of a food supplement is fairly straightforward, the definition of a nutraceutical remains hazy, straddling the line between food, supplements, and medications. The phrase was invented by Stephen DeFelice, a syncretic neologism of the words nutrient and pharmaceutical. The notion has been offered as a new approach to food science, with the potential use characterized as extending beyond diet but before the development of pharmaceuticals (Santini et al., 2018).

Honey is used worldwide due to its high nutritional content, which is favorable to human health. Traditionally, the Egyptians, Greeks, Romans, and Chinese have utilized honey to cure wounds and ailments of the gut, including gastric ulcers. Cough, sore throat, and earaches have also been treated with it. The honey's active constituents, including glucose, fructose, flavonoid, polyphenols, and organic acids, contribute to its therapeutic activity (Keskin et al., 2021). Many in vitro studies in cell cultures in human and animal research indicated that honey polyphenols protect against various diseases, including CVS disease, diabetes, cancer, neurodegenerative, lung, and liver diseases (Hossen et al., 2017a). Recent experimental studies support honey as a novel antidiabetic agent that may be potentially significant for managing diabetes and its complications and being an antidiabetic agent (Bobiş et al., 2018).

Propolis is sometimes referred to as "bee glue," a generic term for the resinous substance gathered by bees from various types of plants (Maroof et al., 2020). The word "propolis" comes from the Greek word, which means "defense" for "pro" and "city or community" for "polis" (Rojczyk et al., 2020). Propolis functions in sealing holes and cracks or reconstruction of the beehive. It is also used for smoothing the inner surface of the beehive, retaining the hive's internal temperature (35°C), preventing weathering and invasion by predators (Pasupuleti et al., 2017; Wakgari and Yigezu, 2021). Polyphenols, esters, flavonoids, terpenes, vitamins, minerals, and certain enzymes have all been reported to be present in propolis (Maroof and Gan, 2020). Due to its antiseptic, antiinflammatory, antioxidant, antibacterial, antimycotic, antifungal, antiulcer, anticancer, and immunomodulatory qualities, propolis and its preparations are used to treat many diseases (Cornara et al., 2017). Propolis has also been shown to control oxidative stress, the formation of AGEs, and adipose tissue inflammation in animal and cellular models, all of which lead to insulin resistance or insulin secretion abnormalities. As a result, propolis therapy has the potential to prevent diabetes sequelae such as nephropathy, retinopathy, foot ulcers, and nonalcoholic fatty liver diseases (Kitamura, 2019).

Royal jelly is a white, viscous jelly-like material produced by the worker bees' hypopharyngeal and mandibular glands. It is also regarded as a "superfood" since it is consumed only by the queen bee. When the honeybee larvae hatch, they are fed with royal jelly, which helps to nourish the brood (Ramanathan et al., 2018). Besides being the queen bee's exclusive diet in her entire life cycle, royal jelly is the only nutrient supplied to immature young larvae in their first 2–3 days of development (Ahmad et al., 2020). The primary ingredient in royal jelly is royalactin which enables a larva to develop further into a queen bee (Maleszka, 2018). Compared to other bees, the royal jelly superfood is the main reason for the queen bee's longevity (Ali and Kunugi, 2020). To date, royal jelly is commonly utilized as a dietary nutritional complex to treat various chronic illnesses due to its antibacterial, antitumor, antiallergy, antiinflammatory, and conferring immunomodulatory effects (Guo et al., 2021). Additionally, royal jelly has been demonstrated to lower fasting BG and serum glycosylated hemoglobin levels while increasing insulin concentrations (Pourmoradian et al., 2014).

Bee pollen is a mass of pollen grains in the hive after transfer to the hive (pollination), storage and fermentation. Bee pollen is often known as the "essence of flowers" because the bees collect the most abundant nutrients from various stamens (Li et al., 2017). In fact, pollen is another important bee product high in bioactive compounds, including flavonoids (Ares et al., 2018). Pollen provides the only protein source available to the bees in nature, which is critical to developing the progeny and the overall growth of the bee colonies (Anđelković et al., 2014). Bee pollen has several biological benefits, including as an antioxidant, is cardioprotective, hepatoprotective, and has antiinflammatory properties (Li et al., 2018). Additionally, bee pollen polysaccharide has been shown to ameliorate hepatic steatosis and insulin resistance (Li et al., 2017). For the production of bee pollen, pollen is first collected by the foraging bees and is transported to the hive, where it is packed into honeycomb cells (Fuenmayor et al., 2014). The bees further enrich the pollen during the packing process, contributed by the rich digestive enzymes and organic acids found in the bees' salivary gland secretions (Deveza et al., 2015). Under anaerobic conditions, lactic fermentation of pollen induced by the *Lactobacillus* bacteria occurs spontaneously in the honeycombs. Pollen sheaths dissolve in the process of pollen transformation into bee bread and are more easily ingested in this form. As a result of enzymatic transformations, fermentation to bee bread not only prevents the pollen from losing its characteristics but also produces new components. During the fermentation process, pollen proteins are broken down into peptides and amino acids (Kieliszek et al., 2018), making the lactic acid concentration in bee bread sixfold higher than pollen (Nagai et al., 2004). The presence of lactic acid preserves the bee bread, allowing more extended storage of the resulting product (Kieliszek et al., 2018). Bee bread possesses therapeutic properties such as antioxidant, antiinflammatory, antibacterial, and antidiabetic, to name a few (Mohammad et al., 2021).

Bee venom is a colorless, bitter liquid with a pH of 4.5–5.5 (Hossen et al., 2017b).The venom is mainly produced in the abdomen of the worker bees and originates from a mixture of both acidic and basic secretions (Hussein et al., 2019). It is a complex mixture of proteins that includes enzymes (which catalyze specific reactions), peptides (which contain two or more amino acids), and other low molecular components like carbohydrates (2% of venom dry weight), phospholipids (5% of venom dry weight), amino acids (1% of venom dry weight), and minerals (3%–4% of venom dry weight) (Hussein et al., 2019). Bee venom consists mostly of proteins and peptides, where their combination is referred to as bee venom toxins (Kurek-Górecka et al., 2021). Arthritis, frozen shoulder, disorders of the central and peripheral nervous systems (CNS and PNS), skin illnesses, heart and blood system disorders, cancer, ulcer, colitis, and neuritis are well treated with bee venom proteins; particularly a substance called melittin (Hussein et al., 2019). Recent research in animal models has also confirmed that the bee venom can considerably reduce BG levels via insulin secretion and glucose uptake (Mousavi et al., 2012; Elkotby et al., 2018).

This chapter emphasizes the components and characteristics of bee products relevant to diabetes mellitus (DM) and their use in DM management. Recent scientific data from both preclinical and clinical trials are discussed to evaluate bee products' potential in the treatment of DM.

Preclinical studies on the antidiabetic potential of bee products

Blood glucose, Hb1Ac, and lipid profiles

There is a myth that sweet products such as honey are harmful to diabetes and should be avoided. However, science has proven otherwise. In fact, honey has a hypoglycemic impact in diabetic rats induced by streptozotocin (STZ) and alloxan (Adesoji and Oluwakemi, 2008; Omotayo et al., 2010). Honey supplementation has been reported to enhance serum insulin levels and high-density lipoprotein (HDL) cholesterol in STZ-induced diabetic rats while lowering fructosamine, low-density lipoprotein (LDL), very low-density lipoprotein (VLDL), total cholesterol and triglycerides (TG) (Erejuwa et al., 2011a, 2016). Additionally, honey has also been shown to lower blood pressure and improve hepatic dysfunction (Asaduzzaman et al., 2015).

Erejuwa and colleagues offered more data supporting honey's therapeutic effects in DM in one of their investigations. STZ-induced diabetic rats were administered with glibenclamide or metformin either singly or together with honey to determine if honey would synergize or antagonize the effects of antidiabetic medications. Diabetic rats administered with antidiabetic agents in combination with honey had significantly lower BG, fructosamine, bilirubin, VLDL cholesterol, and TG concentrations than those receiving glibenclamide or metformin alone (Erejuwa et al., 2011a), indicating the potential of honey in ameliorating diabetes.

The findings were further confirmed (Nasrolahi et al., 2012) in a study where metformin treatment with honey administration to diabetic rats caused a higher increase in HDL cholesterol and significantly lower levels of hyperglycemia, TG, bilirubin, VLDL, LDL, and total cholesterol. Another recent study that investigated the effect of honey in STZ–nicotinamide-induced diabetic rats (another model ofT2DM), as opposed to prior studies that employed alloxan- or STZ-induced diabetic rats (a model of type 1 DM), found that honey significantly raised serum insulin and HDL cholesterol levels while lowering BG, total cholesterol, TG, and LDL cholesterol levels (Aziz et al., 2017). It was noteworthy that the studies mentioned above employed a variety of honey samples from numerous countries, indicating the potential of honey from various geographical regions in ameliorating diabetes-induced metabolic derangements and is not limited only to a single variety of honey (Erejuwa, 2014).

In another study, the glycemic increase in rats following honey administration from different localities was further explored and compared. Nondiabetic and alloxan-induced diabetic rats were administered five Nigerian honey as Abakaliki honey (ABH), Calabar (CAH), Ezzamgbo (EZH), Lokoja (LOH), Okitipupa (OKH), and a single Malaysian (MAH) honey at 0.5 g/kg each. Except for ABH and OKH, none of the honey samples caused a substantial increase in BG levels in diabetic rats. When the honey samples were compared, there was also no significant variation in peak BG values or the area under the curve (AUC) except for CAH honey which showed a similar percent change and fluctuation in BG as seen with glibenclamide. These changes were more noticeable in diabetic rats than nondiabetic rats, indicating that diabetic individuals are more vulnerable to the harmful effects of adulterated honey and that further research in diabetic subjects is needed (Erejuwa et al., 2021).

Another study in diabetic rats suggested that the hypoglycemic effect of honey is dose-dependent (Omotayo et al., 2010). Metformin combined with honey is more effective at lowering serum glucose levels, while the administration of metformin and glibenclamide combined with honey is more successful at lowering serum glucose than glibenclamide alone (Syarifuddin et al., 2020). Other studies have shown that stevia and honey can help lower blood sugar levels. However, more recent studies have suggested that raising the honey dose to approximately 3 g/kg of the body weight causes the hypoglycemic effect to be lost. In diabetic rats administered with 3 g/kg of honey, other favorable benefits on metabolic derangements such as lowering LDL, non-HDL cholesterols, coronary risk index, CVS risk index and the atherogenic index, as well as enhanced HDL cholesterol, were lost (Erejuwa et al., 2016).

Honey has been shown in certain studies to improve metabolism-related indicators in nondiabetic animals and its positive benefits in animal models of DM. Increased body weight, poor glucose tolerance and lipid abnormalities are all risk factors for DM.The body weight and epididymal fat weight, as well as leptin, TG levels, and food intake, were all significantly reduced in rats fed with honey (20%) for 33 days. However, sucrose-fed rats did not have the same positive effects (Nemoseck et al., 2011). Nondiabetic rats fed a honey-containing diet had higher serum vitamin E and fructose levels, as well as lower TG, with no effect seen on glucose levels (Busserolles et al., 2002). Honey given for 52 weeks was shown to dramatically enhance HDL besides lowering HbA1c values in nondiabetic rats. Nevertheless, these effects were not seen in rats fed with either a sugar-free or sucrose-containing diet (Chepulis and Starkey, 2008).

In another related study, although the trial was only 6 weeks long, only body weight, but not HbA1c or food consumption, was significantly reduced in honey-fed rats (Chepulis, 2007).The lack of effect on HbA1c may be attributed to the study's short duration. However, increased body weight was recorded without a substantial drop in BG (Adesoji and Oluwakemi, 2008). Nevertheless, honey's positive benefits on glucose, leptin, and insulin in diabetic rats have not been demonstrated, according to a recent study (Al Aamri and Ali, 2017).

A substantial number of studies on propolis' medicinal benefits discussed BG levels and lipid profiles. Water or ethanol extracts of propolis from China administered by oral gavaging reversed the alloxan-induced increase in FBG. Furthermore, both extracts reversed the effects of alloxan on total cholesterol, LDL-C, VLDL-C, and HDL-C (Okutan et al., 2005). In rats pretreated with alloxan, oral administration of Nigerian propolis lowered FBG levels after 2 weeks and partially decreased HbA1c after 6 weeks. Additionally, the propolis treatment also raised HDL-C levels while lowering LDL-C and cholesterol levels (Oladayo, 2016).

In another study from China on STZ-pretreated rats, propolis extract in gum tragacanth marginally but considerably lowered FBG and Hb1Ac (Shi et al., 2019). Overall, oral administration of Chinese and Brazilian propolis lowered BG levels after one (Chinese) or 3 weeks (Brazilian) after treatment. HbA1c levels were concurrently and dramatically decreased in the Chinese propolis-treated group, while Brazilian propolis exhibited a trend toward reduction in the values. Although Chinese propolis did not lower total cholesterol levels, Brazilian propolis did so marginally (17%), while propolis did not affect the blood triglycerides, LDL-C, or HDL-C levels (Zhu et al., 2011a).

Propolis gathered from Malaysia, Mexico, Iran, Egypt, Saudi Arabia, Turkey, and Croatia had similar effects on BG levels, despite the botanical origins being diverse (Fuliang et al., 2005; Oršolić et al., 2012; Al Ghamdi et al., 2015; Usman et al., 2018; Sameni et al., 2016; Nna et al., 2018). In a dose-dependent way, Indonesian propolis reduced BG levels (Rifa'i and Widodo, 2014). Matsushige et al. reported that pretreatment with a water extract of Brazilian Yukari propolis (200 mg/kg/day, i.e., for 1 week) significantly reduced the STZ-induced increase in FBG. In contrast, a similar amount of methanol extract did not (Matsushige et al., 1996).

Propolis was also found to improve BG and lipid profiles in T2DM models. A high-fat, high-fructose diet combined with a low-dose STZ injection resulted in significant hyperglycemia, especially following 1 month of administration. In another study, Taiwanese green propolis ethanolic extract (184 or 920 mg/kg/day) (Chen et al., 2018) lowered rats' FBG levels. In HFD and low-dose STZ-treated diabetic rats, sonication-encapsulated Chinese propolis also reduced FBG levels. Moreover, the encapsulated propolis decreased triglyceride levels in the blood but not the total or LDL-C levels (Li et al., 2012).

Propolis has also been shown to have therapeutic effects on the blood parameters or indices in studies using a monogenic T2DM model. Aoi et al. reported that feeding Otsuka Long Evans Tokushima Fatty Rat (OLETF) rats a 0.1% or 0.5% Brazilian propolis-containing diets for 8 weeks reduced BG levels while leaving plasma free fatty acids and β-hydroxybutyrate unchanged (Aoi et al., 2013). Intraperitoneal injections of Brazilian green propolis ethanol extract (100 mg/kg, twice a week for 12 weeks) resulted in a 2/3-fold reduction in BG levels (Kitamura et al., 2013).

Royal jelly also has the potential to affect glucose levels. In a study, royal jelly (100 mg/kg) dissolved in water (1 mL) daily for 6 weeks resulted in a significant decrease in FBG levels and serum HDL-C level of STZ-induced diabetic rats (Ghanbari et al., 2016). In another study, administration of royal jelly slightly decreased BG levels, albeit in a nondose–dependent manner (Fujii et al., 1990). In another study, administering a high-fat diet with 5% lyophilized royal jelly powder to healthy male mice for 17 weeks resulted in significantly improved hyperglycemia and hyperlipidemia compared to the control group, which received only HFD (Yoneshiro et al., 2018).

Polysaccharides are broken down into glucose by intestinal enzymes (α-amylase and α-glucosidase), then delivered into the body cells. The impaired function of these enzymes may affect glucose levels. Therefore, glycemic control can be achieved by α-amylase and α-glucosidase inhibitors. Nevertheless, synthetic drugs have adverse side effects such as liver problems, abdominal pain, flatulence, and kidney tumors; thus, natural inhibitors are required to keep BG levels in check (Khalifa et al., 2021). Significant α-amylase inhibition was observed in bee pollen aqueous-ethanol extracts (IC_{50} 4.51 mg/mL), which was more powerful than the control (acarbose) (IC_{50} 6.52 mg/mL). In comparison to the control's IC_{50} of 11.30 mg/mL, bee pollen water extracts inhibited α-glucosidase with the lowest IC_{50} seen at 0.60 mg/mL, indicating that bee pollen can reduce blood sugar levels acting as a natural α-glucosidase inhibitor (Daudu, 2019). In another study, treating STZ-induced diabetic male Wistar rats with bee pollen suspension at 100 mg/kg/day for 4 weeks resulted in a noticeable reduction in high fasting and post-prandial serum glucose levels (Mohamed et al., 2018) confirming the potential of the bee pollen suspension.

In T2DM Zucker fatty (ZDF) rats, daily treatment with bee bread at 700 mg/kg resulted in a considerable reduction in glucose levels (Capcarova et al., 2019). Bee bread has also been shown to confer hypoglycemic, hypolipidemic and hepatoprotective effects in STZ-induced diabetic rats (Lyoussi et al., 2021). In another study, bee venom was tested on alloxan-induced diabetic male rats to determine if it had any hypoglycemic impact. Compared to the diabetic group, the levels of glucose, triglycerides and total cholesterol in the treated group were significantly lower (Mousavi et al., 2012), confirming its potential.

Insulin deficiency

Insulin deficiency may be absolute (type 1 DM) or relative (type 2 DM). Evidence in the literature reveals that the effect of honey on insulin secretion varies and may be dependent on the levels of circulating insulin. In contrast to a comparable glucose–fructose solution, honey consumption in healthy individuals has been linked to reduced insulin, C-peptide levels, and significantly lower serum glucose levels after 1 h (Münstedt et al., 2008). Accordingly, honey was less effective on serum glucose, C-peptide and insulin values than the honey-comparable glucose–fructose solution in healthy subjects. Hence, the higher BG levels might be related to the insulin response following consumption of comparable glucose–fructose or glucose solution. Honey raises plasma insulin and C-peptide in individuals with impaired glucose tolerance or diabetes, linked to better glucose tolerance and lower hyperglycemia (Al-Waili, 2003).

Honey consumption causes a significant increase in C-peptide levels in both nondiabetic and diabetic subjects compared to the consumption of sucrose or glucose (Abdulrhman et al., 2013a). Honey's capacity to stimulate the secretion of C-peptide and insulin from the pancreas of diabetic patients has been confirmed in additional investigations (Abdulrhman et al., 2013b; Al-Waili, 2004). Honey can stimulate both healthy and diseased β-cells to boost the production of C-peptide and insulin. In fact, honey's ability to stimulate insulin secretion in β-cells will be advantageous to people with insulin insufficiency.

Honey supplementation raised serum insulin levels in STZ-induced diabetic rats with insulin shortage. On the other hand, honey did not affect insulin levels in nondiabetic rats (Erejuwa et al., 2011a). Overall, these findings imply that honey can stimulate insulin release from the pancreatic β-cells should a higher level of insulin be required for glucose utilization or absorption (as in diabetes). Several reports have assessed the effects of propolis on serum insulin levels in STZ-induced diabetic models. In a study, STZ-induced insulin insufficiency was partially prevented by ethanolic extract of Malaysian propolis (300 mg/kg/day) (Nna et al., 2018), which was also similarly seen by using ethanol extracts of Mexican and Saudi Arabian propolis (Al Ghamdi et al., 2015; Rivera-Yañez et al., 2018) as well as administration of propolis from Mexico (300 mg/kg/day for 15 days). Additionally, propolis from Mexico interfered with STZ's ability to eliminate β-islet-cells. Propolis inhibits β-cell destruction due to STZ-elicited pathways since many flavonoids scavenge ROS, while flavonoids like quercetin may protect β-cells from oxidative damage (El Rabey et al., 2017). Moreover, quercetin

(15 mg/kg/day, i.p. for 4 weeks) partially reverses STZ-induced insulin deficit in rats (Coskun et al., 2005).

Pancreatic β-cells increase insulin release in the presence of arginine, which targets for insulin secretion (AITS) complex in the endoplasmic reticulum of β-cells (Umeda et al., 2015). On pancreatic cell line (NIT-1), Brazilian, but not Chinese propolis, replicates the actions of arginine. In another study, Brazilian propolis (0.01%) exhibited a more significant effect on insulin secretion than arginine and drupanin, implying that Brazilian propolis contains unknown greater insulin secretion stimulants and propolis treatment raised circulating insulin levels in lean mice in tandem with a decrease in BG levels (Umeda et al., 2015).

In another study, treating STZ-induced diabetic male Wistar rats with bee pollen suspension (100 mg/kg/day) for 4 weeks dramatically improved the reduced serum insulin concentration in diabetic rats where the bee pollen suspension appeared to be very potent, producing a percentage rise of 90.85% when compared to the diabetic control (Mohamed et al., 2018). In animal models, recent research has also revealed that melittin, a bee venom protein, can drastically lower BG levels via insulin secretion and glucose uptake (Mousavi et al., 2012; Khulan et al., 2015; Prakash and Bhargava, 2014).

Oxidative stress and insulin resistance

Changes in oxidative stress levels are linked to insulin resistance. Unlike insulin deficiency which is only seen in type 2 diabetes (T2DM), insulin resistance affects both types 1 and 2 diabetes. Feedback mechanisms raised fasting insulin levels due to insulin insensitivity, especially in the early stages of T2DM (Maciejczyk et al., 2019; Apostolopoulou et al., 2018; Fazakerley et al., 2018). In in vitro investigations, honey ameliorates insulin resistance. Honey extracts and quercetin have been used to investigate the impact of hyperglycemia-induced insulin resistance in HIT-T15 cells. Insulin resistance was characterized by decreased insulin levels and protein kinase B (PKB) expression, elevated levels of NF-ҚB, mitogen-activated protein kinases (MAPK) as well as insulin receptor substrate 1 (IRS-1) serine phosphorylation (ser307). By lowering the expression of NF-ҚB, MAPK and IRS-1 serine phosphorylation, pretreatment with honey and quercetin extracts improved insulin resistance while boosting insulin levels and PKB expression (Batumalaie et al., 2013).

A recent study investigating the effect of honey on a high-fat high-sucrose diet-induced insulin resistance in rats showed that honey supplementation ameliorated impaired glucose tolerance and enhanced insulin sensitivity (Ding et al., 2019). The phenomenon is attributed to oxidative stress that leads to insulin insufficiency and resistance. Therefore, honey supplementation in diabetes patients can help overcome pancreatic oxidative stress while improving insulin deficit (Eguchi et al., 2021; Terzo et al., 2020).

Honey consumption may also reduce oxidative stress in some organs, including the skeletal muscles, adipose tissues, and hepatic cells. Thus, honey consumption may assist diabetes patients in enhancing both insulin sensitivity and secretion. Overall, honey can help protect the pancreas from oxidative stress or damage, minimize islet of Langerhans damage and enhance the size and quantity of islet cells (Erejuwa et al., 2010a,b, 2011b). Honey supplementation lowered the levels of inflammatory markers such as tumor necrosis factor-alpha (TNF-α), inhibitor of nuclear factor kappa-B kinase subunit beta (IKK-β), interleukin 1 beta (IL-β) and nuclear factor-kappa B (NFĶ-B) in the pancreatic islets in STZ–nicotinamide-induced diabetic rats (Aziz et al., 2017). Given the importance of the pancreas in BG regulation, honey's positive effects on the pancreas may increase insulin synthesis and/or secretion while alleviating insulin shortage.

Twelve weeks of administration of propolis ethanolic extract (200 mg/kg/day) from China decreased serum ROS and reactive nitrogen species (RNS) (Shi et al., 2019) in STZ-pretreated diabetic mice. When compared to administration of a similar dose of propolis (100 or 200 mg/kg/day) from China for 8 weeks, Zhu et al. found that Brazilian propolis (100 or 200 mg/kg/day for 8 weeks) exert a more noticeable restorative effect on the blood nitric oxide synthase (NOS), SOD, and malondialdehyde (MDA) levels in STZ-treated rats. The studies also reported that Brazilian propolis had more positive effects on hepatic SOD, GSH, and MDA, as well as on the renal MDA in T1DM model rats (Zhu et al., 2011a,b). Similarly, in STZ-pretreated rats, administration of Brazilian propolis ethanol extract (100–300 mg/kg/day) for 40 days reduced serum as well as renal and pancreatic MDA levels, with an increase in serum and renal SOD, GSH and CAT levels, as well as pancreatic SOD levels seen (El-Sayed et al., 2009; Abo-Salem et al., 2009).

Administration of Mexican propolis (300 mg/kg/day) for 15 days increased the activities of SOD, CAT, and GPx in the pancreas of STZ-induced (130 mg/kg) diabetic CD1 mice (Rivera-Yañez et al., 2018). In the pancreas of STZ-treated (60 mg/kg) rats, treatment with Malaysian propolis (300 mg/kg/day) increased SOD, CAT, GPx, GSH, GST, glutathione reductase (GR), and the overall antioxidant activity, whereas MDA level was considerably reduced. Diabetic rats treated with propolis showed lower levels of IL-1, TNF-α, RelA, and cleaved caspase 3 in the pancreas, which is consistent with the notion that ROS causes inflammation and apoptosis (Nna et al., 2018).

Administration of propolis ethanolic extract (183.9 or 919.5 mg/kg/day) from Taiwan for 8 weeks has also been shown to protect β-cells from oxidative stress in STZ/HFD model mice (Chen et al., 2018). However, oral gavaging of Iranian propolis (200 mg/kg/day) to STZ-induced diabetic rats for 6 weeks restored SOD, GPx, and ferric-reducing antioxidant capacity (FRAP) in the kidney while decreasing renal MDA concentration (Sameni et al., 2016). Furthermore, in alloxan-treated rats, administration of Nigerian propolis (200 or 300 mg/kg/day) for 28 days ameliorated blood antioxidant indices (Babatunde et al., 2015).

ROS levels were similarly reduced in the bone marrow, spleen, blood, and liver of diabetic mice treated with Saudi Arabian propolis (100 mg/kg/day) for 1 month. Additionally, in STZ-induced diabetic mice, methanolic extract of propolis purchased from a local herbal medicine shop in Saudi Arabia improved antioxidant enzyme abnormalities and MDA build-up in the serum and kidney (Al Ghamdi et al., 2015), confirming the potential of propolis in enhancing glucose control. A recent study reported that a diet containing 0.5% Brazilian propolis ethanol extract caused a 78% drop in plasma insulin in OLETF rats, implying restoration of insulin sensitivity (Aoi et al., 2013). In OLETF rats, oral administration of Brazilian propolis (100 or 300 mg/kg) for 4 weeks improved blood insulin levels and homeostatic model assessment-insulin resistance (HOMA-IR) according to a similar study (Zamami et al., 2010). In another study, Li et al. found that administration of encapsulated propolis from China (50–200 mg/kg/day) for 10 weeks reduced fasting serum insulin (from 40 to 44 to 56 IU/mL) and insulin action index. In the euglycemic hyperinsulinemic glucose clamp test, the Chinese propolis (100 mg/kg)-treated group demonstrated a substantial improvement in insulin sensitivity compared to the control group (Li et al., 2012). According to these findings, administration of poplar bud (50 or 100 mg/kg/day for 4 weeks), a key ingredient in Chinese propolis, reduced blood insulin levels in HFD- and STZ-induced T2DM model mice by more than 25%. Notably, both doses of poplar bud extract had a stronger effect on blood insulin levels than that reported for metformin (100 mg/kg/day) (Peng et al., 2018).

Royal jelly's antioxidative qualities may ameliorate insulin resistance and have even been reported to be beneficial in treating hypertension due to insulin resistance. For 8 weeks, treatment with royal jelly (100 and

300 mg/kg) improved insulin resistance, reduced hyperglycemia and alleviated diet-induced obesity in fructose-drinking rats (Zamani et al., 2012). Similarly, treatment with bee pollen raised the activities of SOD, CAT, and GSH-Px and improved the immune status parameters of STZ-treated (65 mg/kg) rats (Mohamed et al., 2013). Diet supplementation with royal jelly for 4 weeks in rats reduced BG levels and serum lipids and may be an effective adjuvant with oral hypoglycemic medications (Rezk, 2018). Another study reported that administration of royal jelly for 4 weeks alleviated hyperglycemia and decreased the expression of the glucose-6-phosphatase (G-6-Pase) gene in KK-Ay mice (Yoshida et al., 2016).

Bee venom dissolved in phosphate-buffered saline PBS (200 μg/kg) administered by subcutaneous injection to (STZ)-induced type I diabetic mouse model (1) rescues the prolonged-expression of proinflammatory cytokines IL-1b, IL-6, and TNF-a (2) upregulates the expression of TGF-β and vascular endothelial growth factor (VEGF) while (3) significantly restoring master regulator activating transcription factor-3 (ATF-3) and iNOS-mediated oxidative stress (Badr et al., 2016).

Preclinical studies of bee products on diabetic complications

Diabetes produces hyperglycemia, which increases ROS formation and AGEs. Through positive feedback loops, both ROS and AGEs accumulate more in cells (Volpe et al., 2018). Damaged cells eventually lose their ability to retain functional integrity, leading to cell death (Burgos-Morón et al., 2019). Cell abnormalities in diabetic individuals become obvious because ROS generation and AGE-RAGE ligation are enhanced in certain types of cells, including the endothelial, kidney and retinal epithelial cells (Kang and Yang, 2020). Insulin resistance is also caused by an excess of carbohydrates and lipids, resulting in the malfunction of energy metabolism-competent organs such as the liver, skeletal muscle, and adipose tissues (Kitamura, 2019). As a result, dyslipidemia combined with hyperglycemia creates blood circulatory problems, a common sign of geriatric disorders, including atherosclerosis, stroke, myocardial infarction, liver cancer, and Alzheimer's disease (Beckman et al., 2002). The phrase "diabetic complications" is used in this context to refer to a wide range of disorders. This section focuses on diabetes complications that are more strictly characterized, encompassing diabetic nephropathy, retinopathy and neuropathy. Fig. 3.1 summarizes the general pathway in the progression of diabetic microvascular complications.

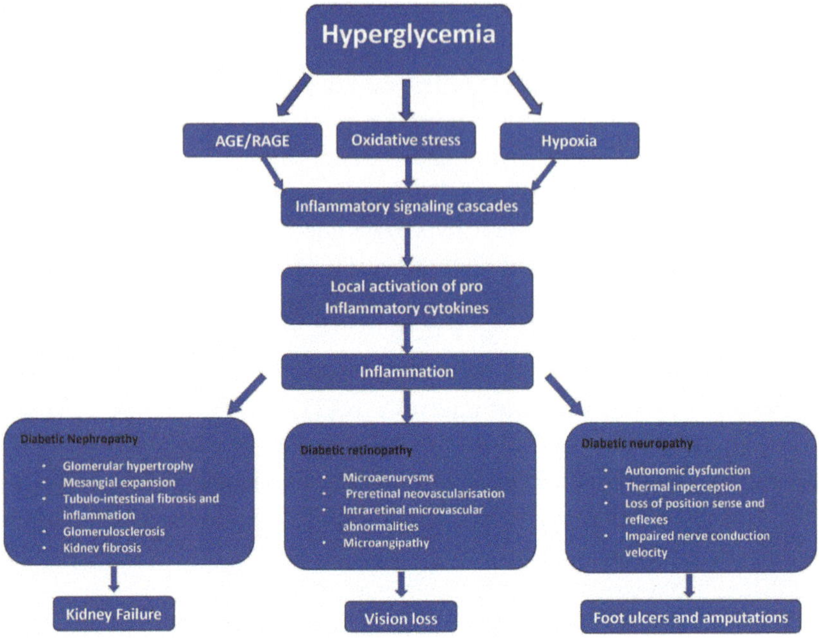

Fig. 3.1 General pathway in the progression of diabetic microvascular complications.

Diabetic nephropathy

One of the most common consequences of DM is diabetic nephropathy. In many cases, histopathology and blood biochemical investigations are conducted to determine the disease condition. Normally, the levels of blood urea nitrogen (BUN), serum creatinine and proteinuria are regularly checked (Zhu et al., 2011a; Granados-Pineda et al., 2018; Kang et al., 2017). Continuous high BG destroys kidney cells and, as a result, lowers glomeruli, in addition to causing a high blood pressure contributed by obesity (Sandino et al., 2021). AGEs rapidly build in the kidneys as a result of diabetic nephropathy, eventually reaching twofold above normal levels in diabetic patients without the end-stage renal disease (Kitamura, 2019). AGEs have been shown to play a role in diabetic nephropathy in a variety of animal models. AGE administration in STZ-treated rats resulted in the advancement of diabetic nephropathy (Tesch et al., 2014; Matsui et al., 2017). In T1DM rats, transgenic overexpression of the methylglyoxal-detoxifying enzyme glyoxalase-1 reduced early renal dysfunction markers such as urine albumin, osteopontin, kidney-inflammation-molecule1, and nephrin, which are all in agreement with these findings (Brouwers et al., 2014). AGEs decrease podocyte autophagy in db/db mice, which is required for optimal kidney function. AGE-induced autophagy in podocytes is mediated by mTOR hyperactivation and subsequent suppression of proautophagic transcription factor EB nuclear translocation (TFEB) (Zhao et al., 2018). Given the importance of AGEs in diabetic nephropathy, modulators of AGE formation or RAGE activation are likely to be employed as medications to treat the condition. Therefore, monitoring AGE or AGE-elicited signaling can be utilized to assess nephropathy progression in experimental animal models (Kitamura, 2019).

The effects of AGEs on the kidney have already been extensively studied using isolated renal cells. Mesangial cells are stimulated by AGEs to create fibronectin, collagen IV, and TGF-1, all of which are important in diabetic kidney fibrosis (Zhou et al., 2004; Tuttle et al., 2005; Huang et al., 2013). Protein kinase C (PKC), extracellular signal-regulated kinase (ERK), and sphingosine kinase 1 (SK1) are thought to be involved in AG*E*-induced fibronectin and TGF-1 production in mesangial cells (Tuttle et al., 2005; Huang et al., 2013; Chen et al., 2017). Another significant target cell type for AGEs is podocytes, predictors of proteinuria. AGEs activate the transcription factor forkhead box protein O4 (FoxO4), which causes apoptosis, block migration and cell adhesion in a neuropilin-1-dependent manner (Chuang et al., 2007; Bondeva et al., 2011). Furthermore, by activating NF-B and zinc finger E-box-binding homeobox 2 (Zeb2) in podocytes, AGEs induce epithelial-mesenchymal transition (EMT), followed by an increase in glomerular permeability (Kumar et al., 2016).

In diabetics, the epithelial cells in the proximal uriniferous tubule transdifferentiate into myofibroblasts, contributing to fibrosis in chronic kidney disease progression (Kang et al., 2017; Hills and Squires, 2011). Therefore, the expression profile of cellular markers can be used to assess epithelial cell transdifferentiation into myofibroblasts. The specific epithelial markers include E-cadherin, zonula occludens-1 (ZO-1), cytokeratin, connective tissue growth factor (CTGF), tissue inhibitor of metalloprotease-2 (TIMP-2), and collagen IV. In contrast, mesenchymal/myofibroblast markers include smooth muscle actin (-SMA), fibroblast-specific protein-1 (FSP-1), N-cadherin, and vimentin. On the other hand, Snail and membrane type 1-matrix metalloprotease, mesenchymal/myofibroblast markers include smooth muscle actin (-SMA), fibroblast-specific protein-1 (FSP-1), N-cadherin, and vimentin (MT1-MMP) (Kang et al., 2017; Hills and Squires, 2011).

Microalbuminuria is a symptom of diabetic nephropathy (DN) caused by albumin leakage through a compromised glomerular membrane. Elevated amounts of urea and creatinine indicate impaired renal function. Honey administration has been reported to significantly reduced blood creatinine while increasing serum albumin and the albumin/globulin ratio in diabetic rats (Cozma et al., 2012). Several inflammatory mechanisms, including NF-kB activation, have been linked to DN. Honey reduces inflammatory cell infiltration into the kidneys of cisplatin-induced nephrotoxicity patients, including cytokine and chemokine production (Hamad et al., 2015), indicating its protective function on the kidneys.

Several studies have shown that propolis slows the progression of nephropathy in diabetic animals. Moreover, there were reported beneficial effects of propolis on DN by causing changes in antioxidative enzyme activity and lipid peroxidation (Oršolić et al., 2012; Zhu et al., 2011b; Abo-Salem et al., 2009; Gong et al., 2017). Oxidative stress is one of the propolis' most important therapeutic targets in diabetic nephropathy. Zhu et al. investigated the therapeutic benefits of administering crude extracts of Chinese or Brazilian propolis (100 mg/kg) to STZ-induced diabetic rats for 8 weeks on nephropathy. The blood urine nitrogen (BUN), serum creatinine, and urine albumin excretion rate (UAER) were all considerably reduced by Brazilian propolis. Chinese propolis, on the other hand, caused some changes in UAER (Zhu et al., 2011a). Conversely, the same research reported that Chinese propolis restored STZ-induced histological alterations more effectively than Brazilian propolis. Additionally, Chinese propolis inhibited mesangial cell growth and vacuolization of renal tubular epithelial cells, but Brazilian propolis merely improved mesangial cell responsiveness (Zhu et al., 2011b), indicating that there are variations in the activity of the propolis depending on its origin.

According to another study, administration of Brazilian propolis extract (100–300 mg/kg/day) to mice for 40 days restored kidney weight, BUN and blood creatinine levels (Abo-Salem et al., 2009). Similarly, oral administration of Saudi Arabian propolis reduced BUN, serum creatinine, uric acid and urine creatinine (El Rabey et al., 2017). Propolis therapy enhanced the histological pictures of the renal cortex in STZ-treated rats, as well as serum sodium and potassium ions levels. Furthermore, Iranian propolis extract (100 or 200 mg/kg/day) administered for 6 weeks reduced STZ-induced kidney atrophy, glomerular enlargement, and basement membrane thickening (Sameni et al., 2016), indicating the protective effects of propolis on the kidneys.

Following gentamicin-induced nephrotoxicity in adult male Wistar rats, administration of royal jelly (100 mg/kg) for 10 days reduced renal tissue injury. Additionally, consuming 100 mg/kg royal jelly for 3 days and the subsequently administration of royal jelly + gentamicin for 7 days protected gentamicin-induced kidney damage (Alaraj, 2020). Royal jelly (100 mg/kg) has also been reported to confer a positive nephroprotective effects with decreased plasma creatinine (CREA) and blood urea nitrogen (BUN) levels after the injury induced by ethylene glycol exposure in male Sprague-Dawley rats (Aksoy and Aslan, 2017). In another study, when the effect of Mongolian honey bee venom (*Apis mellifera*) on renal dysfunction was investigated in alloxan-induced diabetic Chinchilla rabbits, plasma creatinine and BUN were reduced by 19.8% and 4.8%, respectively (Ts et al., 2015), again confirming its potential in protecting the kidneys.

Diabetic retinopathy

In diabetic retinas, ROS is created in the mitochondria by NADPH Oxidase (Nox) (Santiago et al., 2018). In fact, oxidative damage was plainly visible in the retinas of STZ-induced diabetic Lewis rats (Du et al., 2003). ROS activates various metabolic pathways, including the AGE-RAGE and PKC pathways, resulting in an increase in intracellular ROS buildup (Santiago et al., 2018). Additionally, ROS promotes leukocyte adherence to endothelial cells by raising the expression of intracellular adhesion molecule (ICAM), in addition to causing cell death in pericytes and endothelial cells (Kitamura, 2019; Santiago et al., 2018). Due to the infiltrations, leukocytes generate low-grade inflammation, presumably as a result of sensing apoptotic cell death.

Furthermore, the local synthesis of proinflammatory cytokines such as TNF- and IL-1 activates NF-B, causing a worsening of inflammation. The blood–retina–barrier (BRB) eventually breaks down due to the loss of the tight junction proteins ZO-1 and occlusion (Santiago et al., 2018). BRB is determined in rat models by injecting Evans blue into the tail vein and investigating the extravasation of Evans blue in the retina (Shi et al., 2019) where in alloxan-induced diabetes mouse, enhanced cell death of pericytes and endothelial cells can be histologically confirmed (Mizutani et al., 1996). Furthermore, the notable loss of capillary pericytes leads to the production of microaneurysms and the thickening of the retinal basement membrane (Barot et al., 2013).

Aqueous flow in the retina is reduced (15%) in T1DM patients with no indication of microvascular problems where the severity of diabetic retinopathy was related to the length and magnitude of the hyperglycemic exposure (Ojieh et al., 2015; Jenkins et al., 2015). According to a recent study, the intraocular pressure (IOP) of patients with uncontrolled diabetes was considerably higher than the IOP of patients with controlled diabetes (Shi et al., 2019). In another study, the concentrations of glucose and albumin and oxidative stress biomarkers (GSH and MDA) in the eye humor of honey-fed male and female Wistar rats were measured and compared with the fructose-fed group. Honey was unrelated to the likelihood of increased ocular pressure occurring as a result of an elevated glucose concentration. A significantly higher concentration of GSH in aqueous humor and a stronger protection against oxidative stress was found in rats fed with honey than in the fructose-fed group (Ojieh et al., 2015).

The effects of Chinese propolis ethanol extract (200 mg/kg/day; p.o.) for 12 weeks on retinopathy in STZ-induced diabetic mice were studied by Shi and colleagues. Propolis prevented retinal membrane thickening by counteracting diabetes-induced photoreceptor cell death, accompanied by a small drop in FBG and Hb1Ac and a nearly complete restoration of serum MDA, ROS and RNS. As a result, propolis lowered BRB permeability in diabetic rats. In addition, propolis restored the expression of occludin and ZO-1 in the retina. Similarly, Chinese propolis appears to protect against oxidative stress-induced BRB breakdown through improving retinal cell tight junction dysfunction (Shi et al., 2019).

Delayed wound healing

Diabetic neuropathy predisposes to chronic diabetic wound ulcers (Deng et al., 2021) where it has been reported that one out of every four diabetics will develop a diabetic wound ulcer at some point in their lives (Antohe and Popa, 2021; Armstrong et al., 2020). According to estimates, a lower limb has to be amputated every 20 s in the world, due to DM (Aalaa et al., 2012). Hyperglycemia damages endothelial cells in the skin and mucosa, similar to other diabetic problems. The hemostasis, proliferation and remodeling phases are all influenced by the considerable restoration of VEGF expression. VEGF induces platelet aggregation and causes blood clots to protect the wound environment, stimulates endothelial cell proliferation, activates angiogenesis, regulates collagen fiber formation, and turns collagen fibers into scars (Kurek-Górecka et al., 2021; Ellis et al., 2018). Since VEGF production in endothelial cells and fibroblasts is downregulated in diabetic individuals and animals, wound remodeling is significantly disrupted (Kitamura, 2019; Maezawa et al., 2015). Moreover, diabetic individuals display aberrant production of ECM in fibrocytes through interference with signals of growth factors such as TGF-β, fibroblast growth factor (FGF), and hepatocyte growth factor (HGF) (Kitamura, 2019; An et al., 2021).

TGF-β is a growth factor generated by platelets early in the healing process that drives the chemotaxis of inflammatory and immune cells to wounds, increases extracellular matrix deposition and influences the creation of granulation tissue (Kurek-Górecka et al., 2021; Makowski et al., 2021). TGF-β also plays an important part in the remodeling process. TGF-β encourages reepithelization by promoting the replacement of collagen type III with collagen type I (Kurek-Górecka et al., 2021; Mori et al., 2016). Furthermore, β-defensin-2(BD2) is an inflammatory, reepithelization and angiogenesis factor. BD2 enhances keratinocyte migration and proliferation by increasing the production of proinflammatory cytokines (Kurek-Górecka et al., 2021; Koeninger et al., 2020). Macrophages are also important in the healing process because they participate in phagocytosis and impact the release of mediators such as TGF-β, PDGF, FGF, Il-1, Il-6, and TNF-α, which regulate inflammation and govern epithelialization and angiogenesis (Kurek-Górecka et al., 2021; Raziyeva et al., 2021). The reduced activity of enzymes such as glutathione peroxidase (GSH-Px), manganese superoxide dismutase (MnSOD), and catalase is linked to a slowed healing process in diabetic wounds (Kurek-Górecka et al., 2021; Farzaei et al., 2018). According to many studies, oxidative stress and the high production of ROS slow down the healing process (Kurek-Górecka et al., 2021; Johnson et al., 2021).

Honey has antibacterial properties that inhibits *Staphylococcus aureus* (*S. aureus*), a prevalent Gram-positive bacterium linked to wound infection (Almasaudi et al., 2017). Other antibiotic-resistant species, such as methicillin-resistant *S. aureus* (MRSA) (Almasaudi et al., 2017) and several Gram-negative bacteria involved in wound infection such as *Pseudomonas aeruginosa* (*P. aeruginosa*), are also particularly susceptible to honey (Fernandes et al., 2020). Honey employs a bactericidal activity against wound biofilm-embedded microbes such as *P. aeruginosa* and *S. aureus* (Al-Kafaween et al., 2021).

Honey's wound-healing effects are also extended to the human fibroblasts. Honey improved fibroblast wound closure by increasing IL-4, IL-6, and IL-8 levels (McLoone et al., 2016). Its hypoxia-modulating impact may help to speed wound healing in addition to its antioxidant properties. Due to these properties, honey inhibits the generation of ROS, raises the expression of the hypoxia-inducible factor 1 (HIF-1) marker, while promoting cell proliferation and differentiation in hypoxic wound healing (Memar et al., 2018).

On histological investigation, topical administration of Manuka honey to diabetic rats' excision wounds resulted in considerable wound healing due to early epithelialization with well-formed keratinized squamous epithelium and normal collagen tissue around hair follicles (Gill et al., 2019). In STZ-DM rats, the effects of topical administration of mad honey on temporal wound healing were examined. Animals from the mad honey group had significantly lower MDA levels, TNF-α and MMP-9 expression, as well as a higher antioxidant enzyme activity and IL-10 expression. Finally, the topical use of mad honey promotes diabetic wound healing in diabetic rats (Malkoç et al., 2020). In another investigation on diabetic rats, honey showed promise for wound treatment when compared to amikacin (Eyarefe et al., 2014).

McLennan et al. reported that topical use of propolis ("Honey Spring" product; 20 L/wound) in STZ-induced diabetic rats, improved delayed reepithelization and epithelial closure of a scratched lesion (McLennan et al., 2008). Propolis increased macrophage infiltration while reducing neutrophil accumulation in the lesions, preventing tissue damage from ROS. Propolis did, in fact, reduce the production of myeloperoxidase (MPO) while increasing the number of proliferative cells in the lesions. Therefore, propolis appears to protect macrophages from neutrophil attacks while also promoting tissue healing (McLennan et al., 2008).

A bacterial cellulose membrane with 1% adsorbed Brazilian red propolis yielded similar results. As compared to ethanol or butanol extracts, STZ-treated rats had better lesion closure after treatment with the bacterial membrane with absorbed propolis. Edema, hyperemia, and exudation were also significantly decreased by the butanol extract-containing membrane, whereas edema, hyperemia, and exudation were significantly reduced by the ethanol extract-containing membrane. Furthermore, both extract-containing membranes encouraged wound contraction as well as increased TGF-β induction. Propolis-containing membranes, on the other hand, boosted MPO activity and TNF levels in the lesions. As a result, appropriate neutrophil accumulation in the lesions may contribute to a fast wound healing in this paradigm (Picolotto et al., 2019).

Other studies focused on propolis' ability to increase ECM synthesis. In STZ-induced diabetic mice, "Honey spring" propolis restored abnormal TGF-β expression. Propolis increased type I collagen formation in the wound while inhibiting MMP9 growth, which was harmful (Hozzein et al., 2015). An oral traumatic ulcer model was also treated with propolis extracts. Artificial sores on the lower lip of STZ-treated Wistar rats were treated with an Indonesian propolis-absorbed hydroxypropyl methylcellulose gel where fibroblast growth factor-2 (FGF-2) and VEGF were considerably induced in the lesions of the propolis-treated group after 3 days compared to the control (Ernawati and Puspa, 2018; Puspasari et al., 2018). Based on these data, the number of fibroblasts in propolis-treated group was significantly higher (Puspasari et al., 2018). Additionally, in STZ-induced diabetic mice, poly(vinyl alcohol)-blended cellulose membranes with absorbed Brazilian propolis and vitamin C improved back injury repair compared to control membranes (Voss et al., 2018). In another study, royal jelly improved the rate of dermal wound healing in diabetic rats by increasing the migration of cultured human dermal fibroblasts, improving granulation tissue production and decreasing the healing time of desquamated skin lesions (Fujii et al., 1990), again confirming the potential of honey products on wound healing.

Subcutaneous administration of bee venom to diabetic mice (50 μL for 15 days) increased TGF-and VEGF expressions. Additionally, bee venom accelerates the healing process in experimental animals by impairing caspase-3, -8 and -9 activities. Bee venom activates endothelium progenitors cells in tissues by decreasing caspase-3, -8 and -9 activities and also reduced the expression of ATF-3 and iNOS to lower the amount of free radicals and minimize oxidative stress. Furthermore, subcutaneous administration of bee venom restores the levels of the proinflammatory cytokines including IL-1, IL-6, and TNF-α (Badr et al., 2016).

In another study, subcutaneous administration of bee venom (50 μL) into the wounded area of diabetic mice for 15 days enhanced the levels of collagen types I and β-defensin-2 (BD2) in diabetic wounds. Moreover, bee venom therapy reduced apoptotic macrophages, resulting in an increase in the phagocytic index. Bee venom also stimulates angiogenesis by reactivating Ang-1/Tie-2 signaling while increasing the production of Nrf2 and BD2 (Hozzein et al., 2015).

Amin and Abdel-Raheem (2014) reported that a wound dressing made of polyvinyl alcohol (PVA), chitosan and bee venom improves wound healing in diabetic rats. The researchers reported that wound dressings containing bee venom at 3% and 4% resulted in wound closure rates of 84 and 91%, respectively, following 14 days of treatment. Furthermore, the scar was free of contamination and was clean and smooth. Bee venom-loaded patch raises glutathione levels in diabetic rats' wound tissues. Additionally, bee venom-loaded hydrogel therapy can provide significant benefits in the event of skin lesions caused by hyperglycemia (Amin and Abdel-Raheem, 2014).

Clinical studies involving bee products

There is a significant biochemical distinction between animal models and humans, thus limiting the utilization of animal models in examining the therapeutic effects of bee products in human. Clinical trials on bee products are required before they can be administered to patients. We have presented an overview of human clinical data on the impact of bee products on diabetes and diabetic complications in this section.

Honey consumption decreased GI and peak incremental index (PII) in patients and controls in a case-control cross-sectional study involving 20 children and adolescents with type 1 diabetes as well as in 10 healthy children and adolescents as controls. Honey consumption increased C-peptide levels in both groups. Nevertheless, the rise in C-peptide following honey ingestion was significant only in the control group compared to sucrose or glucose (Abdulrhman et al., 2011) indicating the potential of honey in diabetes.

Sadeghi et al. found that 50 g/day honey combined with a weight-maintenance diet increased HbA1c and decreased waist circumference in T2DM patients after 8 weeks compared to individuals who solely followed a weight-maintenance diet. On the other hand, FBS, insulin, Homeostatic Model Assessment for Insulin Resistance (HOMA-IR), quantitative insulin sensitivity check index (QUICKI), weight, BMI, hip circumference and waist-to-hip ratio remained unchanged (Sadeghi et al., 2019). Honey was associated with reduced GI and PII in a comparable case-control cross-sectional investigation with a larger sample size (50 type 1 DM and 30 non-DM participants) than sucrose consumption. In contrast to the previous study, honey consumption resulted in a significant increase in C-peptide levels in both diabetic and nondiabetic patients compared to that for sucrose or glucose group (Abdulrhman et al., 2013a). Another study reported that honey generated a significantly lower rise in BG in both healthy and diabetic individuals and higher increases in insulin and C-peptide levels in diabetic patients but substantially lower insulin and C-peptide elevations in nondiabetic persons when compared to dextrose or sucrose administrations.

Honey consumption over 15 days boosted HDL cholesterol while lowering BG, total cholesterol, LDL cholesterol, TG, CRP, and homocysteine (Hcys) in people with high TG, cholesterol, or CRP levels (Al-Waili, 2004). A 12-week honey intervention reduced HbA1c and resulted in significant reductions in fasting BG as well as elevations in fasting and 2 h postprandial C-peptide in a randomized crossover clinical research involving T1DM

($n = 20$) patients with 10% HbA1c. Honey supplementation also lowered TG, LDL cholesterol and total cholesterol levels while minimizing subscapular skin fold thickness. Additionally, honey withdrawal resulted in a considerable reduction in triceps skin fold thickness and midarm circumference, on top of some of the previously described positive metabolic benefits (Abdulrhman et al., 2013b).

In healthy volunteers ($n = 12$) and T2DM patients ($n = 16$), the effect of intrapulmonary honey (60% solution) administration on random and fasting BGs, plasma insulin, and C-peptide were examined. Honey inhalation increased plasma insulin and C-peptide levels, dramatically reduced hyperglycemia during glucose tolerance while reducing random and fasting BG levels in healthy people and diabetic patients (Al-Waili, 2003). Another study which evaluated honey and glucose tolerance in 24 impaired glucose tolerant patients (whose parent or mother had a history of type 2 DM) discovered that all subjects had significantly lower BG values after honey ingestion than after glucose consumption (Erejuwa, 2019).

Honey exerted a considerably lower BG than glucose in another set of six participants with mild T2DM (Agrawal et al., 2007). Nazir and colleagues evaluated the glycemic reactions of 97 T2DM patients to natural honey (30 and 75 g) and glucose (75 g). Based on their study, the glycemic response to 30 g of honey was much lower at 2 h than that who received 75 g of honey or glucose. The glycemic reaction to 75 g of honey was substantially lower at 2 h than that for 75 g of glucose, indicating that honey can lower the glycemic effect. The finding was confirmed in another study where honey has been found to have a BG-lowering impact in some subjects (10.7%) who ingested 30 g of honey (Nazir et al., 2014).

Yaghoobi et al. who conducted a randomized controlled trial to study the effect of honey versus sucrose on body weight, BG, lipid abnormalities and CRP in overweight or obese adults found that honey induced a slight decrease in the body and fat weights. The data revealed that honey reduced BG, TG, CRP, total, and LDL cholesterols, while increasing HDL cholesterol in individuals with normal values. In subjects with elevated values, honey ingestion decreased TG, CRP, total, and LDL cholesterols (Yaghoobi et al., 2008). Since obesity and overweight are key risk factors for the development of diabetes, these findings suggest that honey may slow the progression of prediabetes and obesity to overt T2DM.

Munstedt et al. reported that a solution containing 75 g honey lowered blood LDL-C levels only in women but not in man. Additionally, there were no changes in TG and TC when honey was compared to a solution containing 75 g glucose and fructose (Münstedt et al., 2009) after 14 days. Raatz et al. found that daily consumption of 50 g honey, sucrose, or high-fructose corn syrup elevated TG and insulin levels in both glucose-tolerant and intolerant patients (Raatz et al., 2015). In another study, Rasad et al. found that 80 g/day honey, dissolved in 250 mL water, reduced FBS serum levels while not affecting blood pressure in healthy young participants after 30 days (Rasad et al., 2014) after accounting for confounders such as age, physical activity, overall dietary intake. The findings were further strengthened by that conducted by Despland et al. who evaluated the effects of Robinia honey consumption, high free glucose and fructose diet and a low fructose diet (control group) in healthy normal-weight men (Despland et al., 2017). During the intervention, all study groups were given a weight-maintenance diet. They found that a diet containing 25% of total energy from honey or pure fructose-glucose reduced postprandial BG and insulin somewhat, whereas postprandial TG remained unchanged (Despland et al., 2017) as compared to the control group.

Another randomized controlled trial that investigated the effects of honey consumption for 8 weeks in T2DM individuals found that honey reduced blood sugar and hyperlipidemia while increasing HbA1c (Bahrami et al., 2009). The HbA1c finding contradicted what had been observed in prior investigations as well as that in a recent study where honey supplementation for 4 months resulted in significant reductions in HbA_{1c} and total cholesterol levels in T2DM patients (Enginyurt et al., 2017). The discrepancies in the findings may be explained by Erejuwa, who found various characteristics contributing to high HbA1c in T2DM patients which include utilization of different honey type (which had abnormally low fructose: glucose ratio) and using high honey doses (Erejuwa, 2012).

Abdulrhman explored the use of honey as a sole alternative agent for the treatment of DM in an 11-year case study and found that the T2DM patient had persistent hyperglycemia and hyperlipidemia (Abdulrhman, 2013) indicating that long term effects of honey should also be investigated. In a separate clinical trial including 20 T2DM patients, honey administration has been found to considerably lower body weight and BMI but had no favorable impact on hyperglycemia or lipid abnormalities. Additionally, honey supplementation elevated blood sugar levels in these patients without causing diabetic ketoacidosis or hyperglycemic hyperosmolarity (Abdulrhman, 2016). Nevertheless, since honey was used as the sole treatment for DM in these two investigations, the results must be regarded with caution since the patients stopped using their antidiabetic medications for a long time (0.42–13.5 years) which may pose a risk hazard to their health with 50% of the subjects took honey alone for at least a year. In a study from Malaysia, 30 g/day stingless bee honey did not affect FBS, lipid profiles, blood pressure, or BMI in patients with impaired fasting glucose (IFG) after 30 days compared to a control group (Rashid et al., 2019).

Unfortunately, honey's effects were not compared with that for placebo in the majority of the clinical trials included in this chapter which may result in a significant bias and have an impact on both the internal and external validities of the studies. Although one may argue that comparison with a placebo is more relevant for subjective outcomes, the lack of comparison may contribute to an overestimation or underestimation of the intervention's actual influence on objective outcomes (Hróbjartsson et al., 2011).

Another flaw in most research was that results were reported without considering confounders such as baseline characteristics, total energy consumption, physical activity and BMI. Additionally, the total energy intake is the key confounder often ignored by many researchers, which may further skew the findings. Honey is a nutrient-dense functional food with 64 cal per tablespoon (about 20 g). As a result, adding honey to one's diet, especially in large doses with no calorie substitution, no changes in daily calorie consumption or in the physical activity, might increase body weight and fat mass due to the higher calorie intake, all of which may impact the metabolic parameters (Akhbari et al., 2021). Only two of the included clinical trials (Sadeghi et al., 2019; Despland et al., 2017) advocated a weight-maintenance diet in addition to honey consumption with most studies not disclosing the data after controlling for confounding factors. Nevertheless, Despland et al. attempted to clarify this issue and reported that after tailored weight maintenance meals, a diet containing 25% of total caloric intake had no negative impacts on glucose and TG levels (Despland et al., 2017).

Propolis supplementation has been shown in multiple studies to improve a variety of blood parameters in T2DM patients. For example, in a 90-day double-blind, placebo-controlled clinical trial, Iranian propolis (1 g/day; $n = 50$) significantly decreased HbA1c, blood insulin, HOMA-IR (insulin resistant), HOMA- levels when compared to the placebo group ($n = 44$) (Zakerkish et al., 2019). Propolis also boosted HDL-C levels in the blood but did not affect the body weight, BMI, blood triglyceride levels, total cholesterol, LDL-C, or VLDL-C levels (Samadi et al., 2017). Additionally, the levels of C-reactive protein (CRP) and tumor necrosis factor (TNF-α) in the blood were reduced by 30% and 16%, respectively, following propolis administration (Zakerkish et al., 2019). A study on Iranian propolis further confirmed that there was a reduction in systemic inflammation in T2DM patients. Similarly, another study in Iran on a pill containing propolis (900 mg/day) administered for 12 weeks on 30 individuals significantly lowered FBG, HbA1c, total blood cholesterol and LDL-C, while having no effect on blood triglyceride levels, HDL–C, VLDL-C, or insulin resistance indices in T2DM patients (Samadi et al., 2017).

In another study, administration of propolis capsules from Iran (1.5 g/day for 8 weeks, $n = 30$) reduced FBG, blood insulin, HbA1C, and insulin resistance as compared to placebo ($n = 30$) (Afsharpour et al., 2019). Iranian propolis has been demonstrated to have positive glycemic effects in T2DM patients, despite showing some discrepancies in the lipid markers and insulin indices across several studies. Additionally, Iranian propolis raised the overall antioxidant activity, SOD, GPx and CAT, as well as oxidized LDL and fructosamine levels in T2DM patients (Afsharpour et al., 2019; Hesami et al., 2019) suggesting that propolis' antidiabetic properties are mainly contributed by its antioxidant activity.

The therapeutic benefits of Chinese propolis in T2DM patients are less than those of Iranian propolis, based on a basic comparison of data from prior studies (Zakerkish et al., 2019; Samadi et al., 2017; Afsharpour et al., 2019; Hesami et al., 2019; Gao et al., 2018). When compared to the control group ($n = 30$), supplementation with Chinese propolis (900 mg/day) for 18 weeks exerted no effect on serum glucose, glycosylated hemoglobin, insulin, FRAP, MDA, SOD, GPx, IL-1, and TNF- levels in T2DM patients (Gao et al., 2018). On the other hand, there was increased serum GSH (from 2.2 to 7.4 g/L) that was accompanied by a decrease in lactate dehydrogenase (LDH) (Gao et al., 2018). Nevertheless, supplementation of Chinese propolis limited the tissue damage in T2DM patients via accumulation of circulating botanical antioxidants, as blood polyphenols and flavonoids rose.

Similar like the Chinese propolis, clinical experiments employing Brazilian green propolis have also revealed a minor effect on blood parameters (Fukuda et al., 2015; Zhao et al., 2016). When compared to the placebo group ($n = 39$), administration of Brazilian propolis extract (226.8 mg/day) for 8 weeks did not result in any significant improvement in HOMA-IR, FBG, HbA1c, blood insulin, total cholesterol, HDL–C, LDL-C, triglyceride, remnant-like particle cholesterol, TNF-, IL-6, CRP, and urinary albumin excretion in T2DM patients (Fukuda et al., 2015). Furthermore, there were no discernible variations in glucose metabolism markers such as BG, glycosylated hemoglobin, insulin, aldose reductase, or adiponectin in T2DM patients who received Brazilian green propolis (900 mg/day, $n = 32$)

(Zhao et al., 2016). Treatment with Brazilian green propolis for 18 weeks, on the other hand, resulted in significant increase in GSH and total polyphenol in the blood, as well as decrease in circulating carbonyls, LDH and TNF-α (Zhao et al., 2016). In a clinical trial for diabetic individuals, commercially available propolis (BioPropolis) was used in T2DM patients with periodontitis. The patients were given either BioPropolis (400 mg/day, $n = 26$) or placebo ($n = 24$) capsules for 3 or 6 months. Propolis reduced fasting plasma glucose and HbA1c levels at both time points as compared to placebo (El-Sharkawy et al., 2016).

The effects of propolis on tissue damage caused by T2DM have been assessed using blood indicators. Iranian propolis (1 g/day) for 90 days significantly reduced BUN, AST, and ALT levels in T2DM patients, showing that Iranian propolis has hepato- and renoprotective properties (Zakerkish et al., 2019). As a result, the estimated glomerular filtration rate (eGFR) was maintained during propolis administration, although the eGFR was dramatically reduced in placebo-treated patients (Zakerkish et al., 2019). Similarly, administration of Brazilian propolis (226.8 mg/day) for 8 weeks kept blood uric acid levels stable, whereas the parameters for the placebo group increased during the experimental period indicating that the propolis ameliorated the decline in eGFR in T2DM patients (Fukuda et al., 2015). Furthermore, "BioPopolis" (400 mg/day) improved periodontal measures like probing pocket depth and clinical attachment, which coincided with a decrease in blood N6-(carboxymethyl)-L-lysine, a major AGE (El-Sharkawy et al., 2016).

The effects of propolis topical treatment on T2DM patients' foot ulcers were also investigated. In comparison to the control group ($n = 84$), topical treatment of "Honey spring" propolis on diabetic foot ulcer of T2DM patients ($n = 24$) for 6 weeks reduced the ulcerative area by 41% (Henshaw et al., 2014). At 4, 5, and 7 weeks after treatment was started, the number of patients whose ulcers had healed was considerably higher in the propolis-treated group than in the control group, indicating that propolis improved wound healing (Henshaw et al., 2014). Additionally, propolis also lowered the bacterial load. Propolis reduced the amount of active MMP-9 in wound fluid following debridement (Henshaw et al., 2014), thereby promoting wound healing by reducing bacterial load and MMP-9-induced inflammation. According to these findings, topical application of a 5% propolis water extract-containing ointment reduced the size of ulcer 1 week following the start of propolis treatment ($n = 20$) when compared to the use of a control ointment ($n = 20$) (Afkhamizadeh et al., 2018). Compared to the control group (discharge, 15%; erythema, 15%), the number of patients with ulcer discharge and erythema was also lower in the propolis-treated group (discharge, 5%; erythema, 0%) (Afkhamizadeh et al., 2018).

More recently, the effects of propolis spray from Chile on diabetic foot ulcers have also been investigated (Mujica et al., 2019). The propolis spray reduced the average ulcer size and promoted fibrosis (Mujica et al., 2019). Patients who received propolis spray on the other hand, had higher levels of tissue GSH as well as a higher GSH/glutathione disulfide ratio, indicating that propolis improved local antioxidant activity (Mujica et al., 2019). Propolis treatment decreased TNF levels while producing a net increase in IL-10 concentration in the ulcer site, indicating propolis' antiinflammatory potential (Mujica et al., 2019). Overall, topical propolis therapy enhances wound healing in T2DM patients with foot ulcers.

A considerable number of human studies investigating on the effects of royal jelly supplementation on blood lipids and glycemic indicators have also been performed. In a recent metaanalysis, royal jelly was confirmed to be an effective and relatively safe alternative for blood cholesterol regulation (Hadi et al., 2018). Nevertheless, the studies were diverse in terms of the clinical scenario. Two studies, including in T2DM individuals, found substantial benefits (Pourmoradian et al., 2014; Khoshpey et al., 2016). However, three groups who had healthy participants (Guo et al., 2007), subjects with dry eye (Inoue et al., 2017) and swimmers (Saritas et al., 2011), did not show any significant improvements in glycemia. As a result, it is reasonable to conclude that royal jelly is more effective in diabetics having higher glucose levels. Following royal jelly consumption, diabetic patients demonstrated more significant decrease in both FPG and HbA1c levels although these changes were not significant.

In a study by Khoshpey et al. (Khoshpey et al., 2016), royal jelly caused an 11.70 unit reduction in FPG in diabetic patients, which was statistically nonsignificant due to the large confidence interval seen. However, HbA1c levels improved with a substantially narrower confidence interval, although the *P*-value remained nonsignificant. Based on these findings, it is plausible that royal jelly has a more marked effect on blood glucose levels in individuals with impaired glycemia although more well-designed trials are required to confirm this hypothesis. Royal jelly ingestion can boost overall antioxidant capacity in T2DM considerably (Khoshpey et al., 2016). Pourmandian et al. found significant improvements in oxidative stress indicators (MDA, SOD, and GSH–PX) in their clinical trial (Pourmoradian et al., 2014). The dosage of royal jelly supplements does not appear to significantly impact the findings, as no apparent association was seen when different doses are used. One of the most important elements regulating glycemia and insulin responses is dietary intake. Nevertheless, except for the study by Khoshpey et al. (Khoshpey et al., 2016), no research have adjusted for the study groups' food intakes, which could potentially add to bias to the findings.

To our knowledge, only a single clinical trial on the antidiabetic potential of bee pollen has been published to date. The study has demonstrated improvement in a number of blood parameters in insulin-independent diabetes patients. According to the study, twice a day bee pollen consumption (250 mg/kg) for 21–28 days reduced BG levels by 22.3%. Additionally, blood cholesterol levels dropped significantly (Senyuk et al., 2016). To the best of our knowledge, antidiabetic potential of bee venom and bee bread in humans have not been investigated. The findings on the effects of bee products on metabolic derangements in clinical studies are presented in Table 3.1.

Conclusion

Current preclinical research in the field of "Apitherapy" products in diabetes suggests that honey, propolis, royal jelly, bee pollen, bee bread, and bee venom are safe for human consumption in diabetic individuals. Based on animals and human studies, honey produces a substantially lower glycemic reaction than all other frequently accessible sugars and sweeteners and therefore may be used as a sugar substitute in diabetic diets. Honey stimulates insulin production from pancreatic β-cells in both healthy and diabetic rats/subjects. Nevertheless, due to limited and conflicting data, it is unclear whether honey can have hypoglycemic/antihyperglycemic and/or antihyperlipidemic benefits in diabetic individuals. Honey's antioxidant and antiinflammatory properties, on the other hand, are undeniable. In addition to oral administration, honey can be applied topically on diabetic wounds to achieve debridement and to hasten wound healing. Since resistance to honey has not been reported, applying topical honey for diabetic wounds may be a better option in ameliorating resistance.

Propolis has been shown to control oxidative stress, the build-up of AGEs and adipose tissue inflammation in preclinical studies, all of which lead to insulin resistance or insulin secretion abnormalities. Therefore, propolis therapy may help reduce diabetes complications including nephropathy, retinopathy, foot ulcers, and nonalcoholic fatty liver disease. To date, a variety of clinical research on the health effects of propolis on BG levels,

Table 3.1 Summary of findings on the effects of bee products on metabolic derangements in clinical studies.

Bee product	Location	Study design	Gender	Mean age	Sample size at baseline	Disease background
Honey	Iran	RCT (cross-over)	Both	57.5	Total ($n=53$) Honey ($n=27$) Placebo ($n=26$)	T2DM
	Malaysia	RCT	Both	Honey (52.8),Placebo (50.4)	Total ($n=60$) Kelulut honey ($n=30$) Control ($n=30$)	Impaired fasting glucose (IFG)
	Turkey	RCT	Both	18–80	Total ($n=64$) Intervention/ patient ($n=32$) Control/ healthy ($n=32$)	T2DM (using metformin)
	North Dakota	RCT (cross-over)	Both	Glucose tolerant: 38.9 Glucose intolerant: 52.1	Total ($n=89$) Glucose tolerant ($n=52$) Glucose intolerant ($n=37$)	Glucose tolerant, glucose in tolerant.

Sample size at end	Dose (duration)	Findings		Reference
		Index	Effect	
Total ($n=42$) Honey ($n=18$) Control ($n=24$)	50 g/day for 8 weeks (4 weeks washout)	HbA1c, FBG, Weight, BMI, Insulin level, HC, Waist-to-hip ratio, HOMA-IR, HOMA-β, QUICKI	No significant change	Sadeghi et al. (2019)
		WC Waist-to-height ratio	Decrease	
Total ($n=64$) Kelulut honey ($n=27$) Control ($n=27$)	30 g/day for 30 days	FBS TC HDL-C LDL-C BMI SBP DBP	No significant change	Rashid et al. (2019)
Total ($n=64$) Intervention/ patient ($n=32$) Control/ healthy ($n=32$)	5, 15 and 25 g/d for 4 months	HbA1c (all of dose) TC (doses of 5 and 25 g/d), LDL (In all groups).	Decrease	Enginyurt et al. (2017)
		TG, HDL	No significant change	
Total ($n=55$) Glucose tolerant ($n=28$) Glucose intolerant ($n=27$)	50 g carbohydrate from honey, HFCS and sucrose for 2 weeks (2–4 weeks wash out)	Chylomicrons, TG, LDL, Glucose Insulin	Decreased	Raatz et al. (2015)
		Weight BMI Glucose Insulin HOMA_IR-glucose iAUC Insulin iAUC SBP DBP TC HDL-C LDL-C TG	No significant change	

Continued

Table 3.1 Summary of findings on the effects of bee products on metabolic derangements in clinical studies—cont'd

Bee product	Location	Study design	Gender	Mean age	Sample size at baseline	Disease background
	Iran	RCT (double-blind)	M	Honey: 21.53 Sucrose: 24.23	Total (n = 60) Honey (n = 30) Sucrose (n = 30)	Healthy
	Germany	RCT	Both	Honey: 62.6 Honey-comparable sugar solution: 58.8	Total (n = 60) Honey group (n = 30) Honey-comparable sugar solution group (n = 30)	High cholesterol
	Iran	RCT	Both	57.2	Total (n = 54) Honey (n = 28) Control (n = 26)	T2DM
	Iran	RCT	Both	41.2	Total (n = 60) Honey: 40 Control: 20	Obese or overweight
Propolis	Iran	RCT (double-blind)	Both	Propolis (55.4), Placebo (54.86)	Total (n = 100) Propolis (n = 50) Placebo (n = 50)	T2DM

Sample size at end	Dose (duration)	Findings		Reference
		Index	Effect	
Total ($n=60$) Honey ($n=30$) Sucrose ($n=30$)	80 g honey in 250 ml water for 6 weeks	FBS Blood pressure	Decrease No significant change	Rasad et al. (2014)
Total = 60 Honey group: 30 Honey-comparable sugar solution group: 30	75 g of honey solution or a honey-comparable sugar solution once daily over 14 days.	LDL (Only Women), TG, LDL, HDL and TG	No significant change	Münstedt et al. (2009)
Total ($n=48$) Honey ($n=25$) Control ($n=23$)	First 2 weeks: 1 g/kg/d Second 2 weeks: 1.5 g/kg/d Third 2 weeks: 2 g/kg/d Last 2 weeks: 2.5 g/kg/d	Bodyweight TG TC LDL-C FBS LDL/HDL HDL-C HbA1c	Decrease Increase	Bahrami et al. (2009)
Total ($n=55$) Honey: 38 Control: 17	70 g/250 ml tap water for 30 days	Normal variable: in subjects with normal values: BMI and FBG↓ Patients with TG > 150 mg/dl Bodyweight Body fat TC LDL-C HDL-C	Decrease Increase	Yaghoobi et al. (2008)
Total ($n=94$) Propolis ($n=50$) Control ($n=44$)	500 mg twice daily for 90 days	HbA1c Insulin, HOMA-IR, HOMA-β, TNF-α, BUN, ALT, AST, IL-1β,	Decrease	Zakerkish et al. (2019)

Continued

Table 3.1 Summary of findings on the effects of bee products on metabolic derangements in clinical studies—cont'd

Bee product	Location	Study design	Gender	Mean age	Sample size at baseline	Disease background
	Iran	RCT (double-blind)	Both	Propolis (51.3), Placebo (56.07)	Total ($n=66$) Propolis: $n=33$ Placebo: $n=23$	T2DM
	Iran	RCT (double-blind)	Both	Propolis (51.81), Placebo (49.05)	Total ($n=62$) Propolis: $n=31$ Placebo: $n=31$	T2DM
	China	NRCT (double-blind)	Both	Propolis (57.7), Placebo (60.6)	Total ($n=61$) Propolis: $n=31$ Placebo: $n=30$	T2DM
	Brazil	RCT (double-blind)	Both	Propolis (63.7), Placebo (62.9)	Total ($n=82$) Propolis: $n=41$ Placebo: $n=41$	T2DM

Sample size at end	Dose (duration)	Findings		Reference
		Index	Effect	
		2 hp., ALP, CRP, Body weight, BMI, FBG, Triglyceride, Total-C, LDL-C, VLDL-C, Creatinine, Uric acid, IL-6.	No significant change	
		HDL-C	Increase	
Total ($n=57$) Propolis: $n=30$ Placebo: $n=27$	300 mg propolis pills, three times a day (12 weeks)	FBG, HbA1c, BMI	Decrease	Samadi et al. (2017)
		Total-C, LDL-C, Bodyweight, TG's, HDL−C, VLDL-C	No significant change	
		Insulin	Increase	
Total ($n=60$) Propolis: $n=30$ Placebo: $n=30$	500 mg propolis capsules, three times a day (8 weeks)	FBG, 2 hp., Insulin, HOMA-IR, HbA1c, Fructosamine Ox-LDL	Decrease	Afsharpour et al. (2019), Hesami et al. (2019)
		Bodyweight, BMI	No significant change	
		TAC, SOD, GPx, CAT	Increase	
Total ($n=55$) Propolis: $n=25$ Placebo: $n=30$	900 mg daily for 18 weeks	BG, Glycosylated hemoglobin, Insulin	No significant change	Gao et al. (2018)
Total ($n=80$) Propolis: $n=41$ Placebo: $n=39$	Propolis tablets 226.8 mg, /day for 8 weeks	HOMA-IR, FBG, HbA1c, Insulin, Uric acid, Total-C, LDL-C, RLP-C, Triglyceride, TNF-α, IL-6, CRP	No significant change	Fukuda et al. (2015)

Continued

Table 3.1 Summary of findings on the effects of bee products on metabolic derangements in clinical studies—cont'd

Bee product	Location	Study design	Gender	Mean age	Sample size at baseline	Disease background
	Brazil	RCT (double-blind)	Both	Propolis (59.5), Placebo (60.8)	Total (n=70) Propolis: n=35 Placebo: n=35	T2DM
	Egypt	RCT (double-blind)	Both	Propolis (48.9), Placebo (51.2)	Total (n=50) Propolis: n=26 Placebo: n=24	T2DM
Royal jelly	Japan	RCT (double-blind)	Both	Royal jelly (39.0), Placebo (36.9)	Total (n=15) Royal jelly: n=7 Placebo: n=8	Healthy adults
	Iran	RCT (double-blind)	Both	Royal jelly (51.78), Placebo (53.13)	Total (n=50) Royal jelly: n=25 Placebo: n=25	T2DM
	Iran	RCT (double-blind)	Female	Royal jelly (51.7), Placebo (51.4)	Total (n=50) Royal jelly: n=25 Placebo: n=25	T2DM
	Japan	RCT (double-blind)	Both	Royal jelly (29.57), Placebo (37.0)	Total (n=43) Royal jelly: n=21 Placebo: n=22	Patients who complained of subjective symptoms of dry eye
	Turkey	RCT (double-blind)	Male	Aged 18 to 25 years (Mean not given)	Total (n=40) Royal jelly: n=30 Placebo: n=10	Healthy (trained swimmers)
Bee pollen	Ukraine	–	Both	Aged 44 to 76 years (Mean not given)	Total (n=50) Royal jelly: n=30 Placebo: n=22	Insulin-independent DM (IIDM)

Sample size at end	Dose (duration)	Findings		Reference
		Index	Effect	
Total ($n=65$) Propolis: $n=33$ Placebo: $n=32$	Brazilian propolis in capsules at the dose of 900 mg daily for 18 weeks.	Carbonyls, LDH, TNF-α	Decreased	Zhao et al. (2016)
		BG, Glycosylated hemoglobin, Insulin, FRAP, SOD, GPx, MDA, Aldose reductase, Adiponectin, Ox-LDL	No significant change	
		GSH, Polyphenols, IL-1β, IL-6	Increased	
Total ($n=50$) Propolis: $n=26$ Placebo: $n=24$	400 mg/day (3 and 6 months)	HbA1c, Plasma glucose, CML,	Decreased	El-Sharkawy et al. (2016)
Total ($n=15$) Royal jelly: $n=7$ Placebo: $n=8$	6 g royal jelly daily for 4 weeks	TC, LDL-C.	Decreased	Guo et al. (2007)
		FBG, TG and HDL-C	No significant change	
Total ($n=46$) Royal jelly: $n=23$ Placebo: $n=23$	1000 mg royal jelly 3 times a day for 8 weeks	FPG	Decrease	Khoshpey et al. (2016)
		ApoA-I	Increase	
Total ($n=41$) Royal jelly: $n=21$ Placebo: $n=20$	1000 mg royal jelly daily for 8 weeks	FPG, HbA1c, MDA	Decreased	Pourmoradian et al. (2014)
		GPx and SOD activities	Increased	
Total ($n=41$) Royal jelly: $n=19$ Placebo: $n=22$	1200 mg royal jelly daily (6 tabs/day) for 8 weeks	FBG, TG, TC, LDL-C and HDL-C	No significant change	Inoue et al. (2017)
Total ($n=40$) Royal jelly: $n=30$ Placebo	(a) 500 mg royal jelly daily (b) 1 g royal jelly daily (c) 2 g royal jelly daily for 4 weeks	FBG, TG, TC, LDL-C and HDL-C	No significant change	Saritas et al. (2011)
–	250 mg/kg of body weight twice a day for 21–28 days	BG, TC, TG	Decreased	Senyuk et al. (2016)

lipid profiles, and diabetic foot ulcer care have been published. Nevertheless, these clinical investigations have not provided an extensive explanation of the processes behind propolis' therapeutic actions, despite implying potential propolis applications in antidiabetic therapy. However, as the number of human trials grow, studies employing animal models and cultured cells will become more important.

To examine the potential health-related features of royal jelly, many preclinical investigations and clinical trials have been done. Hypoglycemia, antihyperlipidemia, and nephroprotection have been observed in several animal investigations. In addition, a rising number of scientific investigations have been conducted to examine the effects of royal jelly consumption on various metabolic indicators in humans. Nevertheless, trials with longer durations and controls for relevant confounders, particularly nutrition and physical activity, are required. Based on the evidences, we believe that the effects of royal jelly on blood glucose levels be more prominent in diabetic patients. Nevertheless, more research is recommended to draw more conclusive findings.

Although preclinical studies have recommended the use of bee pollen, bee bread, and bee venom in the management of DM, the lack of clinical data on the effectiveness and efficacy of these bee products in metabolic derangements as well as the challenges in DM cannot be overstated in this era of evidence-based medicine. In the absence of sufficient clinical data, the use of bee pollen, bread, and venom in the treatment of DM will continue to be met with suspicion. This phenomenon will almost certainly reduce the chances of the bee products acquiring a more widespread acceptance among diabetic patients and healthcare professionals.

Most significantly, clinical investigations evaluating the effectiveness and utility of antidiabetic drug combinations with bee products and antidiabetic drugs alone on hyperglycemia, hyperlipidemia, and diabetic sequelae are vital and are urgently needed. This will necessitate the use of well-designed, randomized, placebo-controlled investigations and trials with a large enough sample size. It is hypothesized that bee products, whether they have hypoglycemic/antihyperglycemic or antihyperlipidemic properties, can help delay the onset of diabetes complications by reducing oxidative stress and inflammation. As patients' metabolic illnesses progress slowly, the therapeutic effects of bee products in humans should be studied over a longer duration. Furthermore, in human investigations, various aspects such as medical history, genetic inheritance and living environment should be taken into account.

References

Aalaa, M., Malazy, O.T., Sanjari, M., Peimani, M., Mohajeri-Tehrani, M.R., 2012. Nurses' role in diabetic foot prevention and care; a review. J. Diabetes Metab. Disord. 11 (1), 1–6.

Abdulrhman, M., 2013. Honey therapy in a patient volunteer with type 2 diabetes mellitus: case report. J. Clin. Trials 3 (148). 2167-0870.

Abdulrhman, M.A., 2016. Honey as a sole treatment of type 2 diabetes mellitus. Endocrinol. Metab. Syndr. 5 (232). 2161-1017.

Abdulrhman, M., El-Hefnawy, M., Hussein, R., Abou, E.-G.A., 2011. The glycemic and peak incremental indices of honey, sucrose and glucose in patients with type 1 diabetes mellitus: effects on C-peptide level—a pilot study. Acta Diabetol. 48 (2), 89–94.

Abdulrhman, M., El Hefnawy, M., Ali, R., Hamid, I.A., Abou El-Goud, A., Refai, D., 2013a. Effects of honey, sucrose and glucose on blood glucose and C-peptide in patients with type 1 diabetes mellitus. Complement. Ther. Clin. Pract. 19 (1), 15–19.

Abdulrhman, M.M., El-Hefnawy, M.H., Aly, R.H., Shatla, R.H., Mamdouh, R.M., Mahmoud, D.M., Mohamed, W.S., 2013b. Metabolic effects of honey in type 1 diabetes mellitus: a randomized crossover pilot study. J. Med. Food 16 (1), 66–72.

Abo-Salem, O.M., El-Edel, R.H., Harisa, G.E., El-Halawany, N., Ghonaim, M.M., 2009. Experimental diabetic nephropathy can be prevented by propolis: effect on metabolic disturbances and renal oxidative parameters. Pak. J. Pharm. Sci. 22 (2), 205–210.

Adesoji, F., Oluwakemi, A., 2008. Differential effect of honey on selected variables in alloxan-induced and fructose-induced diabetic rats. Afr. J. Biomed. Res. 11 (2), 191–196.

Adwas, A.A., Elsayed, A., Azab, A.E., Quwaydir, F.A., 2019. Oxidative stress and antioxidant mechanisms in human body. J. Appl. Biotechnol. Bioeng. 6 (1), 43.

Afkhamizadeh, M., Aboutorabi, R., Ravari, H., Fathi Najafi, M., Ataei Azimi, S., Javadian Langaroodi, A., Yaghoubi, M.A., Sahebkar, A., 2018. Topical propolis improves wound healing in patients with diabetic foot ulcer: a randomized controlled trial. Nat. Prod. Res. 32 (17), 2096–2099.

Afsharpour, F., Javadi, M., Hashemipour, S., Koushan, Y., 2019. Propolis supplementation improves glycemic and antioxidant status in patients with type 2 diabetes: a randomized, double-blind, placebo-controlled study. Complement. Ther. Med. 43, 283–288.

Agrawal, O.P., Pachauri, A., Yadav, H., Urmila, J., Goswamy, H.M., Chapperwal, A., Bisen, P.S., Prasad, G.B., 2007. Subjects with impaired glucose tolerance exhibit a high degree of tolerance to honey. J. Med. Food 10 (3), 473–478.

Ahmad, S., Campos, M.G., Fratini, F., Altaye, S.Z., Li, J., 2020. New insights into the biological and pharmaceutical properties of royal jelly. Int. J. Mol. Sci. 21 (2), 382.

Ahmed, F., Uddin, A.J., 2020. Functional food-a new notion in modern food culture and a Hope during Covid-19. Int. J. Bus. Soc. Sci. Res. 8 (3), 61–62.

Akhbari, M., Jabbari, M., Ayati, M.H., Namazi, N., 2021. The effects of Oral consumption of honey on key metabolic profiles in adult patients with type 2 diabetes mellitus and nondiabetic individuals: a systematic review of clinical trials. Evid. Based Complement. Alternat. Med. 2021, 6666832.

Aksoy, L., Aslan, Z., 2017. Nephroprotective and antioxidative effects of royal jelly on ethylene glycol induced nephropathy in rats. Ank. Univ. Vet. Fak. Derg. 64 (4), 241–248.

Al Aamri, Z.M., Ali, B.H., 2017. Does honey have any salutary effect against streptozotocin-induced diabetes in rats? J. Diabetes Metab. Disord. 16 (1), 1–6.

Al Ghamdi, A.A., Badr, G., Hozzein, W.N., Allam, A., Al-Waili, N.S., Al-Wadaan, M.A., Garraud, O., 2015. Oral supplementation of diabetic mice with propolis restores the proliferation capacity and chemotaxis of B and T lymphocytes towards CCL21 and CXCL12 by modulating the lipid profile, the pro-inflammatory cytokine levels and oxidative stress. BMC Immunol. 16 (1), 1–4.

Alaraj, M., 2020. Royal Jelly and Aliskiren mutually annul their protective effects against gentamicin-induced nephrotoxicity in rats. Vet. World 13 (12), 2658.

Ali, A.M., Kunugi, H., 2020. Royal Jelly as an intelligent anti-aging agent—a focus on cognitive aging and Alzheimer's disease: a review. Antioxidants 9 (10), 937.

Al-Kafaween, M.A., Hilmi, A.B.M., Al-Jamal, H.A.N., Jaffar, N., Al-Sayyed, H., Zahri, M.K., 2021. Effects of selected Malaysian Kelulut honey on biofilm formation and the gene expression profile of *Staphylococcus aureus*, *Pseudomonas aeruginosa* and *Escherichia coli*. Jordan J. Pharm. Sci. 14 (1), 9–26.

Almasaudi, S.B., Al-Nahari, A.A., El Sayed, M., Barbour, E., Al Muhayawi, S.M., Al-Jaouni, S., Azhar, E., Qari, M., Qari, Y.A., Harakeh, S., 2017. Antimicrobial effect of different types of honey on *Staphylococcus aureus*. Saudi J. Biol. Sci. 24 (6), 1255–1261.

Al-Waili, N., 2003. Intrapulmonary administration of natural honey solution, hyperosmolar dextrose or hypoosmolar distill water to normal individuals and to patients with type-2 diabetes mellitus or hypertension: their effects on blood glucose level, plasma insulin and C-peptide, blood pressure and peaked expiratory flow rate. Eur. J. Med. Res. 8 (7), 295–303.

Al-Waili, N.S., 2004. Natural honey lowers plasma glucose, C-reactive protein, homocysteine and blood lipids in healthy, diabetic and hyperlipidemic subjects: comparison with dextrose and sucrose. J. Med. Food 7 (1), 100–107.

Amin, M.A., Abdel-Raheem, I.T., 2014. Accelerated wound healing and anti-inflammatory effects of physically cross linked polyvinyl alcohol–chitosan hydrogel containing honey bee venom in diabetic rats. Arch. Pharm. Res. 37 (8), 1016–1031.

An, T., Chen, Y., Tu, Y., Lin, P., 2021. Mesenchymal stromal cell-derived extracellular vesicles in the treatment of diabetic foot ulcers: application and challenges. Stem Cell Rev. Rep. 17 (2), 369–378.

Anđelković, B., Jevtić, G., Marković, J., Mladenović, M., Peševa, V., 2014. Quality of honey bee bread collected in spring. In: Proceedings of the International Symposium on Animal Science 2014. Faculty of Agriculture, Belgrade-Zemun (Serbia).

Antohe, I., Popa, A.D., 2021. Ten rules of foot care—the development of a short questionnaire for patients with diabetes. J. Diabetes Mellitus 11 (3), 83–96.

Aoi, W., Hosogi, S., Niisato, N., Yokoyama, N., Hayata, H., Miyazaki, H., Kusuzaki, K., Fukuda, T., Fukui, M., Nakamura, N., Marunaka, Y., 2013. Improvement of insulin resistance, blood pressure and interstitial pH in early developmental stage of insulin resistance in OLETF rats by intake of propolis extracts. Biochem. Biophys. Res. Commun. 432 (4), 650–653.

Apostolopoulou, M., Gordillo, R., Koliaki, C., Gancheva, S., Jelenik, T., De Filippo, E., Herder, C., Markgraf, D., Jankowiak, F., Esposito, I., Schlensak, M., 2018. Specific hepatic sphingolipids relate to insulin resistance, oxidative stress and inflammation in nonalcoholic steatohepatitis. Diabetes Care 41 (6), 1235–1243.

Ares, A.M., Valverde, S., Bernal, J.L., Nozal, M.J., Bernal, J., 2018. Extraction and determination of bioactive compounds from bee pollen. J. Pharm. Biomed. Anal. 147, 110–124.

Armstrong, D.G., Swerdlow, M.A., Armstrong, A.A., Conte, M.S., Padula, W.V., Bus, S.A., 2020. Five year mortality and direct costs of care for people with diabetic foot complications are comparable to cancer. J. Foot Ankle Res. 13 (1), 1–4.

Asaduzzaman, M., Sohanur Rahman, M., Munira, S., Muedur Rahman, M., Hasan, M., Siddique, M.A., Biswas, S., Belal, M.H., Khatun, M., Khan, M.M., Rahman, M.M., 2015. Effects of honey supplementation on hepatic and cardiovascular disease (CVD) marker in streptozotocin-induced diabetic rats. J. Diabetes Metab. 6 (592), 2.

Aziz, M.S., Giribabu, N., Rao, P.V., Salleh, N., 2017. Pancreatoprotective effects of Geniotrigona thoracica stingless bee honey in streptozotocin-nicotinamide-induced male diabetic rats. Biomed. Pharmacother. 89, 135–145.

Babatunde, I.R., Abdulbasit, A., Oladayo, M.I., Olasile, O.I., Olamide, F.R., Gbolahan, B.W., 2015. Hepatoprotective and pancreatoprotective properties of the ethanolic extract of Nigerian propolis. J. Intercult. Ethnopharmacol. 4 (2), 102.

Badr, G., Hozzein, W.N., Badr, B.M., Al Ghamdi, A., Saad Eldien, H.M., Garraud, O., 2016. Bee venom accelerates wound healing in diabetic mice by suppressing activating transcription factor-3 (ATF-3) and inducible nitric oxide synthase (iNOS)-mediated oxidative stress and recruiting bone marrow-derived endothelial progenitor cells. J. Cell. Physiol. 231 (10), 2159–2171.

Bahrami, M., Ataie-Jafari, A., Hosseini, S., Foruzanfar, M.H., Rahmani, M., Pajouhi, M., 2009. Effects of natural honey consumption in diabetic patients: an 8-week randomized clinical trial. Int. J. Food Sci. Nutr. 60 (7), 618–626.

Barot, M., Gokulgandhi, M.R., Patel, S., Mitra, A.K., 2013. Microvascular complications and diabetic retinopathy: recent advances and future implications. Future Med. Chem. 5 (3), 301–314.

Bartáková, V., Barátová, B., Pácal, L., Ťápalová, V., Šebestová, S., Janků, P., Kaňková, K., 2021. Development of a new risk score for stratification of women with gestational diabetes mellitus at high risk of persisting postpartum glucose intolerance using routinely assessed parameters. Life 11 (6), 464.

Batumalaie, K., Zaman Safi, S., Mohd Yusof, K., Shah Ismail, I., Devi Sekaran, S., Qvist, R., 2013. Effect of gelam honey on the oxidative stress-induced signaling pathways in pancreatic hamster cells. Int. J. Endocrinol. 2013, 367312.

Beckman, J.A., Creager, M.A., Libby, P., 2002. Diabetes and atherosclerosis: epidemiology, pathophysiology and management. JAMA 287 (19), 2570–2581.

Bobiş, O., Dezmirean, D.S., Moise, A.R., 2018. Honey and diabetes: the importance of natural simple sugars in diet for preventing and treating different type of diabetes. Oxidative Med. Cell. Longev. 2018, 4757893.

Bondeva, T., Wojciech, S., Wolf, G., 2011. Advanced glycation end products inhibit adhesion ability of differentiated podocytes in a neuropilin-1-dependent manner. Am. J. Physiol. Renal Physiol. 301 (4), F852–F870.

Brouwers, O., Niessen, P.M., Miyata, T., Østergaard, J.A., Flyvbjerg, A., Peutz-Kootstra, C.J., Sieber, J., Mundel, P.H., Brownlee, M., Janssen, B.J., De Mey, J.G., 2014. Glyoxalase-1 overexpression reduces endothelial dysfunction and attenuates early renal impairment in a rat model of diabetes. Diabetologia 57 (1), 224–235.

Burgos-Morón, E., Abad-Jiménez, Z., Martinez de Maranon, A., Iannantuoni, F., Escribano-López, I., López-Domènech, S., Salom, C., Jover, A., Mora, V., Roldan, I., Solá, E., 2019. Relationship between oxidative stress, ER stress and inflammation in type 2 diabetes: the battle continues. J. Clin. Med. 8 (9), 1385.

Busserolles, J., Gueux, E., Rock, E., Mazur, A., Rayssiguier, Y., 2002. Substituting honey for refined carbohydrates protects rats from hypertriglyceridemic and prooxidative effects of fructose. J. Nutr. 132 (11), 3379–3382.

Capcarova, M., Kalafova, A., Schwarzova, M., Schneidgenova, M., Prnova, M.S., Svik, K., Slovak, L., Kisska, P., Kovacik, A., Brindza, J., 2019. Consumption of bee bread influences glycaemia and development of diabetes in obese spontaneous diabetic rats. Biologia 75, 1–7.

Chen, C., Gong, W., Li, C., Xiong, F., Wang, S., Huang, J., Wang, Y., Chen, Z., Chen, Q., Liu, P., Lan, T., 2017. Sphingosine kinase 1 mediates AGEs-induced fibronectin upregulation in diabetic nephropathy. Oncotarget 8 (45), 78660.

Chen, L.H., Chien, Y.W., Chang, M.L., Hou, C.C., Chan, C.H., Tang, H.W., Huang, H.Y., 2018. Taiwanese green propolis ethanol extract delays the progression of type 2 diabetes mellitus in rats treated with streptozotocin/high-fat diet. Nutrients 10 (4), 503.

Chepulis, L.M., 2007. The effect of honey compared to sucrose, mixed sugars and a sugar-free diet on weight gain in young rats. J. Food Sci. 72 (3), S224–S229.

Chepulis, L., Starkey, N., 2008. The long-term effects of feeding honey compared with sucrose and a sugar-free diet on weight gain, lipid profiles and DEXA measurements in rats. J. Food Sci. 73 (1), H1–H7.

Chuang, P.Y., Yu, Q., Fang, W., Uribarri, J., He, J.C., 2007. Advanced glycation endproducts induce podocyte apoptosis by activation of the FOXO4 transcription factor. Kidney Int. 72 (8), 965–976.

Cornara, L., Biagi, M., Xiao, J., Burlando, B., 2017. Therapeutic properties of bioactive compounds from different honeybee products. Front. Pharmacol. 8, 412.

Coskun, O., Kanter, M., Korkmaz, A., Oter, S., 2005. Quercetin, a flavonoid antioxidant, prevents and protects streptozotocin-induced oxidative stress and β-cell damage in rat pancreas. Pharmacol. Res. 51 (2), 117–123.

Cozma, A.I., Sievenpiper, J.L., Souza, D., Anderberg, R.J., Chiavaroli, L., Ha, V., Wang, D.D., Mirrahimi, A., Matt, E.Y., Carleton, A.J., Di Buono, M., Jenkins, A.L., 2012. Effect of fructose on glycemic control in diabetes: a systematic review and meta-analysis of controlled feeding trials. Diabetes Care 35 (7), 1611–1620.

Daudu, O.M., 2019. Bee pollen extracts as potential antioxidants and inhibitors of α-amylase and α-glucosidase enzymes in vitro assessment. J. Apic. Sci. 63 (2), 315–325.

Deng, L., Du, C., Song, P., Chen, T., Rui, S., Armstrong, D.G., Deng, W., 2021. The role of oxidative stress and antioxidants in diabetic wound healing. Oxidative Med. Cell. Longev. 2021, 8852759.

Despland, C., Walther, B., Kast, C., Campos, V., Rey, V., Stefanoni, N., Tappy, L., 2017. A randomized-controlled clinical trial of high fructose diets from either Robinia honey or free fructose and glucose in healthy normal weight males. Clin. Nutr. ESPEN 19, 16–22.

Deveza, M.V., Keller, K.M., Lorenzon, M.C., Nunes, L.M., Sales, É.O., Barth, O.M., 2015. Mycotoxicological and palynological profiles of commercial brands of dried bee pollen. Braz. J. Microbiol. 46 (4), 1171–1176.

Ding, Y., Xu, M., Lu, Q., Wei, P., Tan, J., Liu, R., 2019. Combination of honey with metformin enhances glucose metabolism and ameliorates hepatic and nephritic dysfunction in STZ-induced diabetic mice. Food Funct. 10 (11), 7576–7587.

Du, Y., Miller, C.M., Kern, T.S., 2003. Hyperglycemia increases mitochondrial superoxide in retina and retinal cells. Free Radic. Biol. Med. 35 (11), 1491–1499.

Eguchi, N., Vaziri, N.D., Dafoe, D.C., Ichii, H., 2021. The role of oxidative stress in pancreatic β cell dysfunction in diabetes. Int. J. Mol. Sci. 22 (4), 1509.

El Rabey, H.A., Al-Seeni, M.N., Bakhashwain, A.S., 2017. The antidiabetic activity of Nigella sativa and propolis on streptozotocin-induced diabetes and diabetic nephropathy in male rats. Evid. Based Complement. Alternat. Med. 2017, 5439645.

Elkotby, D., Hassan, A.K., Emad, R., Bahgat, I., 2018. Histological changes in islets of Langerhans of pancreas in alloxan-induced diabetic rats following Egyptian honey bee venom treatments. Int. J. Pure Appl. Zool. 6 (1), 1–6.

Ellis, S., Lin, E.J., Tartar, D., 2018. Immunology of wound healing. Curr. Dermatol. Rep. 7 (4), 350–358.

El-Sayed, E.S., Abo-Salem, O.M., Aly, H.A., Mansour, A.M., 2009. Potential antidiabetic and hypolipidemic effects of propolis extract in streptozotocin-induced diabetic rats. Pak. J. Pharm. Sci. 22 (2), 168–174.

El-Sharkawy, H.M., Anees, M.M., Van Dyke, T.E., 2016. Propolis improves periodontal status and glycemic control in patients with type 2 diabetes mellitus and chronic periodontitis: a randomized clinical trial. J. Periodontol. 87 (12), 1418–1426.

Enginyurt, O., Cakir, L., Karatas, A., et al., 2017. The role of pure honey in the treatment of diabetes mellitus. Biomed. Res. 28, 3305–3312.

Erejuwa, O.O., 2012. The use of honey in diabetes mellitus: is it beneficial or detrimental? Int. J. Endocrinol. Metab. 10, 444–445.

Erejuwa, O.O., 2014. Effect of honey in diabetes mellitus: matters arising. J. Diabetes Metab. Disord. 13 (1), 1–4.

Erejuwa, O.O., 2019. Honey: profile and features: applications to diabetes. In: Bioactive Food as Dietary Interventions for Diabetes. Academic Press, pp. 461–494.
Erejuwa, O.O., Sulaiman, S.A., Wahab, M.S., Sirajudeen, K.N., Salleh, M.M., Gurtu, S., 2010a. Antioxidant protection of Malaysian tualang honey in pancreas of normal and streptozotocin-induced diabetic rats. Ann. Endocrinol. 71 (4), 291–296.
Erejuwa, O.O., Sulaiman, S.A., Wahab, M.S., Sirajudeen, K.N., Salleh, M.S., Gurtu, S., 2010b. Antioxidant protective effect of glibenclamide and metformin in combination with honey in pancreas of streptozotocin-induced diabetic rats. Int. J. Mol. Sci. 11 (5), 2056–2066.
Erejuwa, O.O., Sulaiman, S.A., Ab Wahab, M.S., Sirajudeen, K.N., Salleh, M.S., Gurtu, S., 2011a. Glibenclamide or metformin combined with honey improves glycemic control in streptozotocin-induced diabetic rats. Int. J. Biol. Sci. 7 (2), 244.
Erejuwa, O.O., Sulaiman, S.A., Wahab, M.S., Salam, S.K., Salleh, M.S., Gurtu, S., 2011b. Comparison of antioxidant effects of honey, glibenclamide, metformin and their combinations in the kidneys of streptozotocin-induced diabetic rats. Int. J. Mol. Sci. 12 (1), 829–843.
Erejuwa, O.O., Nwobodo, N.N., Akpan, J.L., Okorie, U.A., Ezeonu, C.T., Ezeokpo, B.C., Nwadike, K.I., Erhiano, E., Abdul Wahab, M.S., Sulaiman, S.A., 2016. Nigerian honey ameliorates hyperglycemia and dyslipidemia in alloxan-induced diabetic rats. Nutrients 8 (3), 95.
Erejuwa, O.O., Akpan, J.L., Okorie, U.A., Ezeokpo, B.C., Ezeonu, T.C., Nwadike, K.I., Nwobodo, N.N., Erhiano, E., Aja, D.O., Wahab, M.S., Sulaiman, S.A., 2021. Glycaemic responses of honeys in normal and alloxan-induced diabetic rats. Afr. Sci. 21 (1), 201–210.
Ernawati, D.S., Puspa, A., 2018. Expression of vascular endothelial growth factor and matrix metalloproteinase-9 in *Apis mellifera* Lawang propolis extract gel-treated traumatic ulcers in diabetic rats. Vet. World 11 (3), 304.
Eyarefe, O.D., Ologunagba, F.M., Emikpe, B.O., 2014. Wound healing potential of natural honey in diabetic and non-diabetic wistar rats. Afr. J. Biomed. Res. 17 (1), 15–21.
Farzaei, M.H., Zobeiri, M., Parvizi, F., El-Senduny, F.F., Marmouzi, I., Coy-Barrera, E., Naseri, R., Nabavi, S.M., Rahimi, R., Abdollahi, M., 2018. Curcumin in liver diseases: a systematic review of the cellular mechanisms of oxidative stress and clinical perspective. Nutrients 10 (7), 855.
Fazakerley, D.J., Minard, A.Y., Krycer, J.R., Thomas, K.C., Stöckli, J., Harney, D.J., Burchfield, J.G., Maghzal, G.J., Caldwell, S.T., Hartley, R.C., Stocker, R., 2018. Mitochondrial oxidative stress causes insulin resistance without disrupting oxidative phosphorylation. J. Biol. Chem. 293 (19), 7315–7328.
Fernandes, L., Oliveira, A., Henriques, M., Rodrigues, M.E., 2020. Honey as a strategy to fight Candida tropicalis in mixed-biofilms with Pseudomonas aeruginosa. Antibiotics 9 (2), 43.
Fitipaldi, H., McCarthy, M.I., Florez, J.C., Franks, P.W., 2018. A global overview of precision medicine in type 2 diabetes. Diabetes 67 (10), 1911–1922.
Fuenmayor, B.C., Zuluaga, D.C., Díaz, M.C., de Quicazán, C.M., Cosio, M., Mannino, S., 2014. Evaluation of the physicochemical and functional properties of Colombian bee pollen. Rev. MVZ Córdoba 19 (1), 4003–4014.
Fujii, A., Kobayashi, S., Kuboyama, N., Furukawa, Y., Kaneko, Y., Ishihama, S., Yamamoto, H., Tamura, T., 1990. Augmentation of wound healing by royal jelly (royal jelly) in streptozotocin-diabetic rats. Jpn. J. Pharmacol. 53 (3), 331–337.
Fukuda, T., Fukui, M., Tanaka, M., Senmaru, T., Iwase, H., Yamazaki, M., Aoi, W., Inui, T., Nakamura, N., Marunaka, Y., 2015. Effect of Brazilian green propolis in patients with type 2 diabetes: a double-blind randomized placebo-controlled study. Biomed. Rep. 3 (3), 355–360.
Fuliang, H.U., Hepburn, H.R., Xuan, H., Chen, M., Daya, S., Radloff, S.E., 2005. Effects of propolis on blood glucose, blood lipid and free radicals in rats with diabetes mellitus. Pharmacol. Res. 51 (2), 147–152.
Galicia-Garcia, U., Benito-Vicente, A., Jebari, S., Larrea-Sebal, A., Siddiqi, H., Uribe, K.B., Ostolaza, H., Martín, C., 2020. Pathophysiology of type 2 diabetes mellitus. Int. J. Mol. Sci. 21 (17), 6275.

Gao, W., Pu, L., Wei, J., Yao, Z., Wang, Y., Shi, T., Zhao, L., Jiao, C., Guo, C., 2018. Serum antioxidant parameters are significantly increased in patients with type 2 diabetes mellitus after consumption of Chinese propolis: a randomized controlled trial based on fasting serum glucose level. Diabetes Ther. 9 (1), 101–111.

Ghanbari, E., Nejati, V., Khazaei, M., 2016. Improvement in serum biochemical alterations and oxidative stress of liver and pancreas following use of royal jelly in streptozotocin-induced diabetic rats. Cell J. 18 (3), 362.

Gill, R., Poojar, B., Bairy, L.K., Praveen, K.S., 2019. Comparative evaluation of wound healing potential of manuka and acacia honey in diabetic and nondiabetic rats. J. Pharm. Bioallied Sci. 11 (2), 116.

Gong, P., Chang, X., Chen, X., Bai, X., Wen, H., Pi, S., Yang, W., Wang, L., Chen, F., 2017. Metabolomics study of cadmium-induced diabetic nephropathy and protective effect of caffeic acid phenethyl ester using UPLC–Q-TOF-MS combined with pattern recognition. Environ. Toxicol. Pharmacol. 54, 80–92.

Granados-Pineda, J., Uribe-Uribe, N., García-López, P., Ramos-Godinez, M.D., Rivero-Cruz, J.F., Pérez-Rojas, J.M., 2018. Effect of pinocembrin isolated from Mexican brown propolis on diabetic nephropathy. Molecules 23 (4), 852.

Guo, H., Saiga, A., Sato, M., Miyazawa, I., Shibata, M., Takahata, Y., Morimatsu, F., 2007. Royal jelly supplementation improves lipoprotein metabolism in humans. J. Nutr. Sci. Vitaminol. 53 (4), 345–348.

Guo, J., Wang, Z., Chen, Y., Cao, J., Tian, W., Ma, B., Dong, Y., 2021. Active components and biological functions of royal jelly. J. Funct. Foods 82, 104514.

Hadi, A., Najafgholizadeh, A., Aydenlu, E.S., Shafiei, Z., Pirivand, F., Golpour, S., Pourmasoumi, M., 2018. Royal jelly is an effective and relatively safe alternative approach to blood lipid modulation: a meta-analysis. J. Funct. Foods 41, 202–209.

Hamad, R., Jayakumar, C., Ranganathan, P., Mohamed, R., El-Hamamy, M.M., Dessouki, A.A., Ibrahim, A., Ramesh, G., 2015. Honey feeding protects kidney against cisplatin nephrotoxicity through suppression of inflammation. Clin. Exp. Pharmacol. Physiol. 42 (8), 843–848.

Henshaw, F.R., Bolton, T., Nube, V., Hood, A., Veldhoen, D., Pfrunder, L., McKew, G.L., Macleod, C., McLennan, S.V., Twigg, S.M., 2014. Topical application of the bee hive protectant propolis is well tolerated and improves human diabetic foot ulcer healing in a prospective feasibility study. J. Diabetes Complicat. 28 (6), 850–857.

Hesami, S., Hashemipour, S., Shiri-Shahsavar, M.R., Koushan, Y., Haghighian, H.K., 2019. Administration of Iranian Propolis attenuates oxidative stress and blood glucose in type II diabetic patients: a randomized, double-blind, placebo-controlled, clinical trial. Caspian J. Intern. Med. 10 (1), 48.

Hills, C.E., Squires, P.E., 2011. The role of TGF-β and epithelial-to mesenchymal transition in diabetic nephropathy. Cytokine Growth Factor Rev. 22 (3), 131–139.

Hossen, M.S., Ali, M.Y., Jahurul, M.H., Abdel-Daim, M.M., Gan, S.H., Khalil, M.I., 2017a. Beneficial roles of honey polyphenols against some human degenerative diseases: a review. Pharmacol. Rep. 69 (6), 1194–1205.

Hossen, M., Gan, S.H., Khalil, M., 2017b. Melittin, a potential natural toxin of crude bee venom: probable future arsenal in the treatment of diabetes mellitus. J. Chem., 2017.

Hozzein, W.N., Badr, G., Al Ghamdi, A.A., Sayed, A., Al-Waili, N.S., Garraud, O., 2015. Topical application of propolis enhances cutaneous wound healing by promoting TGF-beta/Smad-mediated collagen production in a streptozotocin-induced type I diabetic mouse model. Cell. Physiol. Biochem. 37 (3), 940–954.

Hróbjartsson, A., Kaptchuk, T.J., Miller, F.G., 2011. Placebo effect studies are susceptible to response bias and to other types of biases. J. Clin. Epidemiol. 64 (11), 1223–1229.

Huang, K., Huang, J., Xie, X., Wang, S., Chen, C., Shen, X., Liu, P., Huang, H., 2013. Sirt1 resists advanced glycation end products-induced expressions of fibronectin and TGF-β1 by activating the Nrf2/ARE pathway in glomerular mesangial cells. Free Radic. Biol. Med. 65, 528–540.

Hussein, A.E., El-Ansari, M.K., Zahra, A.A., 2019. Effect of the honeybee hybrid and geographic region on the honey bee venom production. J. Plant Protect. Pathol. 10 (3), 171–176.

Inoue, S., Kawashima, M., Hisamura, R., Imada, T., Izuta, Y., Nakamura, S., Ito, M., Tsubota, K., 2017. Clinical evaluation of a royal jelly supplementation for the restoration of dry eye: a prospective randomized double blind placebo controlled study and an experimental mouse model. PLoS One 12 (1), e0169069.

Jenkins, A.J., Joglekar, M.V., Hardikar, A.A., Keech, A.C., O'Neal, D.N., Januszewski, A.S., 2015. Biomarkers in diabetic retinopathy. Rev. Diabet. Stud. 12 (1–2), 159.

Johnson, J.B., Broszczak, D.A., Mani, J.S., Anesi, J., Naiker, M., 2021. A cut above the rest: oxidative stress in chronic wounds and the potential role of polyphenols as therapeutics. J. Pharm. Pharmacol. https://doi.org/10.1093/jpp/rgab038.

Kang, Q., Yang, C., 2020. Oxidative stress and diabetic retinopathy: molecular mechanisms, pathogenetic role and therapeutic implications. Redox Biol. 37, 101799.

Kang, M.K., Park, S.H., Kim, Y.H., Lee, E.J., Antika, L.D., Kim, D.Y., Choi, Y.J., Kang, Y.H., 2017. Chrysin ameliorates podocyte injury and slit diaphragm protein loss via inhibition of the PERK-eIF2α-ATF-CHOP pathway in diabetic mice. Acta Pharmacol. Sin. 38 (8), 1129–1140.

Katsarou, A., Gudbjörnsdottir, S., Rawshani, A., Dabelea, D., Bonifacio, E., Anderson, B.J., Jacobsen, L.M., Schatz, D.A., Lernmark, Å., 2017. Type 1 diabetes mellitus. Nat. Rev. Dis. Primers 3 (1), 1–7.

Keskin, M., Keskin, Ş., Kolaylı, S., 2021. Health-promoting benefits of honey. In: Preparation of Phytopharmaceuticals for the Management of Disorders. Academic Press, pp. 303–306.

Khalifa, S.A., Elashal, M.H., Yosri, N., Du, M., Musharraf, S.G., Nahar, L., Sarker, S.D., Guo, Z., Cao, W., Zou, X., El-Wahed, A., 2021. Bee pollen: current status and therapeutic potential. Nutrients 13 (6), 1876.

Khoshpey, B., Djazayeri, S., Amiri, F., Malek, M., Hosseini, A.F., Hosseini, S., Shidfar, S., Shidfar, F., 2016. Effect of royal jelly intake on serum glucose, apolipoprotein AI (ApoA-I), apolipoprotein B (ApoB) and ApoB/ApoA-I ratios in patients with type 2 diabetes: a randomized, double-blind clinical trial study. Can. J. Diabetes 40 (4), 324–328.

Khulan, T.S., Ambaga, M., Chimedragcha, C.H., 2015. Effect of honey bee venom (Apis mellifera) on hyperglycemia and hyperlipidemia in alloxan induced diabetic rabbits. J. Diabetes Metab. 6 (3), 507.

Kieliszek, M., Piwowarek, K., Kot, A.M., Błażejak, S., Chlebowska-Śmigiel, A., Wolska, I., 2018. Pollen and bee bread as new health-oriented products: a review. Trends Food Sci. Technol. 71, 170–180.

Kitamura, H., 2019. Effects of propolis extract and propolis-derived compounds on obesity and diabetes: knowledge from cellular and animal models. Molecules 24 (23), 4394.

Kitamura, H., Naoe, Y., Kimura, S., Miyamoto, T., Okamoto, S., Toda, C., Shimamoto, Y., Iwanaga, T., Miyoshi, I., 2013. Beneficial effects of Brazilian propolis on type 2 diabetes in Ob/Ob mice: possible involvement of immune cells in mesenteric adipose tissue. Adipocytes 2 (4), 227–236.

Koeninger, L., Armbruster, N.S., Brinch, K.S., Kjaerulf, S., Andersen, B., Langnau, C., Autenrieth, S.E., Schneidawind, D., Stange, E.F., Malek, N.P., Nordkild, P., 2020. Human β-defensin 2 mediated immune modulation as treatment for experimental colitis. Front. Immunol. 11, 93.

Kumar, S., Krishnan, V., 2017. Phytochemistry and functional food: the needs of healthy life. J. Phytochem. Biochem. 1, 103.

Kumar, P.A., Welsh, G.I., Raghu, G., Menon, R.K., Saleem, M.A., Reddy, G.B., 2016. Carboxymethyl lysine induces EMT in podocytes through transcription factor ZEB2: implications for podocyte depletion and proteinuria in diabetes mellitus. Arch. Biochem. Biophys. 590, 10–19.

Kurek-Górecka, A., Komosinska-Vassev, K., Rzepecka-Stojko, A., Olczyk, P., 2021. Bee venom in wound healing. Molecules 26 (1), 148.

Li, Y., Chen, M., Xuan, H., Hu, F., 2012. Effects of encapsulated propolis on blood glycemic control, lipid metabolism and insulin resistance in type 2 diabetes mellitus rats. Evid. Based Complement. Alternat. Med. 2012, 981896.

Li, X., Gong, H., Yang, S., Yang, L., Fan, Y., Zhou, Y., 2017. Pectic bee pollen polysaccharide from Rosa rugosa alleviates diet-induced hepatic steatosis and insulin resistance via induction of AMPK/mTOR-mediated autophagy. Molecules 22 (5), 699.

Li, Q.Q., Wang, K., Marcucci, M.C., Sawaya, A.C., Hu, L., Xue, X.F., Wu, L.M., Hu, F.L., 2018. Nutrient-rich bee pollen: a treasure trove of active natural metabolites. J. Funct. Foods 49, 472–484.

Lourenço, S.C., Moldão-Martins, M., Alves, V.D., 2019. Antioxidants of natural plant origins: from sources to food industry applications. Molecules 24 (22), 4132.

Lyoussi, B., Bakour, M., El Menyiy, N., El Ghouizi, A., 2021. Hypoglycemic, hypolipidemic and hepato-protective effect of bee bread in streptozotocin-induced diabetic rats. Avicenna J. Phytomed. 11 (4), 343–352.

Maciejczyk, M., Żebrowska, E., Chabowski, A., 2019. Insulin resistance and oxidative stress in the brain: what's new? Int. J. Mol. Sci. 20 (4), 874.

Maezawa, Y., Takemoto, M., Yokote, K., 2015. Cell biology of diabetic nephropathy: roles of endothelial cells, tubulointerstitial cells and podocytes. J. Diabetes Investig. 6 (1), 3–15.

Makowski, L.M., Leffers, M., Waltenberger, J., Pardali, E., 2021. Transforming growth factor-β1 signalling triggers vascular endothelial growth factor resistance and monocyte dysfunction in type 2 diabetes mellitus. J. Cell. Mol. Med. 25 (11), 5316–5325.

Maleszka, R., 2018. Beyond Royalactin and a master inducer explanation of phenotypic plasticity in honey bees. Commun. Biol. 1 (1), 1–7.

Malkoç, M., Yaman, S.Ö., Imamoğlu, Y., İnce, İ., Kural, B.V., Mungan, S., Livaoglu, M., Yıldız, O., Kolaylı, S., Orem, A., 2020. Anti-inflammatory, antioxidant and wound-healing effects of mad honey in streptozotocin-induced diabetic rats. J. Apic. Res. 59 (4), 426–436.

Maroof, K., Gan, S.H., 2020. A review on chemical compositions, biological activity and formulation techniques of Malaysian honey bee and Meliponine Propolis. J. Biol. Act. Prod. Nat. 10 (6), 507–523.

Maroof, K., Lee, R.F., Siow, L.F., Gan, S.H., 2020. Microencapsulation of propolis by spray drying: a review. Dry. Technol., 1–20.

Matsui, T., Higashimoto, Y., Nishino, Y., Nakamura, N., Fukami, K., Yamagishi, S.I., 2017. RAGE-aptamer blocks the development and progression of experimental diabetic nephropathy. Diabetes 66 (6), 1683–1695.

Matsushige, K., Basnet, P., Hase, K., Kadota, S., Tanaka, K., Namba, T., 1996. Propolis protects pancreatic β-cells against the toxicity of streptozotocin (STZ). Phytomedicine 3 (2), 203–209.

McLennan, S.V., Bonner, J., Milne, S., Lo, L., Charlton, A., Kurup, S., Jia, J., Yue, D.K., Twigg, S.M., 2008. The anti-inflammatory agent Propolis improves wound healing in a rodent model of experimental diabetes. Wound Repair Regen. 16 (5), 706–713.

McLoone, P., Warnock, M., Fyfe, L., 2016. Honey: an immunomodulatory agent for disorders of the skin. Food Agric. Immunol. 27 (3), 338–349.

Memar, M.Y., Ghotaslou, R., Samiei, M., Adibkia, K., 2018. Antimicrobial use of reactive oxygen therapy: current insights. Infect. Drug Resist. 11, 567.

Mizutani, M., Kern, T.S., Lorenzi, M., 1996. Accelerated death of retinal microvascular cells in human and experimental diabetic retinopathy. J. Clin. Invest. 97 (12), 2883–2890.

Mohamed, A.M., El Ella, A., Ga, H.I.A., 2013. Antioxidant effect of bee pollen on immune status of hyperglycemic rats. Assiut Vet. Med. J. 59 (139), 107–116.

Mohamed, N.A., Ahmed, O.M., Hozayen, W.G., Ahmed, M.A., 2018. Ameliorative effects of bee pollen and date palm pollen on the glycemic state and male sexual dysfunctions in streptozotocin-induced diabetic wistar rats. Biomed. Pharmacother. 97, 9–18.

Mohammad, S.M., Mahmud-Ab-Rashid, N.K., Zawawi, N., 2021. Stingless bee-collected pollen (bee bread): chemical and microbiology properties and health benefits. Molecules 26 (4), 957.

Mori, H.M., Kawanami, H., Kawahata, H., Aoki, M., 2016. Wound healing potential of lavender oil by acceleration of granulation and wound contraction through induction of TGF-β in a rat model. BMC Complement. Altern. Med. 16 (1), 1.

Mousavi, S.M., Imani, S., Haghighi, S., Mousavi, S.E., Karimi, A., 2012. Effect of Iranian honey bee (Apis mellifera) venom on blood glucose and insulin in diabetic rats. J. Arthropod. Borne Dis. 6 (2), 136.

Mujica, V., Orrego, R., Fuentealba, R., Leiva, E., Zúñiga-Hernández, J., 2019. Propolis as an adjuvant in the healing of human diabetic foot wounds receiving care in the diagnostic and treatment Centre from the regional hospital of Talca. J. Diabetes Res. 2019, 2507578.

Münstedt, K., Sheybani, B., Hauenschild, A., Brüggmann, D., Bretzel, R.G., Winter, D., 2008. Effects of basswood honey, honey-comparable glucose-fructose solution and oral glucose tolerance test solution on serum insulin, glucose and C-peptide concentrations in healthy subjects. J. Med. Food 11 (3), 424–428.

Münstedt, K., Hoffmann, S., Hauenschild, A., Bülte, M., von Georgi, R., Hackethal, A., 2009. Effect of honey on serum cholesterol and lipid values. J. Med. Food 12 (3), 624–628.

Nagai, T., Nagashima, T., Myoda, T., Inoue, R., 2004. Preparation and functional properties of extracts from bee bread. FoodReview 48 (3), 226–229.

Nasrolahi, O., Heidari, R., Rahmani, F., Farokhi, F., 2012. Effect of natural honey from Ilam and metformin for improving glycemic control in streptozotocin-induced diabetic rats. Avicenna J. Phytomed. 2 (4), 212.

Nazir, L., Samad, F., Haroon, W., Kidwai, S.S., Siddiqi, S., Zehravi, M., 2014. Comparison of glycaemic response to honey and glucose in type 2 diabetes. J. Pak. Med. Assoc. 64 (1), 69–71.

Neha, K., Haider, M.R., Pathak, A., Yar, M.S., 2019. Medicinal prospects of antioxidants: a review. Eur. J. Med. Chem. 178, 687–704.

Nemoseck, T.M., Carmody, E.G., Furchner-Evanson, A., Gleason, M., Li, A., Potter, H., Rezende, L.M., Lane, K.J., Kern, M., 2011. Honey promotes lower weight gain, adiposity and triglycerides than sucrose in rats. Nutr. Res. 31 (1), 55–60.

Nna, V.U., Bakar, A.B., Lazin, M.R., Mohamed, M., 2018. Antioxidant, anti-inflammatory and synergistic anti-hyperglycemic effects of Malaysian propolis and metformin in streptozotocin–induced diabetic rats. Food Chem. Toxicol. 120, 305–320.

Oguntibeju, O.O., 2019. Type 2 diabetes mellitus, oxidative stress and inflammation: examining the links. Int. J. Physiol. Pathophysiol. Pharmacol. 11 (3), 45.

Ojieh, A.E., Awire, I.E., Awhin, P.E., Madojemu, H.N., 2015. Changes in biomakers of ocular pressure and oxidative stress in aqueous humour of honey fed rats. Biomed. Pharmacol. J. 4 (1), 47–50.

Okutan, H., Ozcelik, N., Yilmaz, H.R., Uz, E., 2005. Effects of caffeic acid phenethyl ester on lipid peroxidation and antioxidant enzymes in diabetic rat heart. Clin. Biochem. 38 (2), 191–196.

Oladayo, M.I., 2016. Nigerian propolis improves blood glucose, glycated hemoglobin A1c, very low-density lipoprotein and high-density lipoprotein levels in rat models of diabetes. J. Intercult. Ethnopharmacol. 5 (3), 233.

Omotayo, E.O., Gurtu, S., Sulaiman, S.A., Wahab, M.S., Sirajudeen, K.N., Salleh, M.S., 2010. Hypoglycemic and antioxidant effects of honey supplementation in streptozotocin-induced diabetic rats. Int. J. Vitam. Nutr. Res. 80 (1), 74.

Oršolić, N., Sirovina, D., Končić, M.Z., Lacković, G., Gregorović, G., 2012. Effect of Croatian propolis on diabetic nephropathy and liver toxicity in mice. BMC Complement. Altern. Med. 12 (1), 1–6.

Paolino, D., Mancuso, A., Cristiano, M.C., Froiio, F., Lammari, N., Celia, C., Fresta, M., 2021. Nanonutraceuticals: the new frontier of supplementary food. Nano 11 (3), 792.

Pasupuleti, V.R., Sammugam, L., Ramesh, N., Gan, S.H., 2017. Honey, propolis and royal jelly: a comprehensive review of their biological actions and health benefits. Oxidative Med. Cell. Longev. 2017, 1259510.

Peng, S., Wei, P., Lu, Q., Liu, R., Ding, Y., Zhang, J., 2018. Beneficial effects of poplar buds on hyperglycemia, dyslipidemia, oxidative stress and inflammation in streptozotocin-induced type-2 diabetes. J Immunol Res 2018, 7245956.

Picolotto, A., Pergher, D., Pereira, G.P., Machado, K.G., da Silva, B.H., Roesch-Ely, M., Gonzalez, M.H., Tasso, L., Figueiredo, J.G., Moura, S., 2019. Bacterial cellulose membrane associated with red propolis as phytomodulator: improved healing effects in experimental models of diabetes mellitus. Biomed. Pharmacother. (112), 108640.

Pourmoradian, S., Mahdavi, R., Mobasseri, M., Faramarzi, E., Mobasseri, M., 2014. Effects of royal jelly supplementation on glycemic control and oxidative stress factors in type 2 diabetic female: a randomized clinical trial. Chin. J. Integr. Med. 20 (5), 347–352.

Prakash, S., Bhargava, H.R., 2014. Apis cerana bee venom: it's antidiabetic and anti-dandruff activity against malassezia furfur. World Appl. Sci. J. 32 (3), 343–348.

Puspasari, A., Harijanti, K., Soebadi, B., Hendarti, H.T., Radithia, D., Ernawati, D.S., 2018. Effects of topical application of propolis extract on fibroblast growth factor-2 and fibroblast expression in the traumatic ulcers of diabetic Rattus norvegicus. J. Oral Maxillofac. Pathol. 22 (1), 54.

Raatz, S.K., Johnson, L.K., Picklo, M.J., 2015. Consumption of honey, sucrose and high-fructose corn syrup produces similar metabolic effects in glucose-tolerant and-intolerant individuals. J. Nutr. 145 (10), 2265–2272.

Ramanathan, A.N., Nair, A.J., Sugunan, V.S., 2018. A review on royal jelly proteins and peptides. J. Funct. Foods (44), 255–264.

Rasad, H., Dashtabi, A., Khansari, M., Chaboksavar, F., Pahlavani, N., Maghsoudi, Z., Entezari, M.H., 2014. The effect of honey consumption compared with sucrose on blood pressure and fasting blood glucose in healthy young subjects. Glob. J. Med. Res. Stud. 1 (4), 117–121.

Rashid, M.R., Nor Aripin, K.N., Syed Mohideen, F.B., Baharom, N., Omar, K., Md Taujuddin, N.M., Mohd Yusof, H.H., Addnan, F.H., 2019. The effect of Kelulut honey on fasting blood glucose and metabolic parameters in patients with impaired fasting glucose. J. Nutr. Metab. 2019, 3176018.

Raziyeva, K., Kim, Y., Zharkinbekov, Z., Kassymbek, K., Jimi, S., Saparov, A., 2021. Immunology of acute and chronic wound healing. Biomol. Ther. 11 (5), 700.

Rezk, D.M., 2018. A comparative study on the effect of royal jelly on blood glucose and serum lipids in streptozotocin induced diabetic rats. Eur. J. Pharm. Med. Res. 4 (4), 39–44.

Rifa'i, M., Widodo, N., 2014. Significance of propolis administration for homeostasis of CD4+ CD25+ immunoregulatory T cells controlling hyperglycemia. Springerplus 3 (1), 1–8.

Rivera-Yañez, N., Rodriguez-Canales, M., Nieto-Yañez, O., Jimenez-Estrada, M., Ibarra-Barajas, M., Canales-Martinez, M.M., Rodriguez-Monroy, M.A., 2018. Hypoglycaemic and antioxidant effects of propolis of Chihuahua in a model of experimental diabetes. Evid. Based Complement. Alternat. Med. 2018, 4360356.

Rojczyk, E., Klama-Baryła, A., Łabuś, W., Wilemska-Kucharzewska, K., Kucharzewski, M., 2020. Historical and modern research on propolis and its application in wound healing and other fields of medicine and contributions by Polish studies. J. Ethnopharmacol. 262, 113159.

Sadeghi, F., Salehi, S., Kohanmoo, A., Akhlaghi, M., 2019. Effect of natural honey on glycemic control and anthropometric measures of patients with type 2 diabetes: a randomized controlled crossover trial. Int. J. Prev. Med. 10, 3.

Samadi, N., Mozaffari-Khosravi, H., Rahmanian, M., Askarishahi, M., 2017. Effects of bee propolis supplementation on glycemic control, lipid profile and insulin resistance indices in patients with type 2 diabetes: a randomized, double-blind clinical trial. J. Integr. Med. 15 (2), 124–134.

Sameni, H.R., Ramhormozi, P., Bandegi, A.R., Taherian, A.A., Mirmohammadkhani, M., Safari, M., 2016. Effects of ethanol extract of propolis on histopathological changes and anti-oxidant defense of kidney in a rat model for type 1 diabetes mellitus. J. Diabetes Investig. 7 (4), 506–513.

Sami, W., Ansari, T., Butt, N.S., Ab Hamid, M.R., 2017. Effect of diet on type 2 diabetes mellitus: a review. Int. J. Health Sci. 11 (2), 65.

Sandino, J., Luzardo, L., Morales, E., Praga, M., 2021. Which patients with obesity are at risk for renal disease? Nephron 145, 1–9.

Santiago, A.R., Boia, R., Aires, I.D., Ambrósio, A.F., Fernandes, R., 2018. Sweet stress: coping with vascular dysfunction in diabetic retinopathy. Front. Physiol. 9, 820.

Santini, A., Cammarata, S.M., Capone, G., Ianaro, A., Tenore, G.C., Pani, L., Novellino, E., 2018. Nutraceuticals: opening the debate for a regulatory framework. Br. J. Clin. Pharmacol. 84 (4), 659–672.

Saritas, N., Yildiz, K., Büyükipekci, S., Coskun, B., 2011. Effect of different levels of royal jelly on biochemical parameters of swimmers. Afr. J. Biotechnol. 10 (52), 10718–10723.

Senyuk, B., Boreyko, L., Yurnyuk, S., 2016. Correction of clinical and biochemical parameters using bee pollen in patients with diabetes. Unity Sci. 4, 137–139.

Sharma, A., Gupta, P., Prabhakar, P.K., 2019. Endogenous repair system of oxidative damage of DNA. Curr. Chem. Biol. 13 (2), 110–119.

Shi, Y.Z., Liu, Y.C., Zheng, Y.F., Chen, Y.F., Si, J.J., Chen, M.L., Shou, Q.Y., Zheng, H.Q., Hu, F.L., 2019. Ethanol extract of Chinese propolis attenuates early diabetic retinopathy by protecting the blood–retinal barrier in streptozotocin-induced diabetic rats. J. Food Sci. 84 (2), 358–369.

Smokovski, I., 2021. Burden of Diabetes Prevalence. In Managing Diabetes in Low Income Countries. Springer, Cham, pp. 1–12.

Sruthi, C.R., Raghu, K.G., 2021. Advanced glycation end products and their adverse effects: the role of autophagy. J. Biochem. Mol. Toxicol. 35 (4), e22710.

Syarifuddin, S., Hadju, V., Inriasari, R., 2020. Effect of honey variation on blood glucose level in pregnant Wistar rats (*Rattus norvegicus*). Open Access Maced. J. Med. Sci. 8 (T2), 98–103.

Terzo, S., Mulè, F., Amato, A., 2020. Honey and obesity-related dysfunctions: a summary on health benefits. J. Nutr. Biochem. 82, 108401.

Tesch, G., Sourris, K.C., Summers, S.A., McCarthy, D., Ward, M.S., Borg, D.J., Gallo, L.A., Fotheringham, A.K., Pettit, A.R., Yap, F.Y., Harcourt, B.E., 2014. Deletion of bone-marrow-derived receptor for AGEs (RAGE) improves renal function in an experimental mouse model of diabetes. Diabetologia 57 (9), 1977–1985.

Ts, K., Ambaga, M., Ch, C., 2015. The effect of bee venom on kidney function in alloxan induced diabetic rabbits. Mongolian Med. Sci. 172 (2), 82–86.

Tuttle, K.R., Johnson, E.C., Cooney, S.K., Anderberg, R.J., Johnson, E.K., Clifton, G.D., Meek, R.L., 2005. Amino acids injure mesangial cells by advanced glycation end products, oxidative stress and protein kinase C. Kidney Int. 67 (3), 953–968.

Umeda, M., Hiramoto, M., Watanabe, A., Tsunoda, N., Imai, T., 2015. Arginine-induced insulin secretion in endoplasmic reticulum. Biochem. Biophys. Res. Commun. 466 (4), 717–722.

Usman, U.Z., Bakar, A.B., Mohamed, M., 2018. Propolis improves pregnancy outcomes and placental oxidative stress status in streptozotocin-induced diabetic rats. BMC Complement. Altern. Med. 18 (1), 1–6.

van der Pol, A., van Gilst, W.H., Voors, A.A., van der Meer, P., 2019. Treating oxidative stress in heart failure: past, present and future. Eur. J. Heart Fail. 21 (4), 425–435.

Vanessa Fiorentino, T., Prioletta, A., Zuo, P., Folli, F., 2013. Hyperglycemia-induced oxidative stress and its role in diabetes mellitus related cardiovascular diseases. Curr. Pharm. Des. 19 (32), 5695–5703.

Volpe, C.M., Villar-Delfino, P.H., Dos Anjos, P.M., Nogueira-Machado, J.A., 2018. Cellular death, reactive oxygen species (ROS) and diabetic complications. Cell Death Dis. 9 (2), 1–9.

Voss, G.T., Gularte, M.S., Vogt, A.G., Giongo, J.L., Vaucher, R.A., Echenique, J.V., Soares, M.P., Luchese, C., Wilhelm, E.A., Fajardo, A.R., 2018. Polysaccharide-based film loaded with vitamin C and propolis: a promising device to accelerate diabetic wound healing. Int. J. Pharm. 552 (1–2), 340–351.

Wakgari, M., Yigezu, G., 2021. Honeybee keeping constraints and future prospects. Cogent Food Agric. 7 (1), 1872192.

Wang, W., Lo, A.C., 2018. Diabetic retinopathy: pathophysiology and treatments. Int. J. Mol. Sci. 19 (6), 1816.

Yaghoobi, N., Al-Waili, N., Ghayour-Mobarhan, M., Parizadeh, S.M., Abasalti, Z., Yaghoobi, Z., Yaghoobi, F., Esmaeili, H., Kazemi-Bajestani, S.M., Aghasizadeh, R., Saloom, K.Y., 2008. Natural honey and cardiovascular risk factors; effects on blood glucose, cholesterol, triacylglycerole, CRP and body weight compared with sucrose. Sci. World J. 8, 463–469.

Yoneshiro, T., Kaede, R., Nagaya, K., Aoyama, J., Saito, M., Okamatsu-Ogura, Y., Kimura, K., Terao, A., 2018. Royal jelly ameliorates diet-induced obesity and glucose intolerance by promoting brown adipose tissue thermogenesis in mice. Obes. Res. Clin. Pract. 12 (1), 127–137.

Yoshida, M., Hayashi, K., Watadani, R., Okano, Y., Tanimura, K., Kotoh, J., Sasaki, D., Matsumoto, K., Maeda, A., 2016. Royal jelly improves hyperglycemia in obese/diabetic KK-Ay mice. J. Vet. Med. Sci. 79 (2), 16-0458.

Zakerkish, M., Jenabi, M., Zaeemzadeh, N., Hemmati, A.A., Neisi, N., 2019. The effect of Iranian propolis on glucose metabolism, lipid profile, insulin resistance, renal function and inflammatory biomarkers in patients with type 2 diabetes mellitus: a randomized double-blind clinical trial. Sci. Rep. 9 (1), 1.

Zamami, Y., Fujiwara, H., Hosoda, M., Hino, H., Hirai, K., Okamoto, K., Jin, X., Takatori, S., Kawasaki, H., 2010. Ameliorative effect of propolis on insulin resistance in Otsuka long-Evans Tokushima fatty (OLETF) rats. J. Pharm. Soc. Jpn. 130 (6), 833–840.

Zamani, Z., Reisi, P., Alaei, H., Pilehvarian, A.A., 2012. Effect of Royal Jelly on spatial learning and memory in rat model of streptozotocin-induced sporadic Alzheimer's disease. Adv. Biomed. Res. 1, 26.

Zhao, L., Pu, L., Wei, J., Li, J., Wu, J., Xin, Z., Gao, W., Guo, C., 2016. Brazilian green propolis improves antioxidant function in patients with type 2 diabetes mellitus. Int. J. Environ. Res. Public Health 13 (5), 498.

Zhao, X., Chen, Y., Tan, X., Zhang, L., Zhang, H., Li, Z., Liu, S., Li, R., Lin, T., Liao, R., Zhang, Q., 2018. Advanced glycation end-products suppress autophagic flux in podocytes by activating mammalian target of rapamycin and inhibiting nuclear translocation of transcription factor EB. J. Pathol. 245 (2), 235–248.

Zhou, G., Li, C., Cai, L., 2004. Advanced glycation end-products induce connective tissue growth factor-mediated renal fibrosis predominantly through transforming growth factor β-independent pathway. Am. J. Pathol. 165 (6), 2033–2043.

Zhu, W., Chen, M., Shou, Q., Li, Y., Hu, F., 2011a. Biological activities of Chinese propolis and Brazilian propolis on streptozotocin-induced type 1 diabetes mellitus in rats. Evid. Based Complement. Alternat. Med. 2011, 468529.

Zhu, W., Li, Y.H., Chen, M.L., Hu, F.L., 2011b. Protective effects of Chinese and Brazilian propolis treatment against hepatorenal lesion in diabetic rats. Hum. Exp. Toxicol. 30 (9), 1246–1255.

CHAPTER 4

Propolis and the immune system

Beste Özsezen[a] and Sibel Karakaya[b]
[a]Faculty of Medicine, Department of Pediatric Pulmonology, Hacettepe University, Ankara, Turkey
[b]Faculty of Engineering, Department of Food Engineering, Ege University, Izmir, Turkey

Contents

Introduction

Medicinal use of propolis, which has attracted researchers' interest in the last decades due to its beneficial health effects, including immunomodulatory, antimicrobial, antiinflammatory, etc., goes back to ancient times, at least to 300 BC. Propolis has been used for medicinal purposes in many parts of the world especially healing of some lesions of the skin, sores, and ulcers, and as an antiinflammatory agent (Sforcin, 2007). The chemical composition of propolis is quite complex, and more than 300 phytochemicals, including polyphenols, sesquiterpenes, steroids, alcohols, ketones, quinones, coumarins, amino acids, and inorganic substances, have been determined (Ozdal et al., 2019; Touzani et al., 2019). This composition depends on the collection time, location, geographical origin, and plant source (Touzani et al., 2019). For example, the main components of Turkish propolis originated from different parts of Turkey are flavones (apigenin, luteolin), phenolic acids (3,4-dimethoxycinnamic acid, caffeic acid, *p*-coumaric acid, caffeic acid phenethyl ester), flavanols (quercetin, kaempferol), and flavanones (pinobanksin, pinostrobin) (Ozdal et al., 2019). However, these components of Mexican propolis are naringin, naringenin, kaempferol, quercetin, acacetin,

Bee Products and Their Applications in the Food and Pharmaceutical Industries
https://doi.org/10.1016/B978-0-323-85400-9.00008-3

luteolin, pinocembrin, chrysin, epoxypinocembrin chalcone, pinostrobin, izalpinin, cinnamic acid, pinocembrin, caffeic acid, 3,4-dimethoxycinnamic acid, and rhamnetin (Guzmán-Gutiérrez et al., 2018). The main component that exerts immunomodulatory, antiinflammatory, antioxidant, and antimicrobial effects of Brazilian green propolis is artepillin C ((*E*)-3-(4-hydroxy-3,5-bis[3-methylbut-2-enyl]phenyl)prop-2-enoic acid) (Ding et al., 2021).

The pharmacological quality of propolis depends on their composition and the place they are collected, the type of vegetation, season, and species of bee. Although the chemical composition of propolis obtained from different parts of the world was different, most studies have shown the immunomodulatory action of propolis and its phytochemicals (Bueno-Silva et al., 2016; Orsatti et al., 2010; Touzani et al., 2019). In addition, the safe concentration of propolis for humans has been reported as 70 mg/day. No alteration has been found in total lipids, triglycerides, cholesterol, HDL-cholesterol concentrations of rats treated with different concentrations of propolis (1, 3, and 6 mg/kg/day for 30 days), both for water and ethanolic extracts (Sforcin, 2007).

In this chapter, the brief information on the immune system, the relationship between the phytochemicals of propolis and immunity, and the effect of propolis on innate and adaptive immunity are emphasized.

Phytochemicals of propolis responsible for immunomodulatory action

In the last decade, many articles were published on the antioxidant and immunomodulatory action of propolis and/or its bioactive compounds such as phenolics. The immunomodulatory activities of propolis have been attributed to its phenolics, mainly to flavonoids, some phenolic acids (cinnamic acid and 3,5-diprenyl-4-hydroxycinnamic acid), caffeic acid phenethyl ester (CAPE), quercetin, and artepillin C (Farooqui and Farooqui, 2012). The antioxidant activity of propolis has been well studied, and this activity is mainly attributed to CAPE and/or artepillin C contents of propolis. Propolis extract containing CAPE was a more effective antioxidant than propolis extract without CAPE (Russo et al., 2002). However, CAPE is mainly found in propolis originated from the temperate zone. For example, Brazilian green propolis contains only trace amounts of CAPE but contains a considerable amount of artepillin C (Chan et al., 2013). Ozdal et al. (2019) collected propolis samples from various regions of Turkey. They investigated their total phenolics, total flavonoids, phenolic profiles, total antioxidant capacity, and antioxidant activity after in vitro gastrointestinal digestion. They reported that total phenolic and total flavonoid contents of the samples changed in between 10.39 and 19.97 g GAE/100 g sample and 3.07–29.18 g QE/100 g sample, respectively. After in vitro gastrointestinal digestion, the average recovery of the phenolics and flavonoids was reported as 13.1% and 43.4%, respectively. In addition, an increase in the total antioxidant capacity of most of the propolis samples after in vitro gastrointestinal digestion was demonstrated. The antioxidant capacity of ethanolic (PVEE) and hexanic (PVEH) extracts of Brazilian green propolis against DPPH radical were 13.09 and 95.86 μg/mL as IC_{50} values, respectively. Results confirmed that the higher amount of artepillin C was related to the better antioxidant activity (Veiga et al., 2017). Propolis from the Basque country displayed antioxidant activity against ABTS and DPPH radicals and ferric reducing effect with the concentrations of 420–1430 μM Trolox/g, 108–291 mg ascorbic acid/g and 1573–4669 μmol for ABTS radical, DPPH radical, and FRAP, respectively. There was a relationship between total phenols and antioxidant activity (Bonvehí and Gutiérrez, 2011).

In summary, phytochemicals of propolis, especially CAPE and artepillin C, are the main components that provide antioxidant activity and are more probably associated with the immunomodulatory action of propolis.

Overview of the immune response

The main components of the immune system, which is intrinsic to health, are the different types of immune cells and the cytokines that are communication signals. The physiologic function of the immune system is a defense against infectious microorganisms. However, foreign substances can also elicit immune responses. Defense against microorganisms is mediated by the early reactions of innate immunity and the later responses of adaptive immunity.

The innate immune system is evolutionarily conserved among multicellular organisms. It constitutes a rapid defense against invading microorganisms within hours long before the adaptive immune system bearings an antigen-specific response a few days later. Table 4.1 presents the cells of the immune system and their functions. The immune system cells are categorized as antigen-presenting cells (APC), lymphocytes, and effector cells. The cells of the innate immune system are classified as phagocytic cells (neutrophils, macrophages) and natural killer (NK) cells (Table 4.1). In addition, the

Table 4.1 The cells of the immune system and their functions.

Antigen presenting cells		Lymphocytes		Effector cells	
The cells	**Functions**	**The cells**	**Functions**	**The cells**	**Functions**
Dendritic cells	• Stimulatory action on the T cell response • Coordination of both innate and adaptive immunity	B-lymphocytes	Humoral immunity • Production of antibodies • Antigen presentation. • Production of multiple cytokines	T-lymphocytes. T-helper (Th) and T-cytotoxic (Tc)	Th: • Strengthen of the innate immunity via the activation of phagocytes • The stimulation of humoral response by activating the B-lymphocytes Tc: • Killing of the cells containing foreign proteins in their cytoplasm
Macrophages	• Activation of cell mediated immune response • Involving in the effector phase	T- lymphocytes	Cellular immunity • Expression of membrane bound molecules • Secretion of soluble mediators • Modulation of antibody responses. • Activation of innate immune cells	Mononuclear phagocyte system	Phagocytic macrophages: • Expression of cytokines that trigger inflammation • Tissue repair

Follicular dendritic cells	• Stimulatory action on the humoral immune response by presenting antigens to B-lymphocyte	Natural killer cells	• Destroying infected cells via cytolytic activity • Production of a variety of inflammatory cytokines	Granulocytes (neutrophile, eosinophile)	Neutrophiles: • Production of reactive oxygen species, cytokines and chemokines, and secretion of granule components Eosinophiles: • Modulation of lymphocyte recruitment and homeostasis • Presenting antigen to T-cells • Driving Th2 polarization • Innate recognition • Tissue destruction and repair function • Regulation of neural activity

other principal components of the innate immune system are physical and chemical barriers, blood proteins, and cytokines (Abbas et al., 2010).

Monocytes/macrophages, neutrophils, dendritic cells (DCs), and natural killer cells (NK) are the primary effector cells involved in innate immunity. The three main functions of monocytes/macrophages are phagocytosis, antigen presentation to effective T cells, and synthesis of various inflammatory mediators. Phagocytic cells destroy microorganisms by two different mechanisms, oxygen-dependent in which reactive oxygen and nitrogen intermediates are produced and oxygen-independent, in which lysosomal enzymes such as lysozyme and proteases are released. Neutrophils, as phagocytic cells, destroy extracellular microorganisms, whereas monocytes/macrophages destroy cells infected with pathogens (Wolska et al., 2019). The activation of monocytes and macrophages may occur through Toll-like receptor (TLR) dependent or TLR independent pathways. Dendritic cells exhibit essential roles in innate responses to infections and in linking innate and adaptive immune responses. Natural killer cells destroy infected cells via their cytolytic activity. The cells of innate immunity are effective in stopping most infections at an early stage. When macrophages are exposed to inflammatory stimuli, they secrete cytokines such as tumor necrosis factor. Innate immunity exhibits lethal weapons to destroy pathogens with the production of inflammatory cytokines. Although the production of appropriate amounts of cytokines is beneficial for the host, their excessive production can cause toxicity (Duque and Descoteaux, 2014; Ripari et al., 2021). Therefore, modulation of immunity is of vital importance. However, in some instances, innate immune cells cannot recognize some pathogens, and the lymphocytes of the adaptive immune system provide a defense to these pathogens.

Furthermore, the adaptive immune system supports increased protection against subsequent reinfection with the same pathogen (Janeway et al., 2001). Adaptive immunity provides a late (within days) and slow-initiated response against molecules specific to microorganisms that invade tissues (reach the lymphoid organs through epithelial barriers) and other foreign antigens. It constitutes a specific and robust immune response to the host. The cells responsible for the emergence of the adaptive immune response are lymphocytes. B lymphocytes and their products antibodies eliminate extracellular pathogens, while effector T-lymphocytes destroys intracellular pathogens. It has memory, and a faster and more effective immune response occurs in repetitive encounters (Ripari et al., 2021). $CD4^+$ T helper (Th) cells can be functionally differentiated in Th1 and Th2 cells. Th1 cells secrete mainly IFN-γ and IL-2 and inhibit Th2 cell proliferation.

However, Th2 cells secrete mainly IL-4 and IL-10 and inhibit Th1 cells. Activation of Th1 and Th2 cells contribute to cell-mediated immunity and humoral immune response, respectively (Pagliarone et al., 2009a,b). $CD4^+$ T helper cells help macrophages eliminate ingested microorganisms and help B cells produce antibodies (Abbas et al., 2010).

In summary, the immune system has several properties that are of fundamental importance for its normal functions. The innate immune system constitutes the first line of host defense against microorganisms, and molecules produced during innate immune response stimulate adaptive immunity that provides specificity for distinct molecules and an ability to remember and respond to repeated exposures to the same microorganisms (Fig. 4.1).

Immunomodulatory effects on innate and adaptive immunity

The immunomodulatory effect of propolis has been well documented. In vivo and in vitro assays have demonstrated that propolis exhibits modulatory action on a broad spectrum of immune cells, including monocytes/macrophages, neutrophils, lymphocytes, natural killer (NK) cells, and basophile/mastocyte function (Bachiega et al., 2012; Bueno-Silva et al., 2016; Chan et al., 2013; Orsatti et al., 2010; Orsi et al., 2000; Sá-Nunes et al., 2003; Simões et al., 2004) (Table 4.2).

Effects on innate immunity

Effects of propolis on innate immunity can be listed as an increase in phagocytic capacity, activation of NK cells, innate lymphoid cells, and DCs. Both inhibition and activation on mast cells depend on the concentration of propolis (Fig. 4.2).

Effects on phagocytes, NK cells, and cytokines

Several studies have shown that propolis and/or its bioactive compounds can modulate macrophage activation and cause an increase in phagocytic capacity. TNF is a potent proinflammatory cytokine secreted by many innate immune cells and mainly produced by macrophages. Macrophages also produce and secrete other proinflammatory and antiinflammatory cytokines, such as IL-6, IL-10, IL-12, and chemokines. Neutrophils are phagocytic cells and produce a wide range of cytokines such as TGF-α, TNF, IL-6, IL-12, and CXCL2/IL-8 (Lacy and Stow, 2011). Asgharpour et al.

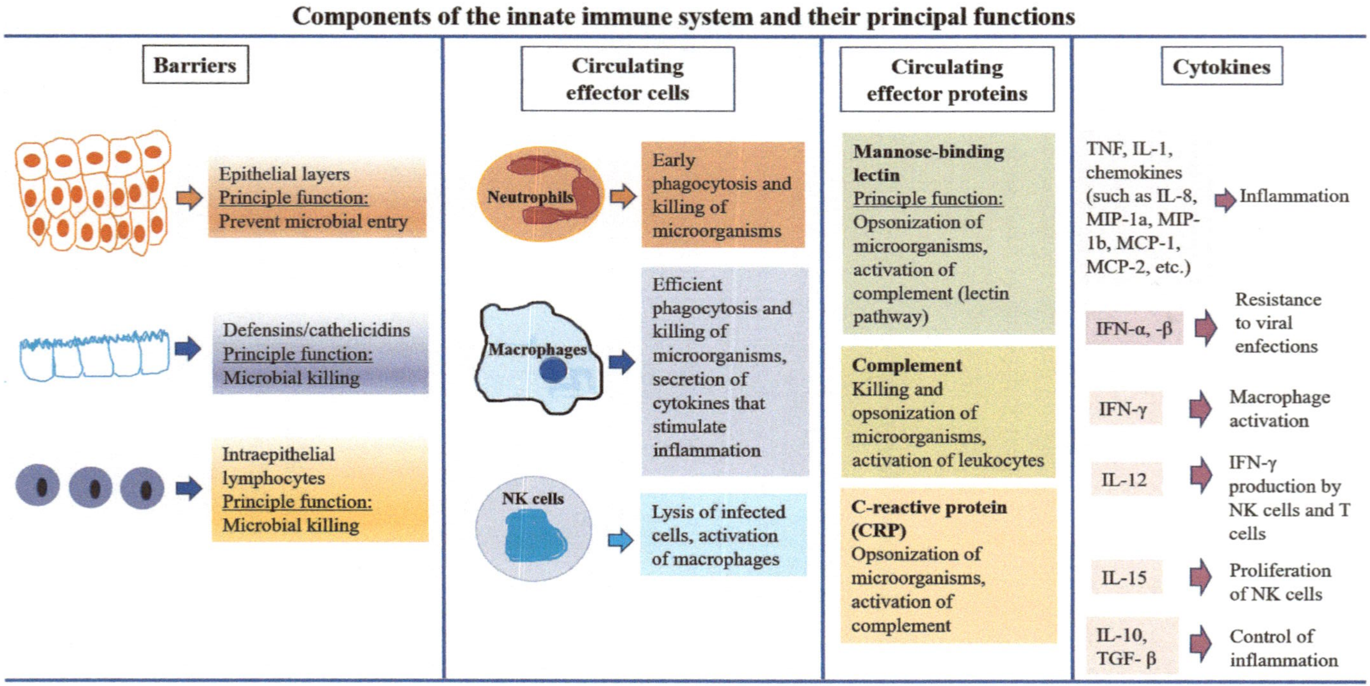

Fig. 4.1 Components of the innate immune system and their principal functions. *IFN*: interferon; *IL*: interleukin; *NK*: natural killer; *TGF*- β: transforming growth factor-β; *TNF*: tumor necrosis factor.

Table 4.2 Recent studies on the immunomodulatory effects of propolis.

Source of propolis	Extraction method	Model	Effects	References
Brazilian propolis	70% EtOH	Male BALB/c mice	Propolis exhibited an immunorestorative role in TLR-4 expression in stressed mice and partially restored TLR2 mRNA expression	Pagliarone et al. (2009a)
Brazilian propolis	70% EtOH	Male BALB/c mice	Propolis did not affect IFN-γ production in acutely stressed mice. However, propolis affected IL-4 production by spleen cells representing the immunorestorative effect	Pagliarone et al. (2009b)
Brazilian propolis	70% EtOH	In vitro splenocyte proliferation assay and Male BALB/c mice	Although an in vitro inhibitory effect on unstimulated splenocyte proliferation (more than 35% at a concentration of 100 μg/mL of propolis extract) was observed, the highest reduction in splenocyte proliferation stimulated with Con A was detected in the samples containing the lowest propolis concentration (5 and 10 μg/mL of propolis extract). In addition, pre-activation of macrophages, lymphocyte proliferation, and stimulation of IFN-γ production were determined in mice treated with propolis (daily 2.5, 5, and 10 mg/kg doses for 3 days intraperitoneally)	Sá-Nunes et al. (2003)
North China	80% EtOH and water	ICR mice and male Wistar rats	Ethanolic extract and a water-soluble fraction of propolis inhibited the activation and differentiation of monocytes/macrophages. Propolis suppressed IL-6 production but not IFN-γ or IL-2 productions	Hu et al. (2005)

Continued

Table 4.2 Recent studies on the immunomodulatory effects of propolis—cont'd

Source of propolis	Extraction method	Model	Effects	References
Turkey	96% EtOH	In vitro method	Immunomodulatory effects, including down-regulation on IDO activity, were observed	Girgin et al. (2009)
Brazilian propolis	30% EtOH	Male BALB/c mice	An increase in basal IL-1β production and TLR-2 and TLR-4 expression in peritoneal macrophages of propolis-treated mice, and upregulation of TLR-2, TLR-4 expressions, and IL-1β, IL-6 productions in the spleen cells of propolis treated cells were observed	Orsatti et al. (2010)
Chinese and Brazilian propolis	Water and ethanol extracts	Antigen-stimulated rat basophilic leukemia (RBL–2H3) cells	The ethanol extract of Chinese propolis exhibited the most potent inhibition on mast cell degranulation. Chrysin, a flavonoid, inhibited IL-4 and MCP-1 production	Nakamura et al. (2010)
Brazilian propolis	30% EtOH	In vitro method	Pro- or antiinflammatory effect of propolis was determined	Bachiega et al. (2012)
Brazilian propolis	30% EtOH	In vitro method (monocytes obtained from healthy blood donors)	Activation of DCs in the presence of LPS. Maintaining the LPS-induced NF-kB activation and TNF-α and IL-6 production at all concentrations of propolis (5, 10, 20, and 40 μg/mL of propolis). Increased expression of TLR-4 and CD86 by DCs	Conti et al. (2016)
Brazilian red propolis	80% EtOH	Male BALB/c mice	Antiinflammatory effect was observed. This effect decreased the neutrophils influx to the inflammatory site by acting through reduced TNF-α, IL-1β, CXCL1/KC and CXCL2/MIP-2 release and decreasing neutrophil chemotaxis by blocking calcium influx	Bueno-Silva et al. (2016)

Not reported	Water-soluble powder of propolis (WPP) extract, artepillin C, *p*-coumaric acid, and drupanin.	C57BL/6 (B6) wild-type (WT) and RAG 2-deficient (*RAG*−/−) mice and BALB/c WT, interferon (IFN)-γ-deficient (*IFN*-$\gamma^{-/-}$), IFN-γ receptor-deficient (*IFN*-$\gamma R^{-/-}$) and $RAG^{-/-}$ mice	Improvement of NK cell cytotoxic activity in orally WPP administrated WT, B6, and BALB/c *RAG*−/− mice. Also, artepillin C and *p*-coumaric acid augmented NK cell cytotoxicity	Takeda et al. (2018)
Propolis from UK, Ghana, Cameroon, and Indonesia	EtOH	In vitro method	Antagonistic effect on TNF-α, IL-1β, and IL-6 proinflammatory cytokines was observed for propolis from Cameroon. The other sources of propolis displayed antagonistic effects on TNF-α and IL-6, but the stimulatory effect on the IL-1β release	Alqarni et al. (2019)
Iranian propolis	Propolis capsules (500 mg)	Human study (T2DM patients)	Decrease in serum hs-CRP and TNF-α levels	Zakerkish et al. (2019)
Propolis powder (LY-009)	Propolis powder (LY-009) containing 8.0% artepillin C	BALB/c mice	Suppression of IgE/antigen-induced production of IL-4, IL-6, and IL-13 in basophils. Inhibition of phosphorylation of FcεRI signaling molecules. Inhibition of IgE-CAI and decrease in ovalbumin-induced intestinal anaphylaxis	Kashiwakura et al. (2020)
Brazilian green propolis	Dry extract of propolis	C57BL/6 female mice (treated with 150 mg/kg of propolis)	Decrease in pulmonary inflammation and mucus production as well as eosinophils and IL-5 in the bronchoalveolar lavage. Increase in the frequency of MDSC and $CD4^+$ $Foxp3^+$ regulatory cells	Piñeros et al. (2020)

Continued

Table 4.2 Recent studies on the immunomodulatory effects of propolis—cont'd

Source of propolis	Extraction method	Model	Effects	References
Chinese propolis, poplar	95% EtOH	Human breast cancer cell line MDA-MB-231	Antiinflammatory effect due to inhibition of inflammatory cytokines was observed	Li et al. (2021)
Brazilian green propolis	EtOH	Male C57BL/6 and BALB/c mice (treated with 50 mg/kg of propolis). In vitro cell culture method (RPMI1640 cells).	Increase in TNFR2 expression in $FoxP3^{+}$ regulatory T cells via the IRF4/cMyc axis Promotion of TNFR2 expression and thus increase in reactivity to TNF- α. Upregulation of TNFR2 by propolis, artepillin C, drupanal, and dihydrokaempferide. Downregulation of CD38 (in vitro result)	Mikami et al. (2021)
Iranian propolis	Propolis capsules (500 mg)	Human study	A significant reduction was detected in TNF-α levels in the group taking propolis capsules	Afsharpour et al. (2017).
Southeastern Brazilian propolis		Human study	Propolis treatment modulated innate immunity by downmodulating the monocytes proinflammatory activity	Conte et al. (2021)
Senegalese propolis	99.5% EtOH	Murine J774.1 macrophage cells	Suppression of LPS-stimulated NO production was determined	Inui et al. (2021)
Southeastern Brazilian brown propolis type from *Apis mellifera*	Ethanol-water extract	Male BALB/c mice	A reducing effect on the inflammatory process was reported	Riberio et al. (2021)

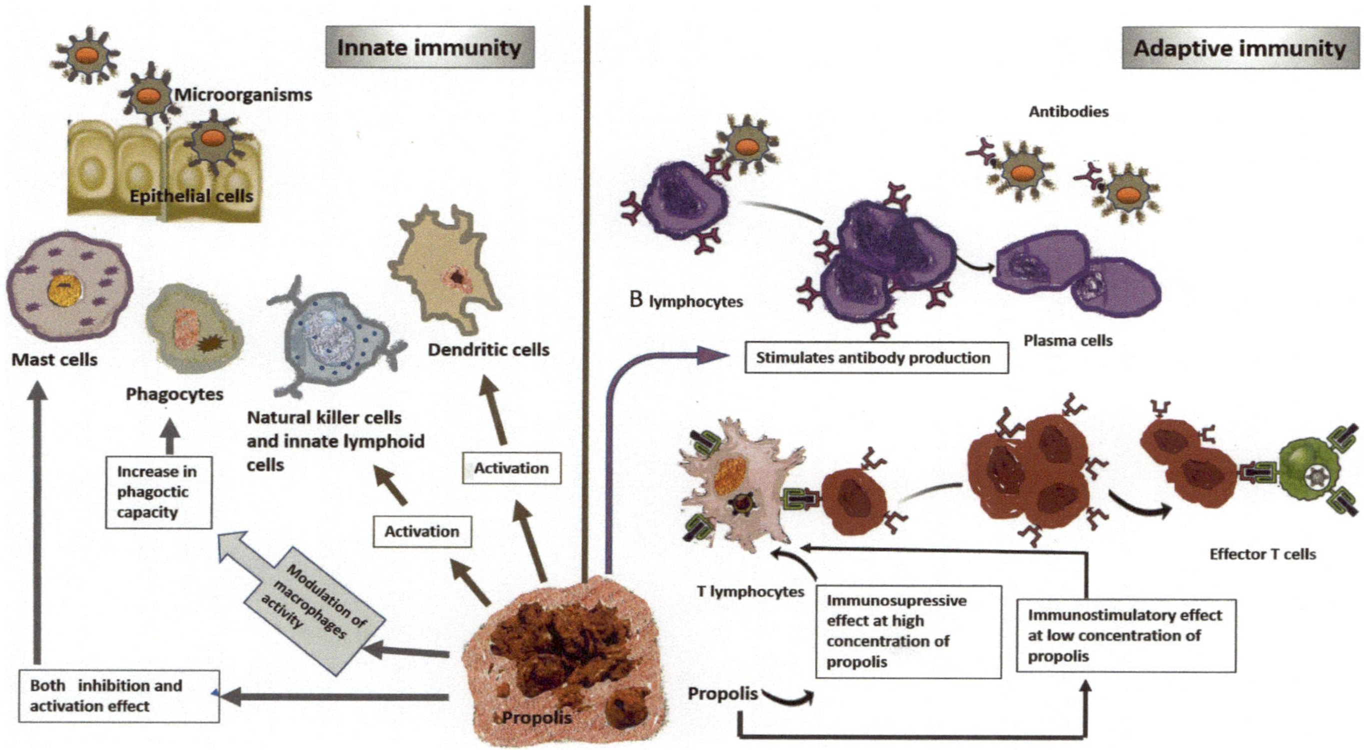

Fig. 4.2 Effect of propolis on innate and adaptive immunity.

(2019) reported that nitric oxide (NO) intermediate metabolite production was not significantly inhibited in bacterial endotoxin lipopolysaccharide-stimulated (LPS) RAW 264.7 cells in the presence of the ethanolic extract of Iranian propolis with the concentration of 15 and 1.5 μg/mL. However, the lowest dose of propolis (0.15 μg/mL) extract inhibited NO production in LPS-challenged cells. Similarly, they reported that reactive oxygen species (ROS) production induced by LPS decreased significantly in the cells treated with 0.15 μg/mL propolis extract. In the case of inflammatory genes expression (increased mRNA expression of IL-1β, IL-6, and Cox-2 in the cells treated with LPS), the significant inhibition on the expression of Cox-2 in all doses of propolis extract and on the expression of IL-1β in 0.15 and 1.5 μg/mL doses of propolis extract were reported. In contrast, the highest dose of propolis extract (15 μg/mL) significantly inhibited IL-6 mRNA expression, whereas inhibitory effects of the low concentrations of propolis extracts were not significant.

The methanolic and water extracts of Netherland propolis inhibited NO production in LPS-activated murine macrophage-like J774.1cells with IC_{50} values of 23.8 and 51.5 μg/mL, respectively (Nagaoka et al., 2003). In addition, they reported that CAPE analogs, CAPE, and cinnamyl caffeate exhibited NO inhibitory activities with IC_{50} values of 13.8, 7.64, and 9.53 μM, respectively. Cytotoxic and NO inhibitory activity of propolis extract encapsulated by complex coacervation to mask the unpleasant bitter taste of propolis were demonstrated. The effect of ethanolic extract of propolis (EEP) on NO production was measured in LPS-induced RAW 264.7 cells. EEP inhibited NO production with a concentration of 0.1 μg/mL. Researchers also indicated that Turkish propolis originating from Sivas, Turkey was superior compared to its counterparts, such as Polish EEP and Brazilian green EEP, in terms of NO inhibition and antiinflammatory effect, respectively (Onbas et al., 2016).

Toll-like receptors play an essential role in the innate immune response by recognizing conserved pathogen-associated molecular patterns such as bacterial lipopolysaccharides, lipopeptides, viral RNA, and DNA. One of the proinflammatory cytokines, IL-1β, is involved in various cellular activities, including cell proliferation, differentiation, and apoptosis. IFN-γ is critical in the protection against viral, some bacterial, and protozoal infections (Table 4.1) (Bao and Cao, 2014; Miller et al., 2009; Zhao et al., 2013). Orsatti et al. (2010) studied the effect of propolis on the expression of Toll-like receptors, TLR 2 and 4, and production of proinflammatory cytokines, IL-1β and IL-6 in propolis administrated (200 mg/kg) BALB/c male mice.

They observed increased TLR-2 and TLR-4 expressions in macrophages and spleen cells of propolis-treated mice than those of the control group. In addition, increased production of proinflammatory cytokines, IL-1β, and IL-6 were determined after TLRs expression.

In the study of Bachiega et al. (2012) immunomodulatory effect of propolis on cytokine production (IL-6, IL-1β, and IL-10) was investigated by peritoneal macrophages obtained from male BALB/c mice in vitro. They found an increased IL-1β production by macrophages incubated with propolis, cinnamic, and coumaric acids at all concentrations (5, 10, 25, and 100 μg/well), representing their immunostimulatory activity. In addition, propolis and its phenolic acids inhibited IL-6 and IL-10 produced by peritoneal macrophages in the presence and absence of LPS. Their LPS protocol was conducted to represent two different conditions indicating the preventive role of the propolis and its constituents by incubating the macrophages with the natural products before applying LPS, and the therapeutic effect of propolis and its constituents by incubating the natural products with LPS-treated macrophages. They found that propolis seemed to be effective in both conditions. However, cinnamic and coumaric acids were only effective in the preventive condition. The dose-dependent response of propolis to reduce proinflammatory cytokines was reported by Xool-Tamayo et al. (2020). The concentrations higher than 100 μg/mL of propolis inhibited cytokine production. In contrast, a concentration of 50 μg/mL of Yucamiel propolis was reported to be enough for high levels of IL-10 and IL-4 production. The authors concluded that a low propolis dose well correlated with in vitro production of IL-10 and IL-4 but a low to moderate in vivo antiinflammatory response in Balb/c male mice at short times (2 h).

Inui et al. (2021) determined the antiinflammatory activity of ethanolic extracts of poplar type, Baccharis-type, Dalbergia-type, and Senegalese propolis using LPS-stimulated NO production in Murine J774.1 macrophage cells. They reported that all the ethanol extracts of propolis exhibited a dose-dependent decrease in LPS-stimulated NO production to <50 μg/mL. In addition, they determined that the eight compounds isolated from Senegalese propolis have high antiinflammatory activity against the LPS-induced expression of inducible NO synthase.

Inhibitory effect of Brazilian green propolis and kaempferol on type 2 diabetes and adiposity by triggering antiinflammatory cytokines, myeloid-derived suppressor cells (MDSCs) was reported (Kitamura et al., 2018). Another study reported that artepillin C of Brazilian green propolis was the compound that activates macrophage phagocytosis. A positive effect on immune function in aged mice after administration of Brazilian green propolis was demonstrated. The authors reported that although production of IL-1β, IL-4, and IFN-γ increased after propolis administration to aged mice, these increased levels were not statistically significant, suggesting that these cytokines were not overproduced at the administrated doses (Gao et al., 2014). However, the overproduction of the cytokines should be carefully considered in terms of harmful inflammatory response or autoimmunity. A recent study reported that propolis inhibited TNF-α, and IL-6 and increased IL-10 production. Moreover, propolis downmodulated LC3 expression induced by LPS. These findings supported that propolis modulates innate immunity by downmodulating the monocytes' proinflammatory activity (Conte et al., 2021).

Natural killer cells have cytotoxic effects, especially against cancer cells or virally infected cells, without requiring prior antigen exposure to mediate their activity. Fundamentally, they recognize infected and/or stressed cells and respond by directly killing these cells and by secreting inflammatory cytokines. Natural killer cells are a significant source of IFN-γ, which activates macrophages to kill ingested microorganisms. They mediate proinflammatory cytokine production, including IFN-γ, TNF, and chemokines (Abbas et al., 2010; Abel et al., 2018). Sforcin et al. (2002) investigated the effect of propolis on the cytotoxic activity of rat NK-like cells using murine lymphoma (Yac-1) as target cells. They found an increased NK activity after administering propolis to rats for 3 days, indicating the short-term effect of propolis on the immune system.

Takeda et al. (2018) used water-soluble powder of propolis (WPP) extract, artepillin C, *p*-coumaric acid, and drupanin on NK cells of C57BL/6 (B6) wild-type (WT) and RAG 2-deficient (*RAG−/−*) mice and BALB/c WT, interferon (IFN)-γ-deficient (*IFN-γ*$^{-/-}$), IFN-γ receptor-deficient (*IFN-γR*$^{-/-}$), and *RAG*$^{-/-}$ mice. Increase in the cytotoxic activity of NK cell after administering 100 mg/kg/day of WPP for 7 days or 200 or 500 mg/kg/day of WPP for 4 days in WT mice. Similarly, orally administrated WPP (200 mg/kg/day for 4 days) caused augment *RAG−/−* mice which indicated WPP induced NK cell activity did not depend on acquired immune cells (T, NKT, and B cells). In addition, they stressed the critical role of IFN-γ in the augmentation of NK cell cytotoxicity by WPP due to administration of WPP did not augment *IFN-γ*$^{-/-}$ or *IFN-γR*$^{-/-}$ BALB/c mice. Furthermore, oral intake of a mixture of artepillin C, *p*-coumaric acid, and drupanin, which are the major components of WPP, displayed similar activity to 200 mg/kg of WPP on NK cells. Likewise, oral intake of individual components, artepillin C, and *p*-coumaric acid augmented NK cell cytotoxicity.

Effects on dendritic cells and cytokines

Innate response to infections and linking innate and adaptive immune responses are to main functions of DCs. They serve a critical function in adaptive immune responses by capturing and displaying microbial antigens to T lymphocytes (Abbas et al., 2010). Adachi et al. (2019), investigated intracellular Ca^{2+} signaling in DCs using conditional calcium biosensor YC3.60 transgenic mice and the effect of oral administration of propolis on DCs. Results indicated that Brazilian propolis could induce Ca^{2+} signaling in immune cells but did not induce the expression of activation markers on lymphocytes and their proliferation. Dendritic cells play an essential role in the pathogenesis of asthma and allergic rhinitis. Impact of CAPE on DCs in the pathogenesis of asthma was reported and inhibitory effect of CAPE on IL-12 p40, IL-12 p70, IL-10 protein expression in mature, healthy human monocyte-derived dendritic cells (MoDCs) and IL-2 p40, IL-10, IP-10 stimulated by crude mite extract. Authors indicated that the mechanism of the cytokine inhibition was related to nuclear factor-κB (NF-κB) signaling, but not the mitogen-activated protein kinase (MAPK) pathway (Wang et al., 2009). Song et al. (2021) investigated the inhibitory effect of galangin, a flavonoid isolated from propolis, on DCs activation and tolerogenic properties. Tolerogenic DCs (tolDCs) can induce the differentiation of immunoregulatory T cells. They demonstrated that exposure to galangin during DC differentiation caused antiinflammatory effects on DCs. Therefore, they suggested that galangin can be a good candidate for preventing or treating inflammatory diseases.

In summary, in vitro and in vivo studies reported a positive impact of ethanolic extracts and aqueous propolis on innate immune cells and cytokine production. However, human studies on the effect of propolis extract on innate immunity are relatively scarce compared to in vitro and in vivo studies. Recently published meta-analysis based on human studies demonstrated sufficient evidence on the positive effect of propolis on IL-6, CRP, and TNF-α. This finding indicates that propolis supplementation exhibits positive effects on chronic inflammation (Shang et al., 2020). However, it should be emphasized that further exploratory studies and large clinical trials are needed to prove the effect of propolis on innate immunity.

Effects on adaptive immunity

Adaptive immunity is composed of two types of immune response called humoral immunity and cell-mediated immunity. Antibodies produced by B lymphocytes mediate humoral immunity, and cell-mediated immunity is mediated by T lymphocytes (Abbas et al., 2010). T and B cells play an important role in adaptive immune defense, and they are another major population of peripheral blood mononuclear cells (PBMC) (Wolska et al., 2019).

The effect of propolis on adaptive immunity can be listed as stimulation of antibody production in B lymphocytes and both immunosuppressive and immunostimulatory effects depending on the concentration of propolis (Fig. 4.2).

The immunosuppressive effect and lymphocyte proliferation stimulating effect of propolis has been demonstrated in several studies (Wolska et al., 2019). It has been reported that whether the suppressive or stimulating effects will occur is related to the concentration of propolis. Márquez et al. (2004) investigated the immunosuppressive activity of CAPE in human T cells and reported that CAPE (10 μM) was a potent inhibitor of receptor-mediated T cell activation. This inhibition was based on the suppression of IL-2 gene transcription and IL-2R expression in stimulated human T cells. Park et al. (2004) demonstrated the immunomodulatory activity of CAPE in female Balb/c mice and found a decrease in cell number, especially T cells. A decrease in thymus weight and/or cellularity of thymus and spleen was observed in the group taking CAPE (5, 10, and 20 mg/kg). However, CAPE did not affect lymphocyte proliferation in B cells while caused an increase in T lymphocyte blastogenesis at the dose of 20 mg/kg. They suggested that CAPE mainly affects T cells in vivo. Ca^{2+} signals regulate proliferation, differentiation, apoptosis, and various transcriptional programs in the cells (T cells, B cells, mast cells) of the immune system. Intracellular Ca^{2+} mobilization is induced by the stimulation of immune receptors, including the B-cell antigen receptor. EEP of Brazilian propolis did not exhibit a remarkable effect on the viability of the splenocytes or the expression of CD86 on B cells and CD69 on T cells. The authors concluded that Brazilian propolis did not affect lymphocyte proliferation in vitro (Adachi et al., 2019). Another study showed that CAPE inhibited IFN-γ and IL-5 secretion and proliferation of human $CD4^+$ T cells from healthy individuals and mite-sensitized asthmatic patients in vitro. These results demonstrated that CAPE was able to inhibit cytokine production and proliferation of T cells, making propolis a good candidate for the treatment of allergic disorders (Wang et al., 2010). Draganova-Filipova et al. (2008) reported that the low concentration of propolis (0.001 mg/mL) stimulated T cell immune response, but the highest concentration of propolis (10 mg/mL) showed a cytotoxic effect. They also indicated that propolis and CAPE induced apoptosis in McCoy-Plovdiv cells and upregulation of the p53 expression, which shows the initial DNA damage. In the meantime, researchers stressed the activation of DNA repair since simultaneous detection of proliferating cell nuclear antigen (PCNA) was observed. They reported that propolis induced apoptosis via activating a p53 dependent pathway, while CAPE activated external apoptosis (p53-independent) mechanisms.

The effect of propolis on the immune system in the stress-induced condition was investigated by measuring Th1/Th2 cytokines production by spleen cells from acutely stressed mice. Stress causes the dysfunction of Th1/Th2 balance. Although there was no significant difference in IL-2 basal production and Con A-stimulated spleen cells, significant inhibition was found in IFN-γ basal and Con A-stimulated production compared to the control group. The amounts of IFN-γ basal were 122.51, 81.45, and 70.90 pg/mL for the control, stress-induced, and propolis treated group, respectively. These amounts determined for IFN-γ in Con A-stimulated condition were 15,068, 6296, and 4933 pg/mL, for the control, stress-induced, and propolis treated group, respectively. In the case of Th2 cytokines production, IL-4 basal production was not determined, but in Con A-stimulated cultures, stress inhibited IL-4 production by spleen cells significantly. However, inhibition was not observed in propolis-treated stressed mice (Pagliarone et al., 2009b).

In summary, propolis can display both immunosuppressive and immunostimulant effects related to the concentration of propolis. Furthermore, CAPE can inhibit cytokine production and proliferation of T cells.

Conclusion

In conclusion, in vitro and in vivo studies showed that propolis might modulate the function of different cells of innate and adaptive immunity and enhance the activity of the cells to fight against infectious agents. Propolis and its constituents, especially CAPE and artepillin C, exhibit a notable immunomodulatory effect and inhibitory and/or stimulatory activity on several cells and components of the immune and inflammatory response.

Although propolis and its constituents have an immunomodulatory effect, their mechanism of action is not completely understood. However, recently published meta-analysis reports based on human studies demonstrated sufficient evidence on the positive effect on IL-6, CRP, and TNF-α, indicating that the propolis supplementation exhibits positive effects on chronic inflammation (Shang et al., 2020). Furthermore, since propolis acts as both pro- and antiinflammatory activity depending on concentration, intake period, and experimental condition, it should be considered that inflammatory mediators are a double-edged sword, with a role to fight against infection but also produce damage to the tissues.

References

Abbas, A.K., Lichtman, A.H., Pillai, S., 2010. Cellular and Molecular Immunology, sixth ed. Saunders Elsevier, Philadelphia (Chapter 1,4).

Abel, A.M., Yang, C., Thakar, M.S., Malarkannan, S., 2018. Natural killer cells: development, maturation, and clinical utilization. Front. Immunol. 9, 1–23.

Adachi, T., Yoshikawa, S., Tezuka, H., Tsuji, N.M., Ohteki, T., Karasuyama, H., Kumazawa, T., 2019. Propolis induces Ca2+ signaling immune cells. Biosci. Microbiota, Food Health 38 (4), 141–149.

Afsharpour, F., Hashemipour, S., Haghighian, H.K., Koushan, Y., 2017. Effects of Iranian Propolis on Glycemic Status Inflammatory Factors and Liver Enzyme Levels in Type 2 Diabetic Patients: A Randomized Double-Blind Placebo-Controlled Clinical Trial. Vol. 3, pp. 9–14.

Alqarni, A.M., Niwasabutra, K., Sahlan, M., Fearnley, H., Fearnley, J., Ferro, V.A., Watson, D.G., 2019. Propolis exerts an anti-inflammatory effect on PMA-differentiated THP-1 cells via inhibition of purine nucleoside phosporylase. Meta 9. https://doi.org/10.3390/metabo9040075.

Asgharpour, F., Moghadamnia, A.A., Motallebnejad, M., Nouri, H.R., 2019. Propolis attenuates lipopolysaccharide-induced inflammatory responses through intracellular ROS and NO levels along with downregulation of IL-1β and IL-6 expressions in murine RAW 264.7 macrophages. J. Food Biochem. 43, 1–10.

Bachiega, T.F., Orsatti, C.L., Pagliarone, A.C., Sforcin, J.M., 2012. The effects of propolis and its isolated compounds on cytokine production by murine macrophages. Phytother. Res. 26 (9), 1308–1313.

Bao, Y., Cao, X., 2014. The immune potential and immunopathology of cytokine-producing B cell subsets: a comprehensive review. J. Autoimmun. 55, 10–23.

Bonvehí, J.S., Gutiérrez, A.L., 2011. Antioxidant activity and total phenolics of propolis form the Basque Country (North Eastern Spain). J. Am. Oil Chem. Soc. 88 (9), 1387–1395.

Bueno-Silva, B., Franchin, M., de Freitas Alves, C., Denny, C., Colón, D.F., Cunha, T.M., Alencar, S.M., Napimoga, M.H., Rosalen, P.L., 2016. Main pathways of action of Brazilian red propolis on the modulation of neutrophils migration in the inflammatory process. Phytomedicine 23 (13), 1583–1590.

Chan, G.C., Cheung, K.W., Sze, D.M., 2013. The immunomodulatory and anticancer properties of propolis. Clin. Rev. Allergy Immunol. 44, 262–273.

Conte, F.L., Santiago, K.B., Conti, B.J., Cardoso, E.O., Oliveira, L.P.G., Feltran, G.S., Zambuzzi, W.F., Golim, M.A., Cruz, M.T., Sforcin, J.M., 2021. Propolis from southern Brazil produced by Apis mellifera affects innate immunity by modulating cell marker expression cytokine production and intracellular pathways in human monocytes. J. Pharm. Pharmacol. 73, 135–144.

Conti, B.J., Santiago, K.B., Cardoso, E.O., Freire, P.P., Carvalho, R.F., Golim, M.A., Sforcin, J.M., 2016. Propolis modulates miRNAs involved in TLR-4 pathway, NF-κB activation, cytokine production and in the bactericidal activity of human dendritic cells. J. Pharm. Pharmacol. 68, 1604–1612.

Ding, J., Matsumiya, T., Hayakari, R., Shiba, Y., Kawaguchi, S., Seya, K., Ueno, K., Imaizumi, T., 2021. Daily Brazilian green propolis intake elevates blood artepillin C levels in humans. J. Sci. Food Agric. https://doi.org/10.1002/jsfa.11132.

Draganova-Filipova, M.N., Georgieva, M.G., Peycheva, E.N., Miloshev, G.A., Sarafian, V.S., Peychev, L.P., 2008. Effects of propolis and cape on proliferation and apoptosis of Mccoy-Plovid cell line. Folia Med. 50, 53–59.

Duque, G.A., Descoteaux, A., 2014. Macrophage cytokines: involvement in immunity and infectious diseases. Front. Immunol. 5, 1–12.

Farooqui, T., Farooqui, A.A., 2012. Beneficial effects of propolis on human health and neurological diseases. Front. Biosci. (Elite Ed.) 4, 773–793.

Gao, W., Wu, J., Wei, J., Pu, L., Guo, C., Yang, J., 2014. Brazilian green propolis improves immune function in aged mice. J. Clin. Biochem. Nutr. 55 (1), 7–10.

Girgin, G., Baydar, T., Ledochowski, M., Schennach, H., Bolukbasi, D.N., Sorkun, K., Salih, B., Sahin, G., Fuchs, D., 2009. Immunomodulatory effects of Turkish propolis: changes in neopterin release and tryptophan degradation. Immunobiology 214, 129–134.

Guzmán-Gutiérrez, S., Nieto-Camacho, A., Castillo-Arellano, J.I., Huerta-Salazar, E., Hernández-Pasteur, G., Silva-Miranda, M., Argüello-Nájera, O., Sepúlveda-Robles, O., Espitia, C.I., Reyes-Chilpa, R., 2018. Mexican propolis: a source of antioxidants and anti-inflammatory compounds and isolation of a novel chalcone and ε-caprolactone derivative. Molecules 23, 334–350.

Hu, F., Hepburn, H.R., Li, Y., Chen, M., Radloff, S.E., Daya, S., 2005. Effects of ethanol and water extracts of propolis (bee glue) on acute inflammatory animal models. J. Ethnopharmacol. 100, 276–283.

Inui, S., Hosoya, T., Yoshizumi, K., Sato, H., Kumazawa, S., 2021. Phytochemical and anti-inflammatory properties Senegalese propolis and isolated compounds. Fitoterapia 152. https://doi.org/10.1016/j.fitote.2021.104861.

Janeway Jr., C.A., Travers, P., Walport, M., Shlomchik, M.J., 2001. Principles of innate and adaptive immunity. In: Immunobiology: The Immune System in Health and Disease, fifth ed. Garland Science, New York. Available from: https://www.ncbi.nlm.nih.gov/books/NBK27090/.

Kashiwakura, J.-i., Yoshihara, M., Saitoh, K., Kagohashi, K., Sasaki, Y., Kobayashi, F., Inagaki, I., Kitai, Y., Muromoto, R., Matsuda, T., 2020. Propolis suppresses cytokine production in activated basophils and basophil-mediated skin and intestinal allergic inflammation in mice. Allergol. Int. https://doi.org/10.1016/j.alit.2020.11.005.

Kitamura, H., Saito, N., Fujimoto, J., Nakashima, K., Fujikura, D., 2018. Brazilian propolis ethanol extract and its component kaempferol induce myeloid-derived suppressor cells from macrophages of mice in vivo and in vitro. BMC Complement. Altern. Med. 18 (1), 1–11.

Lacy, P., Stow, J.L., 2011. Cytokine release from innate immune cells: association with diverse membrane trafficking pathways. Blood 118, 9–18.

Li, J., Liu, H., Liu, X., Hao, S., Zhang, Z., Xuan, H., 2021. Chinese poplar propolis inhibits MDA-MB-231 cell proliferation in an inflammatory microenvironment by targeting enzymes of the glycolytic pathway. J Immunol Res. https://doi.org/10.1155/2021/6641341.

Márquez, N., Sancho, R., Macho, A., Calzado, M.A., Fiebich, B.L., Muñoz, E., 2004. Caffeic acid phenethyl ester inhibits T-cell activation by targeting both nuclear factor of activated T-cells and NF-κB transcription factors. J. Pharmacol. Exp. Ther. 308 (3), 993–1001.

Mikami, N., Tani, H., Kawakami, R., Sugimoto, A., Sakaguchi, S., Ikuta, T., 2021. Brazilian green propolis promotes TNFR2 expression on regulatory T cells. Food Sci. Nutr. https://doi.org/10.1002/fsn3.2281.

Miller, C.H., Maher, S.G., Young, H.A., 2009. Clinical use of interferon-gamma. Ann. N.Y. Acad. Sci. 1182, 69–79.

Nagaoka, T., Banskota, A.H., Tezuka, Y., Midorikawa, K., Matsushige, K., Kadota, S., 2003. Caffeic acid phenethyl ester (CAPE) analogues: potent nitric oxide inhibitors from the Netherlands propolis. Biol. Pharm. Bull. 26 (4), 487–491.

Nakamura, R., Nakamura, R., Watanabe, K., Oka, K., Ohta, S., Mishima, S., Teshima, R., 2010. Effects of propolis from different areas on mast cell degranulation and identification of the effective components in propolis. Int. Immunopharmacol. 10, 1107–1112.

Onbas, R., Kazan, A., Nalbantsoy, A., Yesil-Celiktas, O., 2016. Cytotoxic and nitric oxide inhibition activities of propolis extract along with microencapsulation by complex coacervation. Plant Foods Hum. Nutr. 71, 286–293.

Orsatti, C.L., Missima, F., Pagliarone, A.C., Bachiega, T.F., Búfalo, M.C., Araújo Jr., J.P., Sforcin, J.M., 2010. Propolis immunomodulatory action in vivo on Toll-like receptors 2 and 4 expression and on pro-inflammatory cytokines production in mice. Phytother. Res. 24 (8), 1141–1146.

Orsi, R.O., Funari, S.R.C., Soares, A.M.V.C., Calvi, S.A., Oliveira, S.L., Sforcin, J.M., Bankova, V., 2000. Immunomodulatory action of propolis on macrophage activation. J. Venomous Anim. Toxins 6, 205–219.

Ozdal, T., Ceylan, F.D., Eroglu, N., Kaplan, M., Olgun, E.O., Capanoğlu, E., 2019. Investigation of antioxidant capacity bioaccessibility and LC-MS/MS phenolic profile of Turkish propolis. Food Res. Int. 122, 528–536.

Pagliarone, A.C., Orsatti, C.L., Bufalo, M.C., Missima, F., Bachiega, T.F., Junior, J.P., Sforcin, J.M., 2009a. Propolis effects on pro-inflammatory cytokine production and toll-like receptor 2 and 4 expression in stressed mice. Int. Immunopharmacol. 9, 1352–1356.

Pagliarone, A.C., Missima, F., Orsatti, C.L., Bachiega, T.F., Sforcin, J.M., 2009b. Propolis effect on Th1/Th2 cytokines production by acutely stressed mice. J. Ethnopharmacol. 125, 230–233.

Park, J.H., Lee, J.K., Kim, H.S., Chung, S.T., Eom, J.H., Kim, K.A., Chung, S.J., Paik, S.Y., Oh, H.Y., 2004. Immunomodulatory effect of caffeic acid phenethyl ester in Balb/c mice. Int. Immunopharmacol. 4, 429–436.

Piñeros, A.R., de Lima, M.H.F., Rodrigues, T., Gembre, A.F., Bertolini, T.B., Fonseca, M.D., Berretta, A.A., Ramalho, L.N.Z., Cunha, F.Q., Hori, J.I., Bonato, V.L.D., 2020. Green propolis increases myeloid duppressor cells and $CD4^+$ $Foxp3^+$ cells and reduces Th2 inflammation in the lungs after allergen exposure. J. Ethnopharmacol. 252. https://doi.org/10.1016/j.jep.2019.112496, 112496.

Riberio, V.P., Simaro, G.V., Mejia, J.A.A., Arruda, C., Bastos, J.K., 2021. Anti-inflammatory and antinociceptive activities of the hydroalcoholic extract and the volatile fraction of southeastern Brazilian brown propolis. Rev. Bras 31, 59–66.

Ripari, N., Sartori, A.A., da Silva Honorio, M., Conte, F.L., Tasca, K.I., Santiago, K.B., Sforcin, J.M., 2021. Propolis antiviral and immunomodulatory activity: a review and perspectives for COVID-19 treatment. J. Pharm. Pharmacol. https://doi.org/10.1093/jpp/rgaa067.

Russo, A., Longo, R., Vanella, A., 2002. Antioxidant activity of propolis: role of caffeic acid phenethyl ester and galangin. Fitoterapia 73, S21–S29.

Sá-Nunes, A., Faccioli, L.H., Sforcin, J.M., 2003. Propolis: lymphocyte proliferation and IFN-γ production. J. Ethnopharmacol. 87, 93–97.

Sforcin, J.M., 2007. Propolis and the immune system: a review. J. Ethnopharmacol. 113, 1–4.

Sforcin, J.M., Kaneno, R., Funari, S.R.C., 2002. Absence of seasonal effect on the immunomodulatory action of Brazilian propolis on natural killer activity. J. Venomous Anim. Toxins 8 (1), 1–6.

Shang, H., Bhagavathula, A.S., Aldhaleei, W.A., Rahmani, J., Karam, G., Rinaldi, G., Clark, C., Salehisahlabadi, A., Yuan, Q., 2020. Effect of propolis supplementation on C-reactive protein levels and other inflammatory factors: a systematic review and meta-analysis of randomized controlled trials. J. King Saud Univ.-Sci. 32, 1694–1701.

Simões, L.M.C., Gregório, L.E., Silva Filho, A.A., Souza, M.L., Azzolini, A.E.C.S., Bastos, J.K., Lucisano-Valim, Y.M., 2004. Effect of Brazilian green propolis on the production of reactive oxygen species by stimulated neutrophils. J. Ethnopharmacol. 94, 59–65.

Song, H.-Y., Kim, W.S., Han, J.M., Seo, H.S., Lim, S.-T., Byun, E.-B., 2021. Galangin treatment during dendritic cell differentiation confers tolerogenic properties in response to lipopolysaccharide stimulation. J. Nutr. Biochem. 87. https://doi.org/10.1016/j.jnutbio.2020.108524, 108524.

Takeda, K., Nagamatsu, K., Okumura, K., 2018. A water-soluble derivative of propolis augments the cytotoxic activity of natural killer cells. J. Ethnopharmacol. 218, 51–58.

Touzani, S., Embaslat, W., Imtara, H., Kmail, A., Kadan, S., Zaid, H., ElArabi, I., Badiaa, L., Saad, B., 2019. In vitro evaluation of the potential use of propolis as a multitarget therapeutic product: physicochemical properties, chemical composition, and immunomodulatory, antibacterial, and anticancer properties. Biomed. Res. Int. 2019, 1–11.

Veiga, R.S., De Mendonça, S., Mendes, P.B., Paulino, N., Mimica, M.J., Lagareiro Netto, A.A., Lira, I.S., Lopez, B.G.-C., Negrao, V., Marcucci, M.C., 2017. Artepillin C and phenolic compounds responsible for antimicrobial and antioxidant activity of green propolis and Baccharis dracunculifolia DC. J. Appl. Microbiol. 122, 911–920.

Wang, L.C., Lin, Y.L., Liang, Y.C., Yang, Y.H., Lee, J., Yu, H.H., Wu, W.M., Chiang, B.L., 2009. The effect of caffeic acid phenethyl ester on the functions of human-monocyte derived dendritic cells. BMC Immunol. 10, 1–13.

Wang, L.C., Chu, K.H., Liang, Y.C., Lin, Y.L., Chiang, B.L., 2010. Caffeic acid phenethyl ester inhibits nuclear factor- κB and protein kinase B signalling pathways and induces caspase-3 expression in primary human CD4+ T cells. Clin. Exp. Immunol. 160, 223–232.

Wolska, K., Górska, A., Antosik, K., SuŁugowska, K., 2019. Immunomodulatory effects of propolis and its components on basic immune cell functions. Indian J. Pharm. Sci. 81 (4), 575–588.

Xool-Tamayo, J., Chan-Zapata, I., Arana-Argaez, V., Villa-de la Torre, F., Torres-Romero, J.C., Araujo-Leon, J.A., Aguilar-Ayala, F.J., Rejon-Peraza, M.E., Castro-Linares, N.C., Vargas-Coronado, R.F., Cauich-Rodriguez, J.V., 2020. In vitro and in vivo anti-inflammatory properties of Mayan propolis. Eur. J. Inflammation 18, 1–11.

Zakerkish, M., Jenabi, M., Zaeemzadeh, N., Hemmati, A.A., Neisi, N., 2019. The effect of Iranian propolis on glucose metabolism, lipid profile, insulin resistance, renal function, and inflammatory biomarkers in patients with type 2 diabetes mellitus: a randomized double-blind clinical trial. Sci. Rep. 9, 7289. https://doi.org/10.1038/s41598-019-43838-8.

Zhao, R., Zhou, H., Su, S.B., 2013. A critical role of interleukin-1β in the progression of autoimmune diseases. Int. Immunopharmacol. 17, 658–669.

CHAPTER 5

Propolis and gastrointestinal tract diseases

Büşra Gültekin Subaşı[a,b], Tugba Ozdal[c], and Esra Capanoglu[a]
[a]Faculty of Chemical and Metallurgical Engineering, Department of Food Engineering, Istanbul Technical University, Maslak, Istanbul, Turkey
[b]Hafik Kamer Ornek Vocational School, Sivas Cumhuriyet University, Sivas, Turkey
[c]Faculty of Engineering and Natural Sciences, Department of Food Engineering, Istanbul Okan University, Tuzla, Istanbul, Turkey

Contents

Introduction

Propolis, also known as "bee glue," is a resinous, waxy material. Honey bees prepare it by combining their saliva and beeswax with plant exudates such as tree buds, sap flows, stems, roots, and barks from plants in the beehive area. The ultimate aim of propolis is to protect beehives by sealing holes, protecting bees from pests and microorganisms, and providing thermal insulation (Bankova et al., 2000; Simone-Finstrom and Spivak, 2010; Sung et al., 2017). The chemical composition and pharmacological behavior of propolis are influenced by geographical area, plant origin, selection season, bee species, and extraction solvents. Despite its diverse nature, documents show that propolis has been used in folk medicine since 300 BC (Ghisalberti, 1979). Over the last years, the research on propolis has been fastened, and it has

Bee Products and Their Applications in the Food and Pharmaceutical Industries
https://doi.org/10.1016/B978-0-323-85400-9.00013-7

gained the attention of researchers due to its superior antioxidant, antiinflammatory, anticancer, antimicrobial effects, among others.

Propolis is also commonly used to treat various diseases, particularly those affecting the gastrointestinal tract, such as oral mucositis, colitis, gastritis, peptic ulcers, inflammatory bowel disease, and gastrointestinal cancers. From minor disturbances to life-threatening severe, even fatal cases, gastrointestinal tract disorders have a wide range of symptoms that are usually palliated by medication; however, seldomly they target the main cause of the disease besides downgrading the damaged tissues and recruiting. Symptom palliation and/or suppression might cause many undesirable side effects, which eventually might require different medication. Propolis has a significant potential to treat these challenging gastrointestinal tract diseases instead of suppressing their symptoms. As a result, this chapter aims to summarize and criticize current research on the use of propolis and its vital active ingredients in treating gastrointestinal disorders and other associated disorders.

Phytochemical constituents of propolis

The phenolic composition of propolis is variable and complex depending on the factors such as the origin of the sample, which is closely related to the hive's flora, the geographic and climatic characteristics at the site, honey bee species, and botanical and geographical differences. The variations in propolis bioactivity are associated with various phenolic constituents (Ahn et al., 2007; Moreno et al., 2000). Propolis from temperate regions, also called poplar-type propolis, contains mainly phenolics: flavonoids aglycone, aromatic acids, and their esters (Bankova et al., 2000). Propolis from tropical regions contains different phenolics: prenylated *p*-coumaric acid derivatives, flavonoids, benzophenones, lignans, and terpenes (Popova et al., 2011). Propolis is the richest product within bee products for its polyphenolic content as it contains about 10% of flavonoids with 38 different flavonoid structures (Haroun, 2006; Suver, 2008).

The polyphenolic fraction of propolis has much more therapeutic effects than the other components (Erdoğan et al., 2011). The extraction efficiency highly depends on the method, and the most common is solid-liquid extraction with ethanol, methanol, or water. Besides, some other substances can also be used for propolis extraction, such as oil (Finger et al., 2013). However, for fractionation purposes, ethanolic extract of propolis (EEP) is the most common way (Tagliacollo and Orsi, 2011). The ethanolic extract of propolis primarily represents the highest phenolic acid and flavonoid sources (Watanabe et al., 2011). These extracts are widely used for different purposes, such as antioxidant capsules, throat sprays, toothpaste, and cosmetics (Coneac et al., 2008; Sarıkaya et al., 2009).

The biological activity of propolis is mainly associated with the existence of flavonoids. Their biochemical effect can be grouped into four activities: binding to biological polymers, binding to heavy metal ions, catalyzing the electron transfer, and the ability to scavenge free radicals (Coneac et al., 2008).

Pinocembrin is the most abundant flavonoid found in propolis. Pinobanksin and chrysin are also the characteristic compounds present in propolis (Jaganathan and Mandal, 2009). Some other important flavonoids are galangin, kaempferol, quercetin, tectocricin, acacetin, and apigenin (Jaganathan and Mandal, 2009; Suver, 2008). Propolis also includes some phenolic acids such as caffeic acid, benzoic acid, hydroxycinnamic acid, cinnamic acid, coumaric acid, fumaric acid, vanillic acid, hydroxybenzoic acid, and salicylic acid. Flavonoids and phenolics reported showing antimicrobial activity are pinocembrin, galangin, pinobanksin, *p*-coumaric acid, and caffeic acid (Ahn et al., 2007; Coneac et al., 2008; Kumazawa et al., 2004; Marcucci et al., 2001).

Effect of propolis on oral mucositis

Oral mucositis is one of the acute complications of chemotherapy and radiotherapy used in cancer treatment. It takes place after 3–5 days and makes a peak at 7–14 days after the initiation of chemotherapy (Al Ibraheemi and Shamoun, 2016). Radiation-induced oral mucositis in patients with head and neck cancer occurs within 2 weeks and peaks between 2 and 6 weeks after therapy begin (Mallick et al., 2016; Stokman et al., 2006). Oral mucositis develops as an epithelial and subepithelial injury through chemotherapy or radiation treatment (RT) and progresses through five stages: (1) the activation stage; (2) the message generation stage; (3) the signaling and reinforcement stage; (4) the ulceration stage; and (5) the curing stage (Mallick et al., 2016; Stokman et al., 2006). Mucositis is characterized by mild erythema, sensitivity, irritation, edema, discomfort, ulceration, and hemorrhage. Several scoring schemes exist to describe oral mucositis (Trotti et al., 2003). To grade the severity of oral mucositis, these scales typically use a 5-grade grading scheme representing 1 for mild, 2 for moderate, 3 or 4 for severe (Stokman et al., 2006). Oral mucositis affects 40%–76% of patients getting high-dose radiotherapy/chemotherapy (Al Ibraheemi and Shamoun, 2016).

Fifty-six percent of chemotherapy patients have moderate-to-severe oral mucositis (grades 3–4), while roughly 34% of traditional RT patients have moderate-to-severe oral mucositis (grades 3–4) (Mallick et al., 2016; Trotti et al., 2003). Severe oral mucositis may result in localized or systemic infections, trouble in feeding and drinking, losing weight or malnutrition, medication discontinuation, or even mortality (Eslami et al., 2016; Tomaževič and Jazbec, 2013).

Recently, research has focused on treating oral mucositis with propolis due to its beneficial effects, including antiinflammatory, antioxidant, and antimicrobial effects. Propolis contains a high concentration of flavonoids, which may have antiinflammatory properties by inhibiting reactive oxygen or nitrogen compounds (Noronha et al., 2014). It is also economic adjuvant in the treatment of RT-induced oral mucositis (Bolouri et al., 2015). It has been reported that CAPE in propolis has been shown to attenuate radiation-induced pulmonary injury in vivo through its high antioxidant capacity (Yildiz et al., 2008). Moreover, propolis has been known to have clinically critical antimicrobial properties (Pappalardo et al., 1991). When propolis was used as an ingredient in mouthwashes, it protected against oral disease, most likely due to its antimicrobial potency. There was no statistically significant difference in the potency of propolis when used as a gel, mouthwash, or buccal paste (Ali and Abdul Rasool, 2011; Chen et al., 2009; Dodwad and Kukreja, 2011).

In the past couple of years, studies about the effectiveness of propolis on periodontal diseases gained momentum, especially about oral mucositis as a side effect of chemotherapy and radiotherapy for cancer treatments, primarily with in vivo studies (López-Valverde et al., 2021). Propolis was shown to effectively minimize and/or postpone radiation-induced mucositis in an animal study. The effectiveness of ethanolic propolis extract on radiation-induced mucositis in rats was investigated (Ghassemi et al., 2010). Bolouri et al. (2015) performed a randomized, triple-blind clinical trial with 20 patients selected randomly to swish and swallow 15 mL of water-based extract of propolis mouth wash three times a day in the case group ($n = 10$) and 15 mL placebo mouth wash in the control group ($n = 10$) and indicated that water-based extract of propolis efficiently prevented and healed radiotherapy-induced mucositis. Dastan et al. (2020) applied propolis mouthwash to patients suffering from oral mucositis and dysphagia due to head and neck radiotherapy. In this 4-week study, patients were asked to rinse their mouths three times a day with a 15 mL propolis mouthwash. After 2 weeks of practice, significant downgrading for the disease severity was reported for both oral mucositis and dysphagia (Dastan et al., 2020). Similar to these results, Nakao and Ueno (2021) evaluated the efficacy of propolis in a human trial. The patients who were suffering from the side effects of head and neck radiotherapy used propolis in the form of an oral moisturizing gel. The patients were asked to use the oral gel for 37 days properly. A significant decrease in oral pain was observed after applying oral propolis gel by the patients who were seriously suffering from pain due to oral mucositis (Nakao and Ueno, 2021). However, all researchers working on the relation of propolis and oral mucositis agreed on the requirement of large-scale clinical trials to have a solid judgment on the exact treatment effect of propolis.

Effect of propolis on gastrointestinal cancers

Due to multistage and malignancy processes at the cellular, genetic and epigenetic levels might induce carcinogenesis, uncontrolled cell proliferation in living organisms. Cancer of various types and its metastasis is the second leading cause of death worldwide, right after cardiovascular diseases (Ahmad et al., 2021). Among distinctive types of cancer, gastrointestinal system elements such as esophageal, gastric, liver, pancreas, gallbladder, intestine, and rectum are vulnerable to carcinogenesis as the third most common and fatal cancer types in the world specifically in developed and developing countries (National Cancer Institute, 2018). As a well-known actuality, treatment methods using chemotherapy do not always eventuate in ultimate cancer cell inhibition besides having the potential to cause complex side effects, leading to the requirement of developing safer pharmaceuticals/therapeutics (Shi et al., 2011). Notably, plant-based bioactive compounds are still being investigated for their anticarcinogenic properties due to their low toxicity and being generally recognized as safe (Vukovic et al., 2018). Other alternative natural materials containing bioactives may also interfere, prevent, diminish or reverse the carcinogenic process in the cell by modulating the signaling pathways and/or activation of the apoptosis (programmed cell death) (Khan et al., 2007; Lopaczynski and Zeisel, 2001;Vukovic et al., 2018).

Mechanism of action of propolis on cancer cells

Owing to containing more than 300 bioactive compounds, propolis is considered an important and promising natural substance for its anticarcinogenic, antiinflammatory, and apoptosis-inducing properties (Amini-Sarteshnizi et al., 2015; Premratanachai and Chanchao, 2014; Xuan et al.,

2011). In the literature, there are several studies on the bioactive compounds of propolis such as caffeic acid, caffeic acid phenethyl ester (CAPE), 3,5-diprenyl-4-hydroxycinnamic acid (Artepillin C), chrysin, nemorosone, galangin, cardanol, isoliquiritigenin, vestitol, neovestitol, and many other phenolics for their antitumor activities as well as apoptosis and antiinflammatory effects (Kasiotis et al., 2017; Pasupuleti et al., 2017; Waly et al., 2017; Watanabe et al., 2011). These biologically active components of propolis are reported to slow down the proliferation and invasion of cancer cells throughout some particular gastrointestinal cell lines: HCT-15/116, HT-29, CT 26, SW-480/620, WiDr, CaCo2, AGS, KYSE-30, NCI-N78 both in vitro and in vivo (Catchpole et al., 2015; Chiu et al., 2020).

Mechanism of action of propolis bioactive compounds on cancerous cells could be varying. Apoptosis signaling pathways could be modulated throughout TRAIL/DR4/5 coupled with/or FasL/Fas death signaling loops as intrinsically and extrinsically causing increased miRNA expression hence decreased targeted gene of colon cancer (MAPK/Erk5) expression (Kumazaki et al., 2014). As another pathway, G_0/G_1 arrests were induced by stimulated Cip1/p21 gene expression for colon cancer cells due to upregulating the tumor-suppressing genes (Shimizu et al., 2005). Toll-like receptor 4 (TLR-4), glycogen synthase kinase-3 beta, and NF-κB signaling pathways could also be downregulated by the bioactives of propolis (Da Silva et al., 2018). Forced apoptosis could be observed on colon cancer cells by inducing its DNA condensation and diminishing the proliferation rate (Kubina et al., 2015). Besides these pathways, there are few indirect inhibitory effects of propolis on the cancer cells, metastasis, and/or its proliferation. Modulating the glycolytic metabolism of cancerous cells, avoiding the accessibility of glucose as a main energy source by targeted cells, could reduce cell proliferation (Patel, 2015). Blocking the vascular endothelial growth factor production by CAPE from propolis could reduce the metastatic capacity of targeted cancerous cells (CT26) (Liao et al., 2003).

Numerous studies have been conducted in the last decade on the biological activities of varying propolis sources against gastrointestinal carcinoma. Cytotoxicity of Portuguese propolis phenolic extracts on HCT15-colon carcinoma cell lines was investigated. It was indicated that the phenolic extracts performed a high level of cytotoxicity and inhibited the growth of mentioned cell line (Calhelha et al., 2014). Similarly, 11 different flavonoids extracted from propolis were examined against human colon cancer cells (HCT-116). Specifically, luteolin was observed to have a remarkable cytotoxic effect on the cell line when applied as 66.86 μM for 72 h, which was the observed IC_{50} values besides having the highest apoptotic potential, 27.13% after 72h incubation. Out of these 11 flavonoids, luteolin and galangin also exhibited a prooxidative effect on HCT-116 cell lines after 24h of pretreatment (Vukovic et al., 2018).

A fractionated phytochemical extract, which mainly consisted of 3-methyl-4-isopropyl phenol from Jordanian propolis, was found to have anticancer activity throughout apoptosis induction and cell cycle arrest at G0/G1 phase on human colorectal adenocarcinoma cell line with an IC_{50} value of 36.7 μg/mL (Abutaha, 2020). On the other hand, the effect of Iranian propolis on gastric tissue carcinogenesis was investigated in an in vivo study. A significant decrease in the number of tumors, nuclear/cytoplasmic ratio, nuclear dispolarity, Beta-catenin, and Bcl-2 protein expressions were observed. Associated with reducing Bcl-2 apoptotic gene expression, a concurrent increase of Bax proapoptotic gene expression was determined as an indicator of induced apoptosis (Alizadeh et al., 2015).

A commercial fraction of European-type propolis was tested for its antiinflammatory and anticarcinogenicity for colon carcinoma using TNF-α, COX-1, and COX-2 markers for antiinflammatory effect and DLD-1 cell viability. The propolis fraction consisted of high levels of dihydro flavonoids pinocembrin, pinobanksin-3-*O*-acetate, dimethylallyl, benzyl, 3-methyl-3-butenyl caffeates. In addition to these phenolic compounds, chrysin, galangin, and CAPE showed a strong antiproliferative activity on cancer cell lines (Catchpole et al., 2015). Furthermore, HCT-116 colon carcinoma cell line, KYSE-30 esophageal squamous cancer, and NCI-N87 gastric carcinoma were also treated with fractionized propolis and significant antiproliferative activity was determined for these gastro-intestinal cancer lines (Catchpole et al., 2015).

One of the most abundant bioactive compounds of propolis, CAPE was analyzed for its cytotoxic, genotoxic and proapoptotic potential on neoplastic gastrointestinal cell lines; HCT-116, HT29, and AGS. Significant cytotoxic, genotoxic and proapoptotic activities of CAPE against these gastrointestinal carcinoma cell lines were reported having a synergetic effect with Irinotecan (CPT-11), a drug used for gastrointestinal cancer treatment, and its metabolite, SN-38 (Gajek et al., 2020). On the other hand, different types of extracts from propolis might have varying activities on carcinoma cells. Bioactive extracts of Portuguese propolis were prepared in three different solvents; hexane, chloroform, and ethanol and their antitumor activities were investigated in an in vitro study. Tumor cell viability, proliferation activity, cell death, and metabolism with glycolytic activity were examined.

Among all types of extracts, chloroform extract was observed to have the highest phenolic content as well as the most active inhibitor in terms of cell viability besides inducing significant reduction for glycolytic metabolism of tumor cells (Valença et al., 2013). In another study, ethanolic extract of Polish propolis was found to have significant antitumor and antiproliferative activities on HCT-116 cell lines for colon cancer (Kubina et al., 2015). Ethanol and water extracts of Brazilian green propolis were also investigated for their effects on the postinitiation phase of inflammation-related colon tumorigenesis in rats. Inflammation-related inducible nitric oxide synthase expression, alpha tumor necrotic factor, kappa B nuclear factor, and glutathione peroxidase-2 decreased with the treatment of rat colorectal tumor by ethanolic propolis extract (Doi et al., 2017).

Research on development of pharmaceutics

Studies aim to develop propolis-based drugs and pharmaceuticals against cancerous cells, mostly focus on encapsulation of desired active substance with appropriate coating methods and agents for targeted drug delivery with controlled release. Phenolic extracts of propolis were encapsulated as nano-in-microencapsulates using chitosan-based wall material, and the capsules were used to treat the human HCT-116 cell line. Results demonstrated that the cytotoxicity effect of encapsulated propolis extract was three times higher than the control sample, free propolis, on tumor cell lines (Elbaz et al., 2016). Due to their potential antigastrointestinal cancer activity, New Zealand propolis resin and tincture were encapsulated within three cyclodextrin structures; γ-CD, α-CD, and β-CD and were applied against four human gastrointestinal cancer cell lines; DLD-1, HCT-116, KYSE-30, and NCI-N87. In a time-dependent manner, encapsulated complexes inhibited the proliferation and exhibited strong antiinflammatory and antioxidant activities on all targeted cell lines (Catchpole et al., 2018). In another study, propolis extract was encapsulated as double layered hydroxide nanoparticles to investigate its inhibitory activity on proapoptotic gene expression of HT-29 cell line. Encapsulated propolis extract treatment significantly increased Bax proapoptotic gene expression while decreased Bcl-2 antiapoptotic gene expression of HT-29 cell line compared to uncoated propolis extract (Azarshinfam et al., 2021).

Antiulcer activity of propolis

Peptic ulcers are among the major gastrointestinal diseases, including gastric and duodenal ulcers (Hopkins et al., 1996). Basically, ulcers can be defined as tissue injury due to the unbalance between highly acidic proteolytic excretions and gastrointestinal mucosa, which is expected to protect the gastrointestinal tissue (Hunt et al., 2015). The primary factors to trigger this unbalance are long-term ingestion of nonsteroidal antiinflammatory drugs (NAID) activity of *Helicobacter pylori*, excessive psychological stress, alcohol consumption, and smoking (Sidahmed et al., 2016). Antisecretory drugs widely used to cure, diminish and control acidic proteolytic excretions are either type-2 histamine receptor antagonist or/and proton pump inhibitors. On the other hand, different narrow-spectrum antibiotics are used against *H. pylori* activities (Shapla et al., 2018). Nevertheless, numerous side effects are inevitable coupled with these drugs and therapeutics, such as vitamin and mineral deficiency, hypersensitivity, arrhythmia, impotence, diarrhea, gynecomastia, bone fractions, hypergastrinemia, and gastric cancer (Da Silva et al., 2018). To avoid undesired side effects of these therapeutics, alternative natural bioactive sources have been investigated in the last decade, and propolis is one of the most attractive compounds among them. Particularly, the bioactive compounds of propolis such as artepillin C, CAPE, and galangin are considered to be the most well-known antagonistic agents to be used for their antiulcer activity.

Mechanism of action of propolis on gastric ulcers

Based on the available literature, these natural agents affect the mechanisms such as vascular endothelial growth factor, macrophage apoptosis, vacuolating cytotoxin A, peptidyl deformylase, cyclooxygenase-2, ornithine decarboxylase, xanthine oxidase with their antiangiogenic, antiapoptotic, and antienzymatic activities to avoid chronic duodenal ulcer, gastric inflammation as well as gastric carcinoma (Baltas et al., 2016; Cui et al., 2013; Shapla et al., 2018).

Recently, numerous studies have been conducted using propolis from different regions of the world to investigate their protective and curing potentials on the gastric ulcer. Besides in vitro studies, several in vivo studies also exhibited the activity of propolis on the ulcerous zone, which was forced to occur with different external stresses. Aqueous extract of propolis was studied against an indomethacin-induced gastric ulcer in irradiated and nonirradiated rats. Reduction in the acid secretion and peptic activity, increased mucin excretion and mucosal prostaglandin E_2, suppression for tumor necrosis factor-alpha, and interleukin-1 beta were observed with the treatment of propolis extract even in irradiated rats (El-Ghazaly et al., 2011). Antiulcer properties of triterpenoid fraction (consisting of lanosterol, α-amyrin, 28-norolean-12-en-3-ol, cycloartenol, 3-epi-α-amyrin, lupeol, and 24-methylenecycloartenol) obtained from Cameroon propolis were investigated on ethanol/HCl induced lesion in rats. Ulcerous zones were significantly reduced (47.25%), and mucus production increased by 25% with a 50 mg/kg propolis extract dose. At the same time, for triterpene fraction treatment, ulcer lesions decreased by 53.18% besides the significant induction of mucus production (Carol et al., 2017). The antiulcer potential of various bioactive isolates in Brazilian green propolis was evaluated against gastritis and gastric ulcer. An application of 30 mg/kg artepillin C, drupanin, aromadendrin-4′-*O*-methyl-ether, and kaempferide prevented ethanol/HCl and indomethacin-induced ulcerous zones. According to the results, aromadendrin-4′-*O*-methyl-ether and kaempferide increased the mucin secretion significantly, and artepillin C reduced the TNF amount. Furthermore, all extracts decreased the total acidity, pepsin activity, and total volume of gastric juice in rats (Costa et al., 2018). The same researchers performed another study focusing on the antioxidant properties and gastroprotective and gastric healing activity of artepillin C and the hydroalcoholic extract of green propolis in rats. Healing of acetic acid-induced chronic ulcer was accelerated by 71% with a daily treatment of 300 mg/kg green propolis extract besides inducing mucin secretion and reducing myeloperoxidase activity in the ulcer lesion. Artepillin C was the most efficient bioactive compound for this healing process, having the highest contribution to green propolis (Costa et al., 2020).

In another study, aspirin-induced gastric mucosa damage was treated with Indonesian propolis in rats. A significant reduction in the damaged zone was observed for propolis administrated rats than the control group (Achadiyani et al., 2020). Hydroalcoholic extracts of red propolis and its isolated compounds vestitol, neovestitol, methylvestitol, medicarpin, and oblongifolin AB were treated for gastric ulcer in mice. Red propolis protected gastric mucosa against ethanol/HCl induced damage and increased mucin production with a prostaglandin secretion dependent manner. On the other hand, methylvestitol and medicarpin were found to have essential roles on the bioactivity of propolis (Boeing et al., 2021). Ethanolic extract of Chihuahua propolis with an application dose of 50 mg/kg was demonstrated to have a gastroprotective effect by decreasing the damage lesions and histological injuries, maintaining the mucosal secretion, and reducing thiobarbituric acid reactive substance and myeloperoxidase activity (Ruiz-Hurtado et al., 2021).

Efficacy of propolis on *H. pylori*

Ethanolic extracts of 15 different propolis were studied for their anti-*H. pylori* and antiurease activities in an in vitro study. The strain J99 of *H. pylori* active zones were significantly diminished with all applied extracts. The authors proposed urease inhibitory activity for *H. pylori* with an application dose ranging between 0.26 and 1.152 mg/mL (IC_{50}) (Baltas et al., 2016). The synergetic effect of propolis extract with an antibiotic (clarithromycin) on *H. pylori* and its gastrointestinal ailments was studied. Synergetic and/or additional activities of propolis and clarithromycin were observed to have the potential to control *H. pylori*-induced gastroduodenal diseases (Nostro et al., 2006). *H. pylori* peptide deformylase enzyme is a vital tool for the survival of the organism, and hence, this enzyme is considered a promising target for anti-*H. pylori* therapeutics. CAPE was indicated to be a significant competitive inhibitor against peptide deformylase with an IC_{50} value of 4.02 μM, and its blockage activity for substrate entrance was postulated as a promising tool for further studies on drug development (Cui et al., 2013). In another study, four major bioactive phenolics of propolis (chrysin, pinocembrin, galangin, and CAPE) were extracted to examine their anti-*H. pylori* effectiveness in vitro. Phenolics presented a minimum inhibitory concentration in a range of 256–1024 μg/mL for two different strains of *H. pylori* (ATCC 43504 and J99), while CAPE was reported as the most active inhibitor among all (Romero et al., 2019).

The effect of hydroalcoholic extract of propolis and its isoflavonoid formononetin was investigated on a drug-induced (indomethacin) ulcer in rats as well as their anti-*H. pylori* activities. It was reported that propolis extract and isoflavonoid formononetin showed a significant synergistic effect on *H. pylori* viability in vivo, and ethanol and indomethacin-induced ulcers were reduced (de Mendonça et al., 2020). In a recent study, Korean propolis was tested on *H. pylori*-induced human adenocarcinoma gastric cell line. Due to the suppression of mitogen-activated protein kinases and nuclear factor kappa B pathway by propolis pretreatment in a dose-dependent manner, significant decreases in *H. pylori*-induced proinflammatory interleukins, tumor necrosis factor-alpha, cyclooxygenase-2, and nitric oxide synthase were observed (Song et al., 2020).

Effect of propolis on inflammatory bowel disease

Noninfectious chronic inflammation is one of the most frequent diseases among all other gastrointestinal tract morbidities, also known as inflammatory bowel disease (Soleimani et al., 2021). "Crohn's disease" and "ulcerative colitis" are subtypes of inflammatory bowel disease, and Crohn's disease typically affects any zone of the gastrointestinal tract. Still, ulcerative colitis is a malady usually for the colon mucosal layer and rectum (Abraham and Cho, 2009). Extensive suffering from inflammatory bowel disease could damage intestinal cells, induce malabsorption, ulceration, bleeding, and many other serious problems. The contribution of environmental conditions to the mucosal immune system could interfere with the progress or the regression of illness even though the etiology and pathogenesis of inflammatory bowel disease have not been completely revealed (Knights et al., 2013).

Since synthetic therapeutics and pharmaceuticals mainly focus on curing the anguished symptoms, alternative natural compounds are of emerging interest to propose a better treatment with non or diminished undesirable side effects (Santana et al., 2017; Wehkamp et al., 2016). With this purpose, propolis was considered a potential candidate to treat inflammatory bowel disease due to its scrutinized positive effects on gut microbiome regulation, inflammatory pathways, and immune response systems (Franchin et al., 2016; Al Hariri, 2019; Soleimani et al., 2021; Xue et al., 2019).

Since 1979, the date on which the first study was conducted about the curative effect of propolis on ulcerative colitis (Danø et al., 1979), researchers have been investigating for efforts to clarify the remedial potential of propolis on inflammatory bowel disease. The inhibitory action mechanism of propolis and its individual phenolic compounds on inflammatory bowel disease has been studied for a couple of decades. The studies on the molecular mechanism of CAPE indicated that it performs a selective inhibitory effect on NF-κB signaling pathways throughout degrading IκB and blocking NF-κB binding to DNA besides inflammation-related pathways of Phosphatidylinositol 3-kinase (PI3K) and monophosphate-activated protein kinase. These blocking and inhibitory actions of CAPE resulted in a change in the disease activity index (DAI) score level and proinflammatory cytokines, colonic myeloperoxidase, and epithelial barrier function (Cho et al., 2014; Tambuwala et al., 2018; Tolba et al., 2016).

Furthermore, CAPE decreased the levels of TNF-α, IL-1β, IFN-γ, and IL-6 in terms of antiinflammatory cytokine activity and induced caspase-3 expression in CD4+ T cells (Khan et al., 2018). Decreased expression for TNF-α, IL-1β, and IL-6 but increment for TGF-β gene expression levels were reported for pinocembrin, a propolis flavonoid, inhibiting phosphorylation of cytokines upstream genes such as IκBα, JNK, ERK1/2, and p38MAPK (Soromou et al., 2012).

Many in vivo and in vitro studies have been conducted, particularly in recent years. All studies focused on the healing effects of propolis over the known mechanisms of inflammatory bowel disease and aimed to improve the life quality of individuals who are suffering from this disorder (Miryan et al., 2020). Hydroalcoholic extract of Brazilian green propolis was used to evaluate its effect on dextran sodium sulfate-induced colitis in mice. The treatment significantly reduced macroscopic colonic injury and mucosal damage as well as re-induced mucin secretion levels. Furthermore, propolis normalized SOD activity and GSH levels in treated mice (Mariano et al., 2018). In another in vivo study, Chinese and Brazilian propolis were evaluated for their healing potential on dextran sodium sulfate-induced colitis in rats. Colitis activity index and colonic tissue damage were observed to be decreased while the resistance against induced colonic oxidative stress increased coupled with diminished malonaldehyde and increased T-AOC levels. Colonic apoptosis was significantly reduced by suppressing IL-1β, IL-6, and MCP-1 inflammatory markers (Wang et al., 2018). The effect of Egyptian propolis was also investigated on acetic acid-induced ulcerative colitis in rats. It was indicated that the ulcerative index, lesion score, oxidative markers, and inflammatory mediators were decreased and the enzymatic activity of COX-1 and COX-2 was inhibited (Atta et al., 2019). In another dextran sodium sulfate-induced colitis case study in mice, Brazilian propolis was used to investigate its effect on intestinal barrier defect and inflammation. Propolis application reduced the weight loss, colon shortening, and TJ protein expression; suppressed the expression of TNF-α, IL-6, and IL-17a inflammatory markers and increased the plasma level of lipopolysaccharide-binding protein (Shimizu and Suzuki, 2019). The protective effect of galangin on induced ulcerative colitis was demonstrated with an in vitro study. The expression levels of TNF-α and IL-6 decreased with galangin exposure in a concentration-dependent manner; hence, the severity of colitis was significantly decreased.

Furthermore, antiinflammatory cytokine (IL-10) levels and SOD were increased, whereas myeloperoxidase, nitrites, and TBARS values were decreased (Sangaraju et al., 2019). In another induced colitis (with dextran sodium sulfate) in vivo study, the effect of Korean propolis was investigated on the severity of illness. According to the assessments of body and spleen weight and colonic length, the severity of colitis was significantly decreased after propolis implementation. Inflammatory cytokine level, loss of goblet cells, and inflammatory markers of antibody-reactivity were induced to decrease with propolis treatment (Hwang et al., 2020).

Conclusion

This chapter is a compilation of the evidence reported by several researchers regarding the preventive and/or treatment effect of propolis and/or its bioactives against gastrointestinal disorders, including oral mucositis, gastrointestinal cancers, ulcerative colitis, gastrointestinal cancers, gastric ulcers, and inflammatory bowel disease. According to the current literature, the positive effects of propolis on the treatment of gastrointestinal diseases are due to its bioactive compounds with antioxidant and antiinflammatory activities. The underpinning mechanism of action is regulated by the repression of certain transcriptional factors and associated proteins. Numerous clinical trials have shown that propolis and its associated compounds are helpful in the treatment of gastrointestinal diseases. Nevertheless, just a few clinical trials have been conducted to demonstrate their efficacy and effectiveness against human ulcers and other pathologies. Future research should concentrate on the possible role of propolis and its constituents, either alone or in conjunction with ongoing conventional therapy, to treat gastrointestinal diseases in humans.

References

Abraham, C., Cho, J.H., 2009. Inflammatory bowel disease. N. Engl. J. Med. 361, 2066–2078.

Abutaha, N., 2020. Apoptotic potential and chemical composition of jordanian propolis extract against different cancer cell lines. J. Microbiol. Biotechnol. 30 (6), 893–902. https://doi.org/10.4014/jmb.1905.05027.

Achadiyani, A., Laksmi, A., Girawan, D., 2020. Propolis of Trigona spp. protects mucosa from aspirin-induced gastric mucosal damage in rats. Althea Med. J. 7 (1), 11–15. https://doi.org/10.15850/amj.v7n1.1677.

Ahmad, F.B., Cisewski, J.A., Miniño, A., Anderson, R.N., 2021. Provisional mortality data — United States, 2020. MMWR Morb. Mortal. Wkly Rep. 70 (14), 519–522. https://doi.org/10.15585/mmwr.mm7014e1.

Ahn, M., Kumazawa, S., Usui, Y., Nakamura, J., Matsuka, M., Zhu, F., Nakayama, T., 2007. Antioxidant activity and constituents of propolis collected in various areas of China. Food Chem. 101, 1383–1392.

Al Hariri, M., 2019. Immune's - boosting agent: immunomodulation potentials of propolis. J. Fam. Community Med. 26 (1), 57–60. https://doi.org/10.4103/jfcm.JFCM.

Al Ibraheemi, A.A., Shamoun, S., 2016. Incidence and risk factors of oral mucositis in patients with breast cancer who receiving chemotherapy in Al-Bashir hospital. Int. J. Hematol. Oncol. Stem Cell Res. 10 (4), 217–223.

Ali, H.S., Abdul Rasool, B.K., 2011. Propolis buccal paste in treatment of aphthous ulceration: formulation and clinical evaluation. Asian J. Pharm. Clin. Res. 4 (4), 29–33.

Alizadeh, A.M., Afrouzan, H., Dinparast-Djadid, N., Sawaya, A.C.H.F., Azizian, S., Hemmati, H.R., Mohagheghi, M.A., Erfani, S., 2015. Chemoprotection of MNNG-initiated gastric cancer in rats using Iranian propolis. Arch. Iran. Med. 18 (1), 18–23.

Amini-Sarteshnizi, N., Mobini-Dehkordi, M., Khosravi-Farsani, S., Teimori, H., 2015. Anticancer activity of ethanolic extract of propolis on AGS cell line. J. HerbMed Pharmacol. 4 (1), 29–34.

Atta, A.H., Mouneir, S.M., Nasr, S.M., Sedky, D., Mohamed, A.M., Atta, S.A., Desouky, H.M., 2019. Phytochemical studies and anti-ulcerative colitis effect of Moringa oleifera seeds and Egyptian propolis methanol extracts in a rat model. Asian Pac. J. Trop. Biomed. 9 (3), 98–108. https://doi.org/10.4103/2221-1691.254603.

Azarshinfam, N., Tanomand, A., Soltanzadeh, H., Rad, F.A., 2021. Evaluation of anticancer effects of propolis extract with or without combination with layered double hydroxide nanoparticles on Bcl-2 and Bax genes expression in HT-29 cell lines. Gene Rep. 23 (September 2020), 101031. https://doi.org/10.1016/j.genrep.2021.101031.

Baltas, N., Karaoglu, S.A., Tarakci, C., Kolayli, S., 2016. Effect of propolis in gastric disorders: inhibition studies on the growth of helicobacter pylori and production of its urease. J. Enzyme Inhib. Med. Chem. 31, 46–50. https://doi.org/10.1080/14756366.2016.1186023.

Bankova, V.S., de Castro, S.L., Marcucci, M.C., 2000. Propolis: recent advances in chemistry and plant origin. Apidologie 31 (1), 3–15.

Boeing, T., Mejía, J.A.A., Ccana-Ccapatinta, G.V., Mariott, M., Melo Vilhena de Andrade Fonseca Da Silva, R.D.C., de Souza, P., Mariano, L.N.B., Oliveira, G.R., da Rocha, I.M., da Costa, G.A., de Andrade, S.F., da Silva, L.M., Bastos, J.K., 2021. The gastroprotective effect of red propolis extract from northeastern Brazil and the role of its isolated compounds. J. Ethnopharmacol. 267 (October), 2020. https://doi.org/10.1016/j.jep.2020.113623.

Bolouri, A.J., Pakfetrat, A., Tonkaboni, A., Aledavood, S.A., Najafi, M.F., Delavarian, Z., Shakeri, M.T., Mohtashami, A., 2015. Preventing and therapeutic effect of propolis in radiotherapy induced mucositis of head and neck cancers: a triple-blind, randomized, placebo-con- trolled trial. Iran. J. Cancer Prev. 8 (15), 4019.

Calhelha, R.C., Falcão, S., Queiroz, M.J.R.P., Vilas-Boas, M., Ferreira, I.C.F.R., 2014. Cytotoxicity of portuguese propolis: the proximity of the in vitro doses for tumor and normal cell lines. Biomed. Res. Int. 2014. https://doi.org/10.1155/2014/897361.

Carol, D., Ngenge, T., Emmanuel, T., Joseph, M., Popova, M., Bankova, V., Vernyuy, T., 2017. GC-MS characterization and antiulcer properties of the triterpenoid fraction from Propolis of the north west region of Cameroon. J. Sci. Res. Rep. 15 (4), 1–18. https://doi.org/10.9734/jsrr/2017/35377.

Catchpole, O., Mitchell, K., Bloor, S., Davis, P., Suddes, A., 2015. Antiproliferative activity of New Zealand propolis and phenolic compounds vs human colorectal adenocarcinoma cells. Fitoterapia 106, 167–174. https://doi.org/10.1016/j.fitote.2015.09.004.

Catchpole, O., Mitchell, K., Bloor, S., Davis, P., Suddes, A., 2018. Anti-gastrointestinal cancer activity of cyclodextrin-encapsulated propolis. J. Funct. Foods 41 (December 2017), 1–8. https://doi.org/10.1016/j.jff.2017.12.023.

Chen, J.L., Fan, X.H., Ma, L.L., 2009. A clinical study of efficacy of propolis treatment for recurrent oral aphthae. Cap. Med. 16 (18), 43–44.

Chiu, H.F., Han, Y.C., Shen, Y.C., Golovinskaia, O., Venkatakrishnan, K., Wang, C.K., 2020. Chemopreventive and chemotherapeutic effect of propolis and its constituents: a mini-review. J. Cancer Prev. 25 (2), 70–78. https://doi.org/10.15430/jcp.2020.25.2.70.

Cho, M.S., Park, W.S., Jung, W.K., Qian, Z.J., Lee, D.S., Choi, J.S., Lee, D.Y., Park, S.G., Seo, S.K., Kim, H.J., Won, J.Y., Yu, B.C., Choi, I.W., 2014. Caffeic acid phenethyl ester promotes anti-inflammatory effects by inhibiting MAPK and NF-κB signaling in activated HMC-1 human mast cells. Pharm. Biol. 52 (7), 926–932. https://doi.org/10.3109/13880209.2013.865243.

Coneac, G., Gafiţanu, E., Hădărugă, D.I., Hădărugă, N.G., Pînzaru, I.A., Bandur, G., Gruia, A., 2008. Flavonoid contents of propolis from the west side of Romania and correlation with the antioxidant activity. Chem. Bull. "POLITEHNICA" Univ. (Timişoara) 53 (67), 1–2.

Costa, P., Almeida, M.O., Lemos, M., Arruda, C., Casoti, R., Somensi, L.B., Boeing, T., Mariott, M., da Silva, R.C.M.V.A.F., Stein, B.P., de Souza, P., Dos Santos, A.C., Bastos, J.K., da Silva, L.M., de Andrade, S.F., 2018. Artepillin C, drupanin, aromadendrin-4′-O-methyl-ether

and kaempferide from Brazilian green propolis promote gastroprotective action by diversified mode of action. J. Ethnopharmacol. 15 (226), 82–89. https://doi.org/10.1016/j.jep.2018.08.006.
Costa, P., Somensi, L.B., da Silva, R.C.M.V.A.F., Mariano, L.N.B., Boeing, T., Longo, B., Perfoll, E., de Souza, P., Gushiken, L.F.S., Pellizzon, C.H., Rodrigues, D.M., Bastos, J.K., de Andrade, S.F., da Silva, L.M., 2020. Role of the antioxidant properties in the gastroprotective and gastric healing activity promoted by Brazilian green propolis and the healing efficacy of Artepillin C. Inflammopharmacology 28 (4), 1009–1025. https://doi.org/10.1007/s10787-019-00649-7.
Cui, K., Lu, W., Zhu, L., Shen, X., Huang, J., 2013. Caffeic acid phenethyl ester (CAPE), an active component of propolis, inhibits helicobacter pylori peptide deformylase activity. Biochem. Biophys. Res. Commun. 435 (2), 289–294. https://doi.org/10.1016/j.bbrc.2013.04.026.
Da Silva, L.M., De Souza, P., Jaouni, S.K.A., Harakeh, S., Golbabapour, S., De Andrade, S.F., 2018. Propolis and its potential to treat gastrointestinal disorders. Evid. Based Complement. Alternat. Med. 2018. https://doi.org/10.1155/2018/2035820.
Danø, P., Møller, E.H., Jarnum, S., 1979. Effect of the natural product propolis on ulcerative colitis and Crohn's disease. Ugeskr. Laeger 141 (28), 1888–1890.
Dastan, F., Ameri, A., Dodge, S., Hamidi Shishvan, H., Pirsalehi, A., Abbasinazari, M., 2020. Efficacy and safety of propolis mouthwash in management of radiotherapy induced oral mucositis; a randomized, double blind clinical trial. Rep. Pract. Oncol. Radiother. 25 (6), 969–973. https://doi.org/10.1016/j.rpor.2020.09.012.
de Mendonça, M.A.A., Ribeiro, A.R.S., de Lima, A.K., Bezerra, G.B., Pinheiro, M.S., de Albuquerque-Júnior, R.L.C., Gomes, M.Z., Padilha, F.F., Thomazzi, S.M., Novellino, E., Santini, A., Severino, P., Souto, E.B., Cardoso, J.C., 2020. Red propolis and its dyslipidemic regulator formononetin: evaluation of antioxidant activity and gastroprotective effects in rat model of gastric ulcer. Nutrients 12 (10), 1–17. https://doi.org/10.3390/nu12102951.
Dodwad, V., Kukreja, B., 2011. Propolis mouthwash: a new beginning. J. Indian Soc. Periodontol. 15 (2), 121–125.
Doi, K., Fujioka, M., Sokuza, Y., Ohnishi, M., Gi, M., Takeshita, M., Kumada, K., Kakehashi, A., Wanibuchi, H., 2017. Chemopreventive action by ethanol-extracted brazilian green propolis on post-initiation phase of inflammation-associated rat colon tumorigenesis. In Vivo 31 (2), 187–197. https://doi.org/10.21873/invivo.11044.
Elbaz, N.M., Khalil, I.A., Abd-Rabou, A.A., El-Sherbiny, I.M., 2016. Chitosan-based nano-in-microparticle carriers for enhanced oral delivery and anticancer activity of propolis. Int. J. Biol. Macromol. 92, 254–269. https://doi.org/10.1016/j.ijbiomac.2016.07.024.
El-Ghazaly, M.A., Rashed, R.R.A., Khayyal, M.T., 2011. Anti-ulcerogenic effect of aqueous propolis extract and the influence of radiation exposure. Int. J. Radiat. Biol. 87 (10), 1045–1051. https://doi.org/10.3109/09553002.2011.602378.
Erdoğan, S., Ateş, B., Durmaz, G., Yılmaz, İ., Seçkin, T., 2011. Pressurized liquid extraction of phenolic compounds from Anatolia propolis and their radical scavenging capacities. Food Chem. Toxicol. 49, 1592–1597.
Eslami, H., Pouralibaba, F., Falsafi, P., Bohluli, S., Najati, B., Negahdari, R., Ghanizadeh, M., 2016. Efficacy of Hypozalix spray and propolis mouthwash for prevention of chemotherapy-induced oral mucositis in leukemic patients: a double-blind randomized clinical trial. J. Dent. Res. Dent. Clin. Dent. Prospects 10 (4), 226–233.
Finger, D., Machado, C.S., Torres, Y.R., Quináia, S.P., Thomaz, A.C.G., Gobbo, A.R., Monteiro, M.C., Ferreira, A.G., Sawaya, A.C.H.F., Eberlin, M.N., 2013. Antifungal bioassay-guided fractionation of an oil extract of propolis. J. Food Qual. 36 (5), 291–301.
Franchin, M., Colón, D.F., Da Cunha, M.G., Castanheira, F.V.S., Saraiva, A.L.L., Bueno-Silva, B., Alencar, S.M., Cunha, T.M., Rosalen, P.L., 2016. Neovestitol, an isoflavonoid isolated from Brazilian red propolis, reduces acute and chronic inflammation: involvement of nitric oxide and IL-6. Sci. Rep. 6, 1–12. https://doi.org/10.1038/srep36401.

Gajek, G., Marciniak, B., Lewkowski, J., Kontek, R., 2020. Antagonistic effects of CAPE (a component of Propolis) on the cytotoxicity and genotoxicity of irinotecan and SN38 in human gastrointestinal cancer cells in vitro. Molecules 25 (685).

Ghassemi, L., Zabihi, E., Mahdavi, R., Seyedmajidi, M., Akram, S., Motallebnejad, M., 2010. The effect of ethanolic extract of propolis on radiation induced mucositis in rats. Saudi Med. J. 31 (6), 622–626.

Ghisalberti, E.L., 1979. Propolis: a review. Bee World 60 (2), 59–84.

Haroun, M.I., 2006. Türkiye'de üretilen bazı çiçek ve salgı ballarının fenolik asit ve flavonoid profilinin belirlenmesi (Doctoral dissertation). Ankara Üniversitesi, Gıda Mühendisliği Anabilim Dalı, Fen Bilimleri Enstitüsü, ANKARA.

Hopkins, R.J., Girardi, L.S., Turney, E.A., 1996. Relationship between helicobacter pylori eradication and reduced duodenal and gastric ulcer recurrence: a review. Ann. Fundeni Hospital 110, 1244–1252.

Hunt, R.H., Camilleri, M., Crowe, S.E., El-Omar, E.M., Fox, J.G., Kuipers, E.J., Malfertheiner, P., McColl, K.E.L., Pritchard, D.M., Rugge, M., Sonnenberg, A., Sugano, K., Tack, J., 2015. The stomach in health and disease. Gut 64 (10), 1650–1668. https://doi.org/10.1136/gutjnl-2014-307595.

Hwang, S., Hwang, S., Jo, M., Lee, C.G., Rhee, K.J., 2020. Oral administration of korean propolis extract ameliorates dss-induced colitis in balb/c mice. Int. J. Med. Sci. 17 (13), 1984–1991. https://doi.org/10.7150/ijms.44834.

Jaganathan, S.K., Mandal, M., 2009. Antiproliferative effects of honey and of ıts polyphenols: a review. J. Biomed. Biotechnol. 1, 1–13.

Kasiotis, K.M., Anastasiadou, P., Papadopoulos, A., Machera, K., 2017. Revisiting greek propolis: chromatographic analysis and antioxidant activity study. PLoS One 12 (1), 1–27. https://doi.org/10.1371/journal.pone.0170077.

Khan, N., Afaq, F., Mukhtar, H., 2007. Apoptosis by dietary factors: the suicide solution for delaying cancer growth. Carcinogenesis 28 (2), 233–239. https://doi.org/10.1093/carcin/bgl243.

Khan, M.N., Lane, M.E., McCarron, P.A., Tambuwala, M.M., 2018. Caffeic acid phenethyl ester is protective in experimental ulcerative colitis via reduction in levels of pro-inflammatory mediators and enhancement of epithelial barrier function. Inflammopharmacology 26 (2), 561–569. https://doi.org/10.1007/s10787-017-0364-x.

Knights, D., Lassen, K.G., Xavier, R.J., 2013. Advances in inflammatory bowel disease pathogenesis: linking host genetics and the microbiome. Gut 62 (10), 1505–1510. https://doi.org/10.1136/gutjnl-2012-303954.

Kubina, R., Kabała-Dzik, A., Dziedzic, A., Bielec, B., Wojtyczka, R.D., Bułdak, R.J., Wyszyńska, M., Stawiarska-Pięta, B., Szaflarska-Stojko, E., 2015. The ethanol extract of polish propolis exhibits anti-proliferative and/or pro-apoptotic effect on HCT 116 colon cancer and Me45 malignant melanoma cells in vitro conditions. Adv. Clin. Exp. Med. 24 (2), 203–212. https://doi.org/10.17219/acem/31792.

Kumazaki, M., Shinohara, H., Taniguchi, K., Yamada, N., Ohta, S., Ichihara, K., Akao, Y., 2014. Propolis cinnamic acid derivatives induce apoptosis through both extrinsic and intrinsic apoptosis signaling pathways and modulate of miRNA expression. Phytomedicine 21 (8–9), 1070–1077. https://doi.org/10.1016/j.phymed.2014.04.006.

Kumazawa, S., Hamasaka, T., Nakayama, T., 2004. Antioxidant activity of propolis of various geographic origins. Food Chem. 84, 329–339.

Liao, H.F., Chen, Y.Y., Liu, J.J., Hsu, M.L., Shieh, H.J., Liao, H.J., Shieh, C.J., Shiao, M.S., Chen, Y.J., 2003. Inhibitory effect of Caffeic acid phenethyl ester on angiogenesis, tumor invasion, and metastasis. J. Agric. Food Chem. 51, 7907–7912.

Lopaczynski, W., Zeisel, S.H., 2001. Antioxidants, programmed cell death, and cancer. Nutr. Res. 21 (1–2), 295–307. https://doi.org/10.1016/S0271-5317(00)00288-8.

López-Valverde, N., Pardal-Peláez, B., López-Valverde, A., Flores-Fraile, J., Herrero-Hernández, S., Macedo-De-sousa, B., Herrero-Payo, J., Ramírez, J.M., 2021. Effectiveness of propolis in the treatment of periodontal disease: updated systematic review with meta-analysis. Antioxidants 10 (2), 1–14. https://doi.org/10.3390/antiox10020269.

Mallick, S., Benson, R., Rath, G.K., 2016. Radiation induced oral mucositis: a review of current literature on prevention and management. Eur. Arch. Otorhinolaryngol. 273 (9), 2285–2293.

Marcucci, M.C., Ferreres, F., Garcı'a-Viguera, C., Bankova, V.S., De Castro, S.L., Dantas, A.P., Paulino, N., 2001. Phenolic compounds from Brazilian propolis with pharmacological activities. J. Ethnopharmacol. 74, 105–112.

Mariano, L.N.B., Arruda, C., Somensi, L.B., Costa, A.P.M., Perondi, E.G., Boeing, T., Mariott, M., da Silva, R.C.M.V.A.F., de Souza, P., Bastos, J.K., de Andrade, S.F., da Silva, L.M., 2018. Brazilian green propolis hydroalcoholic extract reduces colon damages caused by dextran sulfate sodium-induced colitis in mice. Inflammopharmacology 26 (5), 1283–1292. https://doi.org/10.1007/s10787-018-0467-z.

Miryan, M., Alavinejad, P., Abbaspour, M., Soleimani, D., Ostadrahimi, A., 2020. Does propolis affect the quality of life and complications in subjects with irritable bowel syndrome (diagnosed with Rome IV criteria)? A study protocol of the randomized, double-blinded, placebo-controlled clinical trial. Trials 21 (1), 1–10. https://doi.org/10.1186/s13063-020-04615-5.

Moreno, M.I., Isla, M.I., Sampietro, A.R., Vattuone, M.A., 2000. Comparison of the free radical-scavenging activity of propolis from several regions of Argentina. J. Ethnopharmacol. 71 (1–2), 109–114.

Nakao, R., Ueno, T., 2021. Effects of oral moisturizing gel containing propolis following head and neck radiotherapy: randomized controlled pilot trial. BDJ Open 7 (1), 1–7. https://doi.org/10.1038/s41405-021-00068-3.

National Cancer Institute, 2018. SEER Stat Fact Sheets: Colon and Rectum Cancer Colon and Rectum Cancer.

Noronha, V.R., Araujo, G.S., Gomes, R.T., Iwanaga, S.H., Barbosa, M.C., Abdo, E.N., Ferreira e Ferreira, E., Viana Campos, A.C., Souza, A.A., Abreu, S.R., Santos, V.R., 2014. Mucoadhesive propolis gel for pre- vention of radiation-induced oral mucositis. Curr. Clin. Pharmacol. 9 (4), 359–364.

Nostro, A., Cellini, L., Di Bartolomeo, S., Cannatelli, M.A., Di Campli, E., Procopio, F., Grande, R., Marzio, L., Alonzo, V., 2006. Effects of combining extracts (from propolis or Zingiber officinale) with clarithromycin on helicobacter pylori. Phytother. Res. 20 (3), 187–190. https://doi.org/10.1002/ptr.1830.

Pappalardo, G., Siegrist, H.H., Francioli, P., 1991. Evaluation of a new antiseptic based on natural substances. Drugs Exp. Clin. Res. 17 (10–11), 531–535.

Pasupuleti, V.R., Sammugam, L., Ramesh, N., Gan, S.H., 2017. Honey, propolis, and royal jelly: a comprehensive review of their biological actions and health benefits. Oxidative Med. Cell. Longev. 2017. https://doi.org/10.1155/2017/1259510.

Patel, S., 2015. Emerging adjuvant therapy for cancer: propolis and its constituents. J. Diet. Suppl. 13 (3), 245–268. https://doi.org/10.3109/19390211.2015.1008614.

Popova, M., Trusheva, B., Antonova, D., Cutajar, S., Mifsud, D., Farrugia, C., Tsvetkova, I., Najdenski, H., Bankova, V., 2011. The specific chemical profile of Mediterranean propolis from Malta. Food Chem. 126, 1431–1435.

Premratanachai, P., Chanchao, C., 2014. Review of the anticancer activities of bee products. Asian Pac. J. Trop. Biomed. 4 (5), 337–344. https://doi.org/10.12980/APJTB.4.2014C1262.

Romero, M., Freire, J., Pastene, E., García, A., Aranda, M., González, C., 2019. Propolis polyphenolic compounds affect the viability and structure of helicobacter pylori in vitro. Rev. Bras 29 (3), 325–332. https://doi.org/10.1016/j.bjp.2019.03.002.

Ruiz-Hurtado, P.A., Garduño-Siciliano, L., Dominguez-Verano, P., Martinez-Galero, E., Canales-Martinez, M.M., Rodriguez-Monroy, M.A., 2021. Evaluation of the

gastroprotective effects of Chihuahua propolis on indomethacin- induced gastric ulcers in mouse. Biomed. Pharmacother. 137. https://doi.org/10.1016/j.biopha.2021.111345.
Sangaraju, R., Nalban, N., Alavala, S., Rajendran, V., Jerald, M.K., Sistla, R., 2019. Protective effect of galangin against dextran sulfate sodium (DSS)-induced ulcerative colitis in Balb/c mice. Inflamm. Res. 68 (8), 691–704. https://doi.org/10.1007/s00011-019-01252-w.
Santana, M.T., Cercato, L.M., Oliveira, J.P., Camargo, E.A., 2017. Medicinal plants in the treatment of depression: evidence from preclinical studies. Planta Med. 83, 588–614. https://doi.org/10.1055/a-1338-1011.
Sarıkaya, A.O., Ulusoy, E., Öztürk, N., Tunçel, M., Kolaylı, S., 2009. Antioxidant activity and phenolic acid constituents of chestnut (castania sativa mill.) honey and propolis. J. Food Biochem. 33, 470–481.
Shapla, U.M., Raihan, J., Islam, A., Alam, F., Solayman, N., Gan, S.H., Hossen, S., Khalil, I., 2018. Propolis: the future therapy against helicobacter pylori-mediated gastrointestinal diseases. J. Appl. Biomed. 16 (2), 81–99. https://doi.org/10.1016/j.jab.2017.10.007.
Shi, Q., Smith, T.G., Michonski, J.D., Stein, K.D., Kaw, C., Cleeland, C.S., 2011. Symptom burden in cancer survivors 1 year after diagnosis. Cancer 117 (12), 2779–2790. https://doi.org/10.1002/cncr.26146.
Shimizu, Y., Suzuki, T., 2019. Brazilian propolis extract reduces intestinal barrier defects and inflammation in a colitic mouse model. Nutr. Res. 69, 30–41. https://doi.org/10.1016/j.nutres.2019.07.003.
Shimizu, K., Das, S.K., Hashimoto, T., Sowa, Y., Yoshida, T., Sakai, T., Matsuura, Y., Kanazawa, K., 2005. Artepillin C in Brazilian propolis induces G0/G1 arrest via stimulation of Cip1/p21 expression in human colon cancer cells. Mol. Carcinog. 44 (4), 293–299. https://doi.org/10.1002/mc.20148.
Sidahmed, H.M.A., Hashim, N.M., Mohan, S., Abdelwahab, S.I., Taha, M.M.E., Dehghan, F., Yahayu, M., Lian Ee, G.C., Loke, M.F., Vadivelu, J., 2016. Evidence of the gastroprotective and anti-helicobacter pylori activities of β-mangostin isolated from Cratoxylum arborescens (Vahl) blume. Drug Des. Devel. Ther. 10, 297–313. https://doi.org/10.2147/DDDT.S80625.
Simone-Finstrom, M., Spivak, M., 2010. Propolis and bee health: the natural history and significance of resin use by honey bees. Apidologie 41 (3), 295–311.
Soleimani, D., Miryan, M., Tutunchi, H., Navashenaq, J.G., Sadeghi, E., Ghayour-Mobarhan, M., Ferns, G.A., Ostadrahimi, A., 2021. A systematic review of preclinical studies on the efficacy of propolis for the treatment of inflammatory bowel disease. Phytother. Res. 35 (2), 701–710. https://doi.org/10.1002/ptr.6856.
Song, M.Y., Lee, D.Y., Kim, E.H., 2020. Anti-inflammatory and anti-oxidative effect of Korean propolis on helicobacter pylori-induced gastric damage in vitro. J. Microbiol. 58 (10), 878–885. https://doi.org/10.1007/s12275-020-0277-z.
Soromou, L.W., Chu, X., Jiang, L., Wei, M., Huo, M., Chen, N., Guan, S., Yang, X., Chen, C., Feng, H., Deng, X., 2012. In vitro and in vivo protection provided by pinocembrin against lipopolysaccharide-induced inflammatory responses. Int. Immunopharmacol. 14 (1), 66–74. https://doi.org/10.1016/j.intimp.2012.06.009.
Stokman, M.A., Spijkervet, F.K., Boezen, H.M., Schouten, J.P., Roodenburg, J.L., de Vries, E.G., 2006. Preventive intervention possibilities in radiotherapy- and chemotherapy-induced oral mucositis: results of meta-analyses. J. Dent. Res. 85 (5), 690–700.
Sung, S.H., Choi, G.H., Lee, N.W., Shin, B.C., 2017. External use of propolis for oral, skin, and genital diseases: a systematic review and meta-analysis. Evid. Based Complement. Alternat. Med. 2017, 10 (Article ID 8025752).
Suver, M., 2008. Arıcılık ve organik bal üretimi el kitabı: İşletme kurma ve pazarlama. (Marmara Grubu Stratejik ve Sosyal Araştırmalar Vakfı).
Tagliacollo, V.A., Orsi, R.D.O., 2011. Quality of propolis commercialized in the informal market. Cienc. Tecnol. Aliment. 31, 752–757.

Tambuwala, M.M., Kesharwani, P., Shukla, R., Thompson, P.D., McCarron, P.A., 2018. Caffeic acid phenethyl ester (CAPE) reverses fibrosis caused by chronic colon inflammation in murine model of colitis. Pathol. Res. Pract. 214 (11), 1909–1911. https://doi.org/10.1016/j.prp.2018.08.020.

Tolba, M.F., Omar, H.A., Azab, S.S., Khalifa, A.E., Abdel-Naim, A.B., Abdel-Rahman, S.Z., 2016. Caffeic acid phenethyl ester: a review of its antioxidant activity, protective effects against ischemia-reperfusion injury and drug adverse reactions. Crit. Rev. Food Sci. Nutr. 56 (13), 2183–2190. https://doi.org/10.1080/10408398.2013.821967.

Tomaževič, T., Jazbec, J., 2013. A double blind randomized placebo controlled study of propolis (bee glue) effectiveness in the treatment of severe oral mucositis in chemotherapy treated children. Complement. Ther. Med. 21 (4), 306–312.

Trotti, A., Bellm, L.A., Epstein, J.B., Frame, D., Fuchs, H.J., Gwede, C.K., Komaroff, E., Nalysnyk, L., Zilberberg, M.D., 2003. Mucositis incidence, severity and associated outcomes in patients with head and neck cancer receiving radiotherapy with or without chemotherapy: a systematic literature review. Radiother. Oncol. 66 (3), 253–262.

Valença, I., Morais-Santos, F., Miranda-Gonçalves, V., Ferreira, A.M., Almeida-Aguiar, C., Baltazar, F., 2013. Portuguese propolis disturbs glycolytic metabolism of human colorectal cancer in vitro. BMC Complement. Altern. Med. 13. https://doi.org/10.1186/1472-6882-13-184.

Vukovic, N.L., Obradovic, A.D., Vukic, M.D., Jovanovic, D., Djurdjevic, P.M., 2018. Cytotoxic, proapoptotic and antioxidative potential of flavonoids isolated from propolis against colon (HCT-116) and breast (MDA-MB-231) cancer cell lines. Food Res. Int. 106 (December 2017), 71–80. https://doi.org/10.1016/j.foodres.2017.12.056.

Waly, M.I., Al Ajimi, H., Al-Lawati, H.T., Guizani, N.I., Rahman, S.S., 2017. In vivo and in vitro evidence of anticancer effects of Omani propolis against colon cancer. FASEB J. 31, 790.22.

Wang, K., Jin, X., Li, Q., Sawaya, A.C.H.F., Le Leu, R.K., Conlon, M.A., Wu, L., Hu, F., 2018. Propolis from different geographic origins decreases intestinal inflammation and Bacteroides spp. populations in a model of DSS-induced colitis. Mol. Nutr. Food Res. 62 (17), 1–11. https://doi.org/10.1002/mnfr.201800080.

Watanabe, M.A.E., Amarante, M.K., Conti, B.J., Sforcin, J.M., 2011. Cytotoxic constituents of propolis inducing anticancer effects: a review. J. Pharm. Pharmacol. 63 (11), 1378–1386. https://doi.org/10.1111/j.2042-7158.2011.01331.x.

Wehkamp, J., Götz, M., Herrlinger, K., Steurer, W., Stange, E.F., 2016. Chronisch entzündliche darmerkrankungen: morbus crohn und colitis ulcerosa. Dtsch. Arztebl. Int. 113 (5), 72–81. https://doi.org/10.3238/arztebl.2016.0072.

Xuan, H., Zhao, J., Miao, J., Li, Y., Chu, Y., Hu, F., 2011. Effect of Brazilian propolis on human umbilical vein endothelial cell apoptosis. Food Chem. Toxicol. 49 (1), 78–85. https://doi.org/10.1016/j.fct.2010.09.034.

Xue, M., Liu, Y., Xu, H., Zhou, Z., Ma, Y., Sun, T., Liu, M., Zhang, H., Liang, H., 2019. Propolis modulates the gut microbiota and improves the intestinal mucosal barrier function in diabetic rats. Biomed. Pharmacother. 118 (August), 109393. https://doi.org/10.1016/j.biopha.2019.109393.

Yildiz, O.G., Soyuer, S., Saraymen, R., Eroglu, C., 2008. Protective effects of caffeic acid phenethyl ester on radiation induced lung injury in rats. Clin. Invest. Med. 31 (15), 242–247.

CHAPTER 6

Potential of propolis against SARS CoV-2 coronavirus infection

Sevgi Kolayli
Faculty of Science, Department of Chemistry, Karadeniz Technical University, Trabzon, Turkey

Contents

Introduction

Like Polio, HIV, Ebola, and Flu viruses, the current pandemic COVID-19 virus is a member of the RNA virus family known as coronavirus (Maruta and He, 2020). Coronaviruses (family *Coronaviridae*; order *Nidovirales*) are large, enveloped, positive-helical RNA viruses. The Coronavirinae subfamily is divided into four subclasses according to sequence and phylogenetic features—alpha-, beta-, gamma-, and delta-coronaviruses. Gamma- and delta-coronaviruses infect some mammal species and birds but do not cause disease in humans. In contrast, alpha- and beta-coronaviruses cause respiratory infections in humans and gastrointestinal system infections in some animal species. Various coronaviruses with an RNA-enveloped structure are capable of infecting humans (CoVh), two alpha-coronavirus (229E and NL63), and five beta-coronavirus (OC43, HKU1, SARS-CoV, SARS-CoV-2, and MERS-CoV) (Lima et al., 2021; Kumar et al., 2020).

Bee Products and Their Applications in the Food and Pharmaceutical Industries
https://doi.org/10.1016/B978-0-323-85400-9.00006-X

The International Virus Taxonomy Committee renamed the 2019-nCov as severe acute respiratory syndrome (SARS) coronavirus 2 (SARS-CoV-2), and the virus genome sequence revealed 89% similarity and 80% identity with SARS-CoV (Zhai et al., 2020; Zhu et al., 2020).

SARS-CoV-2 is responsible for coronavirus disease 19 (COVID-19) and is the seventh member of the coronavirus family that infects human beings, after SARS coronavirus and the Middle East respiratory syndrome (MERS) coronavirus (Bachevski et al., 2020; Zhu et al., 2020). Compared with other viruses, SARS-CoV-2 has high transmissibility and infectivity and is mainly spread through the respiratory tracts. The virus is transmitted directly or indirectly through mucous membranes, nose, mouth, and eyes. Numerous physical and chemical methods were experimented with to protect against the virus until a vaccine was developed. The mask and social distancing are the most widely used physical protection component but are not by themselves sufficient for complete protection. Many natural products, vitamins, and minerals were recommended to strengthen the immune system, such as vitamins D, vitamin C, Zn, Ca, and Mg, curcumin, and propolis during the pandemic (Arshad et al., 2020; Zildzic et al., 2020).

Propolis is a honeybee resinous mixture obtained by being scraped from beehives in raw form. Honeybees collect it from resin leaking from trees, barks, leaves, and plants, add their secretions, and store it in hives (Bankova et al., 2006). Honeybees use propolis for many different purposes. It protects a broad spectrum like an antiseptic, antimicrobial, antiviral, and insulating agent. It is also extensively employed in traditional and complementary medicine due to its antiseptic, antibacterial, antiviral, anesthetic, and antiinflammatory properties (Akçay et al., 2020; Cuce et al., 2020; Dizman et al., 2020; Król et al., 2013). Pharmacological and biochemical studies of propolis over the last 30 years have shown that it exhibits an extensive range of biological activities. In addition to bee venom and royal jelly, propolis is the most commonly employed apitherapeutic product for antibacterial, antiviral, antiinflammatory, antitumoral, immunomodulator, and hepatoprotective uses. Hydrophilic and hydrophobic extracts obtained from crude propolis are commercially produced in various forms (Rasgele and Kekecoglu, 2018).

This chapter overviews the antiviral properties of propolis by discussing possible pathways in treating and preventing infections by SARS CoV-2 as a safe therapeutic option, which is readily available as a natural supplement.

Antiviral properties of propolis

Propolis consists of approximately 50% resin and balm and 30% wax, the remainder being made up of essential oils and aromatic compounds (Keskin et al., 2019; Kızıltas and Erkan, 2021), although its composition and biological properties depend on the flora of the area where it is collected. The active ingredients of propolis, which contains approximately 300 different organic compounds, are various polyphenols and volatile compounds in the balsamic part. Water, ethanol, glycerin, polypropylene glycol, and vegetable oils are used as extraction solvents for raw propolis. However, 60%–70% ethanol is reported to be the best solvent. A quality propolis extract will have a high polyphenol content (Kızıltas and Erkan, 2021; Özcan et al., 2003; Sforcin and Bankova, 2011).

Polyphenols are herbal aromatic compounds, some 10,000 polyphenolic compounds having been described to date. These natural compounds, which are subclassified as phenolic acids, flavonoids, tannins, procyanidins, and anthocyanins, constitute the active ingredients of propolis (Scalbert et al., 2005). The propolis components extracted with ethanol (70%) are primarily phenolic acid and flavonoids, while common components in most propolis varieties are caffeic acid, coumaric acid, ferulic acid, daidzein, hesperetin, and pinocembrin.

Phenolic acids and flavonoids such as caffeic acid and its esters, coumaric acid, ferulic acid, caffeic acid phenethyl ester (CAPE), rutin, quercetin, hesperetin, pinocembrin, acacetin are common ingredients of raw propolis. Studies have reported low biological activities for these standards, but these activities are combined together. For this reason, propolis is primarily consumed in its fractions or as an extract (Lavigne et al., 2020).

Propolis is a good antimicrobial and antiviral agent. Numerous studies confirm that propolis exhibits antimicrobial activity on more than 600 microbial strains (Przybyłek and Karpiński, 2019). Similarly, propolis has been shown to have an antiviral effect on various DNA and RNA viruses, including HIV, Herpes simplex, HSV-1, HSV-2, p-influenza virus, influenza virus type A, B, adenovirus, avian reovirus, Newcastle virus disease, bovine rotavirus, pseudorabies virus (Bachevski et al., 2020; Bankova et al., 2014; Kumar et al., 2020). The broad antiviral spectrum of propolis suggests that it may also have significant potential against SARS CoV- 2.

The first study on the effects of flavonoids on rhinoviruses was performed in 1981 using a synthetic flavonoid, 2,4-dichloroflavone (Bauer et al., 1981). Later studies have reported that flavonoids possess antiviral potential against DNA and RNA viruses (Debiaggi et al., 1990). Quercetin, one of the five flavonoids which were studied, was reported to be the most effective one, but no trial was carried out in propolis extract (Ishitsuha et al., 1982).

The earliest study involving the use of propolis against the coronavirus family was conducted in 1990. That in vitro study investigated the antiviral effects of five flavonoids (chrysin, kaempferol, quercetin, acacetin, and galangin), the active ingredients of propolis. Quercetin was reported to exhibit dose-dependent antiviral activity (Debiaggi et al., 1990).

Experimental studies on this subject are very limited, but molecular docking modeling studies show that propolis components can be effective against the virus. Some flavonoids of Cistus bee pollen have been shown to target the CoV-2 Spike RBD: ACE-II complex by molecular docking study. In the study, some flavonoids blocked the CoV-2 Spike glycoprotein-Human ACE-II complex consisting of A and B chains by the programmer of visualizing protein in Discovery Studio Visualizer. The results showed that catechin possesses the best inhibition potential among the studied compounds, with high binding energy, followed by pinocembrin, chrysin, and caffeic acid phenethyl ester (CAPE). In the study, these flavonoid ligands, successful docking of all the ligands employed in docking experiments, showed significant binding of the ligands with the target proteins (Güler and Kara, 2020).

In another in silico study, the binding energies of some flavonoids found in Anatolian propolis to angiotensin-converting enzyme (ACE)-related carboxypeptidase (ACE-2) receptors were calculated. The authors concluded that rutin, myricetin, caffeic acid phenethyl ester (CAPE), and hesperetin exhibited high binding inhibition values (K_i) (Guler et al., 2021a).

Possible paths of action of propolis against COVID-19

Several comprehensive studies have shown that ethanolic propolis extracts, in particular, will reduce or alleviate the potential effects of SARS-CoV-2 infection through different mechanisms. However, there have been few comprehensive clinical studies to date directly involving propolis extract and COVID-19. Table 6.1 summarizes the possible mechanisms of action of propolis extracts as a food supplement in combating SARS-CoV-2.

Table 6.1 Possible effect paths and mechanisms of propolis extracts against SARS-CoV-2 infection.

N	Target	Evidence
1	Antiviral effect of propolis	Kwon et al. (2020) Bankova et al. (2014) Kai et al. (2014)
2	Viral-RNA dependent polymerase (RdRP) and spike glycoprotein (SGp)	Gekker et al. (2005) Silva et al. (2019)
3	Angiotensin-converting enzyme 2 (ACE-2) receptor binding	Guler et al. (2021a) Khayrani et al. (2021) Basu et al. (2020) Sekiou et al. (2020) Khayrani et al. (2021)
4	PAK-I (RAC/CDC42-activated kinases)	Berretta et al. (2020) Refaat et al. (2020) Nguyen et al. (2017)
5	Protease inhibitors Transmembrane protease serine 2 (TMPRSS2)	Kumar et al. (2020) Indasari et al. (2019)
6	Immunomodulator effector	
7	Adjuvant	
8	Septic shock	Dallan et al. (2020)
9	Antithrombosis activity	Berretta et al. (2020)

Effects on angiotensin-converting enzyme 2 (ACE-2) receptor binding

The main aim of combating the virus is to prevent the virus from entering the respiratory tract, and masks and social distancing are very effective. The second method of protection is vaccination. The third active way is that various interaction agents disrupt the connection between the virus and the related receptors that target receptor is ACE-2. The main purpose here is to prevent the virus from entering the cell by disconnecting the virus from ACE-2 receptors by any natural or synthetic agents. The agents to be used are either prevented from binding to ACE-2 receptors by binding to the virus spike proteins or binding directly to ACE-2 receptors to prevent the virus from binding. Scientists around the world are still seeking effective drugs to treat COVID-19. In silico, preclinical and clinical studies are still being conducted for this purpose (Güler and Kara, 2020; Guler et al., 2021a, b; Narkhede et al., 2020).

It has been argued that flavonoids from natural herbal compounds effectively treat infections with some RNA viruses, such as HIV, and may exert a similar effect in the treatment of COVID-19. Various molecular docking studies indeed suggest that propolis can be effective for that purpose. Studies have shown that propolis from Anatolia in Turkey may prevent the binding of SARS CoV-2 to ACE-2 with high attachment affinity. Flavonol rutin has the best inhibition potentials among the studied flavonoids, with a high binding energy of −8.97 kcal/mol and *Ki* 0.261 M, followed by myricetin, caffeic acid phenethyl ester (CAPE), hesperetin, and pinocembrin (Guler et al., 2021a). A similar docking study involving propolis from the Sulawesi region of Indonesia investigated five active substances—glyasperin A, broussoflavonol F, sulabiroins A (2*S*)-5,7-dihydroxy-40-methoxy-8-prenylflavanone, and isorhamnetin and showed their high attaching potential to ACE-2 and SARS-CoV-2 complex (Khayrani et al., 2021).

Blocking of protein kinases

Protein kinases are sensitive regulatory gap-junctions. The p21-activated kinases (PAKs) participate in cytoskeleton control networks downstream of Rho-family GTPases. Protein kinases are the most common enzymes/receptors in living organisms, and the activated state of PAK1 kinases are described as pathogenic kinases. They also regulate multiple signal transduction pathways in mammalian cells, such as cell motility, survival and proliferation, the organization and function of the cytoskeleton and extracellular matrix, transcription, and translation (Kichina et al., 2010). They are activated by phosphorylation/dephosphorylation through the tyrosine residue (Lei et al., 2005; Nguyen et al., 2016). PAK1 (RAC/CDC42-activated kinase 1) is the major "pathogenic" kinase, abnormal activation of which causes a wide variety of diseases/disorders including cancers, inflammation, malaria, aging, and viral infections including influenza, HIV, and COVID-19 (Maruta and He, 2020). Many diseases are caused by activated PAK1 kinase, including cancer, inflammatory diseases, diabetes, epilepsy, Alzheimer's, depression, acquired immunodeficiency syndrome, and other viral and bacterial infections. PAK1 blockers are, therefore, used as a therapeutic target, including in COVID-19 (Kichina et al., 2010). Some PAK1 blockers are used in the treatment of coronavirus infection. Remdesivir, chloroquine (CQ), and hydroxychloroquine (HQ) are antiviral agents that have been used in the treatment of infections caused by RNA-dependent viruses. These three synthetic agents, previously used as malaria drugs, are PAK1 blockers and the only antiviral synthetic agents used to treat coronavirus infection (Maruta, 2014; Maruta and He, 2020). Apart from these synthetic blockers, there are also some natural PAK1 blockers. Caffeic acid and its esters are reported to act as PAK1 blockers (Maruta, 2014; Maruta and He, 2020; Nguyen et al., 2016). Since PAK1 (RAC/CDC42 activated kinase 1) is also the major oncogenic kinase, various natural herbal PAK1 blockers such as propolis and curcumin have been used for antioncogenic and antimelanogenic purposes (Nguyen et al., 2016). They inhibit melanin synthesis from tyrosine by causing downregulation of tyrosinase enzymes. Another study reported that Artepillin C, a prenylated derivative of cinnamic acid, one of the active ingredients of propolis, and the active flavonoid of Brazilian green propolis, played a role PAK1 blocker and increased the profile of hair cells (Nguyen et al., 2016). Okinawa propolis extract has been reported to suppress PAK-I dependent growth of lung cancer (A549) and pancreatic cancer cells (Nguyen et al., 2017). This mechanism involves the overexpression of PAK1, a kinase that mediates coronavirus-induced lung inflammation, fibrosis, and immune system suppression (Berretta et al., 2020; Hasan et al., 2020; Kumar et al., 2020).

Inhibition of proteases

Proteases are a broad class of enzymes involved in the hydrolysis of peptide bonds in proteins such as serine, cysteine, asparagines, threonine, and glutamic proteases. They play a critical role in many physiological and pathological processes, including blood coagulation, cell growth and migration, protein catabolism, tissue regulation and development, morphogenesis, inflammation, tumor growth, and metastasis, activation of zymogens, and the formation of pharmacologically active peptides (Ishmael, 2011; Puente et al., 2003). Overexpressions of serine proteases therefore occur in pathological conditions such as bronchitis and asthma. Its catalysis by host cell proteases, including trans membrane protease serine 2 (TMPRSS2), endosomal cathepsin L protease, or human airway trypsin-like protease (HAT), is an essential step for virus fusion to the host cell membrane, infectivity, and propagation (Kumar et al., 2020). In general, protease inhibitors are regarded as valuable drugs for treatments or protection against viral infections (Kumar et al., 2020), and serine protease inhibitors also block virus proliferation.

Ethanolic propolis extracts are effective protease inhibitors that provide good protection by inhibiting serine proteases that cause asthma and bronchitis in the respiratory system. Tracheal protease activity in rats has been shown to increase following exposure to tobacco smoke. In contrast, a significant decrease in tracheal protease activities was observed in animals treated with ethanolic propolis extract of *Trigona* sp. (Indasari et al., 2019). Refaat et al. (2020) showed two major compounds of the propolis samples in a liposomal formulation, rutin and caffeic acid phenethyl ester (CAPE), to exhibit high inhibition against COVID-19 3CL-protease and S1 spike protein in a docking study.

Immunomodulatory activity

The first step in dealing with SARS-CoV-2 is to strengthen the immune system. Various vitamins, especially vitamins C and D, minerals (Zn, Ca, and Mg), and natural products are used for this purpose. The second step involves using antiviral drugs affecting the RNA viruses (for example, favipiravir, a new type of RNA-dependent RNA polymerase inhibitor) (Dong et al., 2020). Complementary and traditional treatments have come to the fore in the fight against COVID-19, with demand for propolis extracts being the highest among all-natural products (Dong et al., 2020). Various complementary and traditional medicine practices have been employed to strengthen the immune system and prevent the virus from binding to the receptors. Bee products and especially propolis have become the most widely consumed food supplements as immune boosters (Chan et al., 2013; Araujo et al., 2012).

The protection potential of propolis against oxidative stress is also very effective in strengthening the immune system (Kolankaya et al., 2002; Singla et al., 2014). Propolis has been shown to exhibit a protective effect in kidney and liver tissue against various toxic agents in rats (El-Guendouz et al., 2020; Kolaylı et al., 2018; Okail et al., 2020; Ulusoy et al., 2016). These studies show that consumption of propolis before the infection produces a critical barrier of protection against oxidative stress.

Adjuvant efficiency

Propolis is a functional natural food that can be used as a natural adjuvant in medicine. The success of many vaccines relies on their association with selected agents, known as adjuvants, to increase the immune response and ensure long-term protection (Ashry and Ahmad, 2012). Adjuvants are agents that increase the formation of the immune response, and they function by different mechanisms. However, all available adjuvants exhibit adverse effects due to their side effects and toxicity.

Preclinical studies show that some natural products have been used as adjuvants, as well as propolis (Cunha et al., 2020). One previous study showed that green propolis has a debilitating effect on the humoral immune response against the bovine herpesvirus type 5 (Fischer et al., 2007). In that study, 40 mg/dose of an ethanolic extract of green propolis was applied together with an inactivated oil vaccine against bovine herpesvirus type 5 (BoHV-5). Propolis resulted in a significant increase ($P < 0.01$) in neutralizing antibody levels compared to bovines that received the same vaccine without propolis. A different study in which propolis was used as an adjuvant showed that ethanolic propolis extract had no side effects when administered together with inactivated *Pasteurella multocida* vaccine and had no effects on liver or kidney functions (Nassar et al., 2013). In a study of Iranian propolis, oral ethanolic propolis extracts were given to experimental animals for 45 days, and no adverse findings were observed in their behavior or clinical findings (Mohammadzadeh et al., 2007). The mechanism of the adjuvant effect of propolis is not yet fully understood, but its immunomodulatory activity suggests that it may have the potential for use as an adjuvant agent.

The adjuvant effect of propolis (5 mg/dose) associated with inactivated Suite herpes virus type 1 (SuHV-1) vaccine was compared with that of aluminum hydroxide in one study, and the propolis extract resulted in higher levels of antibodies. The use of the SuHV-1 vaccine plus propolis alone did not produce significant levels of antibodies. Still, the combination was able to increase the cellular immune response, as evidenced by the increase in mRNA expression to produce IFN-γ (gamma-interferon) (Fischer et al., 2007; Sforcin, 2007).

The most comprehensive study of the clinical use of propolis in COVID-19 patients to date was conducted in Brazil. In that research, Brazilian green propolis as an adjunct treatment agent was administered in two different doses (400 mg/day and 800 mg/day) for 7 days together with standard care to COVID-19 patients. Length of hospital stay decreased, and kidney functions were better in the patient group receiving higher doses of propolis than the control group (Silveira et al., 2021).

Efficiency against septic shock

During the COVID-19 pandemic, a septic shock after prolonged intensive care, causing organ loss and death, has been observed in some patients (Dallan et al., 2020). Studies showed that propolis extracts could also be effective in case of septic shock through clinical animal research. A study of lipopolysaccharide-induced acute septic shock in Wister rats reported significant decreases in pro-inflammatory cytokines (TNF-α, IL-1α, IL-1β, IL-6, IL-4, IL-10, and sICAM-1) following treatment with caffeic acid phenyl ester (10 μmol/kg CAPE) (Korish and Arafa, 2011). Ethylene glycol was used to induce hepatotoxicity and nephrotoxicity in another rat study, and propolis was shown to reduce toxicity (El Menyiy et al., 2016).

Antithrombosis activity

The most widely used drugs in patients during the COVID-19 pandemic are blood thinners (Janardhan et al., 2020). Warfarin and aspirin are the most important anticoagulant agents as blood thinners and are widely used for the primary and secondary prevention of arterial and venous thromboembolic diseases (Li et al., 2006; Chan et al., 2016). Due to its high flavonoid content, propolis exhibits blood-thinning properties (Akbay et al., 2017); however, its potential as a supplement in treating COVID-19 patients requires further research.

Conclusion

In silico, in vitro, and clinical studies involving propolis and coronaviruses have shown that propolis has a high anti-COVID-19 potential. However, further studies are now needed to identify the components responsible and the mechanism involved. In conclusion, propolis is an important natural therapeutic agent in preventing COVID-19 due to its effectiveness deriving from its potential to block the binding of the virus to receptors and its immunity-enhancing effect, adjuvant characteristics, preventing septic shock and antithrombotic properties.

References

Akbay, E., Özenirler, Ç., Çelemli, Ö.G., Durukan, A.B., Onur, M.A., Sorkun, K., 2017. Effects of propolis on warfarin efficacy. Pol. Eur. J. Cardiothorac. Surg. 14 (1), 43–46. https://doi.org/10.5114/kitp.2017.66929.

Akçay, Ç., Birinci, E., Birinci, C., Kolaylı, S., 2020. Durability of wood treated with propolis. Bioresources 15 (1), 1547–1562.

Araujo, M.A., Libério, S.A., Guerra, R.N., Ribeiro, M.N.S., Nascimento, F.R.F., 2012. Mechanisms of action underlying the anti-inflammatory and immunomodulatory effects of propolis: a brief review. Rev. bras. Farmacogn. 22 (1), 208–219. https://doi.org/10.1590/S0102-695X2011005000167.

Arshad, M.S., Khan, U., Sadiq, A., Khalid, W., Hussain, M., Yasmeen, A., Asghar, Z., Rehana, H., 2020. Coronavirus disease (COVID-19) and immunity booster green foods: a mini review. Food Sci. Nutr. 8 (8), 3971–3976. https://doi.org/10.1002/fsn3.1719.

Ashry, e.S.H., Ahmad, T.A., 2012. The use of propolis as vaccine's adjuvant. Vaccine 31 (1), 31–39. https://doi.org/10.1016/j.vaccine.2012.10.095.

Bachevski, D., Damevska, K., Simeonovski, V., Dimova, M., 2020. Back to the basics: propolis and COVID-19. Dermatol. Ther. 33 (4). https://doi.org/10.1111/dth.13780, e13780.

Bankova, V., Popova, M., Trusheva, B., 2006. Plant sources of propolis: an update from a chemist's point of view. Nat. Prod. Commun. 1 (11), 1023–1028. https://doi.org/10.1177/1934578X0600101118.

Bankova, V., Galabov, A.S., Antonova, D., Vilhelmova, N., Di Perri, B., 2014. Chemical composition of propolis extract ACF® and activity against herpes simplex virus. Phytomedicine 21 (11), 1432–1438. https://doi.org/10.1016/j.phymed.2014.04.026.

Basu, A., Sarkar, A., Maulik, U., 2020. Molecular docking study of potential phytochemicals and their effects on the complex of SARS-CoV2 spike protein and human ACE2. Sci. Rep. 10. https://doi.org/10.1038/s41598-020-74715-4, e17699.

Bauer, D.J., Selway, J.W.T., Batchelor, J.F., Tisdale, M., Caldwell, I.C., Young, D.A.B., 1981. 4′, 6-Dichloroflavan (BW683C), a new anti-rhinovirus compound. Nature 292 (5821), 369–370. https://doi.org/10.1038/292369a0.

Berretta, A.A., Silveira, M.A.D., Capcha, J.M.C., De Jong, D., 2020. Propolis and its potential against SARS-CoV-2 infection mechanisms and COVID-19 disease. Biomed. Pharmacother. 131. https://doi.org/10.1016/j.biopha.2020.110622, e110622.

Chan, G.C.F., Cheung, K.W., Sze, D.M.Y., 2013. The immunomodulatory and anticancer properties of propolis. Clin. Rev. Allergy Immunol. 44 (3), 262–273. https://doi.org/10.1007/s12016-012-8322-2.

Chan, N.C., Eikelboom, J.W., Weitz, J.I., 2016. Evolving treatments for arterial and venous thrombosis: role of the direct oral anticoagulants. Circ. Res. 118 (9), 1409–1424. https://doi.org/10.1161/CIRCRESAHA.116.306925.

Cuce, P.M., Kolayli, S., Cuce, E., 2020. Enhanced performance figures of solar cookers through latent heat storage and low-cost booster reflectors. Int. J. Low Carbon Technol. 15 (3), 427–433. https://doi.org/10.1093/ijlct/ctz079.

Cunha, B.C., de Miranda, M.B., Afonso, L.C.C., Abreu, S.R.L., Testasicca, M.C.S., Da Silva, G.R., De Moura, S.A.L., 2020. Brazilian green propolis hydroalcoholic extract as a therapeutic adjuvant to treat *Cutaneous leishmaniasis*. J. Appl. Pharm. Sci. 10 (11), 124–132. https://doi.org/10.7324/JAPS.2020.101117.

Dallan, C., Romano, F., Siebert, J., Politi, S., Lacroix, L., Sahyoun, C., 2020. Septic shock presentation in adolescents with COVID-19. Lancet Child Adolesc. Health 4 (7), e21–e23. https://doi.org/10.1016/S2352-4642(20)30164-4.

Debiaggi, M., Tateo, F., Pagani, L., Luini, M., Romero, E., 1990. Effects of propolis flavonoids on virus infectivity and replication. Microbiologica 13 (3), 207–213.

Dizman, S., Yildiz, O., Boyraci, G.M., Kolayli, S., 2020. Determination of natural radioactivity levels and gamma radiation attenuation coefficients in propolis samples and the study of its antioxidant properties. Radiochim. Acta 108 (10), 829–837. https://doi.org/10.1515/ract-2019-3191.

Dong, L., Hu, S., Gao, J., 2020. Discovering drugs to treat coronavirus disease 2019 (COVID-19). Drug Discov Ther. 14 (1), 58–60. https://doi.org/10.5582/ddt.2020.01012.

El Menyiy, N., Al Waili, N., Bakour, M., Al-Waili, H., Lyoussi, B., 2016. Protective effect of propolis in proteinuria, crystaluria, nephrotoxicity and hepatotoxicity induced by ethylene glycol ingestion. Arch. Med. Res. 47 (7), 526–534. https://doi.org/10.1016/j.arcmed.2016.12.010.

El-Guendouz, S., Zizi, S., Elamine, Y., Lyoussi, B., 2020. Preliminary screening of the possible protective effect of Moroccan propolis against chromium-induced nephrotoxicity in animal model. Vet. World 13 (7), 1327–1333. https://doi.org/10.14202/vetworld.2020.1327-1333.

Fischer, G., Cleff, M.B., Dummer, L.A., Paulino, N., Paulino, A.S., de Oliveira Vilela, C., Campos, F.S., Storch, T., D'Avila Vargas, G., de Oliveira Hübner, S., Vidor, T., 2007. Adjuvant effect of green propolis on humoral immune response of bovines immunized with bovine herpesvirus type 5. Vet. Immunol. Immunopathol. 116 (1–2), 79–84. https://doi.org/10.1016/j.vetimm.2007.01.003.

Gekker, G., Hu, S., Spivak, M., Lokensgard, J.R., Peterson, P.K., 2005. Anti-HIV-1 activity of propolis in CD4+ lymphocyte and microglial cell cultures. J. Ethnopharmacol. 102 (2), 158–163. https://doi.org/10.1016/j.jep.2005.05.045.

Guler, H.I., Tatar, G., Yildiz, O., Belduz, A.O., Kolayli, S., 2021a. Investigation of potential inhibitor properties of ethanolic propolis extracts against ACE-II receptors for COVID-19 treatment by molecular docking study. Arch. Microbiol. 203 (6), 3557–3564. https://doi.org/10.1007/s00203-021-02351-1.

Guler, H.I., Ay Şal, F., Can, Z., Kara, Y., Beldüz, A.O., Çanakçi, S., Kolayli, S., 2021b. Targeting CoV-2 spike RBD and ACE-2 interaction with flavonoids of Anatolian propolis by *in silico* and *in vitro* studies in terms of possible COVID-19 therapeutics. Turk. J. Biol. 45 (4), 530–548. https://doi.org/10.3906/biy-2104-5.

Güler, H.I., Kara, Y., 2020. Targeting CoV-2 spike RBD: ACE-II complex with phenolic compounds from Cistus (*Cistus* L.) bee pollen for COVID-19 treatment by molecular docking study. J. Apither. Nat. 3 (1), 10–23.

Hasan, A., Paray, B.A., Hussain, A., Qadir, F.A., Attar, F., Aziz, F.M., Sharifi, M., Derakhshankhah, H., Rasti, B., Mehrabi, M., Shahpasand, K., Saboury, A.A., Falahati, M., 2020. A review on the cleavage priming of the spike protein on coronavirus by angiotensin-converting enzyme-2 and furin. J. Biomol. Struct. Dyn., 1–13. https://doi.org/10.1080/07391102.2020.1754293.

Indasari, E.N., Marhendra, A.P.W., Wardhana, A.W., 2019. Extract bee propolis (*Trigona* sp) for preventive increase protease activity and defect of trachea histology in rats (*Rattus norvegicus*) exposed to cigarette smoke. IOP Conf. Ser. Earth Environ. Sci. 391. https://doi.org/10.1088/1755-1315/391/1/012048, 012048. IOP Publishing.

Ishitsuha, H., Ohsawa, C., Chiwa, T., Umeda, I., Suhara, Y., 1982. Antipicornavirus flavon Ro 09-1179. Antimicrob. Agents Chemother. 22 (4), 611–616. https://doi.org/10.1128/aac.22.4.611.

Ishmael, F.T., 2011. The inflammatory response in the pathogenesis of asthma. J. Am. Osteopath. Assoc. 111 (11 Suppl. 7), 11–17.

Janardhan, V., Janardhan, V., Kalousek, V., 2020. COVID-19 as a blood clotting disorder masquerading as a respiratory illness: a cerebrovascular perspective and therapeutic implications for stroke thrombectomy. J. Neuroimaging 30 (5), 555–561. https://doi.org/10.1111/jon.12770.

Kai, H., Obuchi, M., Yoshida, H., Watanabe, W., Tsutsumi, S., Park, Y.K., Matsuno, K., Yasukawa, K., Kurokawa, M., 2014. *In vitro* and *in vivo* anti-influenza virus activities of flavonoids and related compounds as components of Brazilian propolis (AF-08). J. Funct. Foods 8, 214–223. https://doi.org/10.1016/j.jff.2014.03.019.

Keskin, M., Keskin, Ş., Kolayli, S., 2019. Preparation of alcohol free propolis-alginate microcapsules, characterization and release property. LWT Food Sci. Technol. 108, 89–96. https://doi.org/10.1016/j.lwt.2019.03.036.

Khayrani, A.C., Irdiani, R., Aditama, R., Pratami, D.K., Lischer, K., Ansari, M.J., Chinnathambi, A., Alharbi, S.A., Almoallim, H.S., Sahlan, M., 2021. Evaluating the potency of Sulawesi propolis compounds as ACE-2 inhibitors through molecular docking for COVID-19 drug discovery preliminary study. J. King Saud Univ. Sci. 33 (2). https://doi.org/10.1016/j.jksus.2020.101297, 101297.

Kichina, J.V., Goc, A., Al-Husein, B., Somanath, P.R., Kandel, E.S., 2010. PAK1 as a therapeutic target. Expert Opin. Ther. Targets 14 (7), 703–725. https://doi.org/10.1517/14728222.2010.492779.

Kızıltas, H., Erkan, C., 2021. The effects of different beehives on propolis production and quality. Food Sci. Technol. 41 (4), 877–883. https://doi.org/10.1590/fst.20120.

Kolankaya, D., Selmanoğlu, G., Sorkun, K., Salih, B., 2002. Protective effects of Turkish propolis on alcohol-induced serum lipid changes and liver injury in male rats. Food Chem. 78 (2), 213–217. https://doi.org/10.1016/S0308-8146(01)00400-9.

Kolaylı, S., Eksi, S., Yazıcı, Z., Yıldız, O., Şahin, H., Can, Z., Karaoglu, S., 2018. Some apitherapeutic properties of chestnut propolis. J. Apither. Nat. 1 (3), 14.
Korish, A.A., Arafa, M.M., 2011. Propolis derivatives inhibit the systemic inflammatory response and protect hepatic and neuronal cells in acute septic shock. Braz. J. Infect. Dis. 15 (4), 332–338.
Król, W., Bankova, V., Sforcin, J.M., Szliszka, E., Czuba, Z., Kuropatnicki, A.K., 2013. Propolis: properties, application, and its potential. Evid. Based Complement. Alternat. Med. 2013 (807578). https://doi.org/10.1155/2013/807578.
Kumar, V., Dhanjal, J.K., Bhargava, P., Kaul, A., Wang, J., Zhang, H., Sundar, D., 2020. Withanone and Withaferin-A are predicted to interact with transmembrane protease serine 2 (TMPRSS2) and block entry of SARS-CoV-2 into cells. J. Biomol. Struct. Dyn., 1–13. https://doi.org/10.1080/07391102.2020.1775704.
Kwon, M.J., Shin, H.M., Perumalsamy, H., Wang, X., Ahn, Y.J., 2020. Antiviral effects and possible mechanisms of action of constituents from Brazilian propolis and related compounds. J. Apic. Res. 59 (4), 413–425. https://doi.org/10.1080/00218839.2019.1695715.
Lavigne, J.P., Ranfaing, J., Dunyach-Rémy, C., Sotto, A., 2020. Synergistic effect of propolis and antibiotics on uropathogenic *Escherichia coli*. Antibiotics 9 (11), 739. https://doi.org/10.3390/antibiotics9110739.
Lei, M., Robinson, M.A., Harrison, S.C., 2005. The active conformation of the PAK1 kinase domain. Structure 13 (5), 769–778. https://doi.org/10.1016/j.str.2005.03.007.
Li, Z., Seeram, N.P., Carpenter, C.L., Thames, G., Minutti, C., Bowerman, S., 2006. Cranberry does not affect prothrombin time in male subjects on warfarin. J. Am. Diet. Assoc. 106 (12), 2057–2061. https://doi.org/10.1016/j.jada.2006.09.012.
Lima, W.G., Brito, J.C., da Cruz Nizer, W.S., 2021. Bee products as a source of promising therapeutic and chemoprophylaxis strategies against COVID-19 (SARS-CoV-2). Phytother. Res. 35 (2), 743–750. https://doi.org/10.1002/ptr.6872.
Maruta, H., 2014. Herbal therapeutics that block the oncogenic kinase PAK1: a practical approach towards PAK1-dependent diseases and longevity. Phytother. Res. 28 (5), 656–672. https://doi.org/10.1002/ptr.5054.
Maruta, H., He, H., 2020. PAK1-blockers: potential therapeutics against COVID-19. Med. Drug Discov. 6. https://doi.org/10.1016/j.medidd.2020.100039, 100039.
Mohammadzadeh, S., Shariatpanahi, M., Hamedi, M., Ahmadkhaniha, R., Samadi, N., Ostad, S.N., 2007. Chemical composition, oral toxicity and antimicrobial activity of Iranian propolis. Food Chem. 103 (4), 1097–1103. https://doi.org/10.1016/j.foodchem.2006.10.006.
Narkhede, R.R., Cheke, R.S., Ambhore, J.P., Shinde, S.D., 2020. The molecular docking study of potential drug candidates showing anti-COVID-19 activity by exploring of therapeutic targets of SARS-CoV-2. Eurasian J. Med. Oncol. 4 (3), 185–195. https://doi.org/10.14744/ejmo.2020.31503.
Nassar, S.A., Mohamed, A.H., Soufy, H., Nasr, S.M., 2013. Protective effect of Egyptian propolis against rabbit pasteurellosis. Biomed. Res. Int. 2013, 1–9. https://doi.org/10.1155/2013/163724. Article ID 163724.
Nguyen, B.C.Q., Taira, N., Maruta, H., Tawata, S., 2016. Artepillin C and other herbal PAK1-blockers: effects on hair cell proliferation and related PAK1-dependent biological function in cell culture. Phytother. Res. 30 (1), 120–127. https://doi.org/10.1002/ptr.5510.
Nguyen, B.C.Q., Yoshimura, K., Kumazawa, S., Tawata, S., Maruta, H., 2017. Frondoside A from sea cucumber and nymphaeols from Okinawa propolis: natural anti-cancer agents that selectively inhibit PAK1 *in vitro*. Drug Discov. Ther. 11 (2), 110–114. https://doi.org/10.5582/ddt.2017.01011.
Okail, H.A., Ibrahim, A.S., Badr, A.H., 2020. The protective effect of propolis against aluminum chloride-induced hepatorenal toxicity in albino rats. J. Basic Appl. Zool. 81 (34), 1–11. https://doi.org/10.1186/s41936-020-00169-9.

Özcan, M., Ceylan, A., Ünver, A., Yetişir, R., 2003. Antifungal effect of pollen and propolis extracts collected from different regions of Turkey. Uludag Bee J. 3 (3), 27–34.

Przybyłek, I., Karpiński, T.M., 2019. Antibacterial properties of propolis. Molecules 24 (11), 2047. https://doi.org/10.3390/molecules24112047.

Puente, X.S., Sanchez, L.M., Overall, C.M., Lopez-Otin, C., 2003. Human and mouse proteases: a comparative genomic approach. Nat. Rev. Genet. 4, 544–558. https://doi.org/10.1038/nrg1111.

Rasgele, P.G., Kekecoglu, M., 2018. Investigation of phenolic content of propolis produced in Yigilca district of Duzce province in Western Black Sea region of Turkey. J. Apither. Nat. 1 (2), 20–28.

Refaat, H., Mady, F.M., Sarhan, H.A., Rateb, H.S., Alaaeldin, E., 2020. Optimization and evaluation of propolis liposomes as a promising therapeutic approach for COVID-19. Int. J. Pharm. 592. https://doi.org/10.1016/j.ijpharm.2020.120028, 120028.

Scalbert, A., Johnson, I.T., Saltmarsh, M., 2005. Polyphenols: antioxidants and beyond. Am. J. Clin. Nutr. 81 (1), 215S–217S. https://doi.org/10.1093/ajcn/81.1.215S.

Sekiou, O., Omar, I., Bouziane, Z., Djemel Bouslama, A., 2020. In-silico identification of potent inhibitors of COVID-19 main protease (Mpro) and angiotensin converting enzyme 2 (ACE2) from natural products. ChemRxiv. https://doi.org/10.26434/chemrxiv.12181404.

Sforcin, J.M., 2007. Propolis and the immune system: a review. J. Ethnopharmacol. 113 (1), 1–14. https://doi.org/10.1016/j.jep.2007.05.012.

Sforcin, J.M., Bankova, V., 2011. Propolis: is there a potential for the development of new drugs? J. Ethnopharmacol. 133 (2), 253–260. https://doi.org/10.1016/j.jep.2010.10.032.

Silva, C.C.F.D., Salatino, A., Motta, L.B.D., Negri, G., Salatino, M.L.F., 2019. Chemical characterization, antioxidant and anti-HIV activities of a Brazilian propolis from Ceará state. Rev. Bras. Farmacogn. 29 (3), 309–318. https://doi.org/10.1016/j.bjp.2019.04.001.

Silveira, M.A.D., De Jong, D., Beretta, A.A., dos Santos Galvão, E.B., Riberio, J.C., Cerqueira-Silva, T., Amorim, T.C., da Conceição, L.F.M.R., Gomes, M.M.D., Teixeria, M.B., de Souza, S.P., dos Santos, M.H.C.A., San Martin, R.L.A., de Oliveira Silva, M., Lírio, M., Moreno, L., Sampaio, J.C.M., Mendonça, R., Ultchak, S.S., Amorim, F.S., Ramos, J.G.R., Batista, P.B.P., da Guarda, S.N.F., Mendes, A.V.A., da Hora Passos, R., 2021. Efficacy of Brazilian green propolis (EPP-AF®) as an adjunct treatment for hospitalized COVID-19 patients: A randomized, controlled clinical trial. Biomed. Pharmacother. 138. https://doi.org/10.1016/j.biopha.2021.111526, 111526.

Singla, S., Kumar, N.R., Kaur, J., 2014. *In vivo* studies on the protective effect of propolis on doxorubicin-induced toxicity in liver of male rats. Toxicol. Int. 21 (2), 191–195. https://doi.org/10.4103/0971-6580.139808.

Ulusoy, H.B., Öztürk, İ., Sönmez, M.F., 2016. Protective effect of propolis on methotrexate-induced kidney injury in the rat. Ren. Fail. 38 (5), 744–750. https://doi.org/10.3109/0886022X.2016.1158070.

Zhai, P., Ding, Y., Wu, X., Long, J., Zhong, Y., Li, Y., 2020. The epidemiology, diagnosis and treatment of COVID-19. Int. J. Antimicrob. Agents 55 (5). https://doi.org/10.1016/j.ijantimicag.2020.105955, 105955.

Zhu, F.C., Li, Y.H., Guan, X.H., Hou, L.H., Wang, W.J., Li, J.X., Wu, S.P., Wang, B.S., Wang, Z., Wang, L., Jia, S.Y., Jiang, H.D., Wang, L., Jiang, T., Hu, Y., Gou, J.B., Xu, S.B., Xu, J.J., Wang, X.W., Wang, W., Chen, W., 2020. Safety, tolerability, and immunogenicity of a recombinant adeno virus type-5 vectored COVID-19 vaccine: a dose-escalation, open-label, non-randomised, first-in-human trial. Lancet 395 (10240), 1845–1854. https://doi.org/10.1016/S0140-6736(20)31208-3.

Zildzic, M., Masic, I., Salihefendic, N., Jasic, M., Hajdarevic, B., 2020. The importance of nutrition in boosting immunity for prevention and treatment COVID-19. Int. J. Biomed. Healthc. 8 (2), 73–79. https://doi.org/10.5455/ijbh.2020.8.73-79.

CHAPTER 7

Effect of propolis on fungi of human clinical interest

Isabella Letícia Esteves Barros[a,b], Jakeline Luiz Corrêa[a,b], Flávia Franco Veiga[a,b], Marcos Luciano Bruschi[c], Melyssa Negri[a,b], and Terezinha Inez Estivalet Svidzinski[a,b]

[a]Postgraduate Program in Health Sciences, Health Sciences Center, State University of Maringá, Maringá, PR, Brazil
[b]Medical Mycology Laboratory, Department of Clinical Analysis and Biomedicine, State University of Maringá, Maringá, PR, Brazil
[c]Laboratory of Research and Development of Drug Delivery Systems, Department of Pharmacy, State University of Maringá, Maringá, PR, Brazil

Contents

Bee Products and Their Applications in the Food and Pharmaceutical Industries
https://doi.org/10.1016/B978-0-323-85400-9.00004-6

A modern portrait of human fungal infections

The fungal kingdom contains more than three million species; approximately 600 of these species cause human disease (Caceres et al., 2020). Over 300 million people are acutely or chronically infected by fungi, leading to death, long-term illness, blindness, psychological problems, and reduced work capacity. Deaths from invasive fungal diseases exceed those from malaria (Worldwide, n.d.). Fungi can attack human tissues in several ways, characterizing different clinical conditions. The most common fungal infections are detailed below. The in vitro antifungal property of the propolis in both planktonic fungal cells and in the ability to inhibit biofilm formation is well known. Furthermore, some pharmaceutical forms involving propolis have already shown efficiency regarding various human fungal infections.

Tegumentar infections

Cutaneous mycoses or dermatomycosis infections are caused by fungi in the epidermis (skin, hair, and nails), specifically in deep areas of the corneal layer. These are the most common human fungal infections, affecting approximately 20%–25% of the global population (Vasconcellos et al., 2013). These fungi are classified in three groups: the most frequent are the dermatophytes fungi (Sav et al., 2018), encompassing three genres (*Trichophyton* sp., *Microsporum* sp., and *Epidermophyton* sp.), the most common being *Trichophyton rubrum*. The yeasts are the second group, above all genus *Candida* (Gupta et al., 2017; Ferreira et al., 2019), specially *C. albicans* and *C. parapsilosis* have long been recognized among the principal dermatomycosis agents (Ferreira et al., 2019; Subramanya et al., 2016). Besides, some nonkeratinophilic filamentous fungi, called nondermatophyte molds (NDM), such as *Aspergillus* spp., *Alternaria alternata,* and *Fusarium* spp. also are able to infect human tissue specially toenails (de Albuquerque Maranhão et al., 2019).

Among all dermatophytosis, ringworm on the nails, formerly called tinea unguium, are the most feared, frequent, cause social discomfort and in many cases pain and itching, present varied clinical lesions (onycholysis, onychodystrophy, subungual hyperkeratosis, and color change). Onychomycosis by *Candida* and *Fusarium*, often cause paronychia, erythema, and edematous changes of the around the nail fold soft tissue, are also described especially in immunocompromised patients (Otašević et al., 2016; Galletti et al., 2015) which are extremely uncomfortable, painful lesions suggesting an inflammatory response.

Mucosal infections

For most individuals, yeasts are part of the normal mucosal microbiota. They are typically suppressed by host defense mechanisms and competitive inhibition from a wide range of other microorganisms. They often cause infection only if an opportunity arises (Rautemaa and Ramage, 2011; Willems et al., 2020).

Oral infections of soft and hard tissue by *Candida* species are well defined and caused by complex biofilms that interact with bacteria and the host (Rautemaa and Ramage, 2011). Yeasts have been isolated from a variety of oral environments, such as: periodontal pockets, enamel, root canals, mucous surfaces, orthodontic appliances, and dentures. Several factors (e.g., humidity, nutrients, presence of commensal bacteria, hyphae growth) influence the development of biofilm, its architecture, and even the virulence of *C. albicans* (Bertolini et al., 2015; O'Donnell et al., 2015). The main oral disorders associated with these yeasts are caries, periodontitis, endodontic and orthodontic infections, angular cheilitis, dental stomatitis, and oral candidiasis. They may even lead to oropharyngeal and respiratory infections (O'Donnell et al., 2015).

Vulvovaginal candidiasis (VVC) is also caused by these yeasts and affects the reproductive tract of immunocompetent women. This common condition has a profoundly negative impact on quality of life and high global incidence (Willems et al., 2020). Symptomatic VVC infection results in inflammation of the vaginal mucosa. It is mainly caused by yeast overgrowth and subsequent epithelial invasion and production of virulence factors such as biofilm. Common symptoms of the disease include itching, pain, burning, and vaginal redness. Often, there is vaginal discharge composed of peeling epithelium, immune cells, yeast, and vaginal fluid (Willems et al., 2020; Rodríguez-Cerdeira et al., 2019).

Candida albicans is the most prevalent species in these infections. However, Non-*C. albicans Candida* (NCAC) (e.g., *C. glabrata*, *C. parapsilosis*, *C. krusei*, and *C. tropicalis*) are also frequently isolated from patients with this disease (Rautemaa and Ramage, 2011; Willems et al., 2020; Stenderup, 1990). *Candida* species are responsible for approximately 134 million mucosal infections per year (Caceres et al., 2020).

Invasive fungal infections (IFIs)

Pathogenic fungi, particularly yeasts, have the ability to gain access to the bloodstream, reaching the internal organs of patients and causing systemic or disseminated fungal infections (Costa-Orlandi et al., 2017). These can be life-threatening, causing high mortality rates in severely immunocompromised patients (especially in nosocomial fungal infections). IFIs are from opportunistic pathogens. These include (in order of frequency) *C. albicans,* NCAC, *Trichosporon*, *Rhodotorula*, *Geotrichum,* and *Saccharomyces* (Gunaydin et al., 2020).

IFIs are an emerging problem worldwide; invasive candidiasis and candidemia are responsible for most cases. This is mainly due to the widespread adoption of aggressive immunosuppressive therapy in certain patient populations and increasing use of invasive devices, such as central venous catheters (Enoch et al., 2017). It is estimated that only half of these infections are detected during the patient's life, which makes this one of the most common causes of overlooked death in intensive care patients (von Lilienfeld-Toal et al., 2019). *Candida* species cause approximately 750,000 life-threatening bloodstream infections per year. They are associated with mortality rates $>40\%$ (Caceres et al., 2020).

Fungal virulence attributes

Most human fungal infections are caused by opportunistic fungi. However, fungal virulence agents actively participate in infection etiopathogenesis. Fungal virulence factors encompass a wide range of attributes (e.g., pathogenic microorganisms). These attributes enhance the abilities of the fungi to evade host defenses and cause disease. Through the formation of biofilms, fungi also have the ability to adhere to medical devices, implanted prostheses, or catheters (Bernard et al., 2020). Biofilms are communities of highly structured microbial cells surrounded by an extracellular matrix. These cell communities adhere to living or inert surfaces. This is considered the most common growth state of many fungal species. The cells that make up biofilms have properties distinct from their planktonic (free-floating) counterparts, such as high resistances to drugs and physical disturbances (Lohse et al., 2018).

It has become clear that most fungal infections are linked to the development of a biofilm on biotic or abiotic surfaces, such as biofilms formed on nails that cause onychomycosis (Gupta and Foley, 2019), biofilms formed on medical devices (Bernard et al., 2020), and biofilms involved in VVC infections (Rodríguez-Cerdeira et al., 2019). Moreover, when organized in a biofilm, fungi gain access to the bloodstream. This is how they reach the internal organs of patients and create disseminated invasive infections. This is especially common in nosocomial infections (Sav et al., 2018).

Although pathogenic fungi, particularly yeasts, are capable of forming biofilms of a single species, it is considerably more common to find two or more bacterial or fungal species that compose a single polymicrobial biofilm (Lohse et al., 2018). *Staphylococcus aureus*, some species of streptococci, and anaerobic Gram-negative and Gram-positive species are examples of polymicrobial biofilms composed of fungi with bacteria (O'Donnell et al., 2015). *C. albicans* and *S. aureus* are two of the most prevalent pathogens. They are responsible for 27% of nosocomial bloodstream infections, according to data in the literature. Polymicrobial biofilms formed by these two microorganisms in implanted medical devices and host tissues are responsible for high rates of morbidity and mortality in hospitalized patients.

Insights on treatment of human fungal infection

The options of antifungal treatment depend on the patient's immune status, the affected area, and the chronicity of predisposing factors. Since fungi are eukaryotic organisms, just like humans, the development of antifungal agents is difficult. Many antifungals cause severe side effects. In addition, the polymicrobial nature of biofilms can have a profound influence on the efficiency of antimicrobial agents and the pathophysiology of infections (Bernard et al., 2020). Three groups of antifungals are effective in the treatment of IFIs: polyenes, azoles, and echinocandins. For cutaneous and mucosal infections, azoles and terbinafine are the only available options (Richardson et al., 2007; Hay, 2018). However, it is important to note that these antifungal agents are not efficient on biofilm fungi forms. In addition, susceptibility to antifungals varies according to the fungal species. Thus, it is relevant to discover new antifungal compounds to improve existing treatments or the development of new antifungal therapy strategies. Propolis is an example of a promising treatment that deserves further examination.

Propolis

Botanical origin

The best indicator of the botanical origin of propolis is the analysis of its chemical composition compared to the probable plant source. The determination of the geographical and vegetal origin and the phenology of the host plant are important factors in quality control and propolis sample standardization for an effective therapeutic application (Park et al., 2002; Santos et al., 2020).

In Brazil, there are several species of plants visited by bees to remove resin. Not all of these have been identified, but "field rosemary" (*Baccharis dracunculifolia*), pine trees (*Pinus* sp. and *Araucaria heterophylla*), aroeira

(*Schinus terebenthifolius*), and eucalyptus (*Eucalyptus* sp.) are some examples (Banskota et al., 2001; Midorikawa et al., 2001; Pamplona, 1997).

Propolis color varies from greenish-red-yellow to dark brown, depending on the origin and age. Propolis has a characteristic odor and spicy taste (Marcucci, 1995; Silva et al., 2019). Red tones predominate in regions where poplar is the main plant source for bees. Pinus causes green tones (König, 1985). Yellow parts in propolis samples are from the plant source and high wax amounts (Bruschi et al., 2002). The southern region of Brazil has a great vegetal diversity for resin removal by bees. In this region, there are five distinct common groups of propolis (Park et al., 2000).

Plant sources from tropical regions are slightly different from those in temperate zones (Bankova et al., 1998a). Fig. 7.1 shows the macroscopic characteristics of a propolis sample from Parana in southern Brazil.

Chemical composition

The chemical composition of propolis is complex and is directly related to the flora from the collection point, which can vary even in different locations in the same region (Marcucci, 1995). The time of year when propolis from a given location is collected can also affect its chemical composition (Bruschi et al., 2002, 2003; Bankova et al., 1998b; Zabaiou et al., 2017).

More than 300 constituents have been identified in propolis (Marcucci, 1995; Bankova et al., 2000). Hiitcheck and Borish in 1907 reported that propolis material consisted of resin (50%), beeswax (30%), volatile principles (e.g., essential and aromatic oils) (10%), pollen (5%), and various other substances (5%), including organic debris, and microelements, such as aluminum, calcium, strontium, iron, copper, manganese, and small amounts of vitamins B1, B2, B6, C, and E. They also found that propolis was soluble in alcohol and melts

Fig. 7.1 Macroscopic characteristics of: (A) *in natura* propolis; (B) crushed (comminuted) propolis.

at approximately 65°C (Bruschi et al., 2016). Ghisalberti (1979) detected the presence of a propolis fraction that was insoluble in water and organic solvents. This was probably formed by polymeric material (Ghisalberti, 1979).

In general, phenolic compounds make up more than 50% of the total weight of propolis. These compounds are related to their biological activity and antifungal properties (Bankova et al., 1994, 1992; Maciejewicz et al., 2001). Aromatic and aliphatic acids, benzoic acid derivatives, esters, and hydroquinone were also found in propolis samples (Vanhaelen and Vanhaelen-Fastré, 1979; Bankova et al., 1995).

Propolis samples from tropical regions have displayed different chemical compositions when compared to samples from temperate zones. Brazilian propolis is the subject of great interest from scientists (Lustosa et al., 2008; Trusheva et al., 2006). Brazilian green propolis is from the southeastern and southern regions of Brazil. In these regions, *Baccharis dracunculifolia* DC (Compositae) is the dominant component source of propolis (Banskota et al., 2001; de Funari et al., 2007; Kumazawa et al., 2004). This vegetable source is rich in phenolic substances; most of these substances are not flavonoids and cannot be found in propolis from North America, Europe, and Asia (Lustosa et al., 2008). Most phenol substances from Brazilian propolis are prenylated phenylpropanoids, such as 3,5-diprenyl-4-hydroxycinnamic acid (artepillin C) (Bruschi et al., 2016; de Funari et al., 2007; Aga et al., 1994; Hayashi et al., 1999; Woisky and Salatino, 1998; Bankova, 2005).

Other compounds have been isolated in samples of Brazilian propolis (3,5-diprenyl-4-hydroxycinnamic acid) and Chinese propolis (octacosanol) (Castaldo and Capasso, 2002). Some prophylactic benzaldehyde derivatives have been found in propolis, such as vanillin and isovaniline (Vanhaelen and Vanhaelen-Fastré, 1979).

More than 30 propolis volatile constituents have been identified. They consist mainly of terpenoids, with a predominance of sesquiterpenoid alcohols and sesquiterpenoid hydrocarbons, accompanied by some monoterpenes (mainly alcohols). Volatiles can be composed of alcohols, aromatics, phenols, aldehydes, ketones, acids (from acetic acid to stearic acid), and esters, that make up the largest group together with terpenoids. Alkanes accompanied by alkylated benzenes and naphthalene, two vinyl esters, and benzofurans were found in Brazilian propolis (Banskota et al., 2001; Bankova et al., 1994).

Propolis often contains a great variety of components that come from plants; some of these components are known for their biological activities that can benefit human health (Bruschi et al., 2002, 2016; Zabaiou et al., 2017; Burdock, 1998).

Antibacterial and antifungal activities are among the most-studied propolis biological activities. Phenolic substances such as flavonones, flavonols, and caffeic acid phenethyl ester are the main groups. Other components include caffeic acid, benzoic acid, and cinnamic acid (Bruschi et al., 2016; Lustosa et al., 2008; Burdock, 1998).

Physicochemical characteristics

As a gum-resin of complex chemical composition, propolis is a compound that displays interesting physicochemical characteristics. It is an amorphous material with a degree of hardness that is dependent on the wax content and temperature. Propolis is a difficult material to remove from human skin, as it can interact strongly with skin oils and proteins.

In nature, propolis is hard and brittle when cold, but soft and supple when heated. Therefore, propolis is normally submitted to sorting and then frozen after collection to facilitate manipulation, comminution, and processing. The fraction of propolis that is insoluble in water and organic solvents contributes to the characteristics of propolis formulations (Bruschi et al., 2016; Soleimanian et al., 2018).

Alone or incorporated in another dosage form, propolis extractive solution is commonly used in therapeutics. Most propolis substances are insoluble in water. Therefore, solvents (e.g., ethanol and propylene glycol) are often used together with water to prepare the extracts.

A strong and unpleasant taste, an aromatic odor, and high ethanol concentration are some of the disadvantages of propolis extract (Burdock, 1998). These disadvantages result in difficulties in packing, transportation, and incorporation of other dosage forms. Patient compliance to therapeutics is also critical to effectiveness (Bruschi et al., 2002). Propolis is a safe drug in low doses. However, doses higher than 15 g/day may cause adverse effects (Bruschi et al., 2016; Burdock, 1998). Standardized propolis preparations tend to be safer and less toxic than many other synthetic biologically active agents.

Products, by-products, and formulations

Pharmaceutical formulations such as creams, ointments, lotions, solutions, tablets, powders, and chewing gums have been developed for tissue regeneration, treatment of burns, neurodermatitis, leg ulcers, psoriasis, herpes simplex, genital pruritus, and against dermatophytes (Bruschi et al., 2002; Burdock, 1998; Kurek-Górecka et al., 2020).

The wax, resins, and gums of propolis aid the development of micro/nanostructured compounds, such as microparticles, nanoparticles, bioadhesive systems, thermoresponsive systems, and films. The use of propolis as an active agent and structuring material is dependent on these compounds (Bruschi et al., 2016; Soleimanian et al., 2018). Innovative systems have been proposed using advanced materials and technologies (e.g., nanotechnology) for the development of new propolis delivery systems. During the propolis extract preparation process, the solvent carries a small amount of substances. Most of the solvent is undissolved and constitutes the residue or propolis by-product that can be used as a source of biologically active substances and excipient for preparation of micro/nanoparticulate systems, films, and semisolids (de Toledo et al., 2015, 2016; Rosseto et al., 2017; de Francisco et al., 2018a, 2018b, 2019; Said Dos Santos et al., 2020; dos Santos et al., 2020).

In this context, several drug delivery systems have been proposed for propolis delivery in antifungal therapies: microparticles (Bruschi et al., 2003, 2006, 2004; de Francisco et al., 2018a, 2018b; Dota et al., 2011; Nori et al., 2011; Durán et al., 2007; de Souza Ferreira et al., 2014), nanoparticles (Roy et al., 2010; Kischkel et al., 2020; Iadnut et al., 2019), films (de Toledo et al., 2015, 2016; Zancanela et al., 2019), liposomes (Aytekin et al., 2020), phytosomes (Permana et al., 2020), niosomes (Arafa et al., 2020), emulsion systems (Said Dos Santos et al., 2020; dos Santos et al., 2020), organogels (de Francisco et al., 2019, 2020), liquid crystalline systems (Bruschi et al., 2008; de Toledo et al., 2018), nanostructured lipid systems (Rosseto et al., 2017; Rassu et al., 2015), and bioadhesive systems (Bruschi et al., 2007; Ceschel et al., 2002; de Lima et al., 2016; de Pereira et al., 2013).

Biological properties of propolis for human health

Pharmacological properties

Propolis has a diverse range of pharmacological properties: antimicrobial (Silva et al., 2019; Aytekin et al., 2020; Touzani et al., 2019), antiprotozoal (da Regueira-Neto et al., 2018; Siheri et al., 2019), antiviral (Yildirim et al., 2016; Silva-Beltrán et al., 2020), antioxidant (Aytekin et al., 2020; Bonamigo et al., 2017; Cavalaro et al., 2019; Nna et al., 2018), antitumor (Silva et al., 2019; Aytekin et al., 2020; Desamero et al., 2019; Salem et al., 2020), antiinflammatory (Nna et al., 2018; Alqarni et al., 2019), and immunomodulatory (Touzani et al., 2019; Alqarni et al., 2019; Piñeros et al., 2020). Various pharmacological properties are dependent on the specific chemical composition of the propolis type (Burdock, 1998; Boisard et al., 2015; Oryan et al., 2018).

Antifungal properties

The main fungal infection pathogens that show susceptibility to propolis include the yeasts *Candida albicans* (Capoci et al., 2015; Tobaldini-Valerio et al., 2016; Corrêa et al., 2020; Sayyadi et al., 2020; Gucwa et al., 2018), *C. glabrata* (Gucwa et al., 2018; Siqueira et al., 2015; Mutlu Sariguzel et al., 2016), *C. tropicalis*, *C. krusei*, *C. parapsilosis* (Nani et al., 2020), *Paracoccidioides brasiliensis* (Murad et al., 2002), *Malassezia* spp. (Deegan, 2018), and *Cryptococcus neoformans* (Thammasit et al., 2018); the filamentous fungi *Aspergillus fumigatus* (Netíková et al., 2013), *A. flavus* (Lorini et al., 2018), *Fusarium solani*, *F. oxysporum*, and *F. subglutinans* (Galletti et al., 2015, 2017; Veiga et al., 2018a); and dermatophytes *Trichophyton rubrum*, *T. tonsurans*, and *T. mentagrophytes* (Siqueira et al., 2009); *Microsporum gypseum* and *M. canis* (Netíková et al., 2013).

Mechanisms of propolis antifungal action

The antifungal activity of propolis has been confirmed in several studies in planktonic cells and biofilms of several fungi (Capoci et al., 2015; Tobaldini-Valerio et al., 2016; Veiga et al., 2018b). Phenolic compounds (e.g., flavonoids) have been considered one of the main biologically active constituents of propolis. They are directly related to the antifungal potential and mechanism of action (Gucwa et al., 2018; Salatino et al., 2005; Salas et al., 2011).

Currently there is evidence that the cell membrane is a possible target for propolis action (Gucwa et al., 2018). However, there is no consensus on these matters, since Pippi et al. (2015), when performing the same membrane depolarization tested by Gucwa et al. (2018), observed that propolis is not complexed with membrane ergosterol, indicating action on the cell wall of *Candida* spp.

The cell wall plays a crucial role in pathogenic fungi, acting as a permeability barrier. It is responsible for the survival and adaptation of the pathogen, as well as the signaling functions in stressful conditions during infection (Latgé et al., 2017). The antifungal activity of propolis is likely due to the ability of polyphenols to form complexes with soluble proteins present in the walls of fungal cells that interrupt chitin synthesis and cause a loss of integrity. The cell wall then becomes disorganized, and the fungal cell undergoes osmotic instability and deformations (Bowman and Free, 2006). As demonstrated by Corrêa et al. (2020), *C. albicans* cells treated with propolis showed a loss of cell wall integrity and deformed surfaces, with ruptures and leakage of intracellular content. This suggests that propolis bioactivity induces damage to the cell wall.

Gucwa et al. (2018) observed that propolis from northern Poland did not directly affect the cell wall of *Candida* spp. The study highlighted the action of propolis on the cell membrane, where they observed a reversal of action by ergosterol and membrane depolarization. This suggests cell walls are only a target of secondary importance for propolis action.

Dimorphism has also been investigated as a propolis mechanism of action. de Castro et al. (2013) evaluated the tolerance of different morphogenetic types of *C. albicans* against propolis. They discovered that the gene MCA1 is important for propolis-mediated cell death. MCA1 encodes a metacaspase responsible for cell death induced by oxidative stress from *C. albicans.* Recently, Corrêa et al. (2020) showed that propolis completely inhibited filamentation and the three morphogenic types (yeast, pseudohyphae, hyphae) of *C. albicans.*

RAS signaling mediates cell death of *Candida* spp. by propolis. This involves cell growth control, differentiation and apoptosis acceleration. The flavonoids present in propolis, when exhibiting pro-oxidant activity, benefit from increased levels of ROS (reactive oxygen species). This promotes mitochondrial dysfunction and has an important role in inducing apoptosis (Hwang et al., 2012).

The antifungal mechanisms of propolis are still unclear due to the synergy of its various components. However, the mechanisms are likely related to damage to the fungal cell membrane and wall.

Propolis cytotoxicity and mutagenic potential

When assessing the potential use of a product, therapeutic and toxicological effects (e.g., cytotoxic determination, mutagenic effects) of a compound are indispensable parameters (Montoro et al., 2012). Veiga et al. (2018b) observed that propolis caused dose-dependent cytotoxicity in HeLa and Vero cell lines. In these cases, the HeLa cell line was more susceptible than the Vero cell line. But when considering the established cytotoxicity threshold, none of the tested propolis concentrations showed a cytotoxic effect on cell lines after 24 h of exposure.

Possible cytotoxic and mutagenic effects of green propolis ethanol extract were studied by Tavares et al. (2006) from the evaluation of the mitotic index and the appearance of chromosomal aberrations in Chinese hamster ovary (CHO) cells. No significant differences were observed in the mitotic index between cultures treated with different concentrations of propolis extract and their respective negative controls and solvents. This demonstrated that the different evaluated concentrations were not cytotoxic. There were also no significant differences in the frequency of chromosomal aberrations between cultures treated with 12.5, 25, or 50 μg/mL of the extract. However, there was a small increase in the frequency of significant chromosomal aberrations with 100 μg/mL concentration.

Topical formulations supplemented with green propolis extract used in the treatment of burns do not have a mutagenic effect, as demonstrated by Senedese et al. (2011). They were evaluated through in vitro tests and micronucleus tests in vivo. However, when tested with 3.6% of the propolis gel, a cytotoxic effect was shown in the in vitro test. Absence of mutagenicity of propolis extract in cells HTC (Hepatoma Tissue Culture) were also observed by Roberto et al. (2016), as well as Turan et al. (2015) who also demonstrated the antimutagenicity of propolis and suggested the protective effect of DNA as a consequence, at least in part, of the antioxidant properties of its main components.

The mutagenicity of propolis was also evaluated by Corrêa et al. (2020). In the concentrations and conditions tested, they found no mutagenic potential and no reading frame mutations, frame shifts, or substitutions of base pairs. The determination of the mutagenicity of a compound with therapeutic activity is essential for its safety assessment. Compounds capable of inducing mutations in the DNA molecule can potentially damage the germ line, leading to problems related to fertility and mutations in future generations. An accumulation of mutations in a cell-dividing gene can lead to cancer (Mortelmans and Zeiger, 2000).

Cytotoxicity and mutagenicity assessments contribute to a better understanding of the action of propolis in the human body and allow for safer use. As previously mentioned, propolis is a substance rich in phenolic compounds. Thus, various researchers have examined the relationship between the biological activity promoted by the samples of propolis and its phytochemicals (e.g., flavonoids, tannins) that contribute significantly to antimutagenic activity and DNA protection. However, the mechanisms of antimutagenicity are still unknown (Horn and Vargas, 2008; Khan et al., 2018). Propolis has several pharmacological properties, but various studies have demonstrated that it has a degree of cytotoxicity and mutagenicity. For this reason, its use should be considered in terms of the concentration used.

Propolis perspectives for prevention or treatment of fungal infections

The pharmacological and nutraceutical benefits of propolis have been widely explored as functional foods, cosmetics, and medicines, among others. Propolis has been an important resource for the prevention and treatment of both oral and systemic diseases (Freires et al., 2016). Propolis is used in topical treatments for its antifungal properties. These treatments are mainly applied to skin, nails, and mucous membranes. Propolis is also widely used as a component in pharmaceutical and cosmetic products, such as antiacne creams, ointments, lotions, body and facial creams, and for oral hygiene formulations (Silva-Carvalho et al., 2015).

Propolis on cutaneous fungal infections

There is a great expectation for using propolis in the onychomycosis treatment. Studies have shown propolis as having cumulative properties, thus it is able to solve the hyperkeratose, dystrophy, and paronychia. It is effective on several fungi agents of onychomycosis, including in biofilm forms (Galletti et al., 2015, 2017; Veiga et al., 2018b). This is promising, as the biofilm structure is significantly more resistant than planktonic cell structures. The presence of biofilms on nails represents a major challenge in onychomycosis treatment (Gupta and Foley, 2019). Propolis reduces the biomass and structure of the biofilm, leaving only a small amount of damaged and viable isolated cells (Galletti et al., 2017).

This reinforces the potential of propolis for topical onychomycosis treatment and prevention of fungi dissemination from the nail. The polarity and chemical compositions of propolis enable good retention and permeation (de Toledo et al., 2015, 2016). These factors allow propolis to permeate the nail, overcoming this important physical barrier.

Transungual drug delivery is an important characteristic of topical antifungal candidates. Topical antifungal treatments are more effective and have faster cure rates than oral antifungal treatments. However, topical antifungal treatments often result in adverse treatment outcomes. This is due to high costs, long-term treatments, ineffective infection management, uncooperative patients, and hepatotoxic and gastrointestinal side effects (Gupta et al., 2017). Our research group has found, in the ex vivo model, that propolis can reach the deep nail layers within 24 h of application. In addition, a cure rate of 56.25% was recorded after 6 months of topical treatment with propolis extract (Veiga et al., 2018b). In the cured patients, there was a total recovery of the anatomical appearance of the nails and an absence of fungi in direct mycological examination. This shows the potential of propolis for nail application.

Propolis could potentially be used by health services without diagnostic laboratory support, as it is an empirical nontoxic treatment with a high chance of success, low drug interaction risk, and easy permeation.

Propolis on mucosal fungal infections

In dentistry, propolis has been used mainly in the composition of dentifrices and mouthwashes (de Carvalho et al., 2019). Several studies in recent years have demonstrated effective action of several propolis extracts against *C. albicans* and other yeasts responsible for this oral infection. Thus, the antifungal potential of propolis suggests that it can be applied as a treatment alternative for mucosal diseases affected by these species.

Propolis extracts are principally used as antimicrobial agents in several dental procedures and applications: oral and maxillofacial surgery, periodontics (controlling of postoperative diseases after gum surgery), endodontics (as a root canal irrigator), orthodontics (minimizing enamel demineralization), restorative dentistry, and oral hygiene (de Carvalho et al., 2019; Šabanović et al., 2019). Propolis is an anticariogenic agent. It can be used as an oral antiseptic for the prevention of caries. Toothpastes and mouthwash solutions containing propolis have shown impressive activity against oral pathogens. The biological activities of propolis are attributed to its variety of chemical constituents (Djais et al., 2020).

A study investigated the antimicrobial properties of six propolis solutions in oral pathogenic microorganisms and evaluated their cytotoxicity in gingival fibroblasts at different dilutions. All solutions showed strong antifungal activity and inhibited all tested *Candida* species. In addition, effective dilutions of propolis solutions have been found to be safe for gingival fibroblasts (Sonmez et al., 2005). Siqueira et al. (2015) evaluated the fungistatic and fungicidal activities of an alcoholic extract of red propolis against different species of *Candida* isolated from patients with cases of chronic periodontitis. Propolis demonstrated better fungistatic and fungicidal properties than fluconazole. All *Candida* isolates were susceptible to the antifungal activity of the propolis extract, but only a few samples were resistant to fluconazole.

In another study, Koo et al. (2000) evaluated the antimicrobial activity of a Brazilian propolis extract against *C. albicans* and other oral pathogens. They also examined the effect of this extract against adherence and formation of water-insoluble glucan. The Brazilian propolis extract caused significant inhibition of all tested microorganisms. It also almost completely inhibited cell adhesion and formation of water-insoluble glucan. A recent study compared the effect of an aqueous extract and ethanolic extract of Iranian propolis with routine antifungal agents against *Candida* species isolated from oral candidiasis lesions in cancer patients. Both extracts of Iranian propolis exhibited greater antifungal activity than the traditional antifungal; the ethanolic extract had a greater effect than the aqueous extract (Sayyadi et al., 2020).

For the treatment of prosthetic stomatitis from *Candida* spp., a multicenter study with older adults tested the efficacy and safety of a mucoadhesive oral gel formulation of Brazilian propolis extract and compared it to miconazole oral gel. Forty patients were randomly allocated to two groups: the control group received the oral miconazole gel, and the study group received a mucoadhesive formulation containing standardized propolis extract. Both treatments resulted in clinical improvements in candidiasis symptoms with a clinical cure rate of 70%. When considering the clinical cure rate, the propolis extract oral gel was just as effective as the miconazole gel. Thus, oral gel from propolis extract could be used as an alternative treatment in older patients (de Pina et al., 2017).

Propolis activity is not only demonstrated for yeasts, but also for bacteria that cause oral infections. Potent antibacterial activity was observed in a study that evaluated the effects of a propolis extract from Pakistan in 35 clinical isolates of anaerobic periodontal pathogens (Shabbir et al., 2016). Nazeri et al. (2019) also noted that a mouthwash with propolis was more effective against oral bacteria than chlorhexidine (CHX) and Listerine. Another study compared the efficiency of an ethanol extract of propolis to CHX in various species of bacteria and *C. albicans*, both in planktonic forms and biofilms of a single species. Both agents inhibited the growth of all species in planktonic forms. When compared to CHX, propolis produced better or equal results against the yeast and bacteria tested (Akca et al., 2016). Thus, propolis extract could potentially be as effective as CHX for oral microorganisms.

Despite the extensive literature evaluating the biological effects of propolis both in vitro and in vivo, few products reach the clinical stage and become commercially available. In addition, there is a lack of clinical trials in the dental field (de Carvalho et al., 2019). The development of new propolis-based dosage forms could significantly reduce the use of synthetic drugs and assist in the treatment of diseases of the oral cavity (Šabanović et al., 2019), as various studies have suggested (Kischkel et al., 2020; Bruschi et al., 2007; Fabri et al., 2011).

For vaginal infections, propolis has shown levels of antifungal activity that are similar to traditional antifungal compounds, regardless of the patient's clinical condition and isolated fungal species (Capoci et al., 2015). The action of an ethanol extract of propolis (PE) and propolis microparticles (PMs) were tested against clinical isolates of VVC yeasts. All yeasts were inhibited by PE and PMs, with little variation, regardless of the yeast species. The results demonstrated the potential application of PMs in VVC therapy and the possible prevention of the occurrence of new symptomatic episodes (Dota et al., 2011).

de Pereira et al. (2013) described the preparation and characterization of mucoadhesive thermoresponsive systems containing propolis for the treatment of VVC. The preparations showed antifungal activity against all tested yeasts, and the data obtained indicated a potentially useful role in the treatment of VVC. Considering such outcomes, these propolis-containing systems deserve clinical evaluation. Berretta et al. (2013) evaluated the in vitro and in vivo efficacy of propolis-based gels against *Candida* species. The results demonstrated the fungicidal action of the propolis extracts on three studied morphotypes (yeast, pseudohyphae, hyphae). Recently, Bonfim et al. (2020) demonstrated the potential use of propolis for VVC treatment. The authors tested a new mucoadhesive thermoresponsive platform for propolis delivery in a preclinical murine model. They found similar and better results than nystatin; results varied with different propolis concentrations. Propolis-based gels in the treatment of Balb/c mice demonstrated antifungal action similar to traditional antifungal treatments (clotrimazole and nystatin), justifying future clinical studies.

Propolis in the prevention of systemic fungal infections and potential use in the coating of medical devices

Few studies evaluate propolis for medical surface treatment and prevention of systemic fungal infections. Gucwa et al. (2018) examined the activity of 50 ethanolic extracts of Polish propolis in a group of 69 *C. albicans* clinical isolates. Most of the extracts showed impressive antifungal activity by eradicating biofilms from several surfaces (e.g., polystyrene, polyvinyl chloride, silicone catheters). Fluconazole and voriconazole had a synergistic effect with the extracts against *C. albicans*, as fluconazole and voriconazole MICs were 256–32 times lower when in the presence of propolis than those with the drug alone.

Extensive metabolism in vivo from commensals or probiotics in the intestine and liver is a factor that influences the bioavailability of natural products in systemic treatments. It can significantly alter the form found in systemic circulation. The development of pharmaceutical formulations that incorporate natural products can reduce side effects, decrease costs, and maximize the benefits of the formulations to avoid the mentioned problems. Possamai et al. (2013) evaluated the effect of a Brazilian propolis extract and its fractions, both isolated and adsorbed to microspheres of polyethylene glycol (PEG), on the activity of human phagocytes against *C. albicans*. Propolis had a stimulating effect on these cells by helping them to combat the fungus. These antifungal properties were significantly enhanced when incorporated into the microspheres. This suggests that when propolis is adsorbed to PEG microspheres, an immunostimulatory effect occurs against human blood phagocytes. Such an effect demonstrates the potential of this system for a variety of natural product therapies.

Conclusion

Propolis bioprospecting has increased in recent years. But obtaining propolis has environmental impacts that researchers need to consider. Thus, as it is a promising product, with different applicability, researchers have been concerned with optimizing its exploitation by applying the principle of 3Rs (reuse, reduce, and refine). The reuse of the by-product generated during the process of obtaining propolis is an example of this and has shown economic potential as much as propolis (de Francisco et al., 2018a, 2018b, 2020). Investments in advanced technologies currently aim to reduce raw material and turn propolis and its main biologically active constituents into safe and efficient pharmacological products. Development of new delivery systems for propolis and its derivatives (e.g., nanotechnology) aims to increase antimicrobial effects and reduce cytotoxicity (Bruschi et al., 2016; Soleimanian et al., 2018; Kischkel et al., 2020; Leo et al., 2017).

However, more research on propolis and its main biologically active constituents is needed, as the mechanisms of its antimicrobial, immunomodulatory, and antimutagenic activities are still unclear (Corrêa et al., 2020; Gucwa et al., 2018; Salatino et al., 2005; Salas et al., 2011; Latgé et al., 2017). New research is underway to reach a better understanding of its active ingredients (Zabaiou et al., 2017; Bruschi et al., 2016; Lustosa et al., 2008; Burdock, 1998). Studies (in vitro and in vivo) focusing on propolis antifungal activity have been increasing in recent years, but this has resulted in few randomized and clinical studies with pharmacological formulations of propolis and its derivatives (Clinicaltrials, 2020a,b,c; Cooper et al., 2020). Thus, we emphasize the urgency of new antifungals, the potential of propolis in this context, and the need for randomized and clinical studies involving propolis. Researching all these benefits requires time and investment, as well as an environmental, social, and economic concern.

References

Aga, H., Shibuya, T., Sugimoto, T., Kurimoto, M., Nakajima, S., 1994. Isolation and identification of antimicrobial compounds in Brazilian propolis. Biosci. Biotechnol. Biochem. 58, 945–946. https://doi.org/10.1271/bbb.58.945.

Akca, A.E., Eralp Akca, A., Akca, G., Topçu, F.T., Macit, E., Pikdöken, L., et al., 2016. The comparative evaluation of the antimicrobial effect of propolis with chlorhexidine against oral pathogens: an *in vitro* study. Biomed. Res. Int. 2016, 1–8. https://doi.org/10.1155/2016/3627463.

Alqarni, A.M., Niwasabutra, K., Sahlan, M., Fearnley, H., Fearnley, J., Ferro, V.A., et al., 2019. Propolis exerts an anti-inflammatory effect on PMA-differentiated THP-1 cells via inhibition of purine nucleoside phosphorylase. Metabolites 9. https://doi.org/10.3390/metabo9040075.

Arafa, M.G., Ghalwash, D., El-Kersh, D.M., Elmazar, M.M., 2020. Publisher correction: Propolis-based niosomes as oromuco-adhesive films: a randomized clinical trial of a therapeutic drug delivery platform for the treatment of oral recurrent aphthous ulcers. Sci. Rep. 10. https://doi.org/10.1038/s41598-020-59349-w.

Aytekin, A.A., Tuncay Tanrıverdi, S., Aydın Köse, F., Kart, D., Eroğlu, İ., Özer, Ö., 2020. Propolis loaded liposomes: evaluation of antimicrobial and antioxidant activities. J. Liposome Res. 30, 107–116. https://doi.org/10.1080/08982104.2019.1599012.

Bankova, V., 2005. Recent trends and important developments in propolis research. Evid. Based Complement. Alternat. Med. 2, 29–32. https://doi.org/10.1093/ecam/neh059.

Bankova, V., Dyulgerov, A., Popov, S., Evstatieva, L., Kuleva, L., Pureb, O., et al., 1992. Propolis produced in Bulgaria and Mongolia: phenolic compounds and plant origin. Apidologie 23, 79–85. https://doi.org/10.1051/apido:19920109.

Bankova, V., Christov, R., Popov, S., Pureb, O., Bocari, G., 1994. Volatile constituents of propolis. Z. Naturforsch. C 49, 6–10. https://doi.org/10.1515/znc-1994-1-202.

Bankova, V., Christov, R., Kujumgiev, A., Marcucci, M.C., Popov, S., 1995. Chemical composition and antibacterial activity of Brazilian propolis. Z. Naturforsch. C 50, 167–172. https://doi.org/10.1515/znc-1995-3-402.

Bankova, V., Boudourova-Krasteva, G., Popov, S., Sforcin, J.M., Cunha Funari, S.R., 1998a. Seasonal variations of the chemical composition of Brazilian propolis. Apidologie 29, 361–367. https://doi.org/10.1051/apido:19980406.

Bankova, V.S., Christov, R., Tejera, A.D., 1998b. Lignans and other constituents of propolis from the canary islands. Phytochemistry 49, 1411–1415. https://doi.org/10.1016/s0031-9422(98)00108-3.

Bankova, V.S., de Castro, S.L., Marcucci, M.C., 2000. Propolis: recent advances in chemistry and plant origin. Apidologie 31, 3–15. https://doi.org/10.1051/apido:2000102.

Banskota, A.H., Tezuka, Y., Kadota, S., 2001. Recent progress in pharmacological research of propolis. Phytother. Res. 15, 561–571. https://doi.org/10.1002/ptr.1029.

Bernard, C., Girardot, M., Imbert, C., 2020. *Candida albicans* interaction with gram-positive bacteria within interkingdom biofilms. J. Mycol. Med. 30. https://doi.org/10.1016/j.mycmed.2019.100909, 100909.

Berretta, A.A., de Castro, P.A., Cavalheiro, A.H., Fortes, V.S., Bom, V.P., Nascimento, A.P., et al., 2013. Evaluation of mucoadhesive gels with propolis (EPP-AF) in preclinical treatment of candidiasis vulvovaginal infection. Evid. Based Complement. Alternat. Med. 2013. https://doi.org/10.1155/2013/641480, 641480.

Bertolini, M.M., Xu, H., Sobue, T., Nobile, C.J., Del Bel Cury, A.A., Dongari-Bagtzoglou, A., 2015. Candida-streptococcal mucosal biofilms display distinct structural and virulence characteristics depending on growth conditions and hyphal morphotypes. Mol. Oral Microbiol. 30, 307–322. https://doi.org/10.1111/omi.12095.

Boisard, S., Le Ray, A.-M., Landreau, A., Kempf, M., Cassisa, V., Flurin, C., et al., 2015. Antifungal and antibacterial metabolites from a French poplar type propolis. Evid. Based Complement. Alternat. Med. 2015. https://doi.org/10.1155/2015/319240, 319240.

Bonamigo, T., Campos, J.F., Alfredo, T.M., Balestieri, J.B.P., Cardoso, C.A.L., Paredes-Gamero, E.J., et al., 2017. Antioxidant, cytotoxic, and toxic activities of propolis from two native bees in Brazil: *Scaptotrigona depilis* and *Melipona quadrifasciata anthidioides*. Oxid. Med. Cell. Longev. 2017, 1–12. https://doi.org/10.1155/2017/1038153.

Bonfim, A.P., Sakita, K.M., Faria, D.R., Arita, G.S., Vendramini, F.A.V.R., Capoci, I.R.G., et al., 2020. Preclinical approaches in vulvovaginal candidiasis treatment with mucoadhesive thermoresponsive systems containing propolis. PLoS One 15. https://doi.org/10.1371/journal.pone.0243197, e0243197.

Bowman, S.M., Free, S.J., 2006. The structure and synthesis of the fungal cell wall. Bioessays 28, 799–808. https://doi.org/10.1002/bies.20441.

Bruschi, M.L., Klein, T., Lopes, R.S., Franco, S.L., Gremião, M.P.D., 2002. Contribution to the quality control protocol of propolis and its extracts. Rev. Ciênc. Farm. 23 (2), 289–306.

Bruschi, M.L., Cardoso, M.L.C., Lucchesi, M.B., Gremião, M.P.D., 2003. Gelatin microparticles containing propolis obtained by spray-drying technique: preparation and characterization. Int. J. Pharm. 264, 45–55. https://doi.org/10.1016/s0378-5173(03)00386-7.

Bruschi, M.L., Lopes, R.S., Franco, S.L., Gremião, M.P.D., 2004. *In vitro* release of propolis from gelatin microparticles prepared by spray-drying technique. Rev. Ciênc. Farm. 25 (2), 79–84.

Bruschi, M.L., Lara, E.H.G., Martins, C.H.G., Vinholis, A.H.C., Casemiro, L.A., Panzeri, H., et al., 2006. Preparation and antimicrobial activity of gelatin microparticles containing propolis against oral pathogens. Drug Dev. Ind. Pharm. 32, 229–238. https://doi.org/10.1080/03639040500466312.

Bruschi, M.L., Jones, D.S., Panzeri, H., Gremião, M.P.D., de Freitas, O., Lara, E.H.G., 2007. Semisolid systems containing propolis for the treatment of periodontal disease: *in vitro* release kinetics, syringeability, rheological, textural, and mucoadhesive properties. J. Pharm. Sci. 96, 2074–2089. https://doi.org/10.1002/jps.20843.

Bruschi, M.L., de Freitas, O., Lara, E.H.G.E., Panzeri, H., Gremião, M.P.D., Jones, D.S., 2008. Precursor system of liquid crystalline phase containing propolis microparticles for the treatment of periodontal disease: development and characterization. Drug Dev. Ind. Pharm. 34, 267–278. https://doi.org/10.1080/03639040701655911.

Bruschi, M.L., de A Pereira, R.R., LMB, d.F., 2016. The use of propolis in micro/nanostructured pharmaceutical formulations. Recent Pat. Drug Deliv. Formul. 10, 130–140. https://doi.org/10.2174/1872211310666151230112616.

Burdock, G.A., 1998. Review of the biological properties and toxicity of bee propolis (propolis). Food Chem. Toxicol. 36, 347–363. https://doi.org/10.1016/s0278-6915(97)00145-2.

Caceres, D.H., Mohd Tap, R., Alastruey-Izquierdo, A., Hagen, F., 2020. Detection and control of fungal outbreaks. Mycopathologia 185, 741–745. https://doi.org/10.1007/s11046-020-00494-1.

Capoci, I.R.G., de Bonfim-Mendonça, P.S., Arita, G.S., de Pereira, R.R.A., Consolaro, M.E.L., Bruschi, M.L., et al., 2015. Propolis is an efficient fungicide and inhibitor of biofilm production by vaginal *Candida albicans*. Evid. Based Complement. Alternat. Med. 2015, 287693. https://doi.org/10.1155/2015/287693.

Castaldo, S., Capasso, F., 2002. Propolis, an old remedy used in modern medicine. Fitoterapia 73 (Suppl. 1), S1–S6. https://doi.org/10.1016/s0367-326x(02)00185-5.

Cavalaro, R.I., da Cruz, R.G., Dupont, S., de Moura Bell, J.M.L.N., De Vieira, T.M.F.S., 2019. *In vitro* and *in vivo* antioxidant properties of bioactive compounds from green propolis obtained by ultrasound-assisted extraction. Food Chem. X 4, 100054. https://doi.org/10.1016/j.fochx.2019.100054.

Ceschel, G.C., Maffei, P., Sforzini, A., Lombardi Borgia, S., Yasin, A., Ronchi, C., 2002. *In vitro* permeation through porcine buccal mucosa of caffeic acid phenetyl ester (CAPE) from a topical mucoadhesive gel containing propolis. Fitoterapia 73 (Suppl. 1), S44–S52. https://doi.org/10.1016/s0367-326x(02)00190-9.

Clinicaltrials, 2020a. Efficacy of Standardized-propolis Extract (EPP-AF®) Gel Formulation as Buccal Antiseptic. https://clinicaltrials.gov/ct2/show/NCT02818803. (accessed November 30, 2020).

Clinicaltrials, 2020b. Efficacy of a Propolis-based Denture Cleanser. https://clinicaltrials.gov/ct2/show/NCT01237782. (accessed November 30, 2020).

Clinicaltrials, 2020c. Patients With Vulvovaginal Candidiasis. ClinicalTrials.gov. https://clinicaltrials.gov/ct2/show/NCT03024502. (accessed November 30, 2020) Full Text View.

Cooper, T.E., Teng, C., Howell, M., Teixeira-Pinto, A., Tong, A., Wong, G., 2020. D-mannose for preventing and treating urinary tract infections. Cochrane Database Syst. Rev. 113, 9. https://doi.org/10.1002/14651858.CD013608.

Corrêa, J.L., Veiga, F.F., Jarros, I.C., Costa, M.I., Castilho, P.F., de Oliveira, K.M.P., et al., 2020. Propolis extract has bioactivity on the wall and cell membrane of *Candida albicans*. J. Ethnopharmacol. 256. https://doi.org/10.1016/j.jep.2020.112791, 112791.

Costa-Orlandi, C., Sardi, J., Pitangui, N., de Oliveira, H., Scorzoni, L., Galeane, M., et al., 2017. Fungal biofilms and polymicrobial diseases. J. Fungi 3, 22. https://doi.org/10.3390/jof3020022.

da Regueira-Neto, M.S., Tintino, S.R., Rolón, M., Coronal, C., Vega, M.C., de Queiroz Balbino, V., et al., 2018. Antitrypanosomal, antileishmanial and cytotoxic activities of Brazilian red propolis and plant resin of *Dalbergia ecastaphyllum* (L) Taub. Food Chem. Toxicol. 119, 215–221. https://doi.org/10.1016/j.fct.2018.04.029.

de Albuquerque Maranhão, F.C., Oliveira-Júnior, J.B., Dos Santos Araújo, M.A., Silva, D.M.W., 2019. Mycoses in northeastern Brazil: epidemiology and prevalence of fungal species in 8 years of retrospective analysis in Alagoas. Braz. J. Microbiol. 50, 969–978. https://doi.org/10.1007/s42770-019-00096-0.

de Carvalho, C., Fernandes, W.H.C., Mouttinho, T.B.F., de Souza, D.M., Marcucci, M.C., D'Alpino, P.H.P., 2019. Evidence-based studies and perspectives of the use of Brazilian green and red propolis in dentistry. Eur. J. Dent. 13, 459–465. https://doi.org/10.1055/s-0039-1700598.

de Castro, P.A., de Castro, P.A., Bom, V.L.P., Brown, N.A., de Almeida, R.S.C., Ramalho, L.N.Z., et al., 2013. Identification of the cell targets important for propolis-induced cell death in *Candida albicans*. Fungal Genet. Biol. 60, 74–86. https://doi.org/10.1016/j.fgb.2013.07.001.

de Francisco, L., Pinto, D., Rosseto, H., Toledo, L., Santos, R., Tobaldini-Valério, F., et al., 2018a. Evaluation of radical scavenging activity, intestinal cell viability and antifungal activity of Brazilian propolis by-product. Food Res. Int. 105, 537–547. https://doi.org/10.1016/j.foodres.2017.11.046.

de Francisco, L.M.B., Pinto, D., Rosseto, H.C., Toledo, L.D.A.S.D., Dos Santos, R.S., Costa, P., et al., 2018b. Development of a microparticulate system containing Brazilian propolis by-product and gelatine for ascorbic acid delivery: evaluation of intestinal cell viability and radical scavenging activity. Food Funct. 9, 4194–4206. https://doi.org/10.1039/c8fo00863a.

de Francisco, L.M.B., Rosseto, H.C., Toledo, L.D.A.S.D., dos Santos, R.S., de Souza Ferreira, S.B., et al., 2019. Organogel composed of poloxamer 188 and passion fruit oil: sol-gel transition, rheology, and mechanical properties. J. Mol. Liq. 289, 111170. https://doi.org/10.1016/j.molliq.2019.111170.

de Francisco, L.M.B., Pinto, D., Rosseto, H.C., Toledo, L.D.A.S.D., Dos Santos, R.S., Da Costa, P.J.C., et al., 2020. Design and characterization of an organogel system containing ascorbic acid microparticles produced with propolis by-product. Pharm. Dev. Technol. 25, 54–67. https://doi.org/10.1080/10837450.2019.1669643.

de Funari, C.S., de Oliveira, F.V., Mathor, M.B., 2007. Analysis of propolis from *Baccharis dracunculifolia* DC. (Compositae) and its effects on mouse fibroblasts. J. Ethnopharmacol. 111, 206–212. https://doi.org/10.1016/j.jep.2006.11.032.

de Lima, G.G., de Souza, R.O., Bozzi, A.D., Poplawska, M.A., Devine, D.M., Nugent, M.J.D., 2016. Extraction method plays critical role in antibacterial activity of propolis-loaded hydrogels. J. Pharm. Sci. 105, 1248–1257. https://doi.org/10.1016/j.xphs.2015.12.027.

de Pereira, R.R.A., Ribeiro Godoy, J.S., Stivalet Svidzinski, T.I., Bruschi, M.L., 2013. Preparation and characterization of mucoadhesive thermoresponsive systems containing propolis for the treatment of vulvovaginal candidiasis. J. Pharm. Sci. 102, 1222–1234. https://doi.org/10.1002/jps.23451.

de Pina, G.M.S., Lia, E.N., Berretta, A.A., Nascimento, A.P., Torres, E.C., Buszinski, A.F.M., et al., 2017. Efficacy of propolis on the denture stomatitis treatment in older adults: a multicentric randomized trial. Evid. Based Complement. Alternat. Med. 2017, 8971746. https://doi.org/10.1155/2017/8971746.

de Souza Ferreira, S.B., de Assis Dias, B.R., Obregón, C.S., Gomes, C.C., de Araújo Pereira, R.R., Ribeiro Godoy, J.S., et al., 2014. Microparticles containing propolis and metronidazole: *in vitro* characterization, release study and antimicrobial activity against periodontal pathogens. Pharm. Dev. Technol. 19, 173–180. https://doi.org/10.3109/10837450.2013.763262.

de Toledo, L.A.S., Bavato, M.I., Rosseto, H.C., Cortesi, R., Bruschi, M.L., 2015. Pharmaceutical films made from the waste material from the preparation of propolis extracts: development and characterization. Braz. J. Pharm. Sci. 51, 847–859. https://doi.org/10.1590/s1984-82502015000400011.

de Toledo, L.A.S., Rosseto, H.C., Dos Santos, R.S., Spizzo, F., Del Bianco, L., Montanha, M.C., et al., 2018. Thermal magnetic field activated propolis release from liquid crystalline system based on magnetic nanoparticles. AAPS PharmSciTech 19, 3258–3271. https://doi.org/10.1208/s12249-018-1163-4.

de Toledo, L.A.S., Rosseto, H.C., Ravani, L., Cortesi, R., Bruschi, M.L., 2016. Waste material of propolis as a film forming agent intended to modify the metronidazole release: preparation and characterization. Curr. Drug Deliv. 13, 1152–1164. https://doi.org/10.2174/1567201813666151113121827.

Deegan, K.R., 2018. Perfil de susceptibilidade de *Malassezia pachydermatis* frente a antifúngicos alopáticos e extratos de própolis vermelha, verde e marrom. Universidade Federal da Bahia, Bahia. https://repositorio.ufba.br/ri/bitstream/ri/29719/1/DISSERTA%c3%87%c3%83O%20Final.pdf.

Desamero, M.J., Kakuta, S., Tang, Y., Chambers, J.K., Uchida, K., Estacio, M.A., et al., 2019. Tumor-suppressing potential of stingless bee propolis in *in vitro* and *in vivo* models of differentiated-type gastric adenocarcinoma. Sci. Rep. 9, 19635. https://doi.org/10.1038/s41598-019-55465-4.

Djais, A.A., Jemmy, P.N., Putri, A.R., Soekanto, S.A., 2020. Description of *Streptococcus mutans*, *Streptococcus sanguinis*, and *Candida albicans* biofilms after exposure to propolis dentifrice by using OpenCFU method. Saudi Dent. J. 32, 129–134. https://doi.org/10.1016/j.sdentj.2019.08.003.

dos Santos, R.S., dos Santos, R.S., Rosseto, H.C., da Silva, J.B., Vecchi, C.F., Caetano, W., et al., 2020. The effect of carbomer 934P and different vegetable oils on physical stability, mechanical and rheological properties of emulsion-based systems containing propolis. J. Mol. Liq. 307. https://doi.org/10.1016/j.molliq.2020.112969, 112969.

Dota, K.F.D., Consolaro, M.E.L., Svidzinski, T.I.E., Bruschi, M.L., 2011. Antifungal activity of Brazilian propolis microparticles against yeasts isolated from vulvovaginal candidiasis. Evid. Based Complement. Alternat. Med. 2011. https://doi.org/10.1093/ecam/neq029, 201953.

Durán, N., Marcato, P.D., Buffo, C.M.S., De Azevedo, M.M.M., Esposito, E., 2007. Poly (epsilon-caprolactone)/propolis extract: microencapsulation and antibacterial activity evaluation. Pharmazie 62, 287–290.

Enoch, D.A., Yang, H., Aliyu, S.H., Micallef, C., 2017. The changing epidemiology of invasive fungal infections. Methods Mol. Biol., 17–65. https://doi.org/10.1007/978-1-4939-6515-1_2.

Fabri, F.V., Cupertino, R.R., Hidalgo, M.M., de Oliveira, R.M.M.W., Bruschi, M.L., 2011. Preparation and characterization of bioadhesive systems containing propolis or sildenafil for dental pulp protection. Drug Dev. Ind. Pharm. 37, 1446–1454. https://doi.org/10.3109/03639045.2011.584387.

Ferreira, E.O., Mendes, I.N.V.F., Monteiro, S.G., Crosara, K.T.B., Siqueira, W.L., de Maria Pedroso Silva de Azevedo, C., et al., 2019. Virulence properties and sensitivity profile of *Candida parapsilosis* complex species and *Kodamaea ohmeri* isolates from onychomycosis of HIV/AIDS patients. Microb. Pathog. 132, 282–292. https://doi.org/10.1016/j.micpath.2019.05.012.

Freires, I.A., de Alencar, S.M., Rosalen, P.L., 2016. A pharmacological perspective on the use of Brazilian red propolis and its isolated compounds against human diseases. Eur. J. Med. Chem. 110, 267–279. https://doi.org/10.1016/j.ejmech.2016.01.033.

Galletti, J., Negri, M., Grassi, F.L., Kioshima-Cotica, É.S., Svidzinski, T.I.E., 2015. *Fusarium* spp. is able to grow and invade healthy human nails as a single source of nutrients. Eur. J. Clin. Microbiol. Infect. Dis. 34, 1767–1772. https://doi.org/10.1007/s10096-015-2410-1.

Galletti, J., Tobaldini-Valerio, F.K., Silva, S., Kioshima, É.S., Trierveiler-Pereira, L., Bruschi, M., et al., 2017. Antibiofilm activity of propolis extract on *Fusarium* species from onychomycosis. Future Microbiol. 12, 1311–1321. https://doi.org/10.2217/fmb-2017-0052.

Ghisalberti, E.L., 1979. Propolis: a review. Bee World 60, 59–84. https://doi.org/10.1080/0005772x.1979.11097738.

Gucwa, K., Kusznierewicz, B., Milewski, S., Van Dijck, P., Szweda, P., 2018. Antifungal activity and synergism with azoles of polish propolis. Pathogens 7. https://doi.org/10.3390/pathogens7020056.

Gunaydin, S.D., Arikan-Akdagli, S., Akova, M., 2020. Fungal infections of the skin and soft tissue. Curr. Opin. Infect. Dis. 33, 130–136. https://doi.org/10.1097/QCO.0000000000000630.

Gupta, A.K., Foley, K.A., 2019. Evidence for biofilms in onychomycosis. G. Ital. Dermatol. Venereol. 154, 50–55. https://doi.org/10.23736/S0392-0488.18.06001-7.

Gupta, A.K., Versteeg, S.G., Shear, N.H., 2017. Onychomycosis in the 21st century: an update on diagnosis, epidemiology, and treatment. J. Cutan. Med. Surg. 21, 525–539. https://doi.org/10.1177/1203475417716362.

Hay, R., 2018. Therapy of skin, hair and nail fungal infections. J. Fungi. 4. https://doi.org/10.3390/jof4030099.

Hayashi, K., Komura, S., Isaji, N., Ohishi, N., Yagi, K., 1999. Isolation of antioxidative compounds from Brazilian propolis: 3,4-dihydroxy-5-prenylcinnamic acid, a novel potent antioxidant. Chem. Pharm. Bull. 47, 1521–1524. https://doi.org/10.1248/cpb.47.1521.

Horn, R.C., Vargas, V.M.F., 2008. Mutagenicity and antimutagenicity of teas used in popular medicine in the salmonella/microsome assay. Toxicol. In Vitro 22, 1043–1049. https://doi.org/10.1016/j.tiv.2007.12.014.

Hwang, I.-S., Lee, J., Jin, H.-G., Woo, E.-R., Lee, D.G., 2012. Amentoflavone stimulates mitochondrial dysfunction and induces apoptotic cell death in *Candida albicans*. Mycopathologia 173, 207–218. https://doi.org/10.1007/s11046-011-9503-x.

Iadnut, A., Mamoon, K., Thammasit, P., Pawichai, S., Tima, S., Preechasuth, K., et al., 2019. *In vitro* antifungal and antivirulence activities of biologically synthesized ethanolic extract of propolis-loaded PLGA nanoparticles against *Candida albicans*. Evid. Based Complement. Alternat. Med. 2019, 1–14. https://doi.org/10.1155/2019/3715481.

Khan, M.S., Qais, F.A., Ahmad, I., Hussain, A., Alajmi, M.F., 2018. Genotoxicity inhibition by Syzygium cumini (L.) seed fraction and rutin: understanding the underlying mechanism of DNA protection. Toxicol. Res. 7, 156–171. https://doi.org/10.1039/c7tx00269f.

Kischkel, B., de Castilho, P.F., de Oliveira, K.M., Rezende, P.S., Bruschi, M.L., Svidzinski, T.I., et al., 2020. Silver nanoparticles stabilized with propolis show reduced toxicity and potential activity against fungal infections. Future Microbiol. 15, 521–539. https://doi.org/10.2217/fmb-2019-0173.

König, B., 1985. Plant sources of propolis. Bee World 66, 136–139. https://doi.org/10.1080/0005772x.1985.11098844.

Koo, H., Gomes, B.P., Rosalen, P.L., Ambrosano, G.M., Park, Y.K., Cury, J.A., 2000. *In vitro* antimicrobial activity of propolis and *Arnica montana* against oral pathogens. Arch. Oral Biol. 45, 141–148. https://doi.org/10.1016/s0003-9969(99)00117-x.

Kumazawa, S., Hamasaka, T., Nakayama, T., 2004. Antioxidant activity of propolis of various geographic origins. Food Chem. 84, 329–339. https://doi.org/10.1016/s0308-8146(03)00216-4.

Kurek-Górecka, A., Górecki, M., Rzepecka-Stojko, A., Balwierz, R., Stojko, J., 2020. Bee products in dermatology and skin care. Molecules 25. https://doi.org/10.3390/molecules25030556.

Latgé, J.-P., Beauvais, A., Chamilos, G., 2017. The cell wall of the human fungal pathogen *aspergillus fumigatus*: biosynthesis, organization, immune response, and virulence. Annu. Rev. Microbiol. 71, 99–116. https://doi.org/10.1146/annurev-micro-030117-020406.

Leo, D.E.V., Cappelli, V., Massaro, M.G., Tosti, C., Morgante, G., 2017. Evaluation of the effects of a natural dietary supplement with cranberry, Noxamicina® and D-mannose in recurrent urinary infections in perimenopausal women. Minerva Ginecol. 69, 336–341. https://doi.org/10.23736/S0026-4784.17.04074-6.

Lohse, M.B., Gulati, M., Johnson, A.D., Nobile, C.J., 2018. Development and regulation of single- and multi-species *Candida albicans* biofilms. Nat. Rev. Microbiol. 16, 19–31. https://doi.org/10.1038/nrmicro.2017.107.

Lorini, A., Wobeto, C., Bonaldo, S.M., de Carvalho Campos Botelho, S., Sinhorin, A.P., 2018. Chemical composition and antifungal activity of propolis on *Aspergillus flavus*. Biosci. J., 1298–1307. https://doi.org/10.14393/bj-v34n5a2018-36601.

Lustosa, S.R., Galindo, A.B., Nunes, L.C.C., Randau, K.P., Rolim Neto, P.J., 2008. Própolis: atualizações sobre a química e a farmacologia. Rev. Bras 18, 447–454. https://doi.org/10.1590/s0102-695x2008000300020.

Maciejewicz, W., Daniewski, M., Bal, K., Markowski, W., 2001. GC-MS identification of the flavonoid aglycones isolated from propolis. Chromatographia 53, 343–346. https://doi.org/10.1007/bf02490438.

Marcucci, M.C., 1995. Propolis: chemical composition, biological properties and therapeutic activity. Apidologie 26, 83–99. https://doi.org/10.1051/apido:19950202.

Midorikawa, K., Banskota, A.H., Tezuka, Y., Nagaoka, T., Matsushige, K., Message, D., et al., 2001. Liquid chromatography-mass spectrometry analysis of propolis. Phytochem. Anal. 12, 366–373. https://doi.org/10.1002/pca.605.

Montoro, A., Soriano, J.M., Barquinero, J.F., Almonacid, M., Montoro, A., Verdú, G., et al., 2012. Assessment *in vitro* of cytogenetic and genotoxic effects of propolis on human lymphocytes. Food Chem. Toxicol. 50, 216–221. https://doi.org/10.1016/j.fct.2011.10.053.

Mortelmans, K., Zeiger, E., 2000. The Ames Salmonella/microsome mutagenicity assay. Mutat. Res. 455, 29–60. https://doi.org/10.1016/s0027-5107(00)00064-6.

Murad, J.M., Calvi, S.A., Soares, A.M.V.C., Bankova, V., Sforcin, J.M., 2002. Effects of propolis from Brazil and Bulgaria on fungicidal activity of macrophages against *Paracoccidioides brasiliensis*. J. Ethnopharmacol. 79, 331–334. https://doi.org/10.1016/s0378-8741(01)00404-4.

Mutlu Sariguzel, F., Berk, E., Koc, A.N., Sav, H., Demir, G., 2016. Antifungal activity of propolis against yeasts isolated from blood culture: in vitro evaluation. J. Clin. Lab. Anal. 30, 513–516. https://doi.org/10.1002/jcla.21889.

Nani, B.D., de Sardi, J.C.O., Lazarini, J.G., Silva, D.R., Massariolli, A.P., Cunha, T.M., et al., 2020. Anti-inflammatory and anti- effects of Brazilian organic propolis, a promising source of bioactive molecules and functional food. J. Agric. Food Chem. 68, 2861–2871. https://doi.org/10.1021/acs.jafc.8b07304.

Nazeri, R., Ghaiour, M., Abbasi, S., 2019. Evaluation of antibacterial effect of propolis and its application in mouthwash production. Front. Dent. 16, 1–12. https://doi.org/10.18502/fid.v16i1.1103.

Netíková, L., Bogusch, P., Heneberg, P., 2013. Czech ethanol-free propolis extract displays inhibitory activity against a broad spectrum of bacterial and fungal pathogens. J. Food Sci. 78, M1421–M1429. https://doi.org/10.1111/1750-3841.12230.

Nna, V.U., Abu Bakar, A.B., Md Lazin, M.R.M.L., Mohamed, M., 2018. Antioxidant, anti-inflammatory and synergistic anti-hyperglycemic effects of Malaysian propolis and metformin in streptozotocin-induced diabetic rats. Food Chem. Toxicol. 120, 305–320. https://doi.org/10.1016/j.fct.2018.07.028.

Nori, M.P., Favaro-Trindade, C.S., de Alencar, S.M., Thomazini, M., de Camargo Balieiro, J.C., Contreras Castillo, C.J., 2011. Microencapsulation of propolis extract by complex coacervation. LWT Food Sci. Technol. 44, 429–435. https://doi.org/10.1016/j.lwt.2010.09.010.

O'Donnell, L.E., Millhouse, E., Sherry, L., Kean, R., Malcolm, J., Nile, C.J., et al., 2015. Polymicrobial *Candida* biofilms: friends and foe in the oral cavity. FEMS Yeast Res. 15. https://doi.org/10.1093/femsyr/fov077.

Oryan, A., Alemzadeh, E., Moshiri, A., 2018. Potential role of propolis in wound healing: biological properties and therapeutic activities. Biomed. Pharmacother. 98, 469–483. https://doi.org/10.1016/j.biopha.2017.12.069.

Otašević, S., Barac, A., Pekmezovic, M., Tasic, S., Ignjatović, A., Momčilović, S., et al., 2016. The prevalence of *Candida* onychomycosis in Southeastern Serbia from 2011 to 2015. Mycoses 59, 167–172. https://doi.org/10.1111/myc.12448.

Pamplona, B., 1997. Própolis: Composição e atividades terapêuticas. Rev. Racine 7 (37), 49–53.

Park, Y.K., Ikegaki, M., Alencar, S.M., 2000. Evaluation of Brazilian propolis by both physicochemical methods and biological activity. Honeybee Sci. 21 (2), 85–90.

Park, Y.K., Alencar, S.M., Scamparini, A.R.P., Aguiar, C.L., 2002. Própolis produzida no sul do Brasil, Argentina e Uruguai: evidências fitoquímicas de sua origem vegetal. Cienc. Rural 32, 997–1003. https://doi.org/10.1590/s0103-84782002000600013.

Permana, A.D., Utami, R.N., Courtenay, A.J., Manggau, M.A., Donnelly, R.F., Rahman, L., 2020. Phytosomal nanocarriers as platforms for improved delivery of natural antioxidant and photoprotective compounds in propolis: an approach for enhanced both dissolution behaviour in biorelevant media and skin retention profiles. J. Photochem. Photobiol. B 205. https://doi.org/10.1016/j.jphotobiol.2020.111846, 111846.

Piñeros, A.R., de Lima, M.H.F., Rodrigues, T., Gembre, A.F., Bertolini, T.B., Fonseca, M.D., et al., 2020. Green propolis increases myeloid suppressor cells and CD4Foxp3 cells and reduces Th2 inflammation in the lungs after allergen exposure. J. Ethnopharmacol. 252. https://doi.org/10.1016/j.jep.2019.112496, 112496.

Pippi, B., Lana, A.J.D., Moraes, R.C., Güez, C.M., Machado, M., de Oliveira, L.F.S., et al., 2015. *In vitro* evaluation of the acquisition of resistance, antifungal activity and synergism of Brazilian red propolis with antifungal drugs on *Candida* spp. J. Appl. Microbiol. 118, 839–850. https://doi.org/10.1111/jam.12746.

Possamai, M.M., Honorio-França, A.C., Reinaque, A.P.B., França, E.L., de Souto, P.C.S., 2013. Brazilian propolis: a natural product that improved the fungicidal activity by blood phagocytes. Biomed. Res. Int. 2013, 541018. https://doi.org/10.1155/2013/541018.

Rassu, G., Cossu, M., Langasco, R., Carta, A., Cavalli, R., Giunchedi, P., et al., 2015. Propolis as lipid bioactive nano-carrier for topical nasal drug delivery. Colloids Surf. B Biointerfaces 136, 908–917. https://doi.org/10.1016/j.colsurfb.2015.10.043.

Rautemaa, R., Ramage, G., 2011. Oral candidosis—clinical challenges of a biofilm disease. Crit. Rev. Microbiol. 37, 328–336. https://doi.org/10.3109/1040841x.2011.585606.

Richardson, M.D., Jones, B.L., Rautemaa, R., 2007. Therapeutic guidelines in systemic fungal infections. Remedica.

Roberto, M.M., Matsumoto, S.T., Jamal, C.M., Malaspina, O., Marin-Morales, M.A., 2016. Evaluation of the genotoxicity/mutagenicity and antigenotoxicity/antimutagenicity induced by propolis and *Baccharis dracunculifolia*, by *in vitro* study with HTC cells. Toxicol. In Vitro 33, 9–15. https://doi.org/10.1016/j.tiv.2016.02.005.

Rodríguez-Cerdeira, C., Gregorio, M.C., Molares-Vila, A., López-Barcenas, A., Fabbrocini, G., Bardhi, B., et al., 2019. Biofilms and vulvovaginal candidiasis. Colloids Surf. B Biointerfaces 174, 110–125. https://doi.org/10.1016/j.colsurfb.2018.11.011.

Rosseto, H.C., de Toledo, L.A.S., de Francisco, L.M.B., Esposito, E., Lim, Y., Valacchi, G., et al., 2017. Nanostructured lipid systems modified with waste material of propolis for wound healing: design, *in vitro* and *in vivo* evaluation. Colloids Surf. B Biointerfaces 158, 441–452. https://doi.org/10.1016/j.colsurfb.2017.07.029.

Roy, N., Mondal, S., Laskar, R.A., Basu, S., Mandal, D., Begum, N.A., 2010. Biogenic synthesis of Au and Ag nanoparticles by Indian propolis and its constituents. Colloids Surf. B Biointerfaces 76, 317–325. https://doi.org/10.1016/j.colsurfb.2009.11.011.

Šabanović, M., Saltović, S., Mujkić, A.A., Jašić, M., Bahić, Z., 2019. Impact of propolis on the oral health. Balk J. Dent. Med. 23, 1–9. https://doi.org/10.2478/bjdm-2019-0001.

Said Dos Santos, R., Vecchi, C.F., Rosseto, H.C., Bassi da Silva, J., Dano, M.E.L., de Castro-Hoshino, L.V., et al., 2020. Emulgels containing carbopol 934P and different vegetable oils for topical propolis delivery: bioadhesion, drug release profile, and ex vivo skin permeation studies. AAPS PharmSciTech 21, 209. https://doi.org/10.1208/s12249-020-01748-3.

Salas, M.P., Céliz, G., Geronazzo, H., Daz, M., Resnik, S.L., 2011. Antifungal activity of natural and enzymatically-modified flavonoids isolated from citrus species. Food Chem. 124, 1411–1415. https://doi.org/10.1016/j.foodchem.2010.07.100.

Salatino, A., Teixeira, E.W., Negri, G., Message, D., 2005. Origin and chemical variation of Brazilian propolis. Evid. Based Complement. Alternat. Med. 2, 33–38. https://doi.org/10.1093/ecam/neh060.

Salem, M.M., Donia, T., Abu-Khudir, R., Ramadan, H., Ali, E.M.M., Mohamed, T.M., 2020. Propolis potentiates methotrexate anticancer mechanism and reduces its toxic effects. Nutr. Cancer 72, 460–480. https://doi.org/10.1080/01635581.2019.1640884.

Santos, L.M., Fonseca, M.S., Sokolonski, A.R., Deegan, K.R., Araújo, R.P., Umsza-Guez, M.A., et al., 2020. Propolis: types, composition, biological activities, and veterinary product patent prospecting. J. Sci. Food Agric. 100, 1369–1382. https://doi.org/10.1002/jsfa.10024.

Sav, H., Baris, A., Turan, D., Altinbas, R., Sen, S., 2018. The frequency, antifungal susceptibility and enzymatic profiles of *Candida* species in cases of onychomycosis infection. Microb. Pathog. 116, 257–262. https://doi.org/10.1016/j.micpath.2018.01.036.

Sayyadi, F., Mahdavi, S., Moghadamnia, A.A., Moslemi, D., Shirzad, A., Motallebnejad, M., 2020. The effect of aqueous and ethanolic extract of Iranian propolis on *Candida Albicans* isolated from the mouth of patients with colorectal malignancy undergone chemotherapy: an *in-vitro* study. Caspian J. Intern. Med. 11, 62–66. https://doi.org/10.22088/cjim.11.1.62.

Senedese, J.M., Rodrigues, A.R., Furtado, M.A., Faustino, V.D., Berretta, A.A., Marchetti, J.M., et al., 2011. Assessment of the mutagenic activity of extracts of brazilian propolis in topical pharmaceutical formulations on mammalian cells *in vitro* and *in vivo*. Evid. Based Complement. Alternat. Med. 2011. https://doi.org/10.1093/ecam/nen049, 315701.

Shabbir, A., Rashid, M., Tipu, H.N., 2016. Propolis, a hope for the future in treating resistant periodontal pathogens. Cureus 8. https://doi.org/10.7759/cureus.682, e682.

Siheri, W., Ebiloma, G.U., Igoli, J.O., Gray, A.I., Biddau, M., Akrachalanont, P., et al., 2019. Isolation of a novel flavanonol and an alkylresorcinol with highly potent anti-trypanosomal activity from libyan propolis. Molecules 24, 1041. https://doi.org/10.3390/molecules24061041.

Silva, F.R.G., Matias, T.M.S., Souza, L.I.O., Matos-Rocha, T.J., Fonseca, S.A., Mousinho, K.C., et al., 2019. Phytochemical screening and *in vitro* antibacterial, antifungal, antioxidant and antitumor activities of the red propolis Alagoas. Braz. J. Biol. 79, 452–459. https://doi.org/10.1590/1519-6984.182959.

Silva-Beltrán, N.P., Balderrama-Carmona, A.P., Umsza-Guez, M.A., Souza Machado, B.A., 2020. Antiviral effects of Brazilian green and red propolis extracts on *Enterovirus surrogates*. Environ. Sci. Pollut. Res. Int. 27, 28510–28517. https://doi.org/10.1007/s11356-019-07458-z.

Silva-Carvalho, R., Baltazar, F., Almeida-Aguiar, C., 2015. Propolis: a complex natural product with a plethora of biological activities that can be explored for drug development. Evid. Based Complement. Alternat. Med. 2015. https://doi.org/10.1155/2015/206439, 206439.

Siqueira, A.B.S., Gomes, B.S., Cambuim, I., Maia, R., Abreu, S., Souza-Motta, C.M., et al., 2009. *Trichophyton* species susceptibility to green and red propolis from Brazil. Lett. Appl. Microbiol. 48, 90–96. https://doi.org/10.1111/j.1472-765x.2008.02494.x.

Siqueira, A.B.S., de Araújo Rodriguez, L.R.N., Santos, R.K.B., Marinho, R.R.B., Abreu, S., Peixoto, R.F., et al., 2015. Antifungal activity of propolis against *Candida* species isolated from cases of chronic periodontitis. Braz. Oral Res. 29, 1–6. https://doi.org/10.1590/1807-3107bor-2015.vol29.0083.

Soleimanian, Y., Goli, S.A.H., Varshosaz, J., Maestrelli, F., 2018. Propolis wax nanostructured lipid carrier for delivery of β sitosterol: effect of formulation variables on physicochemical properties. Food Chem. 260, 97–105. https://doi.org/10.1016/j.foodchem.2018.03.145.

Sonmez, S., Kirilmaz, L., Yucesoy, M., Yücel, B., Yilmaz, B., 2005. The effect of bee propolis on oral pathogens and human gingival fibroblasts. J. Ethnopharmacol. 102, 371–376. https://doi.org/10.1016/j.jep.2005.06.035.

Stenderup, A., 1990. Oral mycology. Acta Odontol. Scand. 48, 3–10. https://doi.org/10.3109/00016359009012728.

Subramanya, S.H., Hamal, D., Nayak, N., Gokhale, S., 2016. Onychomycosis due to *Candida parapsilosis* in a child with ventricular septal defect: an unusual predisposition. Case Rep. Pediatr. 2016, 1–4. https://doi.org/10.1155/2016/7026068.

Tavares, D.C., Mazzaron Barcelos, G.R., Silva, L.F., Chacon Tonin, C.C., Bastos, J.K., 2006. Propolis-induced genotoxicity and antigenotoxicity in Chinese hamster ovary cells. Toxicol. In Vitro 20, 1154–1158. https://doi.org/10.1016/j.tiv.2006.02.009.

Thammasit, P., Iadnut, A., Mamoon, K., Khacha-ananda, S., Chupradit, K., Tayapiwatana, C., et al., 2018. A potential of propolis on major virulence factors of *Cryptococcus neoformans*. Microb. Pathog. 123, 296–303. https://doi.org/10.1016/j.micpath.2018.07.028.

Tobaldini-Valerio, F.K., Bonfim-Mendonça, P.S., Rosseto, H.C., Bruschi, M.L., Henriques, M., Negri, M., et al., 2016. Propolis: a potential natural product to fight *Candida* species infections. Future Microbiol. 11, 1035–1046. https://doi.org/10.2217/fmb-2015-0016.

Touzani, S., Embaslat, W., Imtara, H., Kmail, A., Kadan, S., Zaid, H., et al., 2019. *In vitro* evaluation of the potential use of propolis as a multitarget therapeutic product: physicochemical properties, chemical composition, and immunomodulatory, antibacterial, and anticancer properties. Biomed. Res. Int. 2019, 1–11. https://doi.org/10.1155/2019/4836378.

Trusheva, B., Popova, M., Bankova, V., Simova, S., Marcucci, M.C., Miorin, P.L., et al., 2006. Bioactive constituents of brazilian red propolis. Evid. Based Complement. Alternat. Med. 3, 249–254. https://doi.org/10.1093/ecam/nel006.

Turan, I., Değer, O., Aliyazicioğlu, Y., Demir, S., Kilinç, K., Sümer, A., 2015. Effects of Turkish propolis on expression of hOGG-1 and NEIL-1. Turk. J. Med. Sci. 45, 804–811. https://doi.org/10.3906/sag-1406-98.

Vanhaelen, M., Vanhaelen-Fastré, R., 1979. Propolis.—I. Origin, microscopical investigations, chemical constituents and therapeutical activity (author's transl). J. Pharm. Belg. 34, 253–259.

Vasconcellos, C., Pereira, C.Q.M., Souza, M.C., Pelegrini, A., Freitas, R.S., Takahashi, J.P., 2013. Identification of fungi species in the onychomycosis of institutionalized elderly. An. Bras. Dermatol. 88, 377–380. https://doi.org/10.1590/abd1806-4841.20131884.

Veiga, F.F., Costa, M.I., Cótica, É.S.K., Svidzinski, T.I.E., Negri, M., 2018a. Propolis for the treatment of onychomycosis. Indian J. Dermatol. 63, 515–517. https://doi.org/10.4103/ijd.IJD_365_17.

Veiga, F.F., Gadelha, M.C., da Silva, M.R.T., Costa, M.I., Kischkel, B., de Castro-Hoshino, L.V., et al., 2018b. Propolis extract for onychomycosis topical treatment: from bench to clinic. Front. Microbiol. 9, 779. https://doi.org/10.3389/fmicb.2018.00779.

von Lilienfeld-Toal, M., Wagener, J., Einsele, H., Cornely, O.A., Kurzai, O., 2019. Invasive fungal infection. Dtsch. Arztebl. Int. 116, 271–278. https://doi.org/10.3238/arztebl.2019.0271.

Willems, H.M.E., Ahmed, S.S., Liu, J., Xu, Z., Peters, B.M., 2020. Vulvovaginal candidiasis: a current understanding and burning questions. J Fungi 6. https://doi.org/10.3390/jof6010027.

Woisky, R.G., Salatino, A., 1998. Analysis of propolis: some parameters and procedures for chemical quality control. J. Apic. Res. 37, 99–105. https://doi.org/10.1080/00218839.1998.11100961.

Worldwide Life n.d. http://www.life-worldwide.org/ (accessed November 28, 2020).

Yildirim, A., Duran, G.G., Duran, N., Jenedi, K., Bolgul, B.S., Miraloglu, M., et al., 2016. Antiviral activity of Hatay propolis against replication of *herpes simplex* virus type 1 and type 2. Med. Sci. Monit. 22, 422–430. https://doi.org/10.12659/msm.897282.

Zabaiou, N., Fouache, A., Trousson, A., Baron, S., Zellagui, A., Lahouel, M., et al., 2017. Biological properties of propolis extracts: something new from an ancient product. Chem. Phys. Lipids 207, 214–222. https://doi.org/10.1016/j.chemphyslip.2017.04.005.

Zancanela, D.C., Funari, C.S., Herculano, R.D., Mello, V.M., Rodrigues, C.M., Borges, F.A., et al., 2019. Natural rubber latex membranes incorporated with three different types of propolis: physical-chemistry and antimicrobial behaviours. Mater. Sci. Eng. C Mater. Biol. Appl. 97, 576–582. https://doi.org/10.1016/j.msec.2018.12.042.

CHAPTER 8

Royal jelly and fertility

Róbert Gáspár[a] and Adrienn B. Seres[b]
[a]Albert Szent-Györgyi School of Medicine, Department of Pharmacology and Pharmacotherapy, University of Szeged, Szeged, Hungary
[b]Faculty of Pharmacy, Department of Pharmacodynamics and Biopharmacy, University of Szeged, Szeged, Hungary

Contents

Introduction

Royal jelly is the key nutritional factor for the development and fertility of queen bee. The components of royal jelly ensure the constant egg-laying capability of the queen. The efficacy of royal jelly is so high that the development time from larvae to queen honey bee takes only 16 days (Anton and Grozinger, 2020). One of the main drives of queen fertility is royalactin, a unique protein in royal jelly that has a crucial role in the postembryonic development of honey bees and makes the difference between workers and the queen. While the main queen development focuses on the reproductive functions (ovary development, egg-laying capacity), the development of workers is about the creation of a very complex neurological system to ensure the high capability to memory, learning, and navigation. Royalactin itself is responsible for enhancing the body size and the increased production of the juvenile hormone that stimulates ovary development (Kamakura,

Bee Products and Their Applications in the Food and Pharmaceutical Industries
https://doi.org/10.1016/B978-0-323-85400-9.00003-4

2011). The secretion of the juvenile hormone is regulated by epidermal growth factor receptor that is also overexpressed in the queen by royal jelly (Kunugi and Ali, 2019). Since royal jelly is very rich in methionine and methyl groups having an essential role in phenotype-determining methylation processes, the unique development and fertility of queen honey bee can be explained by these epigenetic processes (Maleszka, 2018). Additionally, royal jelly is free from phenolic compounds like quercetin or *p*-coumaric acid that can induce several cytochrome P enzymes in adult bees responsible for developing worker bees without fertility. The experimental enrichment of royal jelly with *p*-coumaric acid reduces the ovary development of queen bee (Mao et al., 2015).

The insects and the human are very different; thus, theoretically, there is a low chance to find hormones or foods that have the same effect on both species. However, some of the insect hormones have been identified with well-described human actions. Ecdysterone is a natural insect steroid hormone playing a role in molting. In humans, it has an anabolic action increasing the muscle mass and changing the shape of the body, causing a challenge in antidoping endeavors in elite sport (Isenmann et al., 2019). The corticotropin-releasing hormone can be found both in insects and humans, having the same critical role in the stress response (Ketchesin et al., 2017). Even human hormones can act on insects: adipokinetic hormone related to obesity in humans as found to modify the Drosophila metabolism (Gáliková et al., 2017). Hence, it is not surprising that royal jelly has an essential role in maintaining fertility in queen honey bee and has a similar action in other species, even humans.

This chapter discusses the influence of royal jelly on female and male fertility based on scientific evidence generated through various animal experiments, including rodents and farm animals. Also, it reviews the currently available knowledge in clinical trials on humans. In conclusion, future research needs were addressed while paying particular attention to the potential benefits of drone milk.

The effects of royal jelly on fertility in experimental animal models

Most animal experiments investigating the action of royal jelly on fertility were carried out on mice and rats. The fertility studies focused on female and male animals, modeling the trend of reduced human fertility rates in both sexes, especially in industrialized countries (Jensen et al., 2002).

Effect of royal jelly on female fertility in rodents

It is known that estrogen has a crucial role in maintaining female fertility in many species, including rodents. The estrogen receptor gene disrupted female mice become infertile with hypoplastic uteri and hyperemic ovaries (Lubahn et al., 1993). Thus, each compound that has an estrogenic effect may improve the fertility of female rodents. Royal jelly binds to estrogenic receptors, but its affinity is less than that of diethylstilbestrol or phytoestrogens. The estrogenic effect of royal jelly was tested in ovariectomized rats. A single subcutaneous injection of royal jelly (1 g/kg) increased the vascular endothelial growth factor (VEGF) level in the uterus after 6 h of the treatment. Since VEGF is crucial for maintaining a healthy and fertile uterus, the estrogenic effect of royal jelly contributes to the improvement of fertility in female rats (Mishima et al., 2005). In immature rats, royal jelly has a positive action on ovarian follicles and female sex hormones. A 2-week period of daily treatment with royal jelly increased the number of secondary ovarian follicles and the plasma levels of estradiol and progesterone; although these actions were dose-independent, rather a dose efficacy threshold could be determined (100–200 mg/kg). Since the royal jelly treatment also improved the antioxidant parameters (reduced NO level and increased ferric reducing antioxidant power), the fertility promoting effect of royal jelly is the result of a combined estrogenic and antioxidant properties (Ghanbari et al., 2018).

A mice study proved that royal jelly could protect female fertility against the androgenic effect. The animals were treated with anabolic-androgenic oxymetholone for 30 days, and parallel, a group of mice were treated orally with royal jelly. The 1-month study revealed that royal jelly maintained fertility and reduced the number of androgen-arrested embryos (Najafi et al., 2014). This is a piece of further evidence for its estrogenic-antiandrogenic effect and significance in improving fertility rate.

An interesting study proved the central nervous system effect of royal jelly that may have an impact on fertility as well. Female rats were orally fed with royal jelly containing special chow for 7 months. After this long treatment, the pituitary size of the royal jelly group was significantly smaller with a reduction and increase in the expressions of prolactin and thyroid-stimulating hormone β mRNA, respectively. The other pituitary hormones' mRNAs remained unchanged, indicating that the primary function of the pituitary is not modified significantly (Narita et al., 2009). Since the aging process usually enhances the pituitary prolactin production, such a reduction may improve fertility because the low level of prolactin means a higher chance for conception. Additionally, the low thyroid function is also linked to infertility (Banu, 2019); thus, the pituitary action of royal jelly modifies both prolactin and thyroid hormone levels to the direction of improved fertility rate.

Effect of royal jelly on male fertility in rodents

Interestingly much more experiments have been carried out to investigate the effect of royal jelly on male fertility than females in rodents. Major royal jelly proteins (that involve royalactin as well) (Ramanathan et al., 2018) are responsible for the multiple actions of royal jelly, including improvement of fertility in males. Oral exposure of dried major royal jelly protein in 3 different doses was investigated in newborn mice to develop male sexual organs and sex hormones for 35 postnatal days. The low dose (125 mg/kg daily) did not show any significant action, while both the moderate (250 mg/kg daily) and the high (500 mg/kg daily) doses increased the testicular weight and the mean seminiferous tubules diameter. Additionally, the high dose significantly reduced the number of apoptotic sperm cells and reduced the ratio of incomplete spermatogenesis (Shi et al., 2019). These findings suggest that royal jelly has a well-defined sexual maturation increasing effect in male mice, consequently enhancing their fertilizing ability. A rat study proved that the oral or intraperitoneal royal jelly treatment increases the plasma levels of luteinizing hormone and testosterone (Khazaei et al., 2018) as further evidence of the fertility increasing effect of royal jelly in male rodents.

Reduced male fertilizing capacity is frequently linked to diseases. Varicocele is an enlarged testicular vein that is supposed to be responsible for low sperm count and quality, causing reduced male fertility. Most probably, the varicocele-induced oxidative stress is responsible for the harmful processes (Jensen et al., 2017). In a rat varicocele model, the protecting effect of royal jelly was investigated on spermatogenesis against oxidative stress. After the surgical induction of varicocele, the rats were treated with royal jelly (200 mg/kg/day) orally for 6 weeks. The royal jelly treatment restored the testicular total antioxidant capacity to the level of healthy animals, reduced the malondialdehyde level (that is related to oxidative stress), and raised the sperm count and the sperm viability as compared with varicocele-induced rats without treatment. Histopathological analysis revealed that the long-term royal jelly treatment enhanced the Johnsen's score of testicles that assesses the quality and intensity of spermatogenesis; the higher the value, the better the process (Asadi et al., 2019). Another fertility-reducing disease is diabetes mellitus, in which oxidative stress also has a crucial role along with the disturbed metabolic cooperation between the testicular cells that have a negative impact on spermatogenesis (Alves et al., 2013; Omolaoye Temidayo and du Plessis Stefan, 2018). In streptozotocin-induced diabetic adult male rats, the 6-week-long royal jelly treatment (100 mg/kg/day) improved the level of testicular antioxidant enzymes, the Johnsen's score and other spermatogenetic indexes, and the sperm motility, count, and viability (Ghanbari et al., 2015, 2016). These findings confirm that royal jelly has a fertility-preserving effect in disease models, and its antioxidant property has a crucial role in the protective and restoring actions in rodents.

The male fecundity can also be deteriorated by environmental exposition or side effects of anticancer therapy. Cadmium, along with lead and arsenic, are toxic elements impairing sperm quality and male fertility (Tariba Lovaković, 2020). The harmful effect of cadmium and the protective potential of royal jelly were investigated in male rodent models. In an experiment, the male rats were treated with royal jelly for 1 week (100 mg/kg/day). The animals were treated with cadmium to induce testicular dysfunction, and the daily cadmium treatment was maintained for 7 weeks along with the daily administration of royal jelly. The cadmium treatment induced a significant deterioration in testicular function, including reduced testicular weight, testosterone plasma level, sperm motility, and count accompanied with high parameters oxidative stress and inflammatory processes. However, the royal jelly treatment was able to reserve the cadmium-induced deteriorations indicating its protective property in spermatogenesis. Besides the antioxidant activity, it is supposed that amino acid and short-chain fatty acid contents of royal jelly can contribute to testicular protection and the enhancement in sperm motility (Ahmed et al., 2018). Another cadmium-induced testicular dysfunction study was carried out in mice. In that experiment, the cadmium and the parallel royal jelly treatment (250 mg/kg/day) was 7 days long, and the royal jelly administration preceded the cadmium treatment each day. Even during this short experiment, the same cadmium-induced deteriorations were found in mice testicles as that of rats. The protecting efficacy of royal jelly was surprising; in most cases, it restored the cadmium-modified, spermatogenesis reducing parameters to healthy levels. The main explanation of these actions is the antioxidant property of royal jelly (Almeer et al., 2018).

Anticancer therapy is frequently responsible for infertility or subfertility by inducing gonadal dysfunction. As a result, in males, the sperm quality, quantity, and viability are usually reduced along with the plasma testosterone level (Vassilakopoulou et al., 2016). The protective effect of royal jelly was tested in two rat experiments against anticancer drugs. In a bleomycin-induced subfertility model, the animals were treated with bleomycin (10 mg/kg) twice a week for 7 weeks. The gonadally deleterious effect of bleomycin was aimed to be reduced by daily oral treatments with royal jelly (100 mg/kg). At the end of the experimental period, according to the anticipations, bleomycin significantly reduced the sperm count, motility, viability, and plasma testosterone level. It increased the number of immature and DNA-damaged sperms and the level of oxidative stress. The long-term royal jelly treatment did not improve the values as compared with the bleomycin nontreated control animals. However, the bleomycin-induced deteriorations were significantly improved or even restored in the case of each investigated parameter. The supposed primary mechanism of this spermiotoxicity-protecting effect is the antioxidant property of royal jelly (Amirshahi et al., 2014). A similar study had been carried out with another anticancer drug, hydroxyurea. The rats were treated with hydroxyurea for 60 days in two different daily doses (225 and 450 mg/kg), while royal jelly (100 mg/kg) was administered daily, followed by the hydroxyurea treatment. Besides the previously mentioned sperm parameters and sex hormone level measurements, this study emphasized the examination of oxidative and inflammatory processes along with histopathology. The hydroxyurea-induced testicular dysfunction was successfully reduced or restored by royal jelly even in the high-dose hydroxyurea group. Although the parameters of oxidative stress and inflammatory responses (interleukin levels) were also improved, and the royal jelly treatment reduced these values, these parameters remained still higher than normal levels. The histological study of the testicles proved that royal jelly significantly ameliorated the hydroxyurea-induced changes in the histoarchitecture of testicles. Similar to the bleomycin study, the mechanism of the protective effect most possibly is the antioxidant effect of royal jelly with an additional antiinflammatory activity (Tohamy et al., 2019). These results raise the possibility of royal jelly to protect male fertility even during human anticancer therapies.

The rodential effects of royal jelly on fertility are summarized in Table 8.1.

Fertility-improving effect of royal jelly in farm animals

The good reproduction rate is a key issue in animal husbandry. Any natural remedy that positively affects reproduction, even in different species, can readily be accepted both by breeders and customers. Royal jelly belongs to this safe and natural compound group with much evidence among different farm animals.

Table 8.1 The fertility-improving effect of royal jelly in rodents.

Mode of action	Species	Sex	Effect
Estrogenic	Rat	Female	Increased uterine VEGF
Estrogenic, progestogenic, antioxidant	Rat	Female	Maturation of ovarian follicles
Antiandrogenic	Mouse	Female	Increased number of embryos
Reduced prolactin and enhanced TSH mRNA in the pituitary	Rat	Female	Higher chance for conception
Increased plasma LH and testosterone levels	Mouse, rat	Male	Improved sex organ development and better sperm quality
Antioxidant, antiinflammatory	Rat, mouse	Male	Improved sperm quality and protection in varicocele, diabetes mellitus, cadmium, or cytostatic drugs exposition models

LH: luteinizing hormone; *VEGF*: vascular endothelial growth factor.

Chickens

The reproduction of poultries is special via eggs. There is a particular way to improve the reproduction rate and the embryonic development in poultries called in ovo feeding, meaning the addition of external nutrients into embryonated eggs (into air sack or yolk sack) to improve hatchability and broiler performance (Retes et al., 2018). Since royal jelly contains several valuable nutrients, fertile chicken eggs were involved into *in ovo* feeding process. The eggs were treated on the 7th day of incubation with royal jelly in a 0.5 mL volume for each egg. The added royal jelly improved the hatchability of the eggs and increased the relative body weight and relative organs weight (heart, liver, ovaries) of the chicks. Additionally, it significantly enhanced the luteinizing and follicular-stimulating hormone levels of the chicks independently from the site of administration (air or yolk sack).

Interestingly, the ultrafilter royal jelly elicited the opposite effect suggesting that the separation of the components of royal jelly can significantly modify its action. Raw royal jelly seems to be useful in chicken breeding to improve development and increase the fertility of the developing and maturating chicks (Moghaddam et al., 2013). In a very similar study, the plasma testosterone levels of the chicks were also measured after the in ovo feeding with royal jelly, but no change was found (Moghaddam et al., 2014). Although the mechanism of action of royal jelly in chicken has not been clarified, its multispecies estrogenic-antiandrogenic effect might explain the avian action, too.

Rabbits

Male rabbits are susceptible to heat, and high temperature may severely deteriorate their fertility reducing the sperm viability and the plasma level of male sex hormone. In husbandry, the summer heat is one of the main challenges to avoid subfertility or infertility of the bucks (Marai et al., 2008). A study investigated whether royal jelly has any protective effect on the male rabbit's sexual function in a high-temperature environment. For 6 weeks, the rabbits were held in warm circumstances (max. 33°C) and treated orally with different doses of royal jelly (200, 400, or 800 mg/kg) once a week. During the 6 weeks, the heat reduced the testosterone level and the sperm quality and quantity in the control animals. However, the plasma testosterone level gradually increased in the royal jelly-treated rabbits along with the volume of ejaculation, seminal plasma fructose content, and sperm concentration. Additionally, the concentration of abnormal or dead sperms was significantly reduced by royal jelly treatment. Interestingly, the peak improvements were detected at 200 and 400 mg/kg doses. However, the highest dose (800 mg/kg) always induced a weaker improvement in the male fertility determining parameters than the lower doses. In this study, the royal jelly treatment also showed some other beneficial effect on rabbits' health, including an increase in serum protein level and red blood cell count and decreased serum lipid, cholesterol, creatinine, and uric acid levels, and modification in hepatic enzymatic activities (Elnagar, 2010). Thus, the royal jelly not only protects the fertility of male rabbits against heat-induced infertility but also contributes to the maintenance of good general health status that can be deteriorated by high temperature. However, too high doses of royal jelly can moderate its positive effects on fertility and physiologic parameters in bucks.

Buffaloes

Buffalo breeding must be faced with an increasing demand to enhance production. This higher need can be satisfied using in vitro embryo transfer for better reproduction since the possibilities of genetic modification of buffaloes are very limited (Jain et al., 2011). However, the freezing and thawing of the sperm buffalo bull can deteriorate the viability and the fertilizing ability of the semen. The success rate is lower than in the case of cows (Barile, 2012). Thus, the enhanced resistance of buffalo sperm against the freezing-thawing process can contribute to the higher success rate during in vitro fertilization. In a study, the fresh buffalo bull ejaculation was supplemented with royal jelly in different concentrations before freezing. Then the samples were thawed, and sperm parameters were measured. It was found that royal jelly supplementation significantly improved the motility and viability of buffalo sperm along with the integrity of its plasma membrane and acrosome (the latter is a special part in the head of sperm playing a crucial role in the successful binding and fusion to the oocyte). The 0.1% royal jelly supplement was found to be the most efficacious; therefore, such a supplemented ejaculate was used for the in vitro fertilization of buffaloes. Although lower concentration (0.05%) also showed positive action, higher concentration (0.4%) of royal jelly could not enhance the protection of buffalo sperm against thawing. The in vitro fertilizing potential of the royal jelly supplemented cryopreserved sperms was significantly higher than that of sperm without royal jelly; although the pregnancy rate has not been significantly improved after implantation, the number of pregnancies was higher (Shahzad et al., 2016). Considering that the study involved 24 buffalos only, it is possible that if a higher number of buffaloes would have been involved in the process, then it would have resulted in a significant enhancement of successful pregnancies. These findings suggest that royal jelly can protect gametes from the deleterious consequences of freezing and thawing and may contribute to reaching a higher success rate in the in vitro fertilization program of buffaloes and maybe other species. However, the protection mechanism is not clarified yet, but its antioxidant activity might be one explanation of the preserving action.

Sheep

Similar to buffalos, sheep breeding and reproduction face challenges. The main limitation is the seasonality of the sheep reproduction rate, which may cause problems in other ruminants (Rosa and Bryant, 2003). Although the fertility rate of the sheep is tried to be improved with gonadotropins and sex hormones, royal jelly seems to add similar or extra effects to these interventions. In a study, the intravaginal progesterone sponge-treated female sheep (ewes) were treated daily 250 mg royal jelly orally or intramuscularly. The duration of both progesterone and royal jelly treatments was 12 days. The royal jelly treatment did not modify the plasma progesterone level but shortened the intervals to the onset of estrus and increased the number of successful pregnancies after mating with rams. No difference was found between the efficacies of orally or intramuscularly administered royal jelly. As the treatment application is easier, the intramuscular intervention seems more comfortable in sheep husbandry than oral administration in capsules (Husein and Kridli, 2002).

Another experiment investigated the royal jelly efficacy on reproductive functions compared to equine chorionic gonadotropin in ewes. Each animal got a progestogen-containing intravaginal sponge for 12 days. During these 12 days, royal jelly-treated animals got 250, 500, or 750 mg royal jelly orally in capsules. The single gonadotropin intramuscular injection (600 IU) was given on the day of the withdrawal of progestogen sponges to the group that did not get royal jelly treatment. The estrus cycle and progesterone levels of the sheep were followed for 21 days after the last dose of royal jelly and gonadotropin injection. In comparison, their ability to be pregnant was followed for 155 days. The middle and the high doses of royal jelly did not significantly reduce the time interval to the onset of estrus, but the effect of gonadotropin was significant. Royal jelly treatment could not accelerate the increase in plasma progesterone level after the withdrawal of progestogen sponge, but gonadotropin shortened the time necessary for the enhancement. However, the pregnancy and lambing rates and the number of newborn lambs were the highest in groups treated with 500 mg/kg royal jelly (including the gonadotropin treated group). The plasma progesterone level was the highest in pregnant animals in the gonadotropin treated group. Still, each dose of royal jelly produced a significant elevation in pregnant plasma progesterone levels, too (Kridli and Al-Khetib, 2006).

In a similar study, the pregnancy rate increasing effect of royal jelly was investigated. The ewes got 400 mg royal jelly for 12 days along with progesterone delivering intravaginal device. A group of ewes got gonadotropin injection (500 IU) on the day of device removal when the royal jelly administration was also ceased. Then all ewes were mated with rams. The rams were left with them for two consecutive estrus cycles. Besides the gonadotropin, even royal jelly could shorten the interval between 2 estrus cycles that effect was equal to gonadotropin. Furthermore, the pregnancy and the lambing rates, the number of offspring lambs were significantly improved by both royal jelly and gonadotropin; no significant difference was found between them (Husein and Haddad, 2006). Royal jelly even improved the conception rate of ewes after insemination combined with gonadotropin treatment proving their synergistic relationship (Gimenez-Diaz, 2012). These findings suggest that royal jelly can improve the rate and outcomes of pregnancy and may increase the number of estrus cycles contributing to more successful breeding. Unfortunately, the mechanism of action and the pharmacokinetic parameters of royal jelly in sheep have not been clarified yet.

Sheep oocytes were used for the measurement of the effect of royal jelly on maturation in vitro. Royal jelly was added to the maturation media in 3 different concentrations (2.5, 5, and 10 mg/mL). The middle and the high concentrations of royal jelly improved the maturation and the fertilization of the oocyte, enhanced the embryo development and the expression of genes responsible for glucose metabolism. These beneficial actions are due to the enhanced antioxidant enzyme and glycolysis capacities of the oocyte induced by royal jelly. However, the responsible bioactive compounds for these effects in royal jelly have still not been identified (Eshtiyaghi et al., 2016). A similar study with similar results was carried out with goat oocytes (Veshkini et al., 2018). These ovine and goat oocyte experiences are promising and can be translated to the human in vitro fertilization technique.

The effects of royal jelly on the fertility of different farm animals are summarized in Table 8.2.

Table 8.2 The fertility-improving effect of royal jelly in farm animals.

Mode of action	Species	Sex	Effect
Nutritive effect (in ovo feeding), estrogenic	Chicken (including embryonated eggs)	Female	Increased ovary weight, enhanced LH and FSH levels of the chicks, and faster maturation
Increased plasma testosterone	Rabbit	Male	Protection of sexual function against heat
Antioxidant	Buffalo	Male	Preservation of sperms during freezing-thawing process
Unknown, suggested gonadotropin-like effect	Sheep	Female	Shortened estrus cycles, increased pregnancy and lambing rates, and increased plasma progesterone level
Antioxidant, enhanced glycolysis	Sheep (in vitro oocyte)	Female	Improved oocyte fertilization and embryo development

FSH: follicular-stimulating hormone; *LH*: luteinizing hormone.

Royal jelly and human fertility

Although royal jelly has been used for enhancing fertility and sexual force since Old Times, including Egypt, Greece, and China (Abdelnour et al., 2020), well-designed and reliable clinical studies are not available about its fertility effects in women, and only a few studies have been carried out in men. Most of the reports are either commercial materials or publications in nonpeer-reviewed sources. According to the statistics, most women who undergo fertility treatments use royal jelly or other bee products as a supplemental intervention to enhance the chance of the desired pregnancy (Gormack et al., 2015). However, the primary evidence for the royal jelly positive actions on oocytes, zygotes, and pregnancy maintenance gained from cell culture and animal experiments (see chapters above).

Only one study has been published with the involvement of women but aimed to enhance men's fertility. A special mixture of royal jelly and honey (3:100) with bee bread (contains pollen) was tested intravaginally to improve the fertilizing ability of sperm of men suffering from idiopathic asthenozoospermia (male infertility caused by low sperm motility) before or after coitus. The mixture was self-administered from the first day of the last menstruation; the couples freely decided the precoital or postcoital application. The efficacy of this method was compared to the intrauterine insemination technique. The royal jelly-containing mixture gained an 8% pregnancy rate, while the success rate with insemination was less than 3%, only. The local mechanism of the mixture is complex. It was proposed that besides the high fructose and NO contents of the honey, the amino acids, short-chain fatty acids, and the antioxidant attribute of royal jelly contributed to the improved fertilizing ability of sperms that possibly due to the enhanced sperm motility (Abdelhafiz and Muhamad, 2008).

The influence of royal jelly on men's infertility was investigated in a clinical study. The infertile participants received three different oral daily doses (25, 50, and 100 mg) of royal jelly for 3 months. The quarterly treatment significantly increased the plasma luteinizing hormone and testosterone levels, reduced the numbers of sperms moving slowly, and improved the active sperm motility. The testosterone increasing effect of royal jelly is possibly linked to its zinc content (the zinc deficiency may induce low testosterone secretion (Prasad et al., 1996)) along with its gonadotropin level increasing property that molecular mechanism is unknown (Ali et al., 2007).

A sedentary lifestyle, especially watching television for long hours, can significantly reduce the sperm count in men (Priskorn et al., 2016). Sedentary young and healthy men were involved in a study where they got 1000 mg/day raw royal jelly orally for 15 days. The plasma testosterone and body weight measurements before and after the treatment series revealed that royal jelly significantly enhanced (12% increase) the plasma testosterone level compared to the control group (1% increase). However, the body weight and the body mass index were not modified by the treatment than can be explained with the relatively short study period. Unfortunately, the sperm quantity or quality was not measured in this study; therefore, only indirect consequences can be concluded from this study about the royal jelly action on sperms (Tasdogan et al., 2020). Although the testosterone level increasing mechanism of short-term royal jelly treatment was not investigated, it might stimulate the secretion of pituitary gonadotropins that can enhance the sex hormone production in the testicles, as was proved earlier.

The effects of royal jelly on human fertility are summarized in Table 8.3.

Table 8.3 The fertility-improving effect of royal jelly in human.

Mode of action	Sex	Effect
Antioxidant (+ nurturing effect mixed with honey and bee bread)	Man	Precoital or postcoital intravaginal application improved the sperm motility and fertilizing ability
Zinc content, partially unknown, putative gonadotrophic effect	Man	Increased testosterone and LH levels, improved sperm motility

LH: luteinizing hormone.

Beyond royal jelly: The effect of drone milk on fertility

Like royal jelly, drone milk is secreted by the hypopharyngeal and mandibular glands of the worker honey bees. It is a light yellow, slightly sweet, and thick liquid separated from drone brood by extraction to eliminate larvae and pupae during the harvest. It has a low sugar content; therefore, its long-term storage and preservation can be solved by freezing only.

Drone milk preparations are traditionally used in Romania, Russia, and China to regenerate weakened people and improve the fertility of men (Sawczuk et al., 2019). Anecdotally, drone milk is also used for the enhancement of female fertility in Transylvania. Androgens of apilarnil (drone bee larvae) were confirmed on male broilers (Yucel et al., 2011). Since no clinical studies were carried out about the reproductive effects of drone milk, results from rat experiments can show the fertility potential of this bee product.

The effect of drone milk on male rats

The putative androgenic effect of drone milk was tested in castrated rats. The animals were treated orally with 110 mg/kg drone milk daily for 10 days. It was found that the drone milk treatment enhanced the weights of all androgen-sensitive organs (except prostate), and this effect was flutamide (antiandrogen) sensitive. The increase in the weight of seminal vesicles can be partially explained by the increased production of sperms that possibly improve the male rats' fertility. Drone milk also enhanced the plasma testosterone levels of the animals and increased the expression of an androgen-sensitive prostatic protein (SLAP). A Bioactivity-guided fractionation revealed that the active ingredients of drone milk are methyl-oleate and methyl-palmitate in a molar ratio of 4:3. When this fatty acid mixture was administered again to the castrated rats, the androgenic action was detectable, confirming that the combination of methyl-oleate and methyl-palmitate is responsible for the androgenic action (Seres et al., 2014b). However, the molecular mechanism of this effect has not been clarified yet.

The effect of drone milk on female rats

Both estrogenic and progestogenic effects of drone milk have been investigated in female rats. Ovariectomized immature rats were treated orally with 500 mg/kg drone milk daily for 4 days. The drone milk treatment increased the uterine weight of the animals and enhanced the expression of and estrogen-dependent protein (complement component C3). These actions were antiestrogen sensitivities. A bioactivity-guided fractionation revealed that the active ingredient of drone milk is E-dec-2-enedioic acid. This action was comparable with soy isoflavones that are well-known phytoestrogens (Seres et al., 2013). The structure of this fatty acid is very similar to an estrogenic fatty acid found in royal jelly (Suzuki et al., 2008). Since the better estrogenic function may improve fertility, there is a high possibility that drone milk can contribute to female fecundity.

The progestogenic effect of drone milk was investigated in a pregnancy maintenance test using pregnant rats, which were ovariectomized at their gestational day 8. The animals were treated orally with drone milk (500 mg/kg/day) from gestational day eight till 15. Although the drone milk treatment enhanced the expression of a progesterone-dependent ovarian protein (calcitonin receptor-like receptor), this effect was not enough to elicit a significant pregnancy maintenance activity. However, when the drone milk was added together with the weak progestogen spironolactone, their combined effect was significant in the number of fetuses. The active progestogenic ingredient of drone milk has not been identified yet, but it cannot be excluded that the estrogenic compound may have a weak progestogenic effect (Seres et al., 2014a). This light progestogenic effect of drone milk may reduce miscarriages and provide a better maintenance ratio of pregnancies and improve fertility even in humans.

Conclusions

The positive effect of royal jelly on fertility has been confirmed in animal experiments, farm animals, and even humans, although the latter has been proved in men only. The putative mechanism of action of royal jelly includes its antioxidant property in both sexes. In females, its estrogenic-antiandrogenic effect seems to be dominant in the fertility-improving action due to its active ingredients. Additionally, it is supposed that royal jelly has some action on the gonadotroph hormones as well, but the exact pathway of this kind of action is still not clarified. Even royal jelly has beneficial local nurturing and differentiating effects on gametes during in vitro fertilization processes. Surprisingly, the fertility-improving influence of royal jelly has been clearly demonstrated on human sperm; however, the scientific evidence for beneficial effects on human oocytes or conception on women side is still not available and clarified. Although several assisted reproductive centers use royal jelly preparations for women to improve the conception rate, the scientific evidence behind it can still be gained from animal experiments. Besides royal jelly, it is worthy of focusing on other apiary products like drone milk that may significantly contribute to the improvement of fertility in both male and female rats. Since the basic materials and the way of production of these apiary products are similar, there is a high chance that their actions on fertility will also be comparable. However, along with royal jelly, further clinical studies are required to explore their natural potency to enhance human fertility.

References

Abdelhafiz, A.T., Muhamad, J.A., 2008. Midcycle pericoital intravaginal bee honey and royal jelly for male factor infertility. Int. J. Gynecol. Obstet. 101 (2), 146–149. https://doi.org/10.1016/j.ijgo.2007.11.012.

Abdelnour, S.A., et al., 2020. Useful impacts of royal jelly on reproductive sides, fertility rate and sperm traits of animals. J. Anim. Physiol. Anim. Nutr., 1798–1808. https://doi.org/10.1111/jpn.13303. Blackwell Publishing Ltd.

Ahmed, M.M., et al., 2018. Protective potential of royal jelly against cadmium-induced infertility in male rats. Andrologia 50 (5). https://doi.org/10.1111/and.12996, e12996.

Ali, E., et al., 2007. Effect of royal jelly on male infertility. Thi-Qar Med. J. 1 (1), 1–12.

Almeer, R.S., et al., 2018. Royal jelly abrogates cadmium-induced oxidative challenge in mouse testes: involvement of the nrf2 pathway. Int. J. Mol. Sci. 19 (12), 3979. https://doi.org/10.3390/ijms19123979.

Alves, M.G., et al., 2013. Molecular mechanisms beyond glucose transport in diabetes-related male infertility. Biochim. Biophys. Acta Mol. Basis Dis. 1832 (5), 626–635. https://doi.org/10.1016/j.bbadis.2013.01.011.

Amirshahi, T., Najafi, G., Nejati, V., 2014. Protective effect of royal jelly on fertility and biochemical parameters in bleomycin- induced male rats. Iran. J. Reprod. Med. 12 (3), 209–216.

Anton, K., Grozinger, C., 2020. An Introduction to Queen Honey Bee Development. PennState Extension. Available from https://extension.psu.edu/an-introduction-to-queen-honey-bee-development. 03.05.2021.

Asadi, N., et al., 2019. Effect of royal jelly on testicular antioxidant enzymes activity, MDA level and spermatogenesis in rat experimental Varicocele model. Tissue Cell 57, 70–77. https://doi.org/10.1016/j.tice.2019.02.005.

Banu, F., 2019. Correlation of serum prolactin and thyroid hormone in female infertility. J. Nepalgunj Med. Coll. 17 (2), 32–34. https://doi.org/10.3126/jngmc.v17i2.28758.

Barile, V.L., 2012. Technologies related with the artificial insemination in buffalo. J. Buffalo Sci. 1, 139–146. https://doi.org/10.6000/1927-520x.2012.01.02.02.

Elnagar, S.A., 2010. Royal jelly counteracts bucks' 'summer infertility'. Anim. Reprod. Sci. 121 (1–2), 174–180. https://doi.org/10.1016/j.anireprosci.2010.05.008.

Eshtiyaghi, M., et al., 2016. Royal jelly may improve the metabolism of glucose and redox state of ovine oocytes matured in vitro and embryonic development following in vitro fertilization. Theriogenology 86 (9), 2210–2221. https://doi.org/10.1016/j.theriogenology.2016.07.019.

Gáliková, M., et al., 2017. The obesity-related Adipokinetic hormone controls feeding and expression of neuropeptide regulators of Drosophila metabolism. Eur. J. Lipid Sci. Technol. 119 (3), 1600138. https://doi.org/10.1002/ejlt.201600138.

Ghanbari, E., et al., 2015. Study on the effect of royal jelly on reproductive parameters in streptozotocin-induced diabetic rats. Int. J. Fertil. Steril. 9 (1), 113–120. https://doi.org/10.22074/ijfs.2015.4215.

Ghanbari, E., Nejati, V., Khazaei, M., 2016. Antioxidant and protective effects of royal jelly on histopathological changes in testis of diabetic rats. Int. J. Reprod. BioMed. 14 (8), 519–526.

Ghanbari, E., et al., 2018. Royal jelly promotes ovarian follicles growth and increases steroid hormones in immature rats. Int. J. Fertil. Steril. 11 (4), 263–269. https://doi.org/10.22074/ijfs.2018.5156.

Gimenez-Diaz, C., 2012. Improved reproductive response of sheep in intrauterine insemination program with the use of royal jelly. Afr. J. Biotechnol. 11 (61), 12518–12521. https://doi.org/10.5897/ajb12.200.

Gormack, A.A., et al., 2015. Many women undergoing fertility treatment make poor lifestyle choices that may affect treatment outcome. Hum. Reprod. 30 (7), 1617–1624. https://doi.org/10.1093/humrep/dev094.

Husein, M.Q., Haddad, S.G., 2006. A new approach to enhance reproductive performance in sheep using royal jelly in comparison with equine chorionic gonadotropin. Anim. Reprod. Sci. 93 (1–2), 24–33. https://doi.org/10.1016/j.anireprosci.2005.06.012.

Husein, M.Q., Kridli, R.T., 2002. Reproductive responses following royal jelly treatment administered orally or intramuscularly into progesterone-treated Awassi ewes. Anim. Reprod. Sci. 74 (1–2), 45–53. https://doi.org/10.1016/s0378-4320(02)00165-3.

Isenmann, E., et al., 2019. Ecdysteroids as non-conventional anabolic agent: performance enhancement by ecdysterone supplementation in humans. Arch. Toxicol. 93 (7), 1807–1816. https://doi.org/10.1007/s00204-019-02490-x.

Jain, A., et al., 2011. Headway in reproductive biotechniques for genetic improvement of buffaloes. Curr. Trends Biotechnol. Pharm. 5 (3), 1206–1232.

Jensen, T.K., et al., 2002. Poor semen quality may contribute to recent decline in fertility rates. Hum. Reprod. 17 (6), 1437–1440. https://doi.org/10.1093/humrep/17.6.1437.

Jensen, C.F.S., et al., 2017. Varicocele and male infertility. Nat. Rev. Urol. 14 (9), 523–533. https://doi.org/10.1038/nrurol.2017.98.

Kamakura, M., 2011. Royalactin induces queen differentiation in honey bees. Nature 473 (7348), 478–483. https://doi.org/10.1038/nature10093.

Ketchesin, K.D., Stinnett, G.S., Seasholtz, A.F., 2017. Corticotropin-releasing hormone-binding protein and stress: from invertebrates to humans. Stress 20 (5), 449–464. https://doi.org/10.1080/10253890.2017.1322575.

Khazaei, M., Ansarian, A., Ghanbari, E., 2018. New findings on biological actions and clinical applications of royal jelly: a review. J. Diet. Suppl., 757–775. https://doi.org/10.1080/19390211.2017.1363843. Taylor and Francis Ltd.

Kridli, R.T., Al-Khetib, S.S., 2006. Reproductive responses in ewes treated with eCG or increasing doses of royal jelly. Anim. Reprod. Sci. 92 (1–2), 75–85. https://doi.org/10.1016/j.anireprosci.2005.05.021.

Kunugi, H., Ali, A.M., 2019. Royal jelly and its components promote healthy aging and longevity: from animal models to humans. Int. J. Mol. Sci. 20 (19), 4662. https://doi.org/10.3390/ijms20194662. MDPI AG.

Lubahn, D.B., et al., 1993. Alteration of reproductive function but not prenatal sexual development after insertional disruption of the mouse estrogen receptor gene. Proc. Natl. Acad. Sci. U. S. A. 90 (23), 11162–11166. https://doi.org/10.1073/pnas.90.23.11162.

Maleszka, R., 2018. Beyond Royalactin and a master inducer explanation of phenotypic plasticity in honey bees. Commun. Biol. 1, 8. https://doi.org/10.1038/s42003-017-0004-4. Nature Research.

Mao, W., Schuler, M.A., Berenbaum, M.R., 2015. A dietary phytochemical alters caste-associated gene expression in honey bees. Sci. Adv. 1 (7), e1500795. 1 https://doi.org/10.1126/sciadv.1500795.

Marai, I.F.M., Habeeb, A.S.I.A.M., Gad, A.E., 2008. Performance of New Zealand white and Californian male weaned rabbits in the subtropical environment of Egypt. Anim. Sci. J. 79 (4), 472–480. https://doi.org/10.1111/j.1740-0929.2008.00552.x.

Mishima, S., et al., 2005. Royal jelly has estrogenic effects in vitro and in vivo. J. Ethnopharmacol. 101 (1–3), 215–220. https://doi.org/10.1016/j.jep.2005.04.012.

Moghaddam, A., et al., 2013. Effect of royal jelly in ovo injection on embryonic growth, hatchability, and gonadotropin levels of pullet breeder chicks. Theriogenology 80 (3), 193–198. https://doi.org/10.1016/j.theriogenology.2013.04.013.

Moghaddam, A.A., Borji, M., Komazani, D., 2014. Hatchability rate and embryonic growth of broiler chicks following in ovo injection royal jelly. Br. Poultry Sci. 55 (3), 391–397. https://doi.org/10.1080/00071668.2014.921664.

Najafi, G., et al., 2014. Protective role of royal jelly in oxymetholone-induced oxidative injury in mouse testis. Iran. J. Toxicol. 8 (25), 1073–1080.

Narita, Y., et al., 2009. Effects of long-term administration of royal jelly on pituitary weight and gene expression in middle-aged female rats. Biosci. Biotechnol. Biochem. 73 (2), 431–433. https://doi.org/10.1271/bbb.80556.

Omolaoye Temidayo, S., du Plessis Stefan, S., 2018. Diabetes mellitus and male infertility. Asian Pac. J. Reprod. 7 (1), 6–14. https://doi.org/10.4103/2305-0500.220978.

Prasad, A.S., et al., 1996. Zinc status and serum testosterone levels of healthy adults. Nutrition 12 (5), 344–348. https://doi.org/10.1016/S0899-9007(96)80058-X.

Priskorn, L., et al., 2016. Is sedentary lifestyle associated with testicular function? A cross-sectional study of 1,210 men. Am. J. Epidemiol. 184 (4), 284–294. https://doi.org/10.1093/aje/kwv338.

Ramanathan, A.N.K.G., Nair, A.J., Sugunan, V.S., 2018. A review on Royal Jelly proteins and peptides. J. Funct. Foods 44, 255–264. https://doi.org/10.1016/j.jff.2018.03.008.

Retes, P.L., et al., 2018. In ovo feeding of carbohydrates for broilers—a systematic review. J. Anim. Physiol. Anim. Nutr. 102 (2), 361–369. https://doi.org/10.1111/jpn.12807.

Rosa, H.J.D., Bryant, M.J., 2003. Seasonality of reproduction in sheep. Small Rumin. Res. 48 (3), 155–171. https://doi.org/10.1016/S0921-4488(03)00038-5.

Sawczuk, R., Karpinska, J., Miltyk, W., 2019. What do we need to know about drone brood homogenate and what is known. J. Ethnopharmacol. 245. https://doi.org/10.1016/j.jep.2018.10.042, 111581.

Seres, A.B., et al., 2013. Raw drone milk of honey bees elicits uterotrophic effect in rats: evidence for estrogenic activity. J. Med. Food 16 (5), 404–409. https://doi.org/10.1089/jmf.2012.0232.

Seres, A., Ducza, E., Gáspár, R., 2014a. Investigation of gestagenic effect of raw drone milk in rats. A nyers heretej gesztagén hatásainak vizsgálata patkányban, Acta Pharm. Hung. 84 (2), 77–81.

Seres, A.B., et al., 2014b. Androgenic effect of honey bee drone milk in castrated rats: roles of methyl palmitate and methyl oleate. J. Ethnopharmacol. 153 (2), 446–453. https://doi.org/10.1016/j.jep.2014.02.050.

Shahzad, Q., et al., 2016. Royal jelly supplementation in semen extender enhances post-thaw quality and fertility of Nili-Ravi buffalo bull sperm. Anim. Reprod. Sci. 167, 83–88. https://doi.org/10.1016/j.anireprosci.2016.02.010.

Shi, Z., et al., 2019. Freeze-dried royal jelly proteins enhanced the testicular development and spermatogenesis in pubescent male mice. Animals 9 (11), 977. https://doi.org/10.3390/ani9110977.

Suzuki, K.M., et al., 2008. Estrogenic activities of fatty acids and a sterol isolated from royal jelly. Evid. Based Complement. Alternat. Med. 5 (3), 295–302. https://doi.org/10.1093/ecam/nem036.

Tariba Lovaković, B., 2020. Cadmium, arsenic, and lead: elements affecting male reproductive health. Curr. Opin. Toxicol. 19, 7–14. https://doi.org/10.1016/j.cotox.2019.09.005.

Tasdogan, A.M., et al., 2020. The effect of short-term royal jelly supplement on testosterone levels in sedentary and healthy individuals. Prog. Nutr. 22 (1), 275–280. https://doi.org/10.23751/pn.v22i1.8746.

Tohamy, H.G., Gad El-Karim, D.R., El-Sayed, Y.S., 2019. Attenuation potentials of royal jelly against hydroxyurea-induced infertility through inhibiting oxidation and release of pro-inflammatory cytokines in male rats. Environ. Sci. Pollut. Res. 26 (21), 21524–21534. https://doi.org/10.1007/s11356-019-05521-3.

Vassilakopoulou, M., et al., 2016. Anticancer treatment and fertility: effect of therapeutic modalities on reproductive system and functions. Crit. Rev. Oncol. Hematol. 97, 328–334. https://doi.org/10.1016/j.critrevonc.2015.08.002.

Veshkini, A., et al., 2018. Oocyte maturation with royal jelly increases embryo development and reduces apoptosis in goats. Anim. Reprod. 15 (2), 124–134. https://doi.org/10.21451/1984-3143-2017-AR986.

Yucel, B., et al., 2011. The effects of apilarnil (drone bee larvae) administration on growth performance and secondary sex characteristics of male broilers. J. Anim. Vet. Adv. 10 (17), 2263–2266. https://doi.org/10.3923/javaa.2011.2263.2266.

CHAPTER 9

Prevention of side effects from chemoradiotherapy and antitumor potential of royal jelly and its components: A systematic review

Maria G. Campos[a,b], **Ofelia Anjos**[c,d,e], **and Saboor Ahmad**[f]
[a]Observatory of Drug-Herb Interactions/Faculty of Pharmacy, University of Coimbra, Heath Sciences Campus, Coimbra, Portugal
[b]Coimbra Chemistry Centre (CQC, FCT Unit 313) (FCTUC), University of Coimbra, Coimbra, Portugal
[c]Polythecnic Institute of Castelo Branco, Castelo Branco, Portugal
[d]Forest research centre, School of Agriculture (ISA, Instituto Superior de Agronomia), University of Lisbon, Lisbon, Portugal
[e]Center for Plant Biotechnology of Beira interior, Castelo Branco, Portugal
[f]Institute of Apicultural Research/Key Laboratory of Pollinating Insect Biology, Ministry of Agriculture, Chinese Academy of Agricultural Sciences, Beijing, China

Contents

Bee Products and Their Applications in the Food and Pharmaceutical Industries
https://doi.org/10.1016/B978-0-323-85400-9.00007-1

Abbreviations

10-HDA	10-hydroxy-2-decenoic acid
5-FU	5-fluorouracil
ALT	alanine aminotransferase
AST	aspartate aminotransferase
BPA	bisphenol A
CAT	catalase
CBA	cytometric Bead Array
CP	cisplatin
DMSO	dimethyl sulfoxide
FS	fibrosarcoma
GPx	glutathione peroxidase
GSH	glutathione
HepG2	hepatocellular carcinoma cell line
HGs	hypopharyngeal glands
HUVECs	human umbilical vein endothelial cells
IFN-α	interferon-alpha
IL	interleukin
IR	irradiation
I-TASSER	Iterative Threading ASSEmbly Refinement
M-CSF	macrophage colony-stimulating factor
MDA	malondialdehyde
MGs	mandibular
MLKL	mixed lineage kinase domain-like protein
MPO	myeloperoxidase
MRJP2	major royal jelly protein 2
RJ	royal jelly
ROS	reactive oxygen species
RT-PCR	reverse transcription-polymerase chain reaction
SOD	superoxide dismutase
TGF	transformed growth factor
TKIs	tyrosine kinase inhibitors
TNF	tumor necrosis factor
TQ	thymoquinone
VEGF	vascular endothelial growth factor

Introduction

Approximately 10 million people died in 2020 with cancer which is the foremost cause of death worldwide. Generally, this pathology can be described as a rapid formation of abnormal cells that grow beyond their usual borders and can occupy adjoining parts of the body and extend to other organs (metastasis). Breast (2.26 million cases) and lung cancers (2.21 million cases) were the most common in 2020 (in terms of new cases of cancer, but the first only with 31% of deaths contrary to the 81.5% deaths from the last). Colon and rectum had 1.93 million cases with 44.5% of deaths. Prostate, skin, and stomach had almost the same number of new cases in 2020, but the last with 70.5% of casualties (WHO, 2021).

From the data presented above is clear the urgent need to continue the research for new strategies and tools that will provide more efficient treatments for an increase of survivors. First of all, identifying the leading causes is a framework process and will help prevent the disease. However, it is important to be aware of the increase in life span and that the incidence of cancer increases with age due to the combined propensity for low cellular repair machinery as a person grows older. Despite this, all the knowledge acquired each year increases the probability of a better and more successful treatment that usually includes surgery, radiotherapy, and/or chemotherapy (Andres et al., 2020). Despite the improvement of the latest therapies, which induce better survival, we still need better tools, namely, new drugs with high efficacy and low side effects, especially adequate.

Metastasis, the last step of the disease, is the primary cause of death from cancer. Unfortunately, some natural products that induce angiogenesis, cell growth, and decreased apoptosis will contribute to widespread tumor cells in the body and should be avoided. However, many others can do the inverse and were studied among the various possibilities to fight cancer. That's why, among all approved antitumor drugs worldwide (for instance, between 1981 and 2019), details regarding the primary source under research point to a high percentage of natural product structures (animal and plants, mostly). The oldest approved and still used for a broad amount of cancers include Taxol derivatives (Docetaxel and Paclitaxel) previously extracted from *Taxus baccata* L, Vinca alkaloids (Vinblastine, Vinorelbine, Vincristine, and Vindesine) from *Catharanthus roseus* G. Don, or even Etoposide, a semisynthetic derivative of podophyllotoxin (a substance found in the mandrake plant) (Newman and Cragg, 2020). Many other candidate drug entities are currently in clinical trials looking for the highest efficacy and safety. Even though combinatorial techniques have prospered as methods of improving new molecules and have been used successfully in recently approved drugs, in this revision, we are still able to identify new combinatorial compounds, for instance, the constituents of RJ. This product is whitish to a yellow viscous substance with a pungent smell, which is secreted by the mandibular (MGs) and hypopharyngeal glands (HGs) by young nurse bees (Ahmad et al., 2020a; Altaye et al., 2019).

Among the natural resources, which continue under investigation pursuing the main goal to obtain new and better possibilities of successful chemotherapy treatments, RJ and its major peptides and proteins MRJPs (1–9) are well-described (Ahmad et al., 2020b). We will be discussing it in more detail below, not only as a possible antitumoral but also for the development of new strategies for prevention and healing of some of the most common side effects induced by chemotherapy as, for example, eating difficulties, mucositis, liver, and kidney toxicity.

In a previous publication, Ahmad et al. (2020b) pointed to the high biopotential of RJ compounds in different bioactivities, which goes from antibacterial and antifungal activity to animal studies showing the decrease of oxidative stress. The increase of antioxidant properties using the values of catalase (CAT), glutathione peroxidase (GPx), and superoxide dismutase (SOD), and, in humans, the anticancer effects (inhibiting the tumor-induced angiogenesis and activating the immune system) were also evaluated. Following these previous results and among the new approaches in Drug Therapies, RJ is under investigation for various purposes alone and in combinations, for instance, associated with thymoquinone (TQ) against human metastatic breast cancer (Moubarak et al., 2021). Even though the positive results were published, further research is needed to identify the key molecular mechanism for this enhanced antitumor activity.

Very recently, Guo et al. (2021) stated that RJ could be beneficial in preserving human health, and among other bioactivities, they highlight antibacterial, anticancer effect, regulation of immunity, and lifespan. All those indications were beneficial to develop some of the studies that we will discuss in this paper. These findings are beneficial for a comprehensive understanding and use of RJ, making it more effective in maintaining health. Based on the above-referenced research, the objective of this chapter is to give a global overview of RJ and its compounds as antitumor potential new strategies or as coadjutant therapy for certain side-effects induced by chemotherapy. We have also discussed candidate drug entities currently in clinical trials, with some very interesting preliminary reports. This source of novel molecules has a high potential in oncologic treatments. Future research should be made to achieve its complete safety and efficacy, allowing the production of new drugs.

Materials and methods

This systematic review was prepared following the "Preferred Reporting Items for Systematic Reviews and Meta-analysis Protocols (PRISMA) reporting guidelines" updated in 2020.

Data source and search strategy

A literature search was conducted on EMBASE (http://www.embase.com) and MEDLINE (http://www.ncbi.nlm.nih.gov/pubmed) by using the following keywords and medical subject headings (MeSH) terms where available. The search was systematized through the following three sets. The first category (a) includes royal jelly and royal jelly products, (b) the second is mainly about the constituents of royal jelly from the chemical point of view and more broad issues, and the last set (c), the main goal is to get information about bioactivity and health purposes in animal and humans. All essential references were imported in reference management software, namely EndNote X9 (Thomson Reuter CA, USA). The duplicated references were removed from the present book chapter.

(a) "royal jelly" and "royal jelly products"; (b) "royal jelly constituents OR "royal jelly chemical" OR "traditional royal jelly products"; "royal jelly" and "peptide"; "protein" and "royal jelly," "royal jelly" and "nutrients" OR "bioactive compounds"; (c) " royal jelly" and "protein" and "animal"; "royal jelly" and "cancer" OR "protein" and "cancer" OR "peptide" and "antitumor" OR "health benefits" OR "antiangiogenic." The search was extended to references within.

Inclusion and exclusion criteria

The inclusion criteria used to select from the previous articles extracted for the databases and make them eligible for this literature review were:

studies must have research in tumor cell lines;

studies must have research in induced tumors in animals;

studies must have research in oncologic treatment side effects, animals and humans;

studies must have research in clinical trials with oncologic patients;

studies must be experimental;

studies must have been published in the last 20 years. The search was extended to references within, and an older reference could be used if absolutely necessary.

The exclusion criteria of the article that disqualify the study from inclusion in a literature review were:

The study used an observational design;

The study used a qualitative methodology;

Studies those were surveyed and published in conference proceedings or workshops or as book chapters;

The study was published more than 20 years ago;

The study was published in a language other than English.

Selection of articles and quality assessments

The titles and abstracts of selected articles were evaluated on the basis of the above inclusion and exclusion criteria. The titles and abstracts of primary articles were screened by two reviewers independently. Full-text articles were evaluated using eligibility criteria, and studies that met the requirements were included. In the event of a disagreement, a third independent reviewer was consulted, and the matter was resolved by consensus and debate.

Data analysis

Data were summarized, and animal and human information were given to facilitate the critical assessment of the published literature. Furthermore, we discussed current evidence of the possible effects on human health and safety.

Results

Characteristics of selected studies

The RJ and its components tested for their in vitro and in vivo antitumor activities were selected from the studies. Six articles were examined for the preclinical cell trials of antitumor activities of RJ and its components. In addition, 12 studies were investigated for the preclinical animal trials. Five studies examined RJ for drug discovery, and seven studies provided data for preventing the adverse effect of various drugs. It was considered that nine studies investigated the preclinical human trials either for the antitumor effects of RJ and its ingredients or prevention of side effects caused by radio or chemotherapy.

Preclinical cell studies

Breast cancer

Cancer-related deaths in women are often correlated to breast cancer. Nevertheless, the current therapies gave very good results; the search for new molecules continues to drive many scientific groups all over the world. In 2007, Nakaya et al. (2007) from Japan studied the protective effect of RJ in human breast cancer cells affected by bisphenol (BP) A-induced damage. This product is an endocrine disruptor found in various products and has the potential to enhance the proliferation of breast cancer cells. In this study, interesting findings have been achieved. First, the authors checked the power sometimes attributed to RJ in MCF-7 cells and analyzed if changes had been found among heat-treated RJ compared to the administration of intact RJ. No estrogen effect was found, which is good in the case of this kind of tumor. The more relevant effect of RJ was the inhibition of BPA-induced growth promotion on these tumor cells, implying that the active substance in RJ should be heat-stable. The authors explored, for instance, the possibility of using RJ as an antienvironmental hormone activity in a way to prevent the risk of bisphenol A (BPA)-induced breast cancer.

Triple-negative is the most aggressive from the various types of breast cancer, and treatment possibilities are scarce. Most of the time, it resists chemotherapy and metastasizes. For this treatment, the combination of drugs in chemotherapy is a crucial possibility. Moubarak et al. (2021) from the American University of Beirut (Lebanon) used a combination of RJ and thymoquinone (TQ), which is the major bioactive constituent of the essential oil extract from the *Nigella sativa* L. seeds. This plant is also known as black cumin and is often used in traditional medicine like Unani Tibb, Ayurveda, and Siddha. The TQ is a molecule previously studied against various types of cancers which was able, for instance, to inhibit the growth of the resistant MCF-7 breast carcinoma cells. However, for this study, the authors utilized the purified synthetic compound TQ concomitantly with RJ, associating both bioactivities with the aim to prove its synergic cytotoxicity. They used RJ, TQ, and their combinations in the MDA-MB-231 cell line to achieve this goal. Crude RJ was from Rashaya al-Wadi (Lebanon), located at 1200 to 1600 m, above sea level. In the summertime, the main floral sources were *Nasturtium* (Brassicaceae), *Rhus* (Anacardiaceae), *Ulmus* (Ulmaceae), *Rosa* (Rosaceae), and *Eryngium* (Apiaceae), which will contribute to the RJ collected. 20 mg of fresh RJ was dissolved in a solution of 800 μL of dH_2O and 200 μL Dimethyl sulfoxide (DMSO) solution at 37°C for 30 min, and vortex every 10 min to increase and maintain the solution and homogeneity. This solution was also tested alone in FHs 74 Int (human small intestinal cell line) cells and did not show toxic effects in low doses (5 μg/mL). The apoptosis induction, changes in cell proliferation, half-maximal inhibitory concentration, cell death, and cycle regulation were all evaluated by the conventional methodologies for these protocols. For the interaction bioactivity between TQ and RJ, the authors did a *Combination index analysis* using the Chou–Talalay plot, which involves the combination of indices calculated from the mean affected fraction at each drug combination using CompuSyn software. These indices determine the three possibilities of interactions among the products under investigation, antagonism, addition, and synergistic effects. Below 5 μg/mL of RJ only had a mild inhibitory effect on MDA-MB-231 cells and FHs 74 Int. TQ had dose-dependent inhibitory cell viability. The concentration of 15 μgmol/L TQ induced an increase in the pre-G1 population. The association of TQ and RJ gave better results but with the commitment of the cell viability due to the apoptosis induction (Moubarak et al., 2021).

Colorectal adenocarcinoma

Filipič et al. (2015) from Slovenia and Croatia checked the possibility of RJ (0.1 g/10 mL phosphate buffer saline), and especially its bioactive component 10-hydroxy-2-decenoic acid (10-HDA) prepared as 100.0 mμmol/L and human interferon-alpha (HuIFN-alpha N3) (1000 IU/mL), to have antiproliferative activity against the growth of the human colorectal adenocarcinoma cells (CaCo-2). The HuIFN-αN3 (type I interferon) is a multisubtype protein commonly used in oncologic treatments. The authors used the products alone, but they tested a synergy among them and tested different combinations (1:1, 1:2, and 2:1) on CaCo-2 cells. The design of the study involved the verification of the effect on intracellular glutathione (GSH) level (significance for normal redox signaling in cells—high values enhance cell proliferation and low levels induce apoptosis) and oxidative degradation of lipids in cell membranes (cell damage) which was quantified by malondialdehyde (MDA) assay. The best results for the antiproliferative activity (3.8) were verified with RJ and HuIFN-alpha N3 at the ratio of 2:1. For these conditions the GSH was 24.9 ± 2.4 nmol $g^{(-3)}$ of proteins (vs 70.2 ± 3.2 nmol $g^{(-3)}$ in the control) and MDA was 72.3 ± 3.1 nmol $g^{(-3)}$ (vs 23.6 ± 9.1 nmol $g^{(-3)}$ in the control). Briefly, the antiproliferative activity was not very significant, but these findings revealed that GSH was decreased, which is good, despite the MDA increased in the cells. Maybe the products above in this research could contribute to the inhibition of CaCo-2 cells proliferation in vitro, and in the future, be used for developing anticancer treatments. Still, the authors highlight the need for more detailed information about the influence in the tumorigenicity index and tumorigenic potential for these and other tumor cells.

Hepatocellular carcinoma

Liver cancer is one of the various diseases related to this organ, and they are the leading cause of mortality and morbidity worldwide. Basically, uncontrolled fibrosis can progress to cirrhosis, and hepatocellular carcinoma has been the third cause of cancer mortality. Abu-Serie and Habashy (2019) in Egypt purified two proteins from RJ, major RJ protein 2 (MRJP2) and its predicted isoform X1(MRJP2 X1), followed by the in silico assessment

(3D structures and functions) using Iterative Threading ASSEmbly Refinement (I-TASSER) analysis and RAMPAGE program. To achieve the in vitro dual antihepatic damage potency, the authors used CCl4-induced hepatotoxicity hepatocyte culture and silymarin as a standard drug. For the antitumor bioactivity, the cell growth of a human hepatocellular carcinoma cell line (HepG2) comparing with doxorubicin (standard doxorubicin drug for chemotherapy) was studied. The results for this last essay showed that the MRJP2 and its MRJP2 X1 were capable of improving hepatocyte functions by mixed lineage kinase domain-like protein (MLKL) and intracellular reactive species, and also released the necrotic hepatocytes (by 60.4%) via decreasing tumor necrosis factor (TNE)-α. Adding to these properties, they induced caspase-dependent apoptosis and controlled the expression of Bcl-2 and p53 (analyzed by *Real-time quantitative polymerase chain reaction*) in HepG2 cells, showing an important antitumor potential. The authors revealed that treating CCl4-exposed cells with the isolated molecules and silymarin (lower safety compared to the purified RJ proteins) decreased the necrotic cell death and improved hepatocellular activity, confirming a high efficiency for this toxicity test. Doxorubicin had a highly toxic effect in hepatocytes and no significant impact in CCl4-treated cells, as was expected. The MRJP2 and its MRJP2 X1 isolated from RJ showed a promising therapeutic potential for liver diseases, requiring further research.

Lung cancer

Lin et al. (2020) checked the influence of 10-HDA in A549 human lung cancer cells. The results showed that 10-HDA induced cell-cycle arrest, regulating mitochondria-associated apoptosis, and inhibited the proliferation of the tumor cells with no significant toxic effects on normal cells. The 10-HDA inhibited cell migration by regulating transforming growth factor-beta 1 (TGF-beta 1), SNAI1, GSK-3 beta, E-cadherin, N-cadherin, and vimentin. It also regulated nuclear factor kappa B (NF-kappa B) signaling pathways by increasing the expression levels of phosphorylated c-Jun N-terminal kinase, p-p38, and I-kappa B. The apoptosis in the lung cancer cells was induced over-reactive oxygen species (ROS)-mediated MAPK, STAT3, NF-kappa B, and TGF-beta 1 signaling pathways. Thus, these results demonstrate that 10-HDA can be used in further studies of drug discovery targeting human lung cancer.

Vascular endothelial growth factor (VEGF)-induced angiogenesis

The VEGF is a proangiogenic factor fundamental in the good development of healthy people. Still, its potential activity (inducing angiogenesis) should be decreased in oncologic patients to avoid, among other problems, the development of the tumor. Once RJ and its constituents have been studied for potential antitumor purposes, it is noteworthy to evaluate their angiogenic effects. Among the bio compounds in RJ, Izuta et al. (2009) chose the 10-HDA to certify its ability on VEGF-induced proliferation, migration, and tube formation in human umbilical vein endothelial cells (HUVECs). All of them were inhibited with the concentration of 20 M, which implies inhibition of the VEGF-induced angiogenesis. Although these data need further studies, they are nevertheless very promising.

Preclinical animal studies

The investigation with RJ includes, in brief, two types of interventions. One is just looking for the antitumoral effect and another which pretends to be effective in preventing some side effects caused by the actual drugs used in cancer treatments, such as gametogenesis, hepatotoxicity, mucositis, myelosuppression, ototoxicity, nephrotoxicity, and spermiotoxicity. Firstly, the potential of RJ for oncologic drug discovery will be described. After that, the present subtitle will include the research showing the possibility of using this matrix to prevent unwanted adverse effects.

Breast cancer

Oršolić et al. (2005) tested RJ and other products correlated to the beehive against the tumor development and metastasis in murine tumor models as, for instance, spontaneous mammary carcinoma and a methylcholanthrene-induced fibrosarcoma (FS) of cytometric bead array (CBA) mouse. Metastases in the lungs were caused by injecting 10 (5) or 2×10 (5) viable tumor cells intravenously. RJ did not interfere with the appearance of metastases if it was given intraperitoneally or subcutaneously. Still, a strong inhibition on the formation of metastases occurs if its application was simultaneous with the tumor cells. These data are relevant for the control of tumor growth and its metastasis.

Zhang et al. (2017a) investigated the antitumor effect of RJ on 4T1-bearing mice (murine mammary carcinoma 4T1 causes a leukemoid reaction with profound granulocytosis coincident with the production of tumor-derived growth factors). RJ (0.5 and 1.5 g/kg) was given to mice by either the prophylactic therapeutic or therapeutic methods. In the prophylactic group (0.5 and 1.5 g/kg), the weight of the tumor was decreased significantly, and levels of interleukin (IL)-2, interferon-alpha (IFN-α), SOD, and total antioxidant capacity were significantly increased in the serum. In contrast, concentrations of IL-4 and IL-10 were significantly reduced. Further, the effects of antioxidant capacity and glutathione reductase were significantly ameliorated in the liver. Similar activities were significantly enhanced only under a dose of (0.5 g/kg) in the kidney. For the therapeutic group, antitumor effects of RJ were nonsignificant; however, the changes in IL-2, IFN-α, SOD, and antioxidant bioactivity levels in the serum and it changes the glutathione reductase in the liver were similar to those observed in the prophylactic ones. The findings of Zhang et al. (2017b) with the RNA-seq analysis on the effects of RJ on tumor growth in 4T1-bearing mice add more information to the previous research. The authors also conducted a study to get insights into RJ antitumor activities in 4T1-bearing mice but this time using an RNA sequencing system. Fresh RJ (in distilled water) has been administered intragastrically to mice, except in the control group, which had only distilled water. The findings showed 56 differently expressed genes pointedly associated with the immune system process of the 4T1 tumor. The most interesting data was the correlation among RJ and its effect on the expression of these different genes, which probably indicate that they could justify the enhancement in the immune response by RJ-mediation. All these data align with further research to verify if RJ and its components, especially the 10-HDA, decrease breast cancer development growth.

The same research group did a similar investigation, but this time they used Fresh Camellia RJ, which was previously analyzed, and the primary compounds as quality factors were all quantified. A particular highlight for the 10-HAD amount was 1.62 ± 0.05 g/100 g. Mice (BALB/c 32 female) were divided into four groups. The animals were fed with autoclaved food and water. One group was the control; two groups also tacked Camellia RJ in different concentrations (0.5 and 1.5 g/kg body weight per day, respectively). After the 8th day of tumor inoculation, the last group was submitted to standard treatment with cyclophosphamide (20 mg/kg body weight) injected intraperitoneally every day and RJ. All groups were subcutaneously injected into each mouse's right mammary fat pads with 4T1 tumor cell suspension. The results have shown that no significant body weight loss occurred in the mice in the RJ-treated groups. Slight inhibition of the growth of 4T1 breast cancer cells was revealed in serum augmentation ofTNF-a, Immunoglobulin G (IgG), and the reduction of IL-6. The best results were obtained with cyclophosphamide simultaneous with RJ. Among other data about the assessment for possible side effects caused by the Camellia RJ intake, the various liver sections were analyzed. Although the authors showed diffuse infiltration of many inflammatory cells, no apparent metastatic lesions were found (Shuang et al., 2017).

The oxidation process that induces and increases the production of free radicals is one of the key points in tumorigenesis. Malkoç et al. (2017) estimated the effect of RJ (by oral gavage, 100mg/kg/day, for 30days) on the oxidant-antioxidant equilibrium in female Sprague Dawley rats, *N*-methyl-*N*-nitrosourea (50mg/kg, a single dose, i.p.) was used to induce breast cancer. They also compared the data obtained with the assay performed with paclitaxel (a standard antitumor drug) (15mg/kg/week, 3 times, i.p.); one group received paclitaxel RJ. The levels of protein carbonyl and 8-hydroxydeoxyguanosine were higher in this last group, comparing to the untreated group. The levels of MDA were lower, and the enzyme catalase higher in groups witch received paclitaxel alone and in association with RJ. However, the authors emphasize that this antioxidant evidence was insufficient to reduce oxidative stress in breast cancer.

Leukemia

In the study of transplantable tumors of mouse advanced leukemia L1210 and P388 strains and Ehrlich, Sarcoma-180 ascites, and solid tumor strains, some authors used a couple of tactics. One involved a prophylactic-therapeutic, and for that, they gave RJ orally 30days before and 30days after the transplantations of tumor cells, and in the second protocol, a therapeutic approach was used. The intake of RJ (daily dose: 0-control, 10, 100, or 1000mg/kg) occurred during 30days after the transplantations of tumor cells. Different outcomes were obtained in relation to the survival time and the antitumor effect. In advance, leukemia L1210 and P388 strains occurred a short survival period (8–9days); no antitumor effect was verified in the therapeutic experiments. However, for the Sarcoma-180 ascites tumor, the survival time was 16days. The best results were verified with the Ehrlich ascites tumor (22.1 days).

Prevention of side effects—Hepatotoxicity

Some other drugs, cisplatin (CP), cause superoxide anion and hydroxyl radicals (reactive oxygen species—ROS), depleting glutathione levels. The inhibition of the activity of liver antioxidant enzymes would induce damage in this organ. To preserve the integrity of the liver tissue, various authors developed research and will be the base data for further studies in humans (Karadeniz et al., 2011). Although some of the data below discussed involve a specific drug means that similar damage cannot be induced by others used in this treatment. In some instances, extrapolation can be hypotheses for identical results that should be developed in further studies.

Ashry and Elkady (2014) studied the modulation of the hepatotoxicity induced with CP in rats, one of the most active cytotoxic drugs broadly used in many oncologic treatments. The RJ bioactivity explored here was mainly associated with the antioxidative potential against the oxidative stress caused by CP injury in the liver. The essay was performed with male albino rats feed in the gavage with RJ (300 mg/kg/day/orally), and treated for 15 days with CP by a single injection, i.p. (7 mg/kg body weight), against a control group only treated with CP. In the group given RJ, the animals significantly revealed better parameters for the biochemical markers, such as GSH, GPx, and increasing alanine aminotransferase (ALT) and aspartate aminotransferase (AST). According to the histopathological results, the RJ-cisplatin group showed reduced liver tissue lesions and a possible potential to scavenge free radicals produced by CP, which induced hepatotoxicity and inflammatory damages.

In India, Waykar and Alqadhi (2019) studied the protective effect of RJ, and also honey, in liver oxidative stress (hepatotoxicity) caused by the intake of CP (commonly used in cancer treatment). The investigation duration was 15 days, and they used Male Wistar albino rats separated into four groups: control 1 (no intake of RJ or honey, and no cisplatin injection), and control 2 included the animals fed orally with honey (500 mg/kg/day) and RJ (100 mg/kg/day), the cisplatin group (injected 7 mg/kg/day) intraperitoneally/15 days, and the last group with rats injected with cisplatin (7 mg/kg/day) intraperitoneally and fed as control 2. As expected, cisplatin administration caused a significant increase in ROS and MDA levels in liver tissue. On the contrary, the endogenous antioxidants have shown lower levels, for instance, GSH, CAT, SOD, and GPx. The rats supplied with RJ and honey even injected with cisplatin showed improved liver biomarkers for oxidative stress, which implies protection in this organ.

Prevention of side effects—Mucositis

Mucositis is a very common side effect caused by drugs used in oncologic chemotherapy and the radiation applied in certain cancer treatments. If no infection affects the oral mucosa, the healing process will be shorter (2 to 4 weeks); however, it could take more time (6–8 weeks) for patients under radiation therapy. The search for possible infection prevention and a faster healing time of the damaged mucosa stimulate various authors to investigate different possibilities. Among them, RJ shows promising results.

Watanabe et al. (2013) tested RJ film (prepared with chitosan-sodium alginate) on 5-fluorouracil (5-FU)–induced oral mucositis in hamsters.

The alginate allows the preparation of RJ to be sustained in the top of the damaged areas. The methods for healing evaluation include radical-scavenging activity of RJ, the generation of keratinocyte growth factors from human periodontal ligament fibroblasts, among the more commonly used as myeloperoxidase (MPO) activity, microscopic aspects, and reverse transcription-polymerase chain reaction (RT-PCR) for detection of proinflammatory cytokines. This last includes tumor necrosis factor-alpha, interleukin-1 beta, which decreased significantly, shown in the final results in the group of animals that had RJ films (10% and 30%). The myeloperoxidase was also reduced its activity and production. Briefly, the authors state that the antiinflammatory or antioxidative activities of RJ were important in reducing wounds and the healing process. Nevertheless, it is important to consider the effect of chitosan-sodium alginate that avoids external contamination of the mucosa damaged.

At the Kayseri Education and Research Hospital, Department of Radiation Oncology, and at Erciyes University Medical School, Department of Pathology (Kayseri, Turkey), Cihan and Deniz (2014) explored the potential of RJ against radiation-induced oral mucositis in rats. For the essay, they chose 48 adult males Sprague-Dawley rats (8–12 weeks old; weighing 275 ± 35 g). The design included six randomized groups as follow: control without RJ or irradiation (IR) and IR group (total cranium irradiation of 22 Gy as a single dose); other two groups taken only RJ in two different doses (50 and 100 mg/kg, topical administration 7 days which began 30 min before irradiation and continued for a week); and the other two, had the irradiation and RJ in the same doses as previous the ones. The tongue samples were histopathologically analyzed, and the degree of mucositis was the first outcome to be evaluated by the severity of inflammation and ulceration in the samples. Other important parameters, such as animal body mass, blood count, oxidative stress markers, and biochemical values, were checked. The results revealed, as expected, that the IR caused thrombocytopenia, weight loss, a significant increase in MDA levels, and decreased endogenous antioxidant enzymes, for example, SOD and CAT. However, the leucocyte and thrombocyte counts in the groups supplied with 50 mg/kg of RJ were significantly high even under IR. Despite this, the group with double RJ concentration did not show significant differences against radiation-induced toxicities. The global data indicates further investigation with RJ for this IR protection (Cihan and Deniz, 2014).

Some authors tested other options before, for instance, Suemaru et al. (2008) studied a mixture with honey, RJ, and propolis to test its prevention effect on 5-fluorouracil-induced experimental oral mucositis. They induced in hamsters the side effect produced by 5-fluorouracil and mild abrasions, and after the mixture of the products under study was applied topically in the damaged mucosa. The restorative process was analyzed by the size of the mucositis. The interesting results are that the authors also tested honey (1%, 10%, and 100%) and propolis (0.3%, 1%, and 3%) ointments separated, which did not reduce the size of the mucositis, comparing to the control group, where only Vaseline was applied in the mucositis. On the contrary, RJ (3%, 10%, and 30%) ointments meaningfully enhanced the healing process in the oral mucosa in a dose-dependent manner.

Prevention of side effects—Prostatic tissue damage

Abdel-Hafez et al. (2017) assessed the potential protection of RJ on the damage in prostatic tissue caused by chemotherapy used in cancer treatments. In the example essay in this study, cyclophosphamide was used as a standard antitumor drug. Rats were split into groups, and one of them had taken RJ, other cyclophosphamide, and another a concomitant ingestion of RJ and cyclophosphamide. The main results pointed that cyclophosphamide treated rats had an increase in prostatic MDA and C-reactive protein as expected. This data was in line with the decrease in GPx in prostatic tissue, and the histological changes showed marked acinar and stromal prostatic degeneration. These changes reveal that in the animals treated with cyclophosphamide and RJ, this last decreases the cyclophosphamide-induced prostatic tissue toxicity.

Prevention of side effects—Spermiotoxicity

Silici et al. (2009) evaluated the protection of RJ, via its antioxidative activity, on cisplatin-induced spermiotoxicity. They gave a single dose of cisplatin (7 mg/kg i.p) to rats. RJ (50 and 100 mg/kg) was added to gavage daily during the 10 days of the essay. The sperm physiognomies, testicular histologic analysis, testicular tissue oxidative stress grade, and plasma testosterone levels were assessed. The authors found that RJ was able to protect against cisplatin-induced spermiotoxicity significantly (motility and epididymal sperm concentration, as also reductions in weights of testes, epididymis seminal vesicles, and prostate). The group under cisplatin treatment showed the MDA increased in testes in line with the decrease in the enzymes superoxide dismutase, catalase, and glutathione peroxidase. The inverse was found in the group administered with RJ underlining the importance of its antioxidant effect.

Effect of RJ on the human models

As presented above, the antitumor effects of RJ were investigated with different methodologies, which allow further studies in human models. Among the clinical trials available, the more relevant data become from the results obtained when the trials aimed survey of RJ as preventing and reducing some side effects caused by the oncologic treatments as it was already explored in the animal essays.

Clinical trials with RJ for oncologic treatments

Leukemia

Bahoush and Pahlevani (2020) from the Department of Pediatric Hematology and Oncology, Ali-Asghar children's Hospital, Iran University of Medical Sciences (Tehran, Iran) did a single-center, before and after 12 weeks, clinical trial in 26 patients (16 males and 10 females) with childhood acute lymphoblastic leukemia during the maintenance phase of chemotherapy. The immune response blood cells were evaluated following N-chromosome RJ administration (starter dose of 2 g processed new natural N-chromosome RJ before breakfast). The follow-up was made every 2 weeks. Their peripheral white blood cells were raised significantly (beginning with 2510 ± 1192 cells per cubic millimeter and then elevated to 4549 ± 1500 cells per cubic millimeter at the end of the trial) after administering N-RJ and, likewise, they had a positive statistical rapport between them and the absolute neutrophil and absolute lymphocyte count. These results were more protruding at the beginning of the trial ($P < 0.001$) and the last 2 weeks of follow-ups ($P < 0.0005$), It was a relevant fact that none of the patients suffered adverse reactions from the RJ intake. This preliminary information validates that N-chromosome RJ can be, in the near future, explored as an immune-enhancing supplement for oncologic patients. However, further research with more accurate protocols, including bigger groups of patients and standardized RJ products, is necessary.

Effect of RJ on treatable symptoms from oncologic treatments

Anorexia, fatigue, hand-foot syndrome, and hypertension are among the most common side effects of cancer therapy, but also oral mucositis and gastrointestinal events. To help patients undergoing chemotherapy, various researchers and personal clinical search for evidence that could decrease the impact of these severe adverse events; nevertheless, the evidence is good but still scarce. Some of those were discussed above in the preclinical animal studies, and others will be in the text below.

Cancer-related fatigue

In 2016, Mofid et al. (2016) from Iran performed a Double-Blind Randomized Clinical Trial (registered in the Iranian Clinical Trial Registry Center—http://www.irct.ir, with the registration code: IRCT2015081423426N1) which evaluated the effect of processed honey and RJ on cancer-related fatigue on 52 patients with various types of tumors. The choice for RJ involved the content in proteins and free amino acid, fatty acids, minerals, and fat-soluble vitamins, A, D, E, and K. The product that was administered to patients was prepared under standard-state conditions with the permission of the Food and Drug Organization of Ministry of Health & Medical Education of Iran. The amount of 10-HDA was used as a control biomarker. The whole mix was analyzed for proteins (0.62%), lipids (0.04%), total sugars (76.50%), free sugars fructose (47.98%), glucose (23.63%), and sucrose (4.60%), and total ash content (0.06%) (Mofid et al., 2016). The results at the end-point, after 4 weeks of treatment with the mixture prepared with honey and RJ revealed an ameliorating effect for cancer-related fatigue in the patients involved in the trial.

Cancer-related fatigue and anorexia

Araki et al. (2019) from the Department of Urology, Nagasaki University Graduate School of Biomedical Sciences (Japan) performed a randomized, double-blinded, placebo-controlled clinical trial with patients diagnosed with renal cell carcinoma and allocated to start molecularly targeted therapy using tyrosine kinase inhibitors (TKIs). Sunitinib, pazopanib, and axitinib are the usual oral drugs most recommended in first-line therapy in these cases but also, for instance, as neoadjuvant therapy for cytoreductive nephrectomy among other combinations of chemotherapy protocols for various cancers. The authors enrolled 33 patients in Nagasaki University Hospital. In the study, hypertension was the side effect with a high percentage of almost 70%; however, RJ intake did not positively impact that. In relation to fatigue and anorexia, which affected 60.6% and 60.0% of the patients, respectively, the frequencies and severities of both were meaningfully lower in the RJ group. The authors stated the importance of these data, especially in the case of fatigue and anorexia.

Eating difficulty

Many head and neck cancer patients need tube feeding due to eating difficulties which are a common problem from the side effects of radiotherapy treatments. Trachootham et al. (2015) from different Faculties of Dentistry, Nurse, and Nutrition, along with the Maha Vajiralongkorn Cancer Hospital (Pathum Thani, Thailand), to avoid nasogastric feeding, tested a food gel with semisolid texture prepared with RJ (Nutri-jelly). They enrolled 80 head and neck cancer patients for the prospective quasi-randomized controlled trial, but only 74 (50 male and 24 female) filled the inclusion criteria and were divided into two groups. The majority of the tumors were lip/oral cavity/oropharynx or nasopharynx/maxillary sinus. The *health-related quality of life* was measure by questionnaires. The product was supply daily to patients under radiotherapy, with Cobalt-60 radiation (25–35 RT fractions, 5000–7000 cGy), in a 250 mL box corresponding to one serving with 230 kcal (protein, fat, and carbohydrate in the ratio of 20:30:50, respectively). Only five patients in the control group and four in the group test need to do simultaneous chemo-radiotherapy. The results showed that the group supplied with Nutri-jelly maintained the *higher general quality of life* score, for instance, in decreasing swallowing complications and wide-ranging eating difficulties, concerning the control group ($P < 0.0001$). In the cases that still needed tube feeding use, the study group had 13.5% of patients compared to 48.6% in the control group. The self-oral intake was 86.5% against the 51.4% for the patients in control. Despite the promising results, the authors would like to pursue the investigation, not only with cancer patients but also with other kinds of health problems such as dental and mouth diseases, stroke, and Parkinson's disease.

Inflammation

Inflammation is one of the adverse events caused by oncologic treatments. Once RJ can modulate inflammation (Yang et al., 2018), Yuno et al. (2020) from the Department of Urology at Nagasaki University Graduate School of Biomedical Sciences (Japan) performed a randomized controlled trial with 33 metastatic renal cell carcinoma patients ($n = 17$ placebo), that would be treated with first-line TKIs. The goal was to evaluate the decrease of specific adverse events, as cited above, caused by this therapy and the previous oral intake of RJ given to them before treatment. RJ regulates the levels of TNF-α, transforming growth factor (TGF)-β, and IL-6 produced by macrophages that will reduce inflammation. The authors evaluated if RJ alters Macrophage colony-stimulating factor (M-CSF), a crucial macrophage activity regulator. Afterward, they also verify the ratio after and before treatments of M-CSF levels, which is linked to anorexia after 2 weeks and fatigue (evaluation after 2, 4, and 12 weeks).

Mucositis

Various oncologic treatments such as chemo and radiotherapy can induce different adverse effects, such as dysphagia, dysphonia, oral mucositis, pain, or even ulcers. However, the more popular treatments use topical applications of different kinds of products, the need for better solutions.

RJ (made with nectar and pollen from several flowers in Zhejiang, China) was also studied for this purpose, between 2009 and 2010, by Yamauchi et al. (2014) from the Department of Agriculture Otolaryngology, Head and Neck Surgery, Kyorin University School of Medicine. The authors enrolled for a single-blind (physician-blind) clinical trial at the Kyorin University Hospital, Japan, 13 patients diagnosed with head and neck cancer (hypopharyngeal, laryngeal, maxillary sinus, oral cavity, and oropharyngeal) aged between 51 and 84 years (only 1 has a female). The study was performed with conventional chemotherapy drugs used worldwide, as Cisplatin, Nedaplatin, and Docetaxel, with different Regimen intakes. Radiotherapy was also administered (20–40 Gy, at the end of radiation; one-month radiation) and usually caused mucositis in 100% of patients. To prevent such effect, patients were administered three times a day 1 g (measuring spoon) of the ointment made with RJ (3 g/day) during radiation treatment. The evaluation was graded with the *Common Terminology Criteria for Adverse Events version 3.0* (CTCAE). The results reveal that RJ can act as prophylaxis against chemoradiotherapy-induced severity mucositis, decreasing the time progress to G2 and G3 mucositis. No important side effects were verified, such as allergy, irritation, or toxicity during treatment. All patients of both groups completed the planned previous schedule of chemoradiotherapy (Yamauchi et al., 2014).

This data is very important to decrease the suffering from oncologic patients who undergo these treatments but remains the need for more information, as the complete characterization of the compounds in the product and the mechanism of action. Certain bioactivities attributed to RJ can also be involved, for instance, the potential to inhibit proinflammatory cytokine production by activated macrophages, the antibacterial effect that protects wounds from being infected, and the relevant antioxidant power for the mucositis healing process.

The research progress with Erdem and Güngörmüş (2014) that assessed the effect of RJ on oral mucositis stated that the average time for the resolution of oral mucositis was significantly shorter in the group which used the product compared to control. They also suggest a significant healing effect.

Nephrotoxicity

As it was already explained above, chemotherapy induces some side effects and, among them, nephrotoxicity. Osama et al. (2017) from two Faculties of Pharmacy in Egypt did a clinical trial with oncologic patients, randomly divided, with cisplatin chemotherapy. The groups include patients pretreated before the initiation and during cisplatin chemotherapeutic regimen with honey and RJ and a control group only on cisplatin. The main goal of this trial was to evaluate the antioxidant properties of RJ and honey in the protection of cisplatin-induced acute kidney injury. The authors measured serum creatinine and urea levels before and after the chemotherapeutic with cisplatin to achieve this information. They found that these serum levels were lower in the groups supplied with the two hive products, implying possible renal injury protection. However, the values are only preliminary data, and a deeper investigation should be done; once possible, some Cisplatin-RJ could occur and induce bias in the final results.

Inflammation and decrease of immunity

In the Department of Urology, Nagasaki University Graduate School of Biomedical Sciences (Japan), Miyata et al. (2020a, b) also worked to relieve side effects caused by the targeted molecular therapy, such as TKIs, in metastatic renal cell carcinoma. The preliminary data was obtained with a randomized, double-blind clinical trial performed with an 800 mg of RJ/capsule oral intake three times per day (after breakfast, lunch, and dinner) for 3 months. The suppression of the adverse events could be due to the regulation of TNF-α and TGF-β.

Discussion

Cancer is a diverse disease marked by uncontrolled proliferation and a disrupted cell cycle, which results in aberrant cells that infiltrate and spread to other parts of the body (Neagu et al., 2019). To the best of our knowledge, this is the first article that systematically reviews recent evidence associated with the antitumor activities of RJ and its components. Although there is substantial evidence that RJ and its components have antitumor properties, the molecular mechanisms behind this impact are still to be fully explained.

Since Tamura et al. (1987) published information about the antitumoral activity of RJ, many other authors researched the same direction allowing the increase of data. Such data will undoubtedly lead RJ to be a good source of anticancer molecules to be included in new drugs in the near future. In

Miyata and Sakai (2018), various cell and animal studies associated with RJ in preliminary oncologic essays are presented. The data published from these studies present a broad possibility for further investigation. RJ showed the ability to decrease the proliferation of various tumor cells lines in vitro and in vivo assays. The positive impact can be high points in various mechanisms of action, which contribute to oncologic treatments' success. For example, the oxidant metabolites decrease and the increase of antioxidant enzymes levels, associated with the capability to increase the function of the immunologic system and the reducing of angiogenesis.

However, from our point of view, the major clinical evidence indicates an auspicious use for prevention and protection from therapy-induced toxicity. This review described a wide range of these situations, from eating difficulties to hepato- and nephrotoxicity decrease, passing through mucositis, among others. For all of the cited ones, well-proofs were found and pointed to the need to do more robust clinical trials to validate this critical application that can help oncologic patients during the chemotherapy treatments.

All the data collected gave evidence for its antitumoral potential as also the main compound. Regarding the gaps in this field of research, we strongly recommend further investigation, especially using a larger samples size and evaluation of much more parameters as, for instance, the plasma concentration of the drugs (carboplatin, cisplatin, docetaxel, paclitaxel, etc.) used in the assays.

Conclusion

The current outcomes of this systematic review supported the literature evidence of the in vivo and in vitro antitumor effects of RJ and its components, and its ability to inhibit, for instance, the breast, liver, and lung cancer cells, etc., as well as to prevent and/or reduce the impact of some side effect induced by various antitumor drugs such as cisplatin, cyclophosphamide, doxorubicin, etc. We hope that the data presented will encourage other authors to pursue this research, which will lead to an important step for the success of chemo- and radiotherapy, saving so many lives.

Acknowledgments

This research was funded by FCT—Fundação para a Ciência e a Tecnologia (UI0204): UIDB/00313/2020, Center of Chemistry from Faculty of Sciences and Technology of University of Coimbra, Portugal; and Forest Research Centre, a research unit funded by Fundação para a Ciência e a Tecnologia I.P. (FCT), Portugal (UIDB/00239/2020).

References

Abdel-Hafez, S.M.N., Rifaai, R.A., Abdelzaher, W.Y., 2017. Possible protective effect of royal jelly against cyclophosphamide induced prostatic damage in male albino rats; a biochemical, histological and immuno-histo-chemical study. Biomed. Pharmacother. 90, 15–23. https://doi.org/10.1016/j.biopha.2017.03.020. 2017 Jun. Epub 2017 Mar 21 28340377.

Abu-Serie, M.M., Habashy, N.H., 2019. Two purified proteins from royal jelly with in *vitro* dual anti-hepatic damage potency: major royal jelly protein 2 and its novel isoform X1. Int. J. Biol. Macromol. 128, 782–795. https://doi.org/10.1016/j.ijbiomac.2019.01.210. Epub 2019 Feb 1 30711561.

Ahmad, S., Khan, S.A., Khan, K.A., Li, J., 2020a. Novel insight into the development and function of hypopharyngeal glands in honey bees. Front. Physiol. 11, 1853.

Ahmad, S., Campos, M.G., Fratini, F., Altaye, S.Z., Li, J., 2020b. New insights into the biological and pharmaceutical properties of royal jelly. Int. J. Mol. Sci. 21 (2), 382. https://doi.org/10.3390/ijms21020382. 31936187. PMC7014095.

Altaye, S.Z., Meng, L., Li, J., 2019. Molecular insights into the enhanced performance of royal jelly secretion by a stock of honeybee (*Apis mellifera ligustica*) selected for increasing royal jelly production. Apidologie 50, 436–453. https://doi.org/10.1007/s13592-019-00656-1.

Andres, J.M.B., Gyawali, B., Krakauer, E., Trapan, D., 2020. WHO report on cancer. In: Ilbawi, A., Varghese, C. (Eds.), Chapter 05. Cancer Management, pp. 87–104. Executive WHO editors and writers.

Araki, K., Miyata, Y., Ohba, K., Nakamura, Y., Matsuo, T., Mochizuki, Y., Sakai, H., 2019. Oral intake of royal jelly has protective effects against tyrosine kinase inhibitor-induced toxicity in patients with renal cell carcinoma: a randomized, double-blinded, placebo-controlled tria. Medicines 6, 2. https://doi.org/10.3390/medicines6010002.

Ashry, K., Elkady, A.A., 2014. Royal jelly modulates hepatotoxic effect of rats treated with cisplatin. Arab J. Nucl. Sci. Appl. 47 (1), 172–180.

Bahoush, G., Pahlevani, R., 2020. Evaluation of the effectiveness and safety of N-Chromosome Royal Jelly on the number of peripheral white blood cells in patients with acute lymphoblastic leukemia. J. Contemp. Med. Sci. 2413-0516. 6 (6). https://doi.org/10.22317/jcms.v6i6.887. [S.l.]. dec. ISSN.

Cihan, Y., Deniz, K., 2014. The effects of royal jelly against radiation-induced acute oral mucositis. Int. J. Hematol. Oncol. 24 (1), 36–44. https://doi.org/10.4999/uhod.11059.

Erdem, O., Güngörmüş, Z., 2014. The effect of royal jelly on oral mucositis in patients undergoing radiotherapy and chemotherapy. Holist. Nurs. Pract. 28 (4), 242–246. https://doi.org/10.1097/HNP.0000000000000033. 24919094.

Filipič, B., Gradišnik, L., Rihar, K., Šooš, E., Pereyra, A., Potokar, J., 2015. The influence of royal jelly and human interferon-alpha (HuIFN-αN3) on proliferation, glutathione level and lipid peroxidation in human colorectal adenocarcinoma cells in vitro. Arh. Hig. Rada Toksikol. 66 (4), 269–274. https://doi.org/10.1515/aiht-2015-66-2632. 26751858.

Guo, J.Y., Wang, J.Z., Chen, Y., Cao, J., Tian, W., Ma, B., Dong, Y., 2021. Active components and biological functions of royal jelly. J. Funct. Foods 82. https://doi.org/10.1016/j.jff.2021.104514, 104514.

Izuta, H., Chikaraishi, Y., Shimazawa, M., Mishima, S., Hara, H., 2009. 10-Hydroxy-2-decenoic acid, a major fatty acid from royal jelly, inhibits VEGF-induced angiogenesis in human umbilical vein endothelial cells. Evid. Based Complement. Alternat. Med. 6 (4), 489–494. https://doi.org/10.1093/ecam/nem152.

Karadeniz, A., Simsek, N., Karakus, E., Yildirim, S., Kara, A., Can, I., Kisa, F., Emre, H., Turkeli, M., 2011. Royal jelly modulates oxidative stress and apoptosis in liver and kidneys of rats treated with cisplatin. Oxid. Med. Cell. Longev. 2011, 981793. https://doi.org/10.1155/2011/981793. Epub 2011 Aug 1 21904651. PMC3166579.

Lin, X.M., Liu, S.B., Luo, Y.H., Xu, W.T., Zhang, Y., Zhang, T., Xue, H., Zuo, W.B., Li, Y.N., Lu, B.X., Jin, C.H., 2020. 10-HDA induces ROS-mediated apoptosis in A549 human lung cancer cells by regulating the MAPK, STAT3, NF-κB, and TGF-β1 signaling pathways. Biomed. Res. Int. https://doi.org/10.1155/2020/3042636, 3042636. 33376719. PMC7744184.

Malkoç, M., Altay, D., Ahmet, A., Şafak, E., Tuğba, S., Kural, B.K., Hüseyin, U., 2017. The effects of royal jelly on the oxidant-antioxidant system in rats with N-methyl-N-nitrosourea-induced breast cancer. Turkish J. Biochem. 43. https://doi.org/10.1515/tjb-2017-0140.

Miyata, Y., Sakai, H., 2018. Anti-cancer and protective effects of royal jelly for therapy-induced toxicities in malignancies. Int. J. Mol. Sci. 19 (10), 3270. https://doi.org/10.3390/ijms19103270. 30347885. PMC6214034.

Miyata, Y., Araki, K., Ohba, K., Mastuo, T., Nakamura, Y., Yuno, T., Mukai, Y., Otsubo, A., Mitsunari, K., Mochizuki, Y., Sakai, H., 2020a. Oral intake of royal jelly improves anti-cancer effects and suppresses adverse events of molecular targeted therapy by regulating TNF-α and TGF-β in renal cell carcinoma: a preliminary study based on a randomized double-blind clinical trial. Mol. Clin. Oncol. 13 (4), 29. https://doi.org/10.3892/mco.2020.2099. Epub 2020 Jul 20. *Erratum in*: Mol Clin Oncol. 2020 Dec; 13(6): 78 32765876. PMC7403815.

Miyata, Y., Araki, K., Ohba, K., Mastuo, T., Nakamura, Y., Yuno, T., Mukai, Y., Otsubo, A., Mitsunari, K., Mochizuki, Y., Sakai, H., 2020b. *Erratum*: oral intake of royal jelly improves anticancer effects and suppresses adverse events of molecular targeted therapy by regulating TNF-α and TGF-β in renal cell carcinoma: a preliminary study based on a randomized double-blind clinical trial. Mol. Clin. Oncol. 3 (6), 78. https://doi.org/10.3892/mco.2020.2148. Epub 2020 Sep 25. Erratum for: Mol Clin Oncol. 2020 Oct; 13(4): 29 33062268. PMC7549394. CORRIGENDUM DOI:10.3892/mco.2020.2148.

Mofid, B., Rezaeizadeh, H., Termos, A., Rakhsha, A., Mafi, A.R., Taheripanah, T., Ardakani, M.M., Taghavi, S.M., Moravveji, S.A., Kashi, A.S., 2016. Effect of processed honey and royal jelly on cancer-related fatigue: a double-blind randomized clinical trial. Electron. Physician 8 (6), 2475–2482. https://doi.org/10.19082/2475. 27504161. PMC4965196.

Moubarak, M.M., Chanouha, N., Abou Ibrahim, N., Khalife, H., Gali-Muhtasib, H., 2021. Thymoquinone anticancer activity is enhanced when combined with royal jelly in human breast cancer. World J. Clin. Oncol. 12 (5), 342–354. https://doi.org/10.5306/wjco.v12.i5.342.

Nakaya, M., Onda, H., Sasaki, K., Yukiyoshi, A., Tachibana, H., Yamada, K., 2007. Effect of royal jelly on bisphenol A-induced proliferation of human breast cancer cells. Biosci. Biotechnol. Biochem. 71 (1), 253–255. https://doi.org/10.1271/bbb.60453. Epub 2007 17213647.

Neagu, M., Popescu, I.D., Zipeto, D., Tzanakakis, G., Nikitovic, D., Fenga, C., Stratakis, C.A., Spandidos, D.A., Tsatsakis, A.M., 2019. Inflammation and metabolism in cancer cell-mitochondria key player. Front Oncol. 14 (9), 348. https://doi.org/10.3389/fonc.2019.00348. PMCID: PMC6527883.

Newman, D.J., Cragg, G.M., 2020. Natural products as sources of new drugs over the nearly four decades from 01/1981 to 09/2019. J. Nat. Prod. 83 (3), 770–803. https://doi.org/10.1021/acs.jnatprod.9b01285. Epub 2020 Mar 12 32162523.

Oršolić, N., Terzić, S., Šver, L., Basic, I., 2005. Honey-bee products in prevention and/or therapy of murine transplantable tumours. J. Sci. Food Agric. 85, 363–370.

Osama, H., Abdullah, A., Gamal, B., Emad, D., Sayed, D., Hussein, E., Mahfouz, E., Tharwat, J., Sayed, S., Medhat, S., Bahaa, T., Abdelrahim, M.E.A., 2017. Effect of honey and royal jelly against cisplatin-induced nephrotoxicity in patients with cancer. J. Am. Coll. Nutr. 36 (5), 342–346. https://doi.org/10.1080/07315724.2017.1292157. Epub 2017 May 26 28548561.

Shuang, Z., Shao, Q., Shen, Z., Su, S., 2017. Immunomodulatory response of 4T1 murine breast cancer model to camellia royal jelly. Biomed. Res. 28 (3), 1223–1230.

Silici, S., Ekmekcioglu, O., Eraslan, G., Demirtas, A., 2009. Antioxidative effect of royal jelly in cisplatin-induced testes damage. Urology 74 (3), 545–551. https://doi.org/10.1016/j.urology.2009.05.024. Epub 2009 Jul 17 19616287.

Suemaru, K., Cui, R., Li, B., Watanabe, S., Okihara, K., Hashimoto, K., Yamada, H., Araki, H., 2008. Topical application of royal jelly has a healing effect for 5-fluorouracil-induced experimental oral mucositis in hamsters. Methods Find. Exp. Clin. Pharmacol. 30 (2), 103–106. https://doi.org/10.1358/mf.2008.30.2.1159655. 18560624.

Tamura, T., Fujii, A., Kuboyama, N., 1987. Antitumor effects of royal jelly (RJ). Nihon Yakurigaku Zasshi 89 (2), 73–80. Japanese https://doi.org/10.1254/fpj.89.73. 3570105.

Trachootham, D., Songkaew, W., Hongsachum, B., Wattana, C., Changkluengdee, N., Karapoch, J., Thirdsuttironnapumi, S., Meennuch, E., Klaitong, C., Sinthusek, T., Lam-ubol, A., 2015. Nutri-jelly may improve quality of life and decrease tube feeding demand in head and neck cancer patients. Support. Care Cancer 23 (5), 1421–1430. https://doi.org/10.1007/s00520-014-2488-5. Epub 2014 Nov 5 25370890. PMC4376960.

Watanabe, S., Suemaru, K., Takechi, K., Kaji, H., Imai, K., Araki, H., 2013. Oral mucosal adhesive films containing royal jelly accelerate recovery from 5-fluorouracil-induced oral mucositis. J. Pharmacol. Sci. 121 (2), 110–118. https://doi.org/10.1254/jphs.12181fp. Epub 2013 Jan 25 23357874.

Waykar, B.B., Alqadhi, Y.A., 2019. Protective role of honey and royal jelly on cisplatin induced oxidative stress in liver of rat. Int. J. Pharm. Sci. Res. 10 (8), 3898–3904.

WHO, 2021. Cancer. World Health Organization. Available from https://www.who.int/news-room/fact-sheets/detail/cancer. 3 March 2021.

Yamauchi, K., Kogashiwa, Y., Moro, Y., Kohno, N., 2014. The effect of topical application of royal jelly on chemoradiotherapy-induced mucositis in head and neck cancer: a preliminary study. Int. J. Otolaryngol. 2014. https://doi.org/10.1155/2014/974967, 974967. Epub 2014 Oct 21 25400667. PMC4221863.

Yang, Y.C., Chou, W.M., Widowati, D.A., Lin, I.P., Peng, C.C., 2018. 10-hydroxy-2-decenoic acid of royal jelly exhibits bactericide and anti-inflammatory activity in human colon cancer cells. BMC Complement. Altern. Med. 18 (1), 202. https://doi.org/10.1186/s12906-018-2267-9. 29970062. PMC6029378.

Yuno, T., Miyata, Y., Mukae, Y., Otsubo, A., Mitsunari, K., Matsuo, T., Ohba, K., Sakai, H., 2020. Mechanisms underlying the inhibition of tyrosine kinase inhibitor-induced anorexia and fatigue by royal jelly in renal cell carcinoma patients and the correlation between macrophage colony stimulating factor and inflammatory mediators. Med. Sci. (Basel) 8 (4), 43. https://doi.org/10.3390/medsci8040043. 33050250. PMC7712565.

Zhang, S., Shao, Q., Geng, H., Su, S., 2017a. The effect of royal jelly on the growth of breast cancer in mice. Oncol. Lett. 14 (6), 7615–7621. https://doi.org/10.3892/ol.2017.7078. Epub 2017 Sep 27 29344209. PMC575522.

Zhang, S., Nie, H., Shao, Q., kalan Hassanyar, A., Su, S., 2017b. RNA-Seq analysis on effects of royal jelly on tumour growth in 4T1-bearing mice. J. Funct. Foods 36, 459–466.

CHAPTER 10

Royal jelly: Healthy aging and longevity

Huseyin Sahin
Espiye Vocational School, Giresun University, Espiye, Giresun, Turkey

Contents

Introduction

Natural products are indispensable for living organisms due to their bioactive properties. Bee products also have a very effective place among natural products, and their product range is vast. RJ, just one of the bee products, is known as an enormous food source for the continuity of the bee world. This precious food, which is directly related to the bee larval development process, plays the leading role in maintaining a healthy and long life, especially for the queen bee. This advantage has aroused people's curiosity, and scientists have focused studies on this area. These studies have included that if humans can have healthy aging and lifespan with RJ consumption. To get the answer to this question, researchers have used various methodologies and tried to prove the superiority of RJ for a healthy life.

The relevant chapter has tried to present various methodologies subject to antiaging studies supported by the scientific literature and the chemical properties of RJ.

Bee Products and Their Applications in the Food and Pharmaceutical Industries
https://doi.org/10.1016/B978-0-323-85400-9.00002-2

Royal jelly

To feed the larvae, young nurse worker honeybees produce a unique, rich, and acidic compound named royal jelly (RJ), secreted from their mandibular and hypopharyngeal glands. Although the queen bees and worker bees typically have the same genetic structure after hatching, at the end of the larval period, the queen bee exhibits more superior properties in terms of the anatomical, physiological, and morphological structures thanks to being fed with more RJ and at different rates during the larval period. As a result of feeding with RJ until the fifth day of larval life and also ongoing throughout her life, the queen bee becomes resistant to possible diseases and lives up to 5 years; moreover, she can produce eggs the equivalent of her weight (2000–3000 eggs) per day (Fratini et al., 2016). On the other hand, other worker bees easily become infected with diseases due to their weak immunity; furthermore, they can live for only 2–3 months (Fratini et al., 2016) and cannot lay eggs despite being female. It is known that RJ, which has a fluid gel form with a creamy-white to dark-yellow color, is highly rich in proteins, oils, carbohydrates, hormones, vitamins, and minerals (Yucel et al., 2017). Although its constituents can change due to variability in the botanical source, bee species, artificial bee feeding, weather, season, location, and method of processing, its general components include water (50%–60%), protein (18%), carbohydrates (15%), lipids (3%–6%), and other biocompounds (Ali and Kunugi, 2020; Nagai and Inoue, 2004; Yeung and Argüelles, 2019). Especially, trans-10-hydroxy-2-decenoic acid (10-HDA) is a known fatty acid marker for RJ, being present at a considerably high level in the dry weight (Kanelis et al., 2015; Keskin et al., 2020). The largest portion of its dry matter comprises proteins, and the major royal jelly proteins 1–9 (MRJPs 1–9) represent 80% of these proteins (Ali and Kunugi, 2020; Kunugi and Mohammed Ali, 2019). Besides the MRJPs 1–9 found in RJ, minor proteins such as Royalisin, Jelleines, and Aspimin I are seen in small amounts (Ali and Kunugi, 2020). Currently, known proteins in RJ contribute to some bioactive properties (Ali and Kunugi, 2020). Moreover, some studies related to the MRJPs have directly been focused on antisenescence (Jiang et al., 2018; Xin et al., 2016).

The chemical content of RJ, which was determined by chemical analysis and summarized in Table 10.1, has been a matter of curiosity for apitherapy. Apitherapy, a subfield of complementary medicine, deals with bees and their products. Generally, honey, bee pollen, and propolis are well known as the main apitherapeutic products due to having a considerable level of

Table 10.1 The general chemical constituents of royal jelly (Ali and Kunugi, 2020; Yeung and Argüelles, 2019; Kunugi and Mohammed Ali, 2019; Okumus Yukunc, 2019; Sabatini et al., 2009; Xue et al., 2017; Chen et al., 2016; Nagai and Inoue, 2004).

Chemical constitutes	Detailed description
Water	~50%–60%
Protein	~18% (Appendix: Higher than 80% of royal jelly protein consists of MRJPs (1–9))
Carbohydrates	~15%
Lipids	~3%–6% (Appendix: The lipid composition of royal jelly contains approximately 80%–85% fatty acids and the major fatty acids of it are trans-10-hydroxy-2-decenoic acid, 10-hydroxydecanoic acid, and sebacic acid)
Some minerals	Sodium, calcium, potassium, magnesium, iron, zinc, copper, manganese, etc
Some vitamins	Retinol (A), ascorbic acid (C), tocopherol (E), some B vitamins (thiamin, riboflavin, niacin, pantothenic acid, pyridoxine, biotin, nicotinic acid, and folic acid)
Free amino acids	
Other bio compounds	Polyphenols

bioactive compounds such as phenolic acids, flavonoids, anthocyanins, and vitamins; however, besides mentioned products, others, royal jelly, apilarnil, and bee venom, have also many health effects. It has been claimed that all these products protect and strengthen the immune system (Can et al., 2015; Trumbeckaite et al., 2015). Nowadays, these products have gained more scientific attention and have become part of a professional approach to the treatment and prevention of illness (Trumbeckaite et al., 2015). Although apitherapy has a deep historical background and has just taken its place in modern medicinal applications, previous and current studies show us that apitherapy still holds many mysteries. An example of this mystery is the tremendous advantage RJ has shown for the queen bee. RJ, which was consumed extensively by the queen bee during its lifetime, distinguished it positively from other bees in the colony. The fact that queen bees have both superior physiological properties and a stronger immune system, healthy aging, and long lifespan has led people to use RJ more effectively in apitherapy applications. The metabolic differences of human and bee metabolism are clear, but the point of interest for humans is the possibility of having a stronger immunity and especially more involved in healthy aging processes. Therefore, it is highly likely that RJ will maintain its popularity for people.

Aging process

Aging is the sum of physiological, chemical, and biological changes in all living organisms; it is considered an inevitable process. This process can be described by two types: heterogeneous and heterochronic. When different organisms and/or individuals of the same species present varying aging rates, they are considered to display a heterogeneous aging process. On the other hand, if different cells and/or tissues of an organism show asynchrony of aging, this process is defined as heterochronic aging (Carmona and Michan, 2016). Generally, these processes do not move in harmony with the desired healthspan and lifespan because of oxidative stress from various factors and free radicals from biochemical reactions, insufficient antioxidative enzymes against them, age-related and chronic inflammation diseases, malnutrition, and the lack of physical capacity (Kunugi and Mohammed Ali, 2019).

Furthermore, the details of aging, in terms of the changing biological levels of cellular and molecular components, damage to mitochondrial homeostasis, disruption of intracellular communication, cellular senescence, telomere, and epigenetic change, and unchecked genome alterations, must be taken into account (Carmona and Michan, 2016; Kunugi and Mohammed Ali, 2019). However, this natural phenomenon can be improved by promoting cellular homeostasis, increasing stress resistance, and protecting against age-related diseases: This defines healthy aging, which is correlated to increased longevity (Carmona and Michan, 2016). According to the World Health Organization (WHO), healthy aging can be described as improving and protecting the functional ability of the elderly (Beard et al., 2016; Daskalopoulou et al., 2017). Although it is well known that there are multiple causes of aging, biochemical pathways are being investigated with scientific methodologies.

In humans, aging studies are conducted from a very limited perspective for valid reasons, such as the necessity of ethical rules, individual lifestyles, environmental and social factors, and a longer lifespan than other models (Carmona and Michan, 2016; Mitchell et al., 2015). For this reason, a variety of vertebrate and invertebrate models have been evaluated as alternatives in aging studies. According to researchers, the preference is usually in favor of invertebrate models due to their short lifespan, low cost, easier manipulation, and more straightforward experimental design (Carmona and Michan, 2016; Lucanic et al., 2013). Moreover, in vitro studies are also used as alternatives by researchers besides invertebrate models (Lucanic et al., 2013). Especially, in vitro analyses, which are used for the target chemicals in simple eukaryotic cell cultures, have come a long way in terms of revealing strong evidence for endogenous genetic pathways that contribute to lifespan. The cloning of genes using plasmids or viral vectors, then forming mammalian cell cultures for the expression of these genes, is significantly used for therapeutic applications (Dröge, 2005). Furthermore, some of these applications, which are related to the studies of the molecular genetics of aging, present an extra advantage because the recombinant proteins can be directly used—many questions, such as the treatment of some age-related diseases, the effectiveness of the produced drugs, and the activity of investigated chemical agents, can be answered using the clones of the potential gene (Lucanic et al., 2013).

As far as we understand, whatever the mechanisms and models, aging is an inevitable phenomenon; however, it could be delayed with some processes that result in healthy aging. Researchers have been more focused than ever on the new chemical synthesis drugs and, especially, natural products to learn whether they contribute to the lifespan by modulating some biochemical pathways associated with the slowdown of aging (Detienne et al., 2014; Lucanic et al., 2013).

Evaluation of royal jelly studies on healthspan and lifespan

It appears that this valuable natural compound includes many bioactive properties, such as antioxidant, antimicrobial, antitumoral, antiinflammatory, neurotrophic, and the like, and these are supported by scientific research (García et al., 2010; Kohno et al., 2004; Kolayli et al., 2016; Sığ et al., 2019). Because of these properties and the proven pharmacological reports, RJ is considered a commercial food and medicinal product in many countries (Nagai and Inoue, 2004). These properties could be given as a piece of proven evidence to support healthy aging. In this section, some study models using RJ and its products are discussed, evaluating their effects on the healthspan and lifespan, rather than those above mentioned biological properties.

Apis mellifera studies

According to the scientific literature, the queen bee's magnificent health span and long lifespan are associated with RJ's use. To prove the accuracy of this claim, it is the right strategy in the first place to directly use the *Apis mellifera* model without the need for other study models. For example, a comprehensive study was presented by Wang et al. (2014), who investigated the lifespan and gene expression levels of worker bees fed with different protein diets, including pollen and RJ. Results indicated that feeding with pollen and RJ was more effective in survival than feeding only pollen. Especially, it was shown that worker bees fed with pollen and pollen and RJ together had lower expression levels of the *HDAC1* gene, known as the aging gene, than the control (Wang et al., 2014). Yang et al. (2017) found the best lifespan extension was gained using 4% RJ in 50% sucrose solution over a 39-day experimental period. They focused on the isolated RJP60 protein (precipitation with 60% ammonium sulfate) from RJ, as it shows the same degree of the life-extension property with 4% RJ.

Moreover, for colony productivity and crop pollination efficiency, the use of this isolated protein (in sucrose solution) was suggested (Yang et al., 2017). Compared to Yang et al.' (2017) study, Shi et al.'s (2018) study used a wide range of RJ percentage values in 50% sucrose solution, ranging from 0% to 20% RJ showing the effect on the extension of lifespan. At the end of their study, the worker bees fed with 20% RJ exhibited notably enhanced lifespan as well as a higher level of gene expression of *GluRA* and *Nmdar1,* which are considered memory genes (Shi et al., 2018). Another study, which did not test for longevity but investigated some parameters in terms of quality of life (e.g., ovarian development), had impressive results. Besides various feeding combinations, including an RJ-only diet (without honey), worker bees were fed 22% and 40% RJ in honey for 2 weeks. This treatment showed an increase in ovarian development. Additionally, thermal conditions were analyzed for ovarian development, and the optimal incubation condition was highlighted as 34°C for 10 days (Lin and Winston, 1998). Part of these results (that was, ovarian activation) was nearly similar to the report by Altaye et al. (2010), who declared that RJ was the ideal diet for ovarian activation of *Apis mellifera scutellata* (Altaye et al., 2010).

Briefly, these studies have not directly mentioned the biochemical mechanisms related to longevity in *Apis mellifera,* for instance, gene levels. Besides contributing to healthy life parameters, they have indicated that RJ and its products cause long life in honey bees, especially the queen bees.

A nematode model study: *Caenorhabditis elegans*

In recent years, *Caenorhabditis elegans* has become a popular model organism for examining aging studies due to its short-term lifespan and more understood genetic pathways (Honda et al., 2011; Mack et al., 2018; Olsen et al., 2006). Previously, the lifespan studies of *C. elegans* have been only designed using single-gene mutations, which have shown successful results concerning the genetic approach. Besides this genetic approach, the manipulation variety related to aging has increased; however, this nematode has also served to the possible studies regarding this manipulation variety (Olsen et al., 2006). Below are some examples of the aging studies conducted with *C. elegans*.

A study related to RJ and its compounds was performed to demonstrate the mystery of longevity and how RJ affects the biological mechanisms in *C. elegans*. In that study, RJ and protease-treated RJ (pRJ) were used in considering its pharmacological properties. In the first place, before the giving results of fractionation processes applied for advanced analysis, the study claimed that RJ and pRJ were agents of extending lifespan, with the optimal data coming from the tested concentration of 10 μg/mL. In the next stage detailing the fractionation process, the fraction of pRJ.F5 eluted with 30% methanol from the octadecyl-silica column showed a significant increase in lifespan compared to the other fractions; hence, the study focused on this fraction for further analysis of gene levels. Moreover, the study highlighted the correlation between reduced *insulin/IGF-1* signaling and the activity of *DAF-16*, the unique ortholog of the *FOXO* family of transcription factors, in the investigated nematode (Honda et al., 2011). *DAF-16* is a well-known prominent factor for metabolism, promoting lifespan because it is responsible for stress resistance. Unfortunately, the response to stress is decreased with *insulin/IGF* signaling (Henderson and Johnson, 2001; Tatar et al., 2003). Honda et al. (2011) summarized the gene expression levels as follows: pRJ-Fr5 upregulated *dod-3* even though it downregulated *dod-19*, *dao-4*, and *fkb-4*; this situation is in coordination with the lowered reduction of insulin/IGF-1 signaling. Substantially, *fkb-4*, known to encode the homolog FKBP protein, was given attention because FKPB has a close relationship with the target of the rapamycin (TOR) pathway. In that study, the importance of TOR pathway inhibition was emphasized by the references to the relationship of the increasing lifespan in some species, including the nematode (Honda et al., 2011).

A comprehensive study on the samples of RJ and enzyme-treated RJ (eRJ) was conducted by Wang et al. (2015) to reveal the lifespan and healthspan effects in *C. elegans*. They emphasized the pivotal role of insulin/IGF-1 signaling for the modulation of longevity. The study showed the optimal concentration levels of 2 mg/mL for RJ and 1 mg/mL for eRJ. Those concentration levels were also used for the advanced study of molecular mechanisms. The current data of molecular mechanisms related to the application of RJ/eRJ indicated that the lifespan of *C. elegans* correlated with the transcriptional activities of *DAF-16*. Moreover, Wang et al. (2015) suggested that treatment with RJ/eRJ under abnormal conditions such as oxidative stress, irradiation, and heat shock could be effective, as indicated by the stress resistance response modulated by *DAF-16*, *SIR-2.1*, *HCF-1*, and *14-3-3* proteins.

Another study by Japanese researchers investigated whether 10-HDA, a well-known major fatty acid in RJ, is associated with the extension of lifespan in *C. elegans* (Honda et al., 2015). According to their results, 10-HDA caused longevity in conjunction with the downstream process of dietary restriction signaling (Honda et al., 2015). Notably, it is already a well-known scientific reality that some processes, such as cell growth and proliferation, development, metabolism, and aging, are closely related to the TOR pathway (Markaki and Tavernarakis, 2013). Fortunately, Honda and coworkers' study included another significant result in terms of providing evidence for 10-HDA's health span effects, showing a significant survival effect against some stress conditions compared to the control (Honda et al., 2015).

Besides the current knowledge on the improvement of lifespan by royalactin in *Drosophila melanogaster*, a study was performed to understand its role in the *C. elegans* lifespan by focusing on epidermal growth factor (EGF) signaling (Detienne et al., 2014). In the mentioned study, the concentration of 1.6 μg/mL showed the best lifespan-extending effect. Moreover, the researchers also applied some modifications to remove N-linked oligosaccharides from the royalactin. Those strategies were important in proving whether N-linked oligosaccharides rather than plain royalactin causes the extra lifespan-extending effect. The results revealed no effect in the parts of the isolated carbohydrate (Detienne et al., 2014). Epidermal growth factor acts in the growth and differentiation of cells through the binding of its epidermal growth factor receptor (EGFR) to activate the different signaling pathways (Ai et al., 2017). The scientific literature has mentioned a complex network of various genes responsible for encoding many EGF family members in some organisms, such as vertebrates and *Drosophila* (Groenen et al., 1994; Liu et al., 1999; Perrimon and Perkins, 1997). However, it is known that the EGF signaling network in *C. elegans* is simpler. Its genome encodes only the EGF-like growth factor (that is, *LIN-3*) and a homolog of EGFR receptor (that is, *LET-23*) (Mereu et al., 2020; Sundaram, 2013). At this point, Detienne et al. (2014) emphasized the significant role of *LIN-3* and *LET-23* for royalactin-mediated lifespan extension in *C. elegans* after analyzing research outputs.

These current studies focused on the metabolic responses in some situations, such as stress resistance, cell growth, and cell differentiation. It is known that these mechanisms are different from human metabolism; however, their results provide a piece of evidence for why humans should consider RJ consumption in pursuit of a healthy life and, also, healthy aging.

A yeast model study: *Saccharomyces cerevisiae*

Compared to other cell models, studies of aging and RJ using *S. cerevisiae* are somewhat limited; however, *S. cerevisiae* has some significant advantages. According to the studies of Zadrag et al. (2008) and Bitterman et al. (2003), the cells of *S. cerevisiae* could be considered an effective model in aging studies for several reasons, including the limited occurrence of budding. In fact, its short lifespan, well-annotated genome, and easy genetic manipulation are important advantages for these organisms, and thus it is claimed that the results obtained from *S. cerevisiae* could also serve as an effective reference for higher organisms (Bitterman et al., 2003; Zadrag et al., 2008). As mentioned above, there are many studies related to the lifespan of *S. cerevisiae*, but there is a limited number directly associated with RJ. This section focuses on one of the studies written by Jamnik et al. (2007), which yielded significant outputs. This study analyzed the cell energy metabolic activity based on fluorescence measurement by using an indicator, resazurin. Besides cell energy activity, intracellular oxidation was evaluated by the reaction of dichlorofluorescein with possible oxidants. The obtained results showed that RJ's intracellular oxidation was reduced in a dose-dependent manner compared to the control group (Jamnik et al., 2007). The authors also mentioned that RJ affected growth and cell energy metabolic activity, influencing protein expression and the scavenging of ROS (Jamnik et al., 2007). It should also be noted that the fight against ROS products plays a vital role in the healthy aging process for the higher organisms.

A multicellular organism model study: *Drosophila melanogaster*

According to Kamakura (2011), it was quite a problem to learn the current mechanism of caste differentiation of honeybees, owing to the lack of a developed mutant stock. Exactly on this point, it has been known that *D. melanogaster* is a suitable model for developmental and aging studies, including genetic analyses (Kamakura, 2011). Based on this reality, the influences of RJ and its unique compounds on *D. melanogaster* were evaluated, as in the literature discussed below.

Generally, in vitro biological activity of fresh or lyophilized RJ and its compounds has been confirmed by some laboratory studies (Kolayli et al., 2016; Nascimento et al., 2015). However, it might have been a prominent investigative issue to learn whether it can be used in freeze-dried form and have the same effect on lifespan as that of the fresh form. At this point, Kayashima et al. (2012) conducted a study associated with freeze-dried RJ (FDRJ) on the development and physiological efficiency of *Drosophila*. As a result of that study, Kayashima et al. (2012) claimed that FDRJ showed the same biological properties as RJ, reducing the development time of larvae, extending the lifespan of males, and increasing the fecundity of females. Besides, the gene expression levels were investigated, and results showed that the effect was associated with insulin/insulin-like growth factor signaling (*IIS*) and the *TOR* network (Kayashima et al., 2012). Considering the *IIS/TOR* pathway, the expression of *dilp5*, *InR*, and *TOR* was increased in female *Drosophila* fed with FDRJ, but no significant change was observed in males (Kayashima et al., 2012).

Xin et al. (2016) presented a study to show the effect of MRJPs on lifespan as well as nutrition and fecundity in both sexes of *Drosophila* flies. They fed the flies different dietary concentrations of MRJPs, ranging from 1.25% to 5.00%. Surprisingly, the average lifespans of both sexes of *Drosophila* flies were found to be differently affected by the MRJPs. While the best result in terms of the increase of mean lifespan was 13.9% with the 5.0% MRJP diet in males, the lifespan was increased by 20.0% using the 2.50% MRJP diet in females (Xin et al., 2016). After applying MRJP diets, Xin et al. (2016) confirmed that the relationship between the antioxidative effect and lifespan increased due to the upregulation of *CuZn-SOD*. Finally, these findings were explained by a mechanism for lifespan extension; particularly, the researchers focused on the *S6K* gene due to its upregulation in the EGFR-mediating signaling (Xin et al., 2016).

Another study emphasizing RJ's antioxidant properties further investigated its antiaging properties in *Drosophila*. The material in the study was composed of a combination of enzyme-treated RJ (eRJ) and collagen peptide (CP) (Qiu et al., 2020). Although Qui and coworkers did not directly suggest a survival mechanism as in vivo, they concluded that the ERJ-CP was effective in prolonging the lifespan of fruit flies treated with H_2O_2 and paraquat (PQ) via the reduced levels of malondialdehyde (MDA) and protein carbonyl (PCO) (Qiu et al., 2020). In that study, besides the results of MDA and PCO, antioxidant capacity was evaluated in terms of some significant endogenous antioxidant enzymes, such as total superoxide dismutase (T-SOD), glutathione peroxidase (GSH-Px), and catalase (CAT). The activities of these enzymes increased with the intake of ERJ-CP. Thus this combinative product could be considered as a health-promoting agent (Qiu et al., 2020).

Another study discussed the effect of RJ on viability, antioxidants, and especially metabolic modulation of a mutant of *Drosophila*, nprl3 (Cheng et al., 2020). The researchers observed that RJ caused a reduction in ROS levels and concluded that RJ proteins were directly responsible for increasing viability. In fact, the evaluation of TOR complex 1 (TORC1) activity seemed the priority for the study, and it was stated that while the survival rate increased in these mutants treated with RJ, TORC1 activity decreased (Cheng et al., 2020). It should not be forgotten that the regulation of TORC1 is significant for cell metabolism, and its dysregulation (particularly, its rise) is associated with cancer, age-related diseases, and hypertrophy, among others (Cheng et al., 2020; Johnson et al., 2013; Li et al., 2010; Zoncu et al., 2011).

It is known that RJ includes various concentration ranges of vitamins (Bogdanov, 2014), and this topic could be a noteworthy investigation area to find an extension agent of lifespan and healthspan. Here, Gardner (1948) focused on the effects of pantothenic acid, the most abundant vitamin in RJ, and its contribution to longevity in *D. melanogaster*. According to the previous studies, B vitamin complexes, especially pantothenic acid, were detected at higher levels than vitamin C, found in trace amounts in RJ (Bogdanov, 2014; Ciulu et al., 2013). In that study, Gardner (1948) revealed that there was an advantage of pantothenic acid for the average lifespan of the fruit fly cultures treated with the different fractions of RJ and pantothenic acid.

Briefly, all study examples detailed above have indicated that royal jelly and its ingredients are effective as directly and/or indirectly on *Drosophila melanogaster* in terms of quality of life.

Rodent model studies: Rat and mice

Animal model studies are needed to increase human knowledge and show the reasons for some biological and biomedical problems (Andersen and Winter, 2019). Rodents are used as a mammalian model in many experimental types of research. Some rodent studies claim that RJ and its compounds have effective properties, such as improving cognitive deficits, antiinflammation, antifatigue, and others (Chen et al., 2018; Kamakura et al., 2001; You et al., 2019). For example, the scientific literature was presented a different point of view with a rodent study (rat and mice together), which summarized the effects of queen bee acid (10-HDA) on the central nervous system (Weiser et al., 2017). When considering the experimental results of this study, queen bee acid, one of the main components of RJ, showed some positive outputs, such as increased neuron growth, the protection of neurons from damage, and decreased anxiety behaviors (Weiser et al., 2017). These properties could be given as a direct and/or indirect example of the increased quality of life and healthspan.

As far as we know, the number of studies that mention the effect of RJ on the direct lifespan of rodents is relatively few, just like yeasts. This section details only one fully compatible study related to the term *RJ and lifespan,* summarized below.

Working with C3H/HeJ mice, Inoue et al. (2003) found the use of dietary RJ caused a decreasing trend in the formation of overproduced free radicals during metabolism. This fundamental result was correlated with the reduction of 8-hydroxy-2′-deoxyguanosine (8-OHdG) production in the kidney and serum DNA. According to the authors, this reduction seemed to account for the increase in the average survival time partially. Compared to the control group, the survival of RJ treatment groups using the intermediate dose (6 mg/kg/day) and high dose (60 mg/kg/day) was significantly longer (Inoue et al., 2003). Unfortunately, RJ was found not to affect the prolongation of the total lifespan in mice (Inoue et al., 2003).

Conclusion

After examining many types of research that have studied RJ using various aging models, we believe that this chapter provides a more precise answer to why queen bees have a long lifespan. However, they have the same genome as other bees in the colony. As we understand, RJ is generally effective on cell renewal and proliferation and the metabolic functions related to providing vitality in all tissues of the organism and supporting immunity and health. After evaluating the promising results in all these subsections, we agree that learning the effects of RJ on healthspan and lifespan, including current mechanisms in metabolism, may help human health, specifically, healthy aging. However, it is evident that detailed research on the subject of human trials is needed for this. In fact, implementing these studies by separating them according to age category or gender will yield more effective results. Moreover, these studies could be more enlightening if the effects of current metabolic pathways evaluated in human clinical studies are conducted with RJ. The interpretation of changes in metabolic pathways and even gene level in antiaging studies will validate the results.

References

Ai, G., Shao, X., Meng, M., Song, L., Qiu, J., Wu, Y., Zhou, J., Cheng, J., Tong, X., 2017. Epidermal growth factor promotes proliferation and maintains multipotency of continuous cultured adipose stem cells via activating STAT signal pathway in vitro. Medicine 96 (30), e7607. https://doi.org/10.1097/MD.0000000000007607.

Ali, A.M., Kunugi, H., 2020. Apitherapy for age-related skeletal muscle dysfunction (sarcopenia): a review on the effects of royal jelly, propolis, and bee pollen. Foods 9 (10), 1362. https://doi.org/10.3390/foods9101362.

Altaye, S.Z., Pirk, C.W.W., Crewe, R.M., Nicolson, S.W., 2010. Convergence of carbohydrate-biased intake targets in caged worker honeybees fed different protein sources. J. Exp. Biol. 213 (19), 3311–3318. https://doi.org/10.1242/jeb.046953.

Andersen, M.L., Winter, L.M.F., 2019. Animal models in biological and biomedical research - experimental and ethical concerns. An. Acad. Bras. Cienc. 91 (suppl 1), e20170238. https://doi.org/10.1590/0001-3765201720170238.

Beard, J.R., Officer, A., de Carvalho, I.A., Sadana, R., Pot, A.M., Michel, J.-P., Lloyd-Sherlock, P., Epping-Jordan, J.A.E., Peeters, G.M.E.E.(.G.)., Mahanani, W.R., Thiyagarajan, J.A., Chatterji, S., 2016. The world report on ageing and health: a policy framework for healthy ageing. Lancet 387 (10033), 2145–2154. https://doi.org/10.1016/S0140-6736(15)00516-4.

Bitterman, K.J., Medvedik, O., Sinclair, D.A., 2003. Longevity regulation in *Saccharomyces cerevisiae*: linking metabolism, genome stability, and heterochromatin. Microbiol. Mol. Biol. Rev. 67 (3), 376–399. https://doi.org/10.1128/mmbr.67.3.376-399.2003.

Bogdanov, S., 2014. Royal jelly, bee brood: composition, health, medicine: a review. Bee Prod. Sci., 1–35.

Can, Z., Yildiz, O., Sahin, H., Akyuz Turumtay, E., Silici, S., Kolayli, S., 2015. An investigation of Turkish honeys: their physico-chemical properties antioxidant capacities and phenolic profiles. Food Chem. 180, 133–141. https://doi.org/10.1016/j.foodchem.2015.02.024.

Carmona, J.J., Michan, S., 2016. Biology of healthy aging and longevity. Rev. Investig. Clin. 68, 7–16.

Chen, Y.-F., Wang, K., Zhang, Y.-Z., Zheng, Y.-F., Hu, F.-L., 2016. *In vitro* anti-inflammatory effects of three fatty acids from royal jelly. Mediat. Inflamm. 2016, 1–11. https://doi.org/10.1155/2016/3583684 (Article ID: 3583684).

Chen, Y.F., You, M.M., Liu, Y.C., Shi, Y.Z., Wang, K., Lu, Y.Y., Hu, F.L., 2018. Potential protective effect of trans-10-hydroxy-2-decenoic acid on the inflammation induced by Lipoteichoic acid. J. Funct. Foods 45, 491–498. https://doi.org/10.1016/j.jff.2018.03.029.

Cheng, Y., Cai, J., Fu, Y., Feng, C., Hao, Y., Wei, Y., 2020. Royal jelly attenuates metabolic defects in a *Drosophila* mutant with elevated TORC1 activity. Biol. Open 9. https://doi.org/10.1242/BIO.054999, bio054999.

Ciulu, M., Floris, I., Nurchi, V.M., Panzanelli, A., Pilo, M.I., Spano, N., Sanna, G., 2013. HPLC determination of pantothenic acid in royal jelly. Anal. Methods 5 (23), 6682–6685. https://doi.org/10.1039/c3ay41284a.

Daskalopoulou, C., Stubbs, B., Kralj, C., Koukounari, A., Prince, M., Prina, A.M., 2017. Physical activity and healthy ageing: a systematic review and meta-analysis of longitudinal cohort studies. Ageing Res. Rev. 38, 6–17. https://doi.org/10.1016/j.arr.2017.06.003.

Detienne, G., De Haes, W., Ernst, U.R., Schoofs, L., Temmerman, L., 2014. Royalactin extends lifespan of *Caenorhabditis elegans* through epidermal growth factor signaling. Exp. Gerontol. 60, 129–135. https://doi.org/10.1016/j.exger.2014.09.021.

Dröge, A., 2005. Recombinant protein production in mammalian cell culture. In: Encyclopedic Reference of Genomics and Proteomics in Molecular Medicine. Springer, Berlin, Heidelberg, https://doi.org/10.1007/3-540-29623-9_3640.

Fratini, F., Cilia, G., Mancini, S., Felicioli, A., 2016. Royal jelly: an ancient remedy with remarkable antibacterial properties. Microbiol. Res. 192, 130–141. https://doi.org/10.1016/j.micres.2016.06.007.

García, M.C., Silvia Finola, M., Marioli, J.M., 2010. Antibacterial activity of royal jelly against bacteria capable of infecting cutaneous wounds. J. ApiMed. ApiProd. Res. 2 (3), 93–99. https://doi.org/10.3896/IBRA.4.02.3.02.

Gardner, T.S., 1948. The use of *Drosophila melanogaster* as-a screening agent for longevity factors: I. pantothenic acid as a longevity factor in royal jelly. J. Gerontol. 3 (1), 1–8. https://doi.org/10.1093/geronj/3.1.1.

Groenen, L.C., Nice, E.C., Burgess, A.W., 1994. Structure-function relationships for the EGF/TGF-α family of mitogens. Growth Factors 11, 235–257. https://doi.org/10.3109/08977199409010997.

Henderson, S.T., Johnson, T.E., 2001. *Daf*-16 integrates developmental and environmental inputs to mediate aging in the nematode *Caenorhabditis elegans*. Curr. Biol. 11 (24), 1975–1980. https://doi.org/10.1016/S0960-9822(01)00594-2.

Honda, Y., Fujita, Y., Maruyama, H., Araki, Y., Ichihara, K., Sato, A., Kojima, T., Nozawa, Y., Ito, M., Honda, S., 2011. Lifespan-extending effects of royal jelly and its related substances on the nematode *Caenorhabditis elegans*. PLoS One 6 (8). https://doi.org/10.1371/journal.pone.0023527, e23527.

Honda, Y., Araki, Y., Hata, T., Ichihara, K., Ito, M., Tanaka, M., Honda, S., 2015. 10-hydroxy-2-decenoic acid, the major lipid component of royal jelly, extends the lifespan of *Caenorhabditis elegans* through dietary restriction and target of rapamycin signaling. J. Aging Res. 2015 (425261). https://doi.org/10.1155/2015/425261.

Inoue, S.I., Koya-Miyata, S., Ushio, S., Iwaki, K., Ikeda, M., Kurimoto, M., 2003. Royal jelly prolongs the life span of C3H/HeJ mice: correlation with reduced DNA damage. Exp. Gerontol. 38 (9), 965–969. https://doi.org/10.1016/S0531-5565(03)00165-7.

Jamnik, P., Goranovič, D., Raspor, P., 2007. Antioxidative action of royal jelly in the yeast cell. Exp. Gerontol. 42 (7), 594–600. https://doi.org/10.1016/j.exger.2007.02.002.

Jiang, C., Liu, X., Li, C., Qian, H., Chen, D., Lai, C., Shen, L., 2018. Anti-senescence effect and molecular mechanism of the major royal jelly proteins on human embryonic lung fibroblast (HFL-I) cell line. J Zhejiang Univ Sci B 19 (12), 960–972. https://doi.org/10.1631/jzus.B1800257.

Johnson, S.C., Rabinovitch, P.S., Kaeberlein, M., 2013. mTOR is a key modulator of ageing and age-related disease. Nature 493, 338–345. https://doi.org/10.1038/nature11861.

Kamakura, M., 2011. Royalactin induces queen differentiation in honeybees. Nature 473, 478–483. https://doi.org/10.1038/nature10093.

Kamakura, M., Mitani, N., Fukuda, T., Fukushima, M., 2001. Antifatigue effect of fresh royal jelly in mice. J. Nutr. Sci. Vitaminol. 47 (6), 394–401. https://doi.org/10.3177/jnsv.47.394.

Kanelis, D., Tananaki, C., Liolios, V., Dimou, M., Goras, G., Rodopoulou, M.A., Karazafiris, E., Thrasyvoulou, A., 2015. A suggestion for royal jelly specifications. Arh. Hig. Rada Toksikol. 66 (4), 275–284. https://doi.org/10.1515/aiht-2015-66-2651.

Kayashima, Y., Yamanashi, K., Sato, A., Kumazawa, S., Yamakawa-Kobayashi, K., 2012. Freeze-dried royal jelly maintains its developmental and physiological bioactivity in *Drosophila melanogaster*. Biosci. Biotechnol. Biochem. 76 (11), 2107–2111. https://doi.org/10.1271/bbb.120496.

Keskin, M., Özkök, A., Yaylaci Karahalil, F., Kolaylı, S., 2020. What should be the amount of 10- hydroxi-2-decanoic acid (10-HDA) in royal jelly? Mediterr. Agric. Sci. 33 (3), 347–350. https://doi.org/10.29136/mediterranean.698926.

Kohno, K., Okamoto, I., Sano, O., Arai, N., Iwaki, K., Ikeda, M., Kurimoto, M., 2004. Royal jelly inhibits the production of proinflammatory cytokines by activated macrophages. Biosci. Biotechnol. Biochem. 68 (1), 138–145. https://doi.org/10.1271/bbb.68.138.

Kolayli, S., Sahin, H., Can, Z., Yildiz, O., Malkoc, M., Asadov, A., 2016. A member of complementary medicinal food: Anatolian royal jellies, their chemical compositions, and antioxidant properties. J. Evid. Based Complementary Altern. Med. 21 (4), NP43–NP48. https://doi.org/10.1177/2156587215618832.

Kunugi, H., Mohammed Ali, A., 2019. Royal jelly and its components promote healthy aging and longevity: from animal models to humans. Int. J. Mol. Sci. 20 (19), 4662. https://doi.org/10.3390/ijms20194662.

Li, L., Kim, E., Yuan, H., Inoki, K., Goraksha-Hicks, P., Schiesher, R.L., Neufeld, T.P., Guan, K.L., 2010. Regulation of mTORC1 by the Rab and Arf GTPases. J. Biol. Chem. 285 (26), 19705–19709. https://doi.org/10.1074/jbc.C110.102483.

Lin, H., Winston, M.L., 1998. The role of nutrition and temperature in the ovarian development of the worker honey bee (*Apis mellifera*). Can. Entomol. 130 (6), 883–891. https://doi.org/10.4039/Ent130883-6.

Liu, J., Tzou, P., Hill, R.J., Sternberg, P.W., 1999. Structural requirements for the tissue-specific and tissue-general functions of the *Caenorhabditis elegans* epidermal growth factor LIN-3. Genetics 153 (3), 1257–1269.

Lucanic, M., Lithgow, G.J., Alavez, S., 2013. Pharmacological lifespan extension of invertebrates. Ageing Res. Rev. 12 (1), 445–458. https://doi.org/10.1016/j.arr.2012.06.006.

Mack, H.I.D., Heimbucher, T., Murphy, C.T., 2018. The nematode *Caenorhabditis elegans* as a model for aging research. Drug Discov. Today Dis. Model. 27, 3–13. https://doi.org/10.1016/j.ddmod.2018.11.001.

Markaki, M., Tavernarakis, N., 2013. Metabolic control by target of rapamycin and autophagy during ageing - a mini-review. Gerontology 59 (4), 340–348. https://doi.org/10.1159/000348599.

Mereu, L., Morf, M.K., Spiri, S., Gutierrez, P., Escobar-Restrepo, J.M., Daube, M., Walser, M., Hajnal, A., 2020. Polarized epidermal growth factor secretion ensures robust vulval cell fate specification in *Caenorhabditis elegans*. Development 147. https://doi.org/10.1242/DEV.175760, dev175760.

Mitchell, S.J., Scheibye-Knudsen, M., Longo, D.L., De Cabo, R., 2015. Animal models of aging research: implications for human aging and age-related diseases. Annu. Rev. Anim. Biosci. 3, 283–303. https://doi.org/10.1146/annurev-animal-022114-110829.

Nagai, T., Inoue, R., 2004. Preparation and the functional properties of water extract and alkaline extract of royal jelly. Food Chem. 84 (2), 181–186. https://doi.org/10.1016/S0308-8146(03)00198-5.

Nascimento, A.P., Moraes, L.A.R., Ferreira, N.U., Moreno, G.P., Uahib, F.G.M., Barizon, E.A., Berretta, A.A., 2015. The lyophilization process maintains the chemical and biological characteristics of royal jelly. J. Evidence-Based Complementary Altern. Med. 2015, 1–5. https://doi.org/10.1155/2015/825068 (Article ID: 825068).

Okumus Yukunc, G., 2019. Royal jelly: proteins and peptides. J. Apither. Nat. 2 (2), 59–70. https://doi.org/10.35206/jan.679534.

Olsen, A., Vantipalli, M.C., Lithgow, G.J., 2006. Using *Caenorhabditis elegans* as a model for aging and age-related diseases. Ann. N. Y. Acad. Sci. 1067 (1), 120–128. https://doi.org/10.1196/annals.1354.015.

Perrimon, N., Perkins, L.A., 1997. There must be 50 ways to rule the signal: the case of the *Drosophila* EGF receptor. Cell 89, 13–16. https://doi.org/10.1016/S0092-8674(00)80177-4.

Qiu, W., Chen, X., Tian, Y., Wu, D., Du, M., Wang, S., 2020. Protection against oxidative stress and anti-aging effect in *Drosophila* of royal jelly-collagen peptide. Food Chem. Toxicol. 135, 110881. https://doi.org/10.1016/j.fct.2019.110881.

Sabatini, A.G., Marcazzan, G.L., Caboni, M.F., Bogdanov, S., Almeida-Muradian, L.B.D., 2009. Quality and standardisation of royal jelly. J. ApiMed. ApiProd. Res. 1 (1), 1–6. https://doi.org/10.3896/IBRA.4.1.01.04.

Shi, J.-l., Liao, C.-h., Wang, Z.-l., Wu, X.-b., 2018. Effect of royal jelly on longevity and memory-related traits of *Apis mellifera* workers. J. Asia Pac. Entomol. 21 (4), 1430–1433. https://doi.org/10.1016/j.aspen.2018.11.003.

Sığ, A.K., Öz-Sığ, Ö., Güney, M., 2019. Royal jelly: A natural therapeutic?'. Ortadoğu Tıp Derg. 11 (3), 333–341. https://doi.org/10.21601/ortadogutipdergisi.500434.

Sundaram, M.V., 2013. Canonical RTK-Ras-ERK signaling and related alternative pathways. In: The C. elegans Research Community (Ed.), WormBook. WormBook, pp. 1–38. https://doi.org/10.1895/wormbook.1.80.2.

Tatar, M., Bartke, A., Antebi, A., 2003. The endocrine regulation of aging by insulin-like signals. Science 299 (5611), 1346–1351. https://doi.org/10.1126/science.1081447.

Trumbeckaite, S., Dauksiene, J., Bernatoniene, J., Janulis, V., 2015. Knowledge, attitudes, and usage of apitherapy for disease prevention and treatment among undergraduate pharmacy students in Lithuania. J. Evidence-Based Complementary Altern. Med. 2015, 1–9. https://doi.org/10.1155/2015/172502 (Article ID: 172502).

Wang, H., Zhang, S.W., Zeng, Z.J., Yan, W.Y., 2014. Nutrition affects longevity and gene expression in honey bee (*Apis mellifera*) workers. Apidologie 45 (5), 618–625. https://doi.org/10.1007/s13592-014-0276-3.

Wang, X., Cook, L.F., Grasso, L.M., Cao, M., Dong, Y., 2015. Royal jelly-mediated prolongevity and stress resistance in *Caenorhabditis elegans* is possibly modulated by the interplays of DAF-16, SIR-2.1, HCF-1, and 14-3-3 proteins. J. Gerontol. Ser. A Biol. Sci. Med. Sci. 70 (7), 827–838. https://doi.org/10.1093/gerona/glu120.

Weiser, M., Grimshaw, V., Wynalda, K., Mohajeri, M., Butt, C., 2017. Long-term administration of queen bee acid (QBA) to rodents reduces anxiety-like behavior, promotes neuronal health and improves body composition. Nutrients 10 (1), 13. https://doi.org/10.3390/nu10010013.

Xin, X.-x., Chen, Y., Chen, D., Xiao, F., Parnell, L.D., Zhao, J., Liu, L., Ordovas, J.M., Lai, C.-Q., Shen, L.R., 2016. Supplementation with major royal-jelly proteins increases lifespan, feeding, and fecundity in *Drosophila*. J. Agric. Food Chem. 64 (29), 5803–5812. https://doi.org/10.1021/acs.jafc.6b00514.

Xue, X., Wu, L., Wang, K., 2017. Chemical composition of royal jelly. In: Alvarez-Suarez, J. (Ed.), Bee Products—Chemical and Biological Properties. Springer, Cham, pp. 181–190. https://doi.org/10.1007/978-3-319-59689-1_8.

Yang, W., Tian, Y., Han, M., Miao, X., 2017. Longevity extension of worker honey bees (*Apis mellifera*) by royal jelly: optimal dose and active ingredient. PeerJ 5. https://doi.org/10.7717/peerj.3118, e3118.

Yeung, Y.T., Argüelles, S., 2019. Bee products: royal jelly and propolis. In: Nabavi, S.M., Silva, A.S. (Eds.), Nonvitamin and Nonmineral Nutritional Supplements. Academic Press, pp. 475–484, https://doi.org/10.1016/B978-0-12-812491-8.00063-1.

You, M., Pan, Y., Liu, Y., Chen, Y., Wu, Y., Si, J., Wang, K., Hu, F., 2019. Royal jelly alleviates cognitive deficits and β-Amyloid accumulation in APP/PS1 mouse model via activation of the cAMP/PKA/CREB/BDNF pathway and inhibition of neuronal apoptosis. Front. Aging Neurosci. 10, 428. https://doi.org/10.3389/fnagi.2018.00428.

Yucel, B., Topal, E., Kosoglu, M., 2017. Bee products as functional food. In: Waisundara, V., Shiomi, N. (Eds.), Superfood and Functional Food - An Overview of Their Processing and Utilization. IntechOpen, https://doi.org/10.5772/65477.

Zadrag, R., Bartosz, G., Bilinski, T., 2008. Is the yeast a relevant model for aging of multicellular organisms? An insight from the total lifespan of *Saccharomyces cerevisiae*. Curr. Aging Sci. 1 (3), 159–165.

Zoncu, R., Efeyan, A., Sabatini, D.M., 2011. mTOR: from growth signal integration to cancer, diabetes and ageing. Nat. Rev. Mol. Cell Biol. 12, 21–35. https://doi.org/10.1038/nrm3025.

CHAPTER 11

Bee pollen in cosmetics: The chemical point of view

Aleksandar Ž. Kostić[a], Danijel D. Milinčić[a], Živoslav Lj. Tešić[b], and Mirjana B. Pešić[a]
[a]Faculty of Agriculture, Chair of Chemistry and Biochemistry, University of Belgrade, Belgrade, Serbia
[b]Faculty of Chemistry, Chair of Analytical Chemistry, University of Belgrade, Belgrade, Serbia

Contents

Introduction

As "life-giving dust," pollen has been recognized from ancient times (Egypt, Greece), but in the most recent past, it is also recommended as an excellent functional food ingredient (Campos et al., 2010; Kostić et al., 2020a). This description is related to the chemical composition of pollen which can be characterized as "almost perfect" since it contains all essential macro- and micronutrients as well as different phytochemicals. In addition, bee pollen is a source of different antioxidants due to its phenolic and flavonoid content.

Bee pollen can be successfully applied in skin cosmetics production. Due to both lipophilic and hydrophilic properties (mainly influenced by botanical origin), bee pollen can advance emulsions and improve

Bee Products and Their Applications in the Food and Pharmaceutical Industries
https://doi.org/10.1016/B978-0-323-85400-9.00011-3

miscibility/stability in the oil/water system, which is essential for preparing different skin cosmetic products. Since several simple compounds are crucial for the antioxidant defense system in the skin, apart from enzymes such as superoxide dismutases, glutathione peroxidases, catalase, and the thioredoxin system, bee pollen can be applied as a cheap, readily available, and eco-friendly source of it. These advantages are particularly related to the following compounds: quercetin and apigenin and its derivatives, vitamins C and E, and some inorganic antioxidants such as zinc and selenium ions.

In this chapter, a brief review of the possible application of bee pollen in skin cosmetics preparation is made to emphasize an excellent opportunity to use this bee product in the protection and health care of the skin.

Bee pollen as a nutritional and health-promoting agent

According to the most recent literature data, the average chemical composition of pollen can be described as follows: 54.22% of carbohydrates, 21.30% of proteins, 8.75% of fibers, 5.31% of lipids, and 2.91% of ash (Thakur and Nanda, 2020b). Additionally, pollen can be used as an excellent source of several macroelements (potassium, 4157.86 mg/kg; phosphorus, 1751.22 mg/kg; magnesium, 46.97 mg/kg) (Thakur and Nanda, 2020b) and microelements, in particular iron and zinc (Kostić et al., 2015a; Serra Bonvehí and Escolà Jordà, 1997; Thakur and Nanda, 2020b). For instance, the consumption of 10 g (one cup of coffee) of sunflower bee-collected pollen will cover around 10.4% of the daily need for iron, i.e., 4.2% of zinc daily intake for adults (Kostić et al., 2020b). Among bioactive compounds, phenolics stand out as the most important phytochemicals represented in pollen. Moreover, pollen is a good source of several vitamins (0.7% for both water and fat-soluble vitamins such as A, E, D, B1, B2, B6, and C), carotenoids (in particular β-carotene, 0.07%) (Komosinska-Vassev et al., 2015) and some specific compounds such as oleanolic acids, 3-ursolic acid, betulin alcohol (Komosinska-Vassev et al., 2015), betaines (Ares et al., 2020a), and glucosinolates (Ares et al., 2020b). This outstanding diversity of compounds present in pollen enable this material to possess excellent antioxidant properties (Kostić et al., 2019; Mărgăoan et al., 2019) as well as several other health promoting possibilities for humans such as antiinflammatory and anticarcinogenic activities (Mărgăoan et al., 2019).

Bee pollen as a source of antioxidants

Bee pollen is an important source of different antioxidants. Among them, phenolics are the most important, followed by unsaturated fatty acids (UFAs), carotenoids, different vitamins, and some microelements such as zinc and selenium. The chemical contexture of pollen is strongly influenced by botanical origin, also reflected in the antioxidants' content. Particularly, this is expressed in pollen's phenolics due to the amazing diversity of this group of compounds. Thus, it is possible to use a detailed phenolic profile of pollen as a chemotaxonomic tool (Bayram et al., 2021; Campos et al., 1997). Several monofloral pollen samples have been characterized via their phenolic profile (Alimoglu et al., 2021; Almaraz-Abarca et al., 2007; De-Melo et al., 2018a, b; Ecem Bayram, 2021; Kostić et al., 2019). The phenolic profile is important since phytochemicals are quite characteristic of different plants influencing pollen composition as well. Among phenolic compounds, flavonoids are predominant in pollen, followed by phenolic acids (Rzepecka-Stojko et al., 2015). The average phenolic content in pollen is up to 1.6%, with 1.4% of flavonoids, and 0.2% of phenolic acids (Komosinska-Vassev et al., 2015) in total. Among vitamins, the most important antioxidant in bee pollen is ascorbic acid (vitamin C), with an average range of content from 7 to 56 mg/100 g (Thakur and Nanda, 2020b). For vitamins belonging to B group International Honey Commission suggested the following ranges: 0.6 to 1.3 mg/100 g (vitamin B1 or thiamine), 0.6 to 2 mg/100 g (vitamin B2 or riboflavin), 4 to 11 mg/100 g (vitamin B3 or niacin), 0.2 to 0.7 mg/100 g (vitamin B6 or pyridoxine) (Thakur and Nanda, 2020b). Among lipophilic vitamins recent literature review gives average values for tocopherol, the major form of vitamin E (4–32 mg/100 g), as well as β-carotene (1–20 mg/100 g), also known as provitamin A (Thakur and Nanda, 2020b).

In addition to phenolics, carotenoids are also very interesting bioactive phytochemicals regularly present in pollen. The botanical origin of pollen influences its presence and quantities. For several pollen samples from Romania with different botanical origin the range for total carotenoid content was from 63.94 to 527.27 μg/g of dry weight with three different carotenoids quantified on the molecular level: lutein (57.04–476.30 μg/g of dry weight), β-cryptoxanthin (1.31–35.43 μg/g of dry weight), and β-carotene (0.21–18.18 μg/g of dry weight) (Mărgăoan et al., 2014). Among prominent antioxidants, UFAs are also regularly found in bee pollen. Although fatty acid profile is strongly dependent on the botanical origin of bee pollen, it can be stated that the most predominant UFAs are so-called "ω-3 fatty acids." Considering all of them, α-linoleic, linoleic and oleic acids are the most widespread UFAs in bee pollen (Thakur and Nanda, 2020b).

Lipophilic/hydrophilic character of bee pollen

Pollen is the focus of numerous researches because, as a multicomponent natural product, it represents a good source of bioactive components (BCs) such as phenolics, carotenoids, and some vitamins, which provide it with excellent antioxidant properties (Kostić et al., 2020a; Thakur and Nanda, 2020b). However, most BCs have lipophilic character and appear as micro components or in trace amounts. On the other hand, the main constituents of pollen are carbohydrates (13%–55%), proteins (10%–40%), and lipids (1%–13%) (Campos et al., 2008), which significantly contribute to its techno-functional properties (Kostić et al., 2015b) such as emulsifying and foaming properties, oil adsorption capacity, water adsorption capacity, protein and carbohydrate solubility, etc. These properties significantly influence the processing of some final products obtained from bee pollen. Lipid content and fatty acid diversity, as well as solubility of proteins and carbohydrates, significantly affect the emulsifying, foaming, and gelling properties of prepared pollen suspensions (Kostić et al., 2015b). It is vital to evaluate the potential application of pollen as an emulsifier replacer or water/oil adsorber in the formulation of complex food products such as ice cream, mousse, cakes, marshmallows, or high-fat sausages and meatballs (Kostić et al., 2020a), as well as in some other intricate systems such as cosmetic products. Different monofloral (coconut, coriander, rapeseed, and sunflower) (Kostić et al., 2020c; Thakur and Nanda, 2020a) and polyfloral (Kostić et al., 2015b; Thakur and Nanda, 2020a) pollen samples have been shown to possess good emulsifying properties (activity and stability), and oil/water adsorption capacity. According to literature, the emulsifying activity (EAI) of different pollen samples varies from 10.37 (bifloral pollen sample containing predominantly *Sophora* (46%) and *Helianthus annuus* (21%) pollen) to 80.54 (*Helianthus annuus* monofloral pollen) m^2/g, while emulsifying stability (ESI) ranged from 19.6 (monofloral sample containing 72% Fabaceae pollen) to 49.3 min (polyfloral sample containing 18% Asteraceae pollen) (Kostić et al., 2015b, c). Furthermore, values for water and oil adsorption capacity were in the range from 0.47 (coriander monofloral pollen) to 2.25 g/g (sample containing 81% Fabaceae pollen); that is, in the range from 1.00 (sample containing 76% Brassicaceae pollen) to

3.53 g/g (sample containing 42% *Sophora* pollen) (Kostić et al., 2015b, c; Thakur and Nanda, 2020a). Unlike the previously listed properties, pollen has distinct antifoaming properties (Kostić et al., 2015b, c), which is not following the research conducted on Indian bee pollen (Thakur and Nanda, 2020a). These differences can be attributed to the applied methodology and composition of surface-active lipids present in bee pollen samples. Since lipids play a critical role in skin cosmetics preparation, it is extremely important to protect them from oxidation processes (McMullen, 2018) to which they are susceptible. In that sense, pollen can be a good source of different antioxidants able to prevent lipid peroxidation.

The use of pollen in maintaining the health and beauty of the skin dates back to the ancient times of China, precisely because of the numerous chemical constituents of bee pollen that have great potential for application and development in the field of cosmetics (Xi et al., 2018). Bee pollen is most often added to cosmetic preparations in the form of various aqueous, lyophilized, and lipid extracts (0.5%–5%), rich in BCs that are evenly incorporated into the preparation matrix to protect the skin (Kurek-Górecka et al., 2020). However, most cosmetic preparations are in the form of emulsions, suspensions, or gels (Kim et al., 2020), requiring the use of surfactants, which over time cause side effects and skin damage (Xi et al., 2018). Due to the presence of oil-in-water (o/w) or water-in-oil (w/o) emulsion in cosmetics, it is essential to choose adequate antioxidants which will be applied to achieve maximal effect and efficiency (McMullen, 2018). Interestingly, a paradox known as the "polar paradox" in cosmetic products is very useful and should be understood. It claims that polar (hydrophilic) simple antioxidant compounds are a better choice to be applied in the bulk oil system and vice versa (McMullen, 2018). The main reason for this is the accumulation of these compounds in the surface area since they are not miscible with lipophilic compounds in the oil zone, and they will migrate to a more polar area. Therefore, antioxidants will be concentrated in the surface zone where the primary lipid peroxidation processes occur under the influence of atmospheric oxygen (McMullen, 2018). Since pollen contains both hydrophilic and lipophilic antioxidants, it is clear why it can be a good base for different skin products.

The complex structure and good techno-functional properties create the possibility of successful incorporations and applications of micronized bee pollen grains in formulating various skincare and cosmetic formulations as surfactant replacers (Kocot et al., 2018; Kurek-Górecka et al., 2020; Xi et al., 2018). Soluble proteins (Kostić et al., 2015b, c) and polar lipids such as esterified monoglycerides, phospholipids, and lipoproteins (Liang et al., 2013) from bee pollen can form a stable interfacial film between two immiscible phases and act as good emulsifiers (Dalgleish, 2006). Moreover, hydrophobic residues of proteins may affect the increased ability of oil adsorption (Kostić et al., 2015b). Also, proteins can contribute to the formation of a gel structure (Kinsella and Melachouris, 1976), which can be helpful in the development of cosmetic preparations because gel matrixes can affect holding water, oil, and specific odors. In addition, lipids of bee pollen contribute to the formation and functioning of a permeable barrier in the epidermis; they have moisturizing and cleansing properties for the skin, antimicrobial activity and help in the penetration of lipophilic molecules through skin tissues, while polyunsaturated fatty acids (PUFAs) reduce skin sensitivity to radiation and regulate skin lipid metabolism/activity (Xi et al., 2018).

Carbohydrates from bee pollen can act as thickeners or emulsion stabilizers, as well as gelling, moisturizing, and chelating agents. Insoluble polysaccharides easily bind and adsorb water molecules, making them moisturizing agents that are very important for the hydration of dry skin (Xi et al., 2018). Moreover, some sugars and primarily L-fucose or fucose-rich oligo polysaccharides can easily penetrate the dermis and affect the biosynthesis of collagen and fibroblasts (Fodil-Bourahla et al., 2003). Bee pollen grains contain complex polymer sporopollenin that can be of varying degrees of polymerization, crosslinking, and ability to adsorb oil (Ischebeck, 2016; Kostić et al., 2015b). They can influence preserve consistency, protect lipophilic BCs, and retain specific odors in cosmetic formulation. In recent years, amino acids have been recognized as mandatory ingredients in cosmetics because they positively affect changes in the dermis, penetrate deeper into the skin, prevent the accumulation of melanin, and prevent UV rays from causing more severe skin damage (Xi et al., 2018).

As bee pollen is a good source of essential amino acids (Bayram et al., 2021), the application of pollen grains in cosmetics attracts special attention. Bee pollen is often added to antidandruff shampoos because it limits the growth of fungi, has a positive effect on strengthening the hair and the activity of the sebaceous glands (Kurek-Górecka et al., 2020). Possible ways of application and action of pollen in skincare products are suggested in Fig. 11.1.

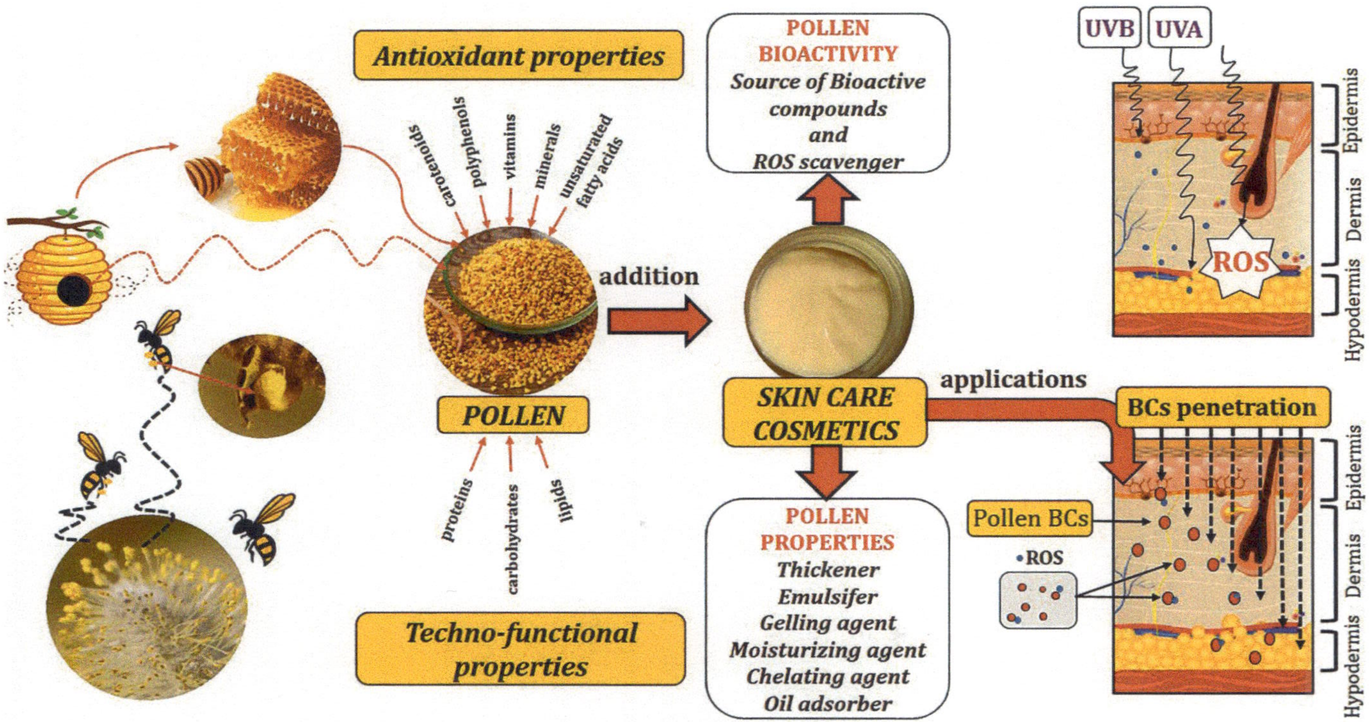

Fig. 11.1 Possible application and ways of action of bee pollen in skin care products.

Skin and skincare cosmetics

As an external and largest organ in our body, skin provides two important features: maintains body temperature and sensation (McMullen, 2018). With all the complexity of the skin structure, there are two main parts/zones: epidermis and dermis. The main component of the epidermis is keratinocyte cells filled with keratin filaments and proteins. The dermis is mainly composed of papillary and reticular dermis providing skin strength with the main cell types: fibroblasts, mast cells, and macrophages (McMullen, 2018). The skin presents the part of our immune system, which is the first physical barrier preventing foreign intruders' direct and free entrance into our body. Both enzymatic and nonenzymatic antioxidants are present in the epidermal antioxidant system. Superoxide dismutases (SODs), glutathione peroxidases (GSH), catalase, the thioredoxin system (Trx), and reductase of NADPH/NADH: Quinone are the primary enzymatic antioxidants of the skin (McMullen, 2018). However, it seems that simple chemical compounds (both lipophilic and hydrophilic) are more important antioxidants in the skin than several enzymatic systems mentioned above, such as SOD and GSH. It is also worth mentioning that most antioxidants are located in the epidermis (McMullen, 2018).

Due to the direct exposure of the skin to light and solar radiation, some changes occur in the dermis causing loss of good mechanical properties of the skin known as photoaging. This phenomenon manifests itself with the appearance of deep furrows in the skin as well as hyperpigmentation. Combined with UV radiation exposure, this can lead to some carcinogenic changes in the skin (McMullen, 2018). UVB light is the most dangerous and detrimental for different chromophores present in the skin, above all DNA and different proteins. In addition, visible and infrared light can induce the production of reactive oxygen species (ROS) and DNA damage in the skin (Hudson et al., 2020; Liebel et al., 2012), causing photoaging. Apart from light, air pollutants (such as different nitrogenous and sulfur oxides, ozone, particulate matters, polycyclic aromatic hydrocarbons (PAHs), tobacco smoke, and volatile organic compounds (VOCs)) can also be the triggers of some adverse effects in the skin such as extrinsic skin aging, inflammation processes, and allergic reactions (McMullen, 2018). To prevent/reduce all the above-mentioned unpleasant processes/changes in the skin and to provide consumers with a safe and healthy product without adding artificial additives, the pharmaceutical industry turned to natural sources, most often plant-based materials (González-Minero and

Bravo-Díaz, 2018). Active components in skincare cosmetics can be hydrophilic or lipophilic, dealing with different limitations: compounds soluble in water cannot penetrate some skin parts. In contrast, compounds soluble in oil usually have an unpleasant "greasy" appearance which is unacceptable for most consumers (Khater et al., 2021). In both cases, pollen can be an excellent source of these compounds. Still, it can also help develop modern systems such as solid lipid nanoparticles, nanocarriers, and nanoemulsions to improve targeted BCs delivery in skin cosmetics (Khater et al., 2021). Such claims are supported by the most recent research which successfully prepared sporopollenin microcapsules from two types of pollen grains (sunflower and chamomile) suitable for possible application as drug delivery systems (Ageitos et al., 2021). In addition, pollen-derived microcapsules, obtained from *Lycopodium clavatum* sporopollenin and date palm pollen grains, were efficiently applied for in vitro and in vivo targeted delivery as a drug (Meligi et al., 2021).

Selected compounds from bee pollen as skincare cosmetics components

Bee pollen can be used as an excellent source of several important antioxidants, as illustrated in Fig. 11.2. Here we summarized selected compounds useful in skincare, emphasizing quercetin and its derivatives and vitamins C and E.

Quercetin and derivatives

Researchers have paid special attention to quercetin, which is known as an exciting polyphenol from the group of flavonols, primarily due to its therapeutic properties and pharmaceutical activity (McMullen, 2018). Quercetin has been shown to have several physiological effects on the skin, such as strong antioxidant activity, ability to scavenge free radicals, prevent depletion of endogenous antioxidants from the skin, protect keratinocytes, inhibit lipid peroxidation and ROS radical formation after skin exposure to UV light (Hatahet et al., 2016). Quercetin reduces the level of malondialdehyde (MDA) and inhibits the formation of ROS radicals in cells exposed to UVB (Liu et al., 2013). An in vivo study involving the topical application of quercetin to hairless mice showed reduced UV-induced myeloperoxidase activity, reduced depletion of glutathione, and a decrease in total connective tissue-damaging proteinases (Casagrande et al., 2006).

Quercetin also has broad antiinflammatory activity and an important role in physiological mechanisms such as wound healing and skin aging

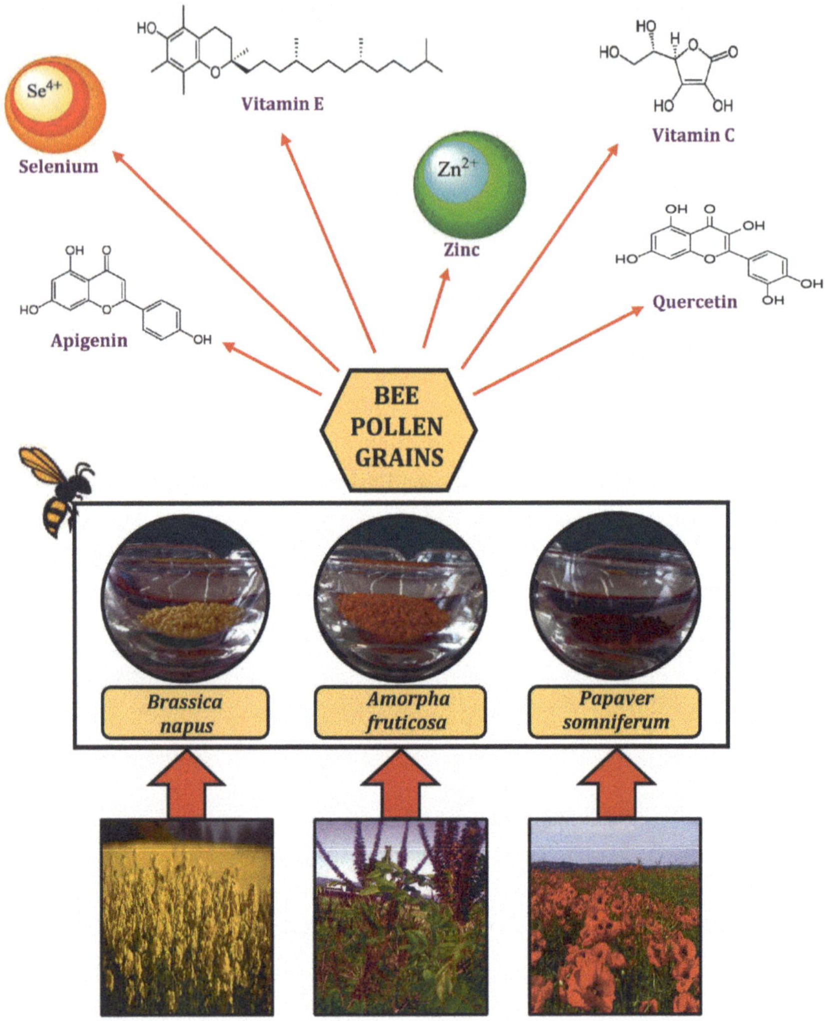

Fig. 11.2 Illustration of bee pollen as a source of important antioxidants.

(Hatahet et al., 2016). Exposure to UV radiation initiates oxidative stress that activates several signal transduction pathways. Nuclear factor-kappa B (NF-κB) is critical because it plays a role in cytokine production and modulation of the immune response (Hatahet et al., 2016). Good antioxidant properties and antiinflammatory activity give an increased possibility to use quercetin to treat chronic wounds. Quercetin suppresses/inhibits uncontrolled inflammation, enables successful skin regeneration and healing of acute wounds.

Furthermore, quercetin as an inhibitor of lipid peroxidation affects skin aging and protects the skin from dehydration; it inhibits tyrosinase and contributes to the whitening effect on the skin (Hatahet et al., 2016). The possible use of quercetin in the suppression of atopic dermatitis due to its antiinflammatory and antioxidant activity and ability to accelerate wound healing has been reported (Beken et al., 2020), as well as its option to be an antimelanogenesis agent (Choi and Shin, 2016). However, the benefits mentioned above of quercetin are limited due to its low hydrophilicity and reduced penetration into the skin (Nan et al., 2018). Therefore, quercetin and quercetin-rich plant extracts require adequate carriers for application in cosmetic preparations, which transfer them through the layers of the skin. In recent years, numerous studies have been conducted to develop suitable formulations for the successful local delivery of quercetin to the skin. Encapsulation of quercetin in microemulsions improves its intradermal delivery (Kitagawa et al., 2009) and reduces histological changes in the skin, such as UVB-Induced connective tissue damage (Vicentini et al., 2010). Quercetin encapsulation in chitosan nanoparticles allows easier passage through the epidermal layer, improves absorption, enhances the inhibitory effect against NF-κB/COX-2 signaling pathway, and alleviates UV-induced edema (Nan et al., 2018).

It has been demonstrated that bee pollen can be a functional and constitutive ingredient in various cosmetic preparations (Kurek-Górecka et al., 2020; Xi et al., 2018). The solid membrane (exine), sporopollenin, and multicomponent nature allow pollen to act as a good "carrier" of numerous bioactive components, such as lipophilic polyphenols and carotenoids (Ageitos et al., 2021; Kostić et al., 2020a) can quickly diffuse into the oil-rich skincare cosmetics and after topical application further in the skin layers. In particular, bee pollen is a rich source of quercetin and its derivatives (Table 11.1).

Given the possibility of using pollen in the formulation of skincare cosmetics, future research should focus on assessing the penetration and activity of quercetin and its derivatives from pollen in the skin layers.

Vitamin C (L-ascorbic acid)

Vitamin C is the main hydrophilic antioxidant for skincare. It helps and reduces hyperpigmentation of skin and changes related to the photoaging process (Ravetti et al., 2019). However, the most recent research revealed the possibility that high vitamin C content in wipes used for skin hygiene of children caused unusual yellow-brown pigmentation of the skin

Table 11.1 The content of quercetin and its derivatives in different bee pollen samples.

Botanical origin of bee pollen	Country	Detected quercetin and its derivatives in pollen samples	The content (mg/kg)	References
Monofloral: *Brassica napus*; *Leonitis leonurus*; *Camellia* sp.; Gallnut bee pollen; *Rosa* sp.; *Zea mays*	China	Quercetin Rutin Isoquercitrin	4.56–7.07 1.52–22.37 4.39–8.41	Duan et al. (2019)
Polyfloral: *Pinus*, *sylvestris*, *Ceratonia siliqua*, *Rosmarinus officinalis*, *Juniperus oxycedrus*, *Thymus vulgaris*, *Bupleurum spinosum*, *Globularia alypum*, and *Cistus landanferus*	Morocco	Quercetin Rutin	48.12 95.36	Laaroussi et al. (2020)
Polyfloral and monofloral: *Cocos nucifera*; *Coriandrum sativum*; *Brassica napus*	India	Quercetin Rutin Identified: Quercetin-3-methyl ether Quercetin-*O*-glucoside Bis-methylated quercetin	31.4–159.4 48.1–248.3	Thakur and Nanda (2021)
Unknown	Brazil	Quercetin Rutin Quercetin-3-*O*-glucopyranoside	36.5–72.3 16.5–54.9 25.5–43.6	Santa Bárbara et al. (2021)
Polyfloral (predominant plant species—*Eschscholzia californica*; *Brassica rapa*; *Azara celastrina*; *Cryptocarya alba*)	Chile	Quercetin	42.9–4099.4	Oyarzún et al. (2020)

Predominant families: Asteraceae, Fabaceae, Plantaginaceae, and Rosaceae	Turkey	Quercetin	3.81–36.32	Bayram et al. (2021)
		Rutin	17.29–58.46	
Helianthus annuus L.	Serbia	Quercetin	26.14–38.96	Kostić et al. (2019)
		Quercetin 3-*O*-galactoside	112.86–128.64	
		Quercetin 3-*O*-rhamnoside	7.00–7.76	
		Rutin	8.64–2.72	
		Quercetin 3-*O*-(2″-hexosyl)hexoside	18.05	
		Quercetin 3-*O*-(6″-acetyl)hexoside	35.12–111.10	
		Quercetin 3-*O*-rhamnoside	7.00–7.76	
		Quercetin 3-*O*-(di-*p*-coumaroyl) rhamnoside	16.84–22.45	
Cynara scolymus	Serbia	Quercetin 3-*O*-glucoside (Isoquercetin)	1.836	Kostić et al. (2021)
		Rutin	4.317	

(Guglielmo et al., 2021). Oral intake of vitamin C is not adequate for skin treatment since it will have low bioavailability in the skin. That is why the best choice for skin treatment with vitamin C is local administration and application directly to the skin surface (Ravetti et al., 2019). Due to high solubility in water, vitamin C is readily available in the skin's cells and acts as a scavenger of free radicals, particularly ROS, produced under the influence of sunlight. During the scavenging process, the vitamin C molecule donates electrons to radical particles stabilizing them. After that, vitamin C is converted in its oxidizing forms (semidehydroascorbate and dehydroascorbate) that are stable and unreactive (Ravetti et al., 2019). At physiological pH, vitamin C is predominantly represented as an anionic form of ascorbate ion. This anion is a strong antioxidant able to neutralize not only ROS but also ozone as well as nitrogen dioxide as air pollutants (McMullen, 2018). Humans are unable to biosynthesize vitamin C so it must be provided from different food stuffs (McMullen, 2018). And here is the situation where bee pollen can help us.

The vitamin C content in bee pollen varied from 70 to 560 mg/kg, providing up to 50% of recommended daily intake for an adult person (Thakur and Nanda, 2020b). It is important how bee pollen is stored to preserve the content of this vitamin. It was suggested that frozen bee pollen is a better source of vitamin C than the dried samples (Estevinho et al., 2019; de Melo and de Almeida-Muradian, 2010). Moreover, the drying process significantly reduced vitamin C content compared to fresh—from 384.8 to 121.1 mg/kg (60 °C) (Isik et al., 2019). However, the botanical origin of pollen significantly affected the content of vitamin C, as is presented in Table 11.2.

Vitamin E (tocopherols)

Tocopherols are composed of a long tail that is crucial for their incorporation into lipid membranes and a ring structure of chromanol that has the ability to donate protons during antioxidant activity. Vitamin E is widely used in the field of dermatology due to its UV protective effects, that is, excellent antioxidant and antiinflammatory properties (McMullen, 2018; Mohd Zaffarin et al., 2020). Inflammation causes enhanced immune responses and the formation of free radicals that affect uncontrolled tissue damage, excessive secretion of eicosanoids, and pro-inflammatory cytokines, which eventually leads to the development of cancer (Mohd Zaffarin et al., 2020). Strong antiinflammatory properties of vitamin E are associated with its ability to inhibit the secretion

Table 11.2 The content of vitamin C in different bee pollen samples influenced by botanical origin.

Botanical/geographical origin of bee pollen	The content (mg/kg)	References
Rhododendron ponticum -Turkey	162.4	Ecem Bayram (2021)
Polyfloral -Turkey	451.5	Kanar and Mazı (2019)
Echium sp.–*Castanea sativa* -Portugal	22.9–214.4	Estevinho et al. (2019)
Unknown -Turkey	384.8	Isik et al. (2019)
Asteraceae, Poaceae, Myrtaceae, Fabaceae, Combretaceae, Euphorbiaceae, Mimosaceae, Salicaceae, Urticaceae -Brazil	9.0–10.6	Negrão and Orsi (2018)
Unknown -Brazil	14–119	de Melo and de Almeida-Muradian (2010)

of inflammation-mediating molecules and pro-inflammatory signaling pathways (nuclear factor kappa beta (NF-κB)) and signal transducer and activator of transcription 3 (STAT-3) (Mohd Zaffarin et al., 2020). Moreover, the antiaging activity of vitamin E is reflected in its ability to enhance synthesis and prevent the breakdown of collagen (McMullen, 2018). Endogenous α-tocopherol is a structural constituent of the skin, found in its various layers. It plays a crucial role in protecting the skin against lipid peroxidation, photoaging, UV-induced ROS radicals, and immunosuppression (McMullen, 2018). However, due to their important role in maintaining skin health, tocopherols also became an almost indispensable ingredient in the formulation of skincare cosmetics. Topically applied vitamin E can easily penetrate the dermal layers, thus preventing products resulting from oxidative stress and photoaging. However, in the production of cosmetic preparations, antioxidant inactive esterified forms of tocopherol are increasingly used (McMullen, 2018).

Another option is to use naturally rich vitamin E extracts or powders. Bee pollen has been recognized as a potential source of vitamin E (De-Melo et al., 2016; Ecem Bayram, 2021; Kanar and Mazı, 2019; Sattler et al., 2015). The content of α-tocopherol in 18 Brazilian bee pollen samples varied from 4.7 (*Andira* monofloral pollen (99%)) to 114 μg/g (*Euterpe* (33%); *Eupatorium* (16%); *Eucalyptus* (15%); *Piper* (15%); and *Machaerium* (10%)).

Similar results were reported for vitamin E content in six analyzed fresh (19.43–43.00 μg/g) and dehydrated (16.27–38.64 μg/g) bee pollen samples (de Melo and de Almeida-Muradian, 2010). Depending on the type of the sample, the process of dehydration of pollen samples was responsible for an average loss of 18.7% of the content of vitamin E. Generally, Brazilian monofloral pollen samples had lower tocopherol content (Sattler et al., 2015), which was in agreement with the result reported for *Rhododendron ponticum* monofloral bee pollen from Turkey: 490 μg/100 g (Ecem Bayram, 2021). It is important to emphasize that by using 25 g of bee pollen rich in vitamin E, up to 20% of daily recommended intake (15 mg/day for adults) can be provided (Sattler et al., 2015). Furthermore, processing fresh polyfloral bee pollen sample (162.35 μg/g) by using different drying methods such as freeze-drying (128.70 μg/g); hot air drying (117.27–132.81 μg/g); vacuum drying (123.79–141.19 μg/g); microwave drying (116.60–131.42 μg/g) and microwave-assisted vacuum drying (114.86–131.85 μg/g), caused the decreased content of vitamin E (Kanar and Mazı, 2019). Significantly lower values for vitamin E content were reported for oven-dehydrated (27.2 and 27.5 μg/g) and lyophilized (37.5–53.7 μg/g) polyfloral bee pollen sample (De-Melo et al., 2016). These differences may be due to the botanical origin and the conditions under which bee pollen was processed. It has been shown that the synergistic effect of antioxidants such as vitamin C and vitamin E in skin preparations effectively protects the skin from sunburn, photoaging, and photocarcinogenesis (McMullen, 2018). By containing both vitamins, bee pollen can be used in skincare cosmetics.

Apigenin and derivatives

As one of the polyphenols (flavone subclass), apigenin is recognized as a good skin therapeutic. It proved to be an inhibitor of UVB-induced cyclooxygenase-2 expression in keratinocyte and chemokine expressions (Lee et al., 2007). To improve bioavailability in dermal treatments, nanoformulation was used to prepare apigenin nanocrystals as novel agents for skin protection from UV light (Al Shaal et al., 2011). There are several reports about bee pollen as a source of apigenin or its derivatives. For instance, in total, artichoke (*C. scolymus*) bee pollen from Serbia contained 3.006 mg/kg of apigenin as well as 1.606 mg/kg of apigenin-7-*O*-glucoside (Kostić et al., 2021), while Indian monofloral bee pollen samples (coconut, coriander, and rapeseed) contained from 4.6 to 30.2 mg/kg of apigenin (Thakur and Nanda, 2021).

Zinc and selenium

Zinc (Zn) and selenium (Se) are micro/trace elements essential for our immune system. For adults (70 kg of body weight) average Zn-content is 1.4–2.3 g, with 6% of total Zn found in the skin (Gupta et al., 2014). In dermatology, it has been used both as elemental/salts form and as topical preparations like zinc oxide, calamine, or zinc pyrithione to prevent skin photo damaging or dandruff formation (Gupta et al., 2014). The importance of Zn as a therapeutic or adjuvant for skincare is reflected in the treatment of different skin infections (warts, cutaneous leishmaniasis, leprosy, herpes genitalis, dermatophytoses, bromhidrosis, pityriasis and versicolor), inflammatory dermatoses (acne vulgaris, rosacea, acne conglobata, folliculitis decalvans, psoriasis, eczemas, ulcers, wounds, and oral aphthosis), hair disorders and mucosa (alopecia, erosive pustular dermatosis of scalp, seborrheic dermatitis, and mucosal lichen planus) as well as some premalignant and malignant dermatoses. In addition, topical Zn application can prevent depigmentation of skin expressed in the form of vitiligo and melasma (Gupta et al., 2014). In the case of Se as a trace element, it is one of the most important mineral antioxidants that is typically inserted into the body via food (plant food, meat, or poultry). This element builds selenocysteine, an amino acid present in thioredoxin reductase that is an enzyme important for the skin antioxidant system (McMullen, 2018). This enzyme provides degradation of hydrogen peroxide and fatty acid hydroperoxides in the skin enabling the renewal of two other important skin antioxidant molecules—vitamin C and lipoic acid (McMullen, 2018).

Bee pollen can be considered as an excellent Zn-source providing up to 15% of the daily need for adults (Kostić et al., 2015a; Serra Bonvehí and Escolà Jordà, 1997) with average Zn content of 3–25 mg/100 g of dry weight (Thakur and Nanda, 2020b). Pollen collected by hands from seven different maize hybrids was a surprisingly excellent Se source (0.069–0.176 mg/kg) covering 12.5%–32% of recommended daily intake for adults, calculated on 100 g of pollen (Kostić et al., 2017). However, bee pollen from Turkey was an even better source of this element, with an average content of 2.56 mg/kg (Altunatmaz et al., 2017).

Conclusion

Besides use in the food sector, bee pollen can be successfully applied in the cosmetic industry. As an excellent source of both hydrophilic (in particular vitamin C, zinc, and selenium ions) and lipophilic antioxidants (in particular vitamin E, quercetin, and apigenin) as well as due to good emulsifying properties, bee pollen can be a possible suitable carrier and base for the preparation of different emulsions applied in skin cosmetics. Also, sporopollenin from pollen can be used to prepare different liposomes for targeted drug delivery in the skin. In that sense, some future research should focus on the development of bee pollen-based skin products through applying nanotechnology and other modern techniques.

References

Ageitos, J.M., Robla, S., Valverde-Fraga, L., Garcia-Fuentes, M., Csaba, N., 2021. Purification of hollow sporopollenin microcapsules from sunflower and chamomile pollen grains. Polymers 13 (13), 2094. https://doi.org/10.3390/polym13132094.

Al Shaal, L., Shegokar, R., Müller, R.H., 2011. Production and characterization of antioxidant apigenin nanocrystals as a novel UV skin protective formulation. Int. J. Pharm. 420 (1), 133–140. https://doi.org/10.1016/j.ijpharm.2011.08.018.

Alimoglu, G., Guzelmeric, E., Yuksel, P.I., Celik, C., Deniz, I., Yesilada, E., 2021. Monofloral and polyfloral bee pollens: comparative evaluation of their phenolics and bioactivity profiles. LWT 142. https://doi.org/10.1016/j.lwt.2021.110973, 110973.

Almaraz-Abarca, N., da Graça Campos, M., Ávila-Reyes, J.A., Naranjo-Jiménez, N., Herrera Corral, J., González-Valdez, L.S., 2007. Antioxidant activity of polyphenolic extract of monofloral honeybee-collected pollen from mesquite (*Prosopis juliflora*, Leguminosae). J. Food Compos. Anal. 20 (2), 119–124. https://doi.org/10.1016/j.jfca.2006.08.001.

Altunatmaz, S.S., Tarhan, D., Aksu, F., Barutçu, U.B., Or, M.E., 2017. Mineral element and heavy metal (cadmium, lead and arsenic) levels of bee pollen in Turkey. Food Sci. Technol. 37 (suppl 1), 136–141. https://doi.org/10.1590/1678-457x.36016.

Ares, A.M., Toribio, L., Nozal, M.J., Martín, M.T., Bernal, J., 2020a. Simultaneous determination of betaines and other quaternary ammonium related compounds in bee pollen by hydrophilic interaction liquid chromatography-mass spectrometry. Microchem. J. 157. https://doi.org/10.1016/j.microc.2020.105000, 105000.

Ares, A.M., Redondo, M., Tapia, J., González-Porto, A.V., Higes, M., Martín-Hernández, R., Bernal, J., 2020b. Differentiation of bee pollen samples according to their intact-glucosinolate content using canonical discriminant analysis. LWT 129. https://doi.org/10.1016/j.lwt.2020.109559, 109559.

Bayram, N.E., Gercek, Y.C., Çelik, S., Mayda, N., Kostić, A.Ž., Dramićanin, A.M., Özkök, A., 2021. Phenolic and free amino acid profiles of bee bread and bee pollen with the same botanical origin—similarities and differences. Arab. J. Chem. 14 (3). https://doi.org/10.1016/j.arabjc.2021.103004, 103004.

Beken, B., Serttas, R., Yazicioglu, M., Turkekul, K., Erdogan, S., 2020. Quercetin improves inflammation, oxidative stress, and impaired wound healing in atopic dermatitis model of human keratinocytes. Pediatr. Allergy Immunol. Pulmonol. 33 (2), 69–79. https://doi.org/10.1089/ped.2019.1137.

Campos, M., Markham, K.R., Mitchell, K.A., da Cunha, A.P., 1997. An approach to the characterization of bee pollens via their flavonoid/phenolic profiles. Phytochem. Anal. 8 (4), 181–185. https://doi.org/10.1002/(SICI)1099-1565(199707)8:4<181::AID-PCA359>3.0.CO;2-A.

Campos, M.G.R., Bogdanov, S., de Almeida-Muradian, L.B., Szczesna, T., Mancebo, Y., Frigerio, C., Ferreira, F., 2008. Pollen composition and standardisation of analytical

methods. J. Apic. Res. 47 (2), 154–161. https://doi.org/10.1080/00218839.2008.11101443.
Campos, M.G.R., Frigerio, C., Lopes, J., Bogdanov, S., 2010. What is the future of bee-pollen? J. ApiProduct ApiMedical Sci. 2 (4), 131–144. https://doi.org/10.3896/IBRA.4.02.4.01.
Casagrande, R., Georgetti, S.R., Verri, W.A., Dorta, D.J., dos Santos, A.C., Fonseca, M.J.V., 2006. Protective effect of topical formulations containing quercetin against UVB-induced oxidative stress in hairless mice. J. Photochem. Photobiol. B Biol. 84 (1), 21–27. https://doi.org/10.1016/j.jphotobiol.2006.01.006.
Choi, M.-H., Shin, H.-J., 2016. Anti-melanogenesis effect of quercetin. Cosmetics 3 (2), 18. https://doi.org/10.3390/cosmetics3020018.
Dalgleish, D.G., 2006. Food emulsions—their structures and structure-forming properties. Food Hydrocoll. 20 (4), 415–422. https://doi.org/10.1016/j.foodhyd.2005.10.009.
de Melo, I.L.P., de Almeida-Muradian, L.B., 2010. Stability of antioxidants vitamins in bee pollen samples. Quim. Nova 33 (3), 514–518. https://doi.org/10.1590/S0100-40422010000300004.
De-Melo, A.A.M., Estevinho, M.L.M.F., Sattler, J.A.G., Souza, B.R., da Freitas, A.S., Barth, O.M., Almeida-Muradian, L.B., 2016. Effect of processing conditions on characteristics of dehydrated bee-pollen and correlation between quality parameters. LWT Food Sci. Technol. 65, 808–815. https://doi.org/10.1016/j.lwt.2015.09.014.
De-Melo, A.A.M., Estevinho, L.M., Moreira, M.M., Delerue-Matos, C., da de Freitas, A.S., Barth, O.M., de Almeida-Muradian, L.B., 2018a. A multivariate approach based on physicochemical parameters and biological potential for the botanical and geographical discrimination of Brazilian bee pollen. Food Biosci. 25, 91–110. https://doi.org/10.1016/j.fbio.2018.08.001.
De-Melo, A.A.M., Estevinho, L.M., Moreira, M.M., Delerue-Matos, C., da de Freitas, A.S., Barth, O.M., de Almeida-Muradian, L.B., 2018b. Phenolic profile by HPLC-MS, biological potential, and nutritional value of a promising food: monofloral bee pollen. J. Food Biochem. 42 (5). https://doi.org/10.1111/jfbc.12536, e12536.
Duan, H., Dong, Z., Li, H., Li, W., Shi, S., Wang, Q., Cao, W., Fang, X., Fang, A., Zhai, K., 2019. Quality evaluation of bee pollens by chromatographic fingerprint and simultaneous determination of its major bioactive components. Food Chem. Toxicol. 134. https://doi.org/10.1016/j.fct.2019.110831, 110831.
Ecem Bayram, N., 2021. Vitamin, mineral, polyphenol, amino acid profile of bee pollen from *Rhododendron ponticum* (source of "mad honey"): nutritional and palynological approach. J. Food Measur. Charact. 15 (3), 2659–2666. https://doi.org/10.1007/s11694-021-00854-5.
Estevinho, L.M., Dias, T., Anjos, O., 2019. Influence of the storage conditions (frozen vs. dried) in health-related lipid indexes and antioxidants of bee pollen. Eur. J. Lipid Sci. Technol. 121 (2), 1800393. https://doi.org/10.1002/ejlt.201800393.
Fodil-Bourahla, I., Bizbiz, L., Schoevaert, D., Robert, A., Robert, L., 2003. Effect of L-fucose and fucose-rich oligo- and polysaccharides (FROP-s) on skin aging: penetration, skin tissue production and fibrillogenesis. Biomed. Pharmacother. 57 (5–6), 209–215. https://doi.org/10.1016/S0753-3322(03)00047-7.
González-Minero, F., Bravo-Díaz, L., 2018. The use of plants in skincare products, cosmetics and fragrances: past and present. Cosmetics 5 (3), 50. https://doi.org/10.3390/cosmetics5030050.
Guglielmo, A., Virdi, A., Vincenzi, C., Zanotti Russo, M., Sechi, A., Patrizi, A., Neri, I., 2021. 'Colouring' wipes phenomenon: a peculiar skin pigmentation induced by ascorbic acid observed during lockdown. Clin. Exp. Dermatol., ced.14606. https://doi.org/10.1111/ced.14606.
Gupta, M., Mahajan, V.K., Mehta, K.S., Chauhan, P.S., 2014. Zinc therapy in dermatology: a review. Dermatol. Res. Pract. 2014, 1–11. https://doi.org/10.1155/2014/709152.

Hatahet, T., Morille, M., Hommoss, A., Devoisselle, J.M., Müller, R.H., Bégu, S., 2016. Quercetin topical application, from conventional dosage forms to nanodosage forms. Eur. J. Pharm. Biopharm. 108, 41–53. https://doi.org/10.1016/j.ejpb.2016.08.011.

Hudson, L., Rashdan, E., Bonn, C.A., Chavan, B., Rawlings, D., Birch-Machin, M.A., 2020. Individual and combined effects of the infrared, visible, and ultraviolet light components of solar radiation on damage biomarkers in human skin cells. FASEB J. 34 (3), 3874–3883. https://doi.org/10.1096/fj.201902351RR.

Ischebeck, T., 2016. Lipids in pollen—they are different. Biochim. Biophys. Acta Mol. Cell Biol. Lipids 1861 (9), 1315–1328. https://doi.org/10.1016/j.bbalip.2016.03.023.

Isik, A., Ozdemir, M., Doymaz, I., 2019. Effect of hot air drying on quality characteristics and physicochemical properties of bee pollen. Food Sci. Technol. 39 (1), 224–231. https://doi.org/10.1590/fst.02818.

Kanar, Y., Mazı, B.G., 2019. Effect of different drying methods on antioxidant characteristics of bee-pollen. J. Food Measur. Charact. 13 (4), 3376–3386. https://doi.org/10.1007/s11694-019-00283-5.

Khater, D., Nsairat, H., Odeh, F., Saleh, M., Jaber, A., Alshaer, W., Al Bawab, A., Mubarak, M.S., 2021. Design, preparation, and characterization of effective dermal and transdermal lipid nanoparticles: a review. Cosmetics 8 (2), 39. https://doi.org/10.3390/cosmetics8020039.

Kim, B., Cho, H.-E., Moon, S.H., Ahn, H.-J., Bae, S., Cho, H.-D., An, S., 2020. Transdermal delivery systems in cosmetics. Biomed. Derm. 4 (1), 10. https://doi.org/10.1186/s41702-020-0058-7.

Kinsella, J.E., Melachouris, N., 1976. Functional properties of proteins in foods: a survey. CRC Crit. Rev. Food Sci. Nutr. 7 (3), 219–280. https://doi.org/10.1080/10408397609527208.

Kitagawa, S., Tanaka, Y., Tanaka, M., Endo, K., Yoshii, A., 2009. Enhanced skin delivery of quercetin by microemulsion. J. Pharm. Pharmacol. 61 (7), 855–860. https://doi.org/10.1211/jpp/61.07.0003.

Kocot, J., Kiełczykowska, M., Luchowska-Kocot, D., Kurzepa, J., Musik, I., 2018. Antioxidant potential of propolis, bee pollen, and royal jelly: possible medical application. Oxid. Med. Cell. Longev. 2018, 1–29. https://doi.org/10.1155/2018/7074209.

Komosinska-Vassev, K., Olczyk, P., Kaźmierczak, J., Mencner, L., Olczyk, K., 2015. Bee pollen: chemical composition and therapeutic application. Evid. Based Complement. Alternat. Med. 2015, 1–6. https://doi.org/10.1155/2015/297425.

Kostić, A.Ž., Pešić, M.B., Mosić, M.D., Dojčinović, B.P., Natić, M.M., Trifković, J.Đ., 2015a. Mineral content of bee pollen from Serbia/Sadržaj minerala u uzorcima pčelinjega peluda iz Srbije. Arch. Ind. Hyg. Toxicol. 66 (4), 251–258. https://doi.org/10.1515/aiht-2015-66-2630.

Kostić, A.Ž., Barać, M.B., Stanojević, S.P., Milojković-Opsenica, D.M., Tešić, Ž.L., Šikoparija, B., Radišić, P., Prentović, M., Pešić, M.B., 2015b. Physicochemical composition and techno-functional properties of bee pollen collected in Serbia. LWT Food Sci. Technol. 62 (1), 301–309. https://doi.org/10.1016/j.lwt.2015.01.031.

Kostić, A.Ž., Kaluđerović, L.M., Dojčinović, B.P., Barać, M.B., Babić, V.B., Mačukanović-Jocić, M.P., 2017. Preliminary investigation of mineral content of pollen collected from different Serbian maize hybrids—is there any potential nutritional value? J. Sci. Food Agric. 97 (9), 2803–2809. https://doi.org/10.1002/jsfa.8108.

Kostić, A.Ž., Milinčić, D.D., Gašić, U.M., Nedić, N., Stanojević, S.P., Tešić, Ž.L., Pešić, M.B., 2019. Polyphenolic profile and antioxidant properties of bee-collected pollen from sunflower (*Helianthus annuus* L.) plant. LWT 112. https://doi.org/10.1016/j.lwt.2019.06.011, 108244.

Kostić, A.Ž., Milinčić, D.D., Barać, M.B., Ali Shariati, M., Tešić, Ž.L., Pešić, M.B., 2020a. The application of pollen as a functional food and feed ingredient—the present and perspectives. Biomol. Ther. 10 (1), 84. https://doi.org/10.3390/biom10010084.

Kostić, A.Ž., Dojčinović, B.P., Nedić, N., Milinčić, D.D., Pantelić, N.Đ., Stanojević, S.P., Barać, M.B., Tešić, Ž.L., Pešić, M.B., 2020b. Content and Nutritional Value of Selected Biogenic Elements in Monofloral Sunflower Bee-Collected Pollen From Serbia. pp. 211–217, https://doi.org/10.1007/978-3-030-40049-1_27.

Kostić, A.Ž., Milinčić, D.D., Trifunović, B.D.Š., Stanojević, S.P., Lević, S., Nedić, N., Nedović, V., Tešić, Ž.L., Pešić, M.B., 2020c. Nutritional and techno-functional properties of monofloral bee-collected sunflower (*Helianthus annuus* L.) pollen. Emirates J. Food Agric., 768. https://doi.org/10.9755/ejfa.2020.v32.i11.2188.

Kostić, A.Ž., Milinčić, D.D., Nedić, N., Gašić, U.M., Špirović Trifunović, B., Vojt, D., Tešić, Ž.L., Pešić, M.B., 2021. Phytochemical profile and antioxidant properties of bee-collected artichoke (*Cynara scolymus*) pollen. Antioxidants 10 (7), 1091. https://doi.org/10.3390/antiox10071091.

Kurek-Górecka, A., Górecki, M., Rzepecka-Stojko, A., Balwierz, R., Stojko, J., 2020. Bee products in dermatology and skin care. Molecules 25 (3), 556. https://doi.org/10.3390/molecules25030556.

Laaroussi, H., Bakour, M., Ousaaid, D., Aboulghazi, A., Ferreira-Santos, P., Genisheva, Z., Teixeira, J.A., Lyoussi, B., 2020. Effect of antioxidant-rich propolis and bee pollen extracts against D-glucose induced type 2 diabetes in rats. Food Res. Int. 138. https://doi.org/10.1016/j.foodres.2020.109802, 109802.

Lee, J.-H., Zhou, H.Y., Cho, S.Y., Kim, Y.S., Lee, Y.S., Jeong, C.S., 2007. Anti-inflammatory mechanisms of apigenin: inhibition of cyclooxygenase-2 expression, adhesion of monocytes to human umbilical vein endothelial cells, and expression of cellular adhesion molecules. Arch. Pharm. Res. 30 (10), 1318–1327. https://doi.org/10.1007/BF02980273.

Liang, M., Zhang, P., Shu, X., Liu, C., Shu, J., 2013. Characterization of pollen by MALDI-TOF lipid profiling. Int. J. Mass Spectrom. 334, 13–18. https://doi.org/10.1016/j.ijms.2012.09.007.

Liebel, F., Kaur, S., Ruvolo, E., Kollias, N., Southall, M.D., 2012. Irradiation of skin with visible light induces reactive oxygen species and matrix-degrading enzymes. J. Invest. Dermatol. 132 (7), 1901–1907. https://doi.org/10.1038/jid.2011.476.

Liu, D., Hu, H., Lin, Z., Chen, D., Zhu, Y., Hou, S., Shi, X., 2013. Quercetin deformable liposome: preparation and efficacy against ultraviolet B induced skin damages *in vitro* and *in vivo*. J. Photochem. Photobiol. B Biol. 127, 8–17. https://doi.org/10.1016/j.jphotobiol.2013.07.014.

Mărgăoan, R., Mărghitaş, L.A., Dezmirean, D.S., Dulf, F.V., Bunea, A., Socaci, S.A., Bobiş, O., 2014. Predominant and secondary pollen botanical origins influence the carotenoid and fatty acid profile in fresh honeybee-collected pollen. J. Agric. Food Chem. 62 (27), 6306–6316. https://doi.org/10.1021/jf5020318.

Mărgăoan, R., Stranţ, M., Varadi, A., Topal, E., Yücel, B., Cornea-Cipcigan, M., Campos, M.G., Vodnar, D.C., 2019. Bee collected pollen and bee bread: bioactive constituents and health benefits. Antioxidants 8 (12), 568. https://doi.org/10.3390/antiox8120568.

McMullen, R.L., 2018. Antioxidants and the Skin. CRC Press, https://doi.org/10.1201/9781315207254.

Meligi, N.M., Dyab, A.K.F., Paunov, V.N., 2021. Sustained *in vitro* and *in vivo* delivery of metformin from plant pollen-derived composite microcapsules. Pharmaceutics 13 (7), 1048. https://doi.org/10.3390/pharmaceutics13071048.

Mohd Zaffarin, A.S., Ng, S.-F., Ng, M.H., Hassan, H., Alias, E., 2020. Pharmacology and pharmacokinetics of vitamin E: nanoformulations to enhance bioavailability. Int. J. Nanomedicine 15, 9961–9974. https://doi.org/10.2147/IJN.S276355.

Nan, W., Ding, L., Chen, H., Khan, F.U., Yu, L., Sui, X., Shi, X., 2018. Topical use of quercetin-loaded chitosan nanoparticles against ultraviolet B radiation. Front. Pharmacol. 9. https://doi.org/10.3389/fphar.2018.00826.

Negrão, A.F., Orsi, R.O., 2018. Harvesting season and botanical origin interferes in production and nutritional composition of bee pollen. An. Acad. Bras. Cienc. 90 (1), 325–332. https://doi.org/10.1590/0001-3765201720150192.

Oyarzún, J.E., Andia, M.E., Uribe, S., Núñez Pizarro, P., Núñez, G., Montenegro, G., Bridi, R., 2020. Honeybee pollen extracts reduce oxidative stress and steatosis in hepatic cells. Molecules 26 (1), 6. https://doi.org/10.3390/molecules26010006.

Ravetti, S., Clemente, C., Brignone, S., Hergert, L., Allemandi, D., Palma, S., 2019. Ascorbic acid in skin health. Cosmetics 6 (4), 58. https://doi.org/10.3390/cosmetics6040058.

Rzepecka-Stojko, A., Stojko, J., Kurek-Górecka, A., Górecki, M., Kabała-Dzik, A., Kubina, R., Moździerz, A., Buszman, E., 2015. Polyphenols from bee pollen: structure, absorption, metabolism and biological activity. Molecules 20 (12), 21732–21749. https://doi.org/10.3390/molecules201219800.

Santa Bárbara, M.F., Moreira, M.M., Machado, C.S., Chambó, E.D., Pascoal, A., de Carvalho, C.A.L., da Silva Sodré, G., Delerue-Matos, C., Estevinho, L.M., 2021. Storage methods, phenolic composition, and bioactive properties of *Apis mellifera* and *Trigona spinipes* pollen. J. Apic. Res. 60 (1), 99–107. https://doi.org/10.1080/00218839.2019.1708595.

Sattler, J.A.G., de Melo, I.L.P., Granato, D., Araújo, E., da de Silva Freitas, A., Barth, O.M., Sattler, A., de Almeida-Muradian, L.B., 2015. Impact of origin on bioactive compounds and nutritional composition of bee pollen from southern Brazil: a screening study. Food Res. Int. 77, 82–91. https://doi.org/10.1016/j.foodres.2015.09.013.

Serra Bonvehí, J., Escolà Jordà, R., 1997. Nutrient composition and microbiological quality of honeybee-collected pollen in Spain. J. Agric. Food Chem. 45 (3), 725–732. https://doi.org/10.1021/jf960265q.

Thakur, M., Nanda, V., 2020a. Exploring the physical, functional, thermal, and textural properties of bee pollen from different botanical origins of India. J. Food Process Eng. 43 (1). https://doi.org/10.1111/jfpe.12935.

Thakur, M., Nanda, V., 2020b. Composition and functionality of bee pollen: a review. Trends Food Sci. Technol. 98, 82–106. https://doi.org/10.1016/j.tifs.2020.02.001.

Thakur, M., Nanda, V., 2021. Screening of Indian bee pollen based on antioxidant properties and polyphenolic composition using UHPLC-DAD-MS/MS: a multivariate analysis and ANN based approach. Food Res. Int. 140. https://doi.org/10.1016/j.foodres.2020.110041, 110041.

Vicentini, F.T.M.C., Fonseca, Y.M., Pitol, D.L., Iyomasa, M.M., Bentley, M.V.L.B., Fonseca, M.J.V., 2010. Evaluation of protective effect of a water-in-oil microemulsion incorporating quercetin against UVB-induced damage in hairless mice skin. J. Pharm. Pharm. Sci. 13 (2), 274. https://doi.org/10.18433/J3830G.

Xi, X., Li, J., Guo, S., Li, Y., Xu, F., Zheng, M., Cao, H., Cui, X., Guo, H., Han, C., 2018. The potential of using bee pollen in cosmetics: a review. J. Oleo Sci. 67 (9), 1071–1082. https://doi.org/10.5650/jos.ess18048.

CHAPTER 12

Pollen and bee bread and liver health

Zaida Zakaria[a], Zaidatul Akmal Othman[a,b], Joseph Bagi Suleiman[c], Victor Udo Nna[d], and Mahaneem Mohamed[a,e]
[a]Department of Physiology, School of Medical Sciences, Universiti Sains Malaysia, Kubang Kerian, Kelantan, Malaysia
[b]Unit of Physiology, Faculty of Medicine, Universiti Sultan Zainal Abidin, Kuala Terengganu, Malaysia
[c]Department of Science Laboratory Technology, Akanu Ibiam Federal Polytechnic, Unwana, Afikpo, Ebonyi State, Nigeria
[d]Department of Physiology, Faculty of Basic Medical Sciences, College of Medical Sciences, University of Calabar, Calabar, Cross River State, Nigeria
[e]Unit of Integrative Medicine, School of Medical Sciences, Universiti Sains Malaysia, Kubang Kerian, Kelantan, Malaysia

Contents

Abbreviations

ACCα acetyl-CoA carboxylase-α
ALD alcoholic liver disease
ALP alkaline phosphatase
ALT alanine aminotransferase
AMPK adenosine monophosphate-activated protein kinase
AST aspartate aminotransferase
CAT catalase
CCl_4 carbon tetrachloride
COL1A1 collagen type I alpha 1 chain
CRP C-reactive protein

Bee Products and Their Applications in the Food and Pharmaceutical Industries
https://doi.org/10.1016/B978-0-323-85400-9.00017-4

DNA	deoxyribonucleic acid
ER	endoplasmic reticulum
FAS	fatty acid synthase
FATP2	fatty acid transport protein 2
FDA	Food and Drug Administration
FFA	free fatty acid
GCMS	gas chromatography–mass spectrometry
HFD	high-fat diet
HGF	hepatocyte growth factor
HO1	heme oxygenase 1
HPLC	high performance liquid chromatography
IL-6	interleukin-6
LCMS	liquid chromatography-mass spectrometry
LPS	lipopolysaccharide
LXRα	liver X receptor α
MDA	malondialdehyde
mTOR	mammalian target of rapamycin
NAFLD	nonalcoholic fatty liver disease
NFκβ	nuclear factor kappa-β
Nrf2	nuclear factor erythroid 2-related factor 2
PCNA	proliferating cell nuclear antigen
PI3K	phosphoinositide 3-kinase
ROS	reactive oxygen species
SCD-1	stearoyl-CoA desaturase-1
SGLT1	sodium-dependent glucose transporter 1
SMA	smooth muscle actin
SOD	superoxide dismutase
SREBP-1c	sterol regulatory element-binding protein-1c
TG	triglyceride
TGF-β	transforming growth factor beta
TLR4	Toll-like receptor 4
TNF-α	tumor necrosis factor-α

Introduction

The use of bee products as complementary medicines and dietary supplements to treat various ailments was long practiced decades ago. According to Nayik et al. (2014), the first evidence revealing the acquisition of bee products by mankind was portrayed on the rock showing a painting of a person who collected honey from the wild bees found in 1919 inside the Cave Spider Spain. The painting is believed to have been made since 8000 to 5000 BC, during Neolithic Age. Today, in addition to honey, there is an increasing interest in researching other bee products, including pollen and bee bread, on their benefits and pharmacological properties for the potential development of nutraceutical and functional food.

The pollen grain is the microscopic structure produced in the anthers of stamen in the angiosperms flower (Stephen, 2014). The pollen grain is collected by the foraging bees from the plant flowers and gathered at their hind legs in the form of pollen loads (moistened with some nectar or honey to bind the grains and form lump) to the beehive. It is then harvested by the beekeepers using a trap set up at the beehive entrance (Barajas et al., 2012). The pollen grains do not come from a single flower. The floral origin of pollen is identified using a scanning electron microscope in terms of their colors, morphologies, and surface textures (Thakur and Nanda, 2020). Although the botanical origin is the primary factor in determining the chemical composition of the pollen, other factors such as geographical origin, weather conditions, climate, type of soil where the pollen is harvested, the time during harvesting has taken place, and even the breed of bees can also influence the pollen composition (Feás et al., 2012). As a result, there is a significant variation in the nutritional components of each pollen that have numerous valuable therapeutic properties (Denisow and Denisow-Pietrzyk, 2016; Kieliszek et al., 2018). The pollen usually has high moisture content and needs to be dried to maintain its nutrients and prevents degradation (Kieliszek et al., 2018). It has been consumed as a health food supplement and even considered medicine from a long time ago (Linskens and Jorde, 1997). Pollen has been demonstrated to have abundant nutritional components such as carbohydrates, proteins, lipids, vitamins, minerals, and phenolic compounds, contributing to its nutritional and therapeutic properties (Li et al., 2018). Previous studies have been reported the valuable properties of pollen, including antioxidant (Morais et al., 2011), hepatoprotective (Eraslan et al., 2009a), antimicrobial (Basim et al., 2006), antifungal (Özcan, 2004), antiradiation (Bevzo and Grygor'eva, 1997), chemopreventive (Pinto et al., 2010), antiinflammatory (Maruyama et al., 2010), and anticancer (Furusawa et al., 1995) activities.

Bee bread or ambrosia refers to the fermented pollen mixed with honey and secretions of the bee's salivary glands and stored in the hive. It is usually served as a source of protein for larvae and young bees by worker bees (Barajas et al., 2012; Vásquez and Olofsson, 2009). The foraging bees deliver the collected pollen loads to the hive and pack directly into the empty cells of the honeycomb, where it is later secured and mixed with wax and honey (Barene et al., 2015). The lactic fermentation process of pollen in the honeycomb is caused by *Lactobacillus* bacteria under anaerobic conditions, which converts pollen into a more preserved state called the "bee bread" (Anderson et al., 2014). The process of breaking down the outer layer of pollen, the exine, which is made of sporopollenin, is essential in the transformation process so that all the pollen nutrients are more accessible to the host. The enzymatic transformation of pollen into bee bread provides pollen protection against the loss of properties and the development of new components (Kieliszek et al., 2018). However, previous data reported a slight difference in its composition compared to pollen (Nagai et al., 2004).

In comparison to pollen, bee bread has been reported to possess higher nutritional value, greater digestibility, and abundant chemical composition (Habryka et al., 2016) due to its fermented form, which makes it easier to be absorbed and assimilated by the body (Barene et al., 2015; Ivanišová et al., 2015). Similar to pollen, the presence of proteins, vitamins, lipids, microelements, phenolic and flavonoid compounds are recognized as the added values to its nutritional and therapeutic properties. Bee bread has demonstrated its antibacterial (Abouda et al., 2011), antioxidant (Baltrušaityte et al., 2007), and anticancer (Sobral et al., 2017) properties and intake of bee bread has shown its medicinal effects on cardiovascular diseases (Othman et al., 2020), male fertility (Suleiman et al., 2020), renal dysfunction (Eleazu et al., 2020), and cancer (Markiewicz-Żukowska et al., 2013) in in vitro and in vivo studies.

The liver is a powerhouse organ as it performs various crucial tasks, ranging from the production of proteins, cholesterol and bile to storing vitamins, minerals, and even carbohydrates. In addition, it also breaks down toxins like alcohol, medications, and natural by-products of metabolism to maintain metabolic homeostasis in an organism. Hence, the liver is the main target for many dysregulated hepatic homeostasis events, leading to hepatic diseases (Casas-Grajales and Muriel, 2015). The liver comprises many cell types; hepatocytes, liver sinusoidal endothelial cells, bile duct cells called cholangiocytes, and hepatic stellate cells. In addition, the liver also contains one of the largest populations of resident macrophages; Kupffer cells, natural killer cells, natural killer T cells, and plasma B cells, as well as other peripheral immune cells such as monocytes and neutrophils (Kim et al., 2021). Liver diseases are becoming the main medical trouble worldwide. A previous study reported that approximately 2 million deaths per year worldwide were attributed to liver diseases where 1 million were due to complications of cirrhosis and the remaining 1 million were from viral hepatitis and hepatocellular carcinoma (Mokdad et al., 2014). Many studies have reported the positive usage of pollen and bee bread in treating liver diseases, including their beneficial effects on liver enzymes, improvements in liver steatosis and fibrosis, and decrease risk of cirrhosis and hepatocellular carcinoma. This chapter hopes to provide readers a general overview of the functional properties of pollen and bee bread related to liver health and the mechanisms on how they act as protective and therapeutic substances are an area of great interest. We also covered the possible adverse effects that might exist upon their application and suggested future research.

Chemical composition and active components of pollen and bee bread

Approximately 200 chemical compounds have been identified in the pollen originating from various plant species (Denisow and Denisow-Pietrzyk, 2016), and the average pollen composition was reviewed by Campos et al. (2008) and compared with the nutritional requirements for an average adult (Table 12.1). The components of pollen ranging from the largest value to the lowest value, such as carbohydrates and sugars, and proteins, followed by lipids and fatty acids, dietary fiber, minerals, and vitamins (Thakur and Nanda, 2020). Other than that, the presence of polyphenols in the pollen, especially flavonoids, adds more to its nutritional value. The pollen composition may differ due to seasonal and regional variations (de Melo et al., 2009).

The chemical composition of bee bread is mainly composed of proteins, free amino acids, carbohydrates, fatty acids, and other bioactive molecules. The average bee bread composition was reviewed by Kieliszek et al. (2018) (Table 12.2). Even though pollen and bee bread are made up of pollen grains, they differ biochemically (Anđelković et al., 2014). The pH value for bee bread is more acidic (pH 3.8–4.3) compared to pollen (pH 4.1–5.9) due to the lactic acid formation during the fermentation process (Vásquez and Olofsson, 2009), and the concentration of lactic acid is sixfold higher than in pollen (Nagai et al., 2004). Bee bread is claimed to have a broader range of sugars and greater reducing sugars compared to pollen (Anđelković et al., 2014; Khalifa et al., 2020). Each bee bread has a different composition which varies from different regions, climatic and seasonal variations, floral origins (Baltrušaityte et al., 2007; Urcan et al., 2017), and soil type (Vásquez and Olofsson, 2009). Kaplan et al. (2016) demonstrated this variation, who reported the contents of protein and lipids in bee bread ranged from 14.8% to 24.3% and 5.9% to 11.5%, respectively, found in bee bread samples *Trifolium* spp. (clover), *Gossypium hirsutum* (cotton), *Castanea sativa* (chestnut), *Citrus* spp. (citrus), and *Helianthus annuus* (sunflower). Another study reported that 15 bee bread samples showed different lipid content ranging from 1.65% to 5.50% (Zuluaga et al., 2015). In addition, Kaplan et al. (2016) also presented 37 fatty acids in the bee bread samples.

Table 12.1 Pollen composition and nutritional requirements as Required Daily Intake (RDI).

Main components	Amount (%)	Average RDI	% RDI for 100 g of pollen
Protein	10–40	50[a]	36–146.67
Lipids	1–13	80[a]	0.67
Carbohydrates (fructose, glucose, sucrose, fiber)	13–55	320[a]	6.67–306.67
Crude fiber	0.3–20	30[a]	2–120
Minerals			
Potassium	400–2000[b]	2000[c]	33.33–180
Phosphorus	80–600[b]	1000[c]	13.33–106.67
Calcium	20–300[b]	1100[c]	3.33–46.67
Magnesium	20–300[b]	350[c]	13.33–153.33
Zinc	3–25[b]	8.5[c]	66.67–526.67
Manganese	2–11[b]	3.5[c]	100–566.67
Iron	1.1–17[b]	12.5[c]	13.33–246.67
Copper	0.2–1.6[b]	1.2[c]	26.67–240
Vitamins			
Ascorbic acid (vitamin C)	7–56[b]	100[c]	13.33–100
β-Carotene (provitamin A)	1–20[b]	0.9[c]	100–4000
Tocopherol (vitamin E)	4–32[b]	13[c]	53.33–440
Niacin (vitamin B_3)	4–11[b]	15[c]	46.67–133.33
Pyridoxin (vitamin B_6)	0.2–0.7[b]	1.4[c]	26.67–86.67
Thiamine (vitamin B_1)	0.6–1.3[b]	1.1[c]	100–213.33
Riboflavin (vitamin B_2)	0.6–2[b]	1.3[c]	80–280
Pantothenic acid	0.5–2[b]	6[c]	13.33–60
Folic acid	0.3–1[b]	0.4[c]	133.33–446.67
Biotin (vitamin H)	0.05–0.07[b]	0.045[c]	200–280

RDI, Required Daily Intake requirements are according to Reports of the Scientific Committee for Food (2010) and Denisow and Denisow-Pietrzyk (2016).
a RDI is given in g/day.
b Amount is given in mg/100 g.
c RDI is given in mg/day.
Pollen composition is according to Campos, M.G.R., Bogdanov, S., de Almeida-Muradian, L.B., Szczesna, T., Mancebo, Y., Frigerio, C., Ferreira, F., 2008. Pollen composition and standardisation of analytical methods. J. Apic. Res. 47 (2), 154–161.

Recently, the active constituents in the pollen and bee bread have become a focus factor as commercial interest, making them a higher research demand among honeybee products. Ideally, methanol is frequently used to concentrate the responsible, active ingredients in the pollen and bee bread other than ethanol, hexane, or chloroform (Bayram et al., 2021; Conte et al., 2017). High-performance liquid chromatography (HPLC) is a standard method used to detect the presence of specific polyphenols

Table 12.2 Bee bread composition and nutritional requirements as Required Daily Intake (RDI).

Main components	Amount (%)	% RDI for 100 g of bee bread
Proteins	14–37	36–146.67
Carbohydrates	24–34	6.67–306.67
Lipids	6–13	0.67
Lactic acid	3.2	–
Cellulose	2.7	–
Flavonoids	n.d.	0.2
Vitamin	n.d.	13.33–466.67
Nucleic acid	n.d.	–

RDI, Required Daily Intake requirements are according to Reports of the Scientific Committee for Food (2010); n.d., no data.

(Ulusoy and Kolayli, 2014). It is widely used due to its capability to perform high separation and relative simplicity. The phenolic compounds are analyzed using modern separation and detection by combination to HPLC with ultraviolet-visible spectroscopy (UV–vis) (Ares et al., 2018) and nuclear magnetic resonance spectroscopy (Li et al., 2015).

Additionally, a few methods have been applied to analyze the presence of compounds, nutrient profiling, and bioactivity mechanisms of investigations and quality control. These included gas chromatography–mass spectrometry (GCMS), liquid chromatography-mass spectrometry (LCMS), and high-resolution mass spectrometry. Besides, as the scientific method evolved, higher advanced technologies are also developed to identify and quantify the polyphenol compounds via combination methods such as HPLC-diode array detection, HPLC-diode array detection-electrospray ionization mass spectrometry (HPLC-DAD-ESI/MS), high-performance liquid chromatography-diode array detection-atmospheric pressure chemical ionization-mass spectrometry (HPLC-DAD-APCI/MS), and ultra-performance liquid chromatography with a linear ion trap-high resolution orbitrap mass spectrometry system (UHPLC-LTQ-orbitrap/MS) (Zhang et al., 2016).

The presence of bioactive compounds in the pollen and bee bread contributes to significant biological effects, making these bee's products valuable supplements. Some of the active compounds like quercetin, naringenin, hesperetin, and epicatechin are directly absorbed in the small intestine and metabolized by the liver to exhibit their biological activity (Spencer et al., 1999). Few of the flavonoids are absorbed by Na-dependent glucose transporter 1 (SGLT1) (Williamson, 2017). The absorption of small metabolites also occurs in the colon, where the gut microflora could degrade the flavonoids into simple-structured phenolic acids, which are further metabolized in the liver (Spencer et al., 2004).

Phenolic compounds are considered the most significant contributor to the high antioxidant activity of natural source products. The phenolic compounds possess dominant numbers of flavonoids and phenolic acids beneficial for mediating biological action for maintaining human health. Divisions of phenolic compounds can be classified into anthraquinones, coumarins and isocoumarins, flavonoids, lignins, naphthoquinones, phenols and phenolic acids, as well as stilbenes (Wollgast and Anklam, 2000). Flavonoids are one of the major polyphenolic classes, derived from benzo-γ-pyrone, and form few subclasses into chalcones, flavan-3-ols, flavanones, flavones, flavonols, isoflavones, and bioflavonoids (Kim et al., 2004).

The bioactive compositions of pollen and bee bread have been studied in many countries. Phenolic acids and flavonoids are among the primary active metabolite compounds found in pollen. From the literature, pollen contains a wide range of phenolic acids such as caffeic acid, ferulic acid, vanillic acid, gallic acid, syringic acid, benzoic acid, chlorogenic acid, tert-cinnamic acid, and *p*-coumaric acid, which are detected from different geographical origins, including China, Brazil, Egypt, and Spain. Besides these, pollen comprises more than 20 flavonoids ranging from apigenin, catechin, epicatechin, hesperetin, kaempferol, luteolin, isorhamnetin, and quercetin (Li et al., 2018).

Serra Bonvehí et al. (2001) identified seven phenolic acid compounds in pollen from Spain, including 3,4-dihydroxybenzoic acid, the ethyl ester of 4-hydroxybenzoic acid, *o*-coumaric acid, *p*-coumaric acid, syringic acid, *trans*-cinnamic acid, and vanillic acid. Additionally, flavonoid isorhamnetin, kaempferol, myricetin, rutin, and quercetin are also detected in that sample. Similar trends of flavonoid contents in the pollen are also observed in the previous report, which comprises kaempferol, quercetin, and isorhamnetin (Maruyama et al., 2010; Tomás-Barberán et al., 1989).

Using the HPLC method, pollen from China has been shown to possess higher composition of rutin followed by benzoic acid, resveratrol, quercetin, protocatechuic acid, cinnamic acid, *p*-hydroxybenzoic acid, *p*-coumaric acid, and kaempferol. Meanwhile, gallic acid and catechin are found in trace amounts (Sun et al., 2017). Similarly, rutin content has also been observed in the pollen from the Baltic region (Kaškonienė et al., 2015). Other than that, Brazillian pollen also contains great quality of active components. The flavonoids isoquercetin, myricetin, tricetin, quercetin, luteolin, selagin, kaempferol, and isorhamnetin are also detected in pollens from Brazil (Freire et al., 2012). According to another study made in Brazil, pollen is reported to possess the following compounds: p-hydroxycinnamic acid, dihydroquercetin,6-β-isorhamnetin, isorhamnetin-3-*O*-(6″-*O*-*E*-pcoumaroyl)-β-D-glucopyranoside, luteolin, and quercetin (Silva et al., 2009). Meanwhile, ferulic acid, *p*-coumaric acid, and rutin are among the dominant phenolic compounds identified in Turkey pollen; however, catechin flavonoid is the only compound not detected in Turkey pollen in addition to rutin and quercetin flavonoid compounds (Ulusoy and Kolayli, 2014). 3,4-Dimethoxycinnamic acid is predominantly found in Egyptian pollen from monofloral sources, namely, maize (*Zea mays*) and gallic acid, catechin, and quercetin, which are also predominantly found in the samples. Other compounds like alpha-catechin, apigenin, caffeic acid, epicatechin, ferulic acid, kaempferol, luteolin, naringenin, rutin, *p*-coumaric acid, syringic acid, and vanillic acid are found in traces (Mohdaly et al., 2015). The main compounds identified in the Spanish pollen are rutin, quercetin, myricetin, and trans-cinnamic acid (Serra Bonvehí et al., 2001). According to Carpes et al. (2013), the presence of rutin in pollen indicates the biological and nutritional quality due to its high antioxidant activity.

Data from bee bread bioactive composition is, however, scarce, hence making it difficult to compare. In a study conducted by Isidorov et al. (2009), bee bread from Poland predominantly possesses kaempferol, isorhamnetin, and *p*-coumaric acid, which are detected using the GCMS method. In another study, flavonoid contents of rutin, naringenin, and quercetin in pollen and bee bread from Georgia are insignificantly found, indicating the contents are almost in similar concentrations (Tavdidishvili et al., 2014). Six samples of bee bread from Portugal possess various flavonoid derivatives, including herbacetin, isorhamnetin, kaempferol, myricetin, and quercetin, as determined with an HPLC method coupled with DAD-MSI (Sobral et al., 2017). LCMS analysis of bee bread from Malaysia demonstrated phenolic and flavonoid compounds ranging from caffeic acid, ferulic acid, apigenin, kaempferol, and isorhamnetin (Othman et al., 2020). The flavonoid contents are also identified in bee bread from Lithuania, namely hyperoside, isorhamnetin, kaempferol, rutin, quercitrin, quercetin, rutin, and vitexin. Rutin has the highest concentration, followed by hyperoside and kaempferol (Čeksterytė et al., 2006).

Literature data are still lacking evidence to demonstrate the differences between pollen and bee bread. In fact, comparative data on the same origin of these samples are required for further study. In another study, comparative bioactive analysis of Turkey pollen and bee bread revealed that the concentrations of caffeic acids, rutin, ethyl gallate, trans-ferulic acid, and myricetin are highly found in the pollen samples compared to bee bread. Both pollen and bee bread contained various concentrations such as caffeic acid, chlorogenic acid, catechin, ethyl gallate, quercetin, isorhamnetin, kaempferol, trans-ferulic acid, myricetin, rutin, and gallic acid. Among these, rutin is the most prominent compound found as high concentrations of rutin are detected in both samples. Certain compounds are not detected in Turkey pollen and bee bread, such as 2-hydroxy-*trans*-cinnamic acid, naringenin, syringic acid, sinapic acid, and propyl gallate. Compared to the pollen sample, bee bread contained a higher amount of 2,5-dihydroxybenzoic acid, kaempferol, protocatechuic acid, and quercetin (Bayram et al., 2021).

On the other hand, a few compounds are present in pollen but not detected in bee bread, such as phlorizin and ethyl gallate compounds. The observed differences might be attributed to the enzymatic reaction following bacterial digestion in bee bread (Tarko et al., 2013). The color variations in the pollen and bee bread are attributed to pigmental compounds such as anthocyanins, carotenoids, and xanthophylls (Naranjo et al., 2004; Owayss et al., 2004). Most of the literature data demonstrated that pollen and bee bread possessed a relatively higher composition of main phenolic compounds abundantly found, such as rutin, caffeic acid, ferulic acid, isorhamnetin, kaempferol, and quercetin. The differences might be contributed by the quantitative value of these compounds depending on flower pollen type carried by the bees and the extraction process. Based on the polyphenolic chemical structures found in the pollen and bee bread, researchers might predict the possible biological activities beneficial for human health. The presence of polymeric polyphenols is attributed to higher antioxidant activity compared to simple monomeric phenolic compounds. The antioxidant activity also depends on the number and localization of hydroxyl and methoxyl groups in the benzene ring and the electron delocalization of the double bonds (Mohdaly et al., 2015). The main phenolic acids and flavonoid compounds in pollen and bee bread are shown in Figs. 12.1 and 12.2, respectively.

Effects of pollen and bee bread on liver enzymes and diseases

Liver enzymes

The liver is a vital organ involves in numerous metabolic processes. An elevation in liver enzymes is recognized as the first manifestation of liver disease. Although they are simply called "liver function test," they represent hepatocyte membrane integrity, cholestasis, hepatocyte or biliary epithelial necrosis, or induction phenomenon rather than liver functional capacity.

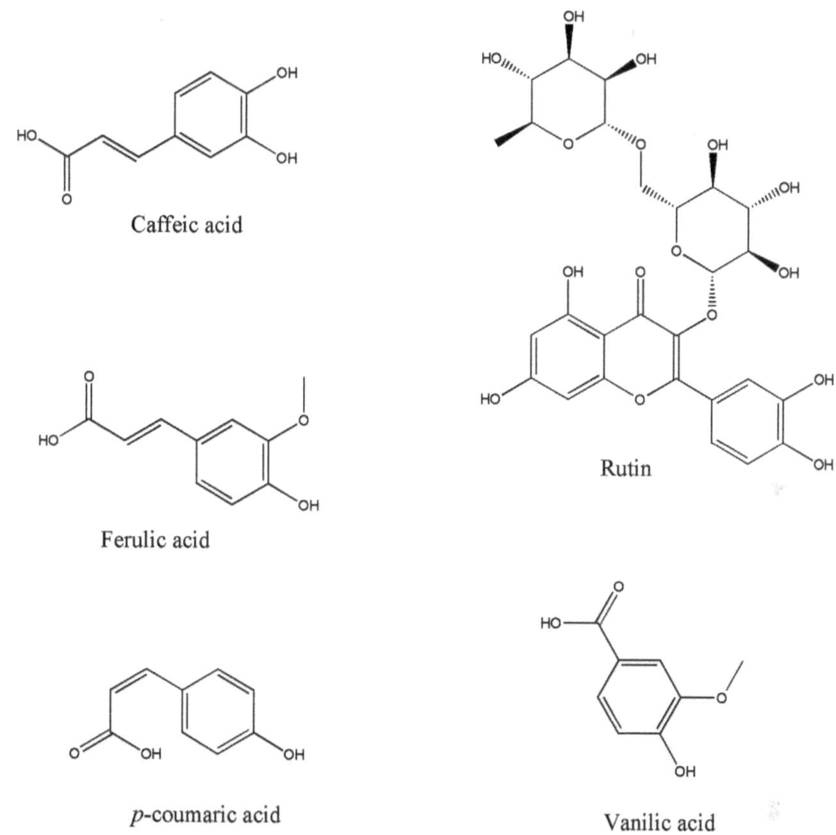

Fig. 12.1 Main phenolic acids in pollen and bee bread.

These enzymes are usually located within the hepatocytes or attached to their plasma membrane. They can be secreted out of the liver into the blood due to the distortion of membrane integrity (via leakage of gap junction), cell necrosis, discharge of membrane-bound enzymes with membrane fragments, or discharge of membrane-bound enzymes from their phosphatidylinositol anchor as soluble fractions (Van Hoof et al., 1997). Many studies have linked the elevations of these enzymes to the increased oxidative stress levels (Marrocco et al., 2017; Kocot et al., 2018).

Previous findings indicated that both pollen and bee bread had been linked with the improvement in liver enzymes. Pollen has been widely used in the poultry industry as a daily diet to improve health performance and shown to improve liver health status in broiler rabbit and chicken by reducing aspartate transaminase (AST) and alanine transferase (ALT) levels

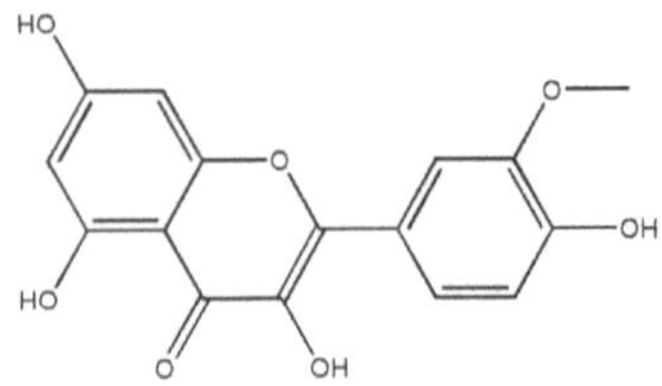

Isorhamnetin

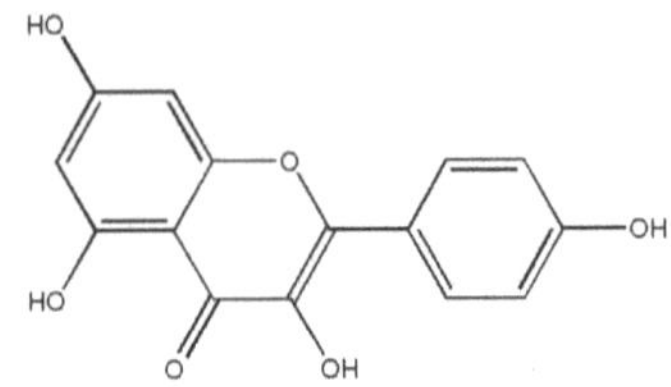

Kaempferol

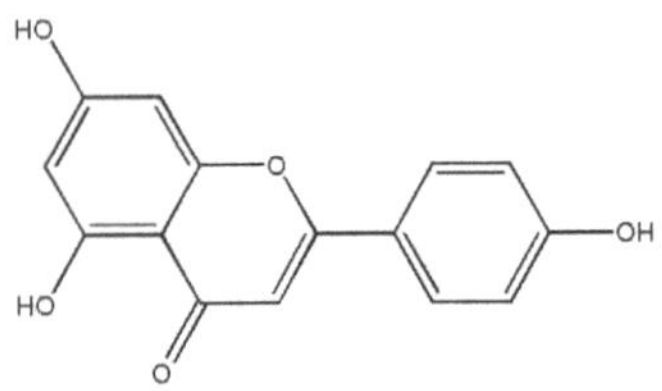

Apigenin

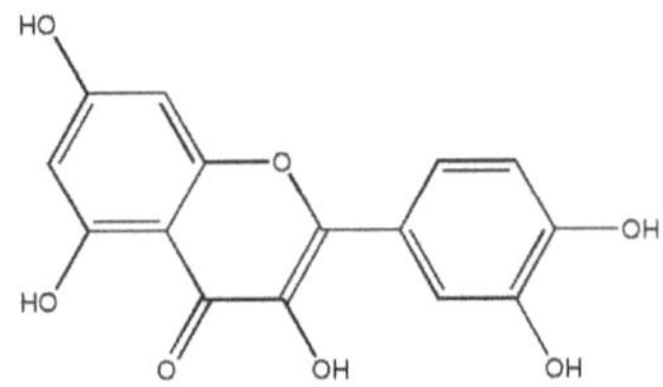

Quercetin

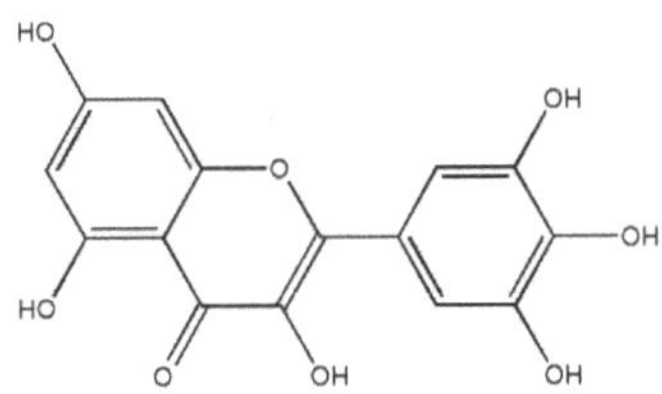

Myricetin

Fig. 12.2 Main flavonoids in pollen and bee bread.

compared to the control group (Attia et al., 2014, 2015). Furthermore, intake of pollen in an autistic rat model (Al-Salem et al., 2020) and pregnant female rabbits (Attia et al., 2011) also improved the mental condition and reproductive performance, respectively, as well as normalizing the levels of liver enzymes. These indicate that pollen can decline hepatic damage and improve liver function by improving the antioxidant activities, scavenging free radicals, and inhibiting free radicals induced-plasma membranes, mitochondrial and endoplasmic reticulum damages in the liver (Cheng et al., 2013). Additionally, improvement in liver oxidative stress along with liver enzymes are also observed in carbaryl-induced hepatic damage rats after pollen administration suggesting its action in preventing the oxidative damage (Eraslan et al., 2009a). Another study has also reported improvements in liver enzymes, malondialdehyde (MDA) level, and hepatic damage in human hepatic cells after intervention with pine pollen (Jin et al., 2015).

Bee bread has been reported to possess many beneficial effects such as antimicrobial, antiaging, antiatherosclerotic, and antiproliferative properties. These specialties of bee bread upgrade them as a liver protector against many liver diseases (Kieliszek et al., 2018). The use of bee bread, like pollen, has been associated with decreases in concentrations of liver enzymes AST, ALT, and ALP. Diabetes mellitus and advanced liver disease are related to each other as the liver regulates blood glucose via the hepatic metabolism of insulin and glucose (Moscatiello et al., 2007). Intake of antioxidant-rich ethyl acetate extract of bee bread normalized plasma AST and ALT levels in streptozotocin-induced diabetic rats as supported by high antioxidant and free radical scavenging activity of bee bread used in the study (Bakour et al., 2021). A previous study using bee bread as a feed additive in maternal goats reduced serum AST and ALT concentrations other than improved the animal digestion, feeding values, and productive performance (Farag et al., 2021).

These hepatoprotective actions of pollen and bee bread are accredited to their high amount of beneficial active compounds and chemical compositions such as phenolic, flavonoid, and glycosides, as well as the presence of vitamins and minerals. For example, administrations of rutin (Caglayan et al., 2019; El-Maadawy et al., 2021), ferulic acid (Sanjeev et al., 2019), vanillic acid (Vinothiya and Ashokkumar, 2017), kaempferol (BinMowyna and AlFaris, 2021), apigenin (Yue et al., 2020), and quercetin (Zhang et al., 2020) exhibited improvement in liver functions in the animal-induced liver injury models. Previous studies have also highlighted the significance of nutrients in pollen and bee bread, such as vitamins and minerals, for liver health. Administrations of these vitamins and minerals have been reported to lower the liver enzymes back to normal levels, including tocopherol (Tokoro et al., 2021), β-carotene (Acar et al., 2018), ascorbic acid (Lee et al., 2018), calcium (Das and Choudhuri, 2020), and magnesium (Karandish et al., 2013). These beneficial actions of active compounds and nutrients denoting their properties in maintaining the integrity of the cell membrane of liver cells.

Hepatitis and hepatotoxicity

Pollen extract has successfully counteracted the consequences of oxidative stress induced by drug toxicity in the rat liver, possibly due to massive sources of natural antioxidants present in the pollen. Wide-ranging and well-documented preclinical studies have been reported the detoxifying actions of pollen. In some studies, the animals were poisoned using organic solvents, specifically carbon tetrachloride (CCl_4) and trichloroethylene, ammonium fluoride, ethionine, and galactosamine, leading to hepatocyte damage and eventually liver steatosis and cirrhosis. These substances increase the levels of liver damage markers such as ALT and AST and ALP in the blood (Komosinska-Vassev et al., 2015). In this regard, administration of pollen demonstrated its detoxification effect by reversing the levels of liver enzymes AST and ALT and ALP back to the physiological level in CCl_4 (Cheng et al., 2013; Yıldız et al., 2013), carbaryl (Eraslan et al., 2009a), and propoxur-induced liver damage animals (Eraslan et al., 2009b). In another study, pollen has been demonstrated to reduce the hepatotoxicity effect as shown by significant decreases in monocyte and lymphocyte count, levels of AST, ALT, ALP, MDA, with significant improvement in superoxide dismutase (SOD) and catalase (CAT) activities in cadmium-induced liver toxicity rats (El Shater et al., 2019). In the experimental comparative studies involving administration of chestnut pollen, honey, propolis, and royal jelly to CCl_4-induced liver injury rats, the authors demonstrated that all the treatment modulations exerted similar efficacy against liver injury by improving the liver enzymes of AST and ALT, as well as reducing the progression of liver necrosis regardless of their differences in antioxidant capacities (Saral et al., 2016).

On the other hand, bee bread has been shown to reduce liver damage in aluminum-induced hepatotoxicity rats by reducing the levels of AST, ALT, C-reactive protein (CRP), and improved the levels of monocyte and lymphocyte counts, indicating its protective effects against hepatotoxicity rat model which could be attributed to its high radical scavenging activity (Bakour et al., 2017).

Hepatitis and hepatotoxicity are the leading risks for developing chronic liver diseases and hepatocellular carcinoma to a great extent. The detoxifying mechanism might be attributed to the role of polyphenols present in the pollen and bee bread. Quercetin phenolic compound might be a good candidate for improving hepatic function and preventing hepatocytes' degradation in CCl_4-induced liver injury in rats (Kalantari et al., 2018).

Additionally, kaempferol, ferulic acid, *p*-coumaric acid, and gallic acid, which can also be found in pollen and bee bread, may serve as beneficial sources to promote hepatoprotective outcomes alleviating liver damage in acetaminophen-induced hepatotoxicity rats (Ahmad and Zeb, 2019).

Nonalcoholic fatty liver disease (NAFLD)

The liver plays a vital role in lipid metabolism involving free fatty acids importation, synthesis, storage, and distribution of lipids and lipoprotein (Mirza, 2011). At normal physiological conditions, the liver carries it functions normally to maintain its low level of lipid; however, as obesity progresses, the abnormal lipid build-up in the liver and reduces its function and may lead to NAFLD, liver failure, and liver carcinoma (Bugianesi et al., 2002; Donnelly et al., 2005). NAFLD has also been linked to an increase in reactive oxygen species (ROS) production, accompanied by increased oxidative stress and low antioxidant enzyme activities in the liver tissue (Rahman et al., 2017). Therefore, prevention and dietary intervention have become people's preferences nowadays. To date, effective and established pharmacological regimens for treating NAFLD are still being explored.

Pollen has been demonstrated to have beneficial effects on lipid metabolism in the liver; it reduces the liver lipid contents, resulting in restriction of lipid deposition (Habryka et al., 2016; Prahastuti et al., 2020). Phenolic compounds in the pollen, which are the secondary plant metabolites, possess strong antioxidant activities, reduced body weight, fasting blood glucose, and the main NAFLD's features such as liver enzymes, lipid deposition, inflammation, oxidative stress, and down-regulated expression of genes related to liver lipid synthesis in the HFD-induced NAFLD mice (Rzepecka-Stojko et al., 2018; Cheng et al., 2019). The antiobesity property of pollen is attributed to its ability to enhance hepatic autophagy and lipase expression via an AMPK/mTOR-mediated signaling pathway, hence reducing NAFLD's progression in an animal model (Li et al., 2017). A previous study has also reported that pollen increased gene expression of Nrf2, which is the key transcription factor for maintaining the redox homeostasis, reducing liver oxidative stress marker MDA, and increasing the antioxidant enzyme activities in the NAFLD animal model (Yan et al., 2021). However, no research has been demonstrated in assessing the potential value of bee bread in NAFLD. Hence, more future studies are warranted to provide more data on this area.

Other than the reported studies on the intakes of these bee's products exclusively, many researchers have also demonstrated data on the beneficial actions of their constituents in the NAFLD treatment. For examples, gallic acid (Chao et al., 2014; Hsu and Yen, 2007), caffeic acid (Kim et al., 2018), quercetin (Li et al., 2013), kaempferol (Chang et al., 2011; Hoang et al., 2019), apigenin (Lv et al., 2019), and ferulic acid (Wang et al., 2015) have shown to improve NAFLD by exerting their protection against hepatic steatosis and others NAFLD-related metabolic diseases. Previous studies demonstrated that the beneficial actions of these bioactive compounds are accredited to their antioxidant and antiinflammatory properties in controlling hepatic lipogenesis. In this regard, the recent studies revealed that the administration of gallic acid inhibited the levels of ROS, HO1, NFκβ, TNF-α, and IL-6 (Fanaei et al., 2021), activated the hepatic AMPK signaling pathway, suppressed the hepatic expression of genes responsible for fatty acid transporters (CD36 and FATP2) and de novo lipogenesis (ACCα, SREBP-1c, and LXRα), and thus prevented the hepatic lipid accumulation in vitro and in vivo of NAFLD models (Tanaka et al., 2020; Sousa et al., 2020). Furthermore, there is also a presence of a link between gut microbiota and NAFLD (Liu et al., 2019) and the proinflammatory lipopolysaccharides (LPS), which are generated by gut microbiota that enhances the hepatic inflammation and TGs accumulation, hence induces NAFLD (Fukunishi et al., 2014). Concerning this, caffeic acid and quercetin have been reported to inhibit the LPS/TLR4 signaling pathway, suppress the levels of hepatic lipogenesis protein expressions such as SREBP-1c, FAS, ACC, and SCD-1, hence reduce the hepatic steatosis and hyperlipidemia in NAFLD animal models (Mu et al., 2021; Porras et al., 2017). Other than the role of gut microbiota in NAFLD, researchers have also revealed the role of endoplasmic reticulum (ER) stress (Malhi and Kaufman, 2011) and autophagy (Singh et al., 2009) in the development of hepatic steatosis. Previous studies established that caffeic acid and quercetin suppressed hepatic steatosis by reducing ER stress and increasing hepatic autophagy in the NAFLD model (Kim et al., 2018; Zhu et al., 2018).

Apart from these bioactive components, there is also the presence of other chemical compositions in the pollen and bee bread, which have added value to their therapeutic properties, including vitamins and minerals. Vitamins and minerals are known for their antioxidant properties, which provide health benefits, and many studies have reported the use of vitamins and minerals in combating NAFLD. A previous study showed that the intake of vitamin C and vitamin E in the NAFLD rats significantly reduced hepatic steatosis (Oliveira et al., 2003). Moreover, Perumpail et al. (2018) reported that the administration of vitamin E exerted its therapeutic effects by increasing the antioxidant activities, reducing the inflammation and apoptosis levels, hence prevented cell injuries and improved hepatic steatosis. Other than that, a cross-sectional study conducted in about 3500 adults in China diagnosed with NAFLD, found that vitamin C consumption reduced the development and progression of NAFLD (Wei et al., 2016). In addition, many studies have also reported the importance of minerals in NAFLD. Intake of dietary calcium exerted its hepatoprotective effects against NAFLD by restoring the hepatic antioxidant and oxidative balance, reducing the hepatic lipogenesis activity (Das and Choudhuri, 2020), and increasing hepatic lipolysis activity (Zhang et al., 2021) in the HFD-induced NAFLD animal model. Similarly, the administration of zinc activated hepatic autophagy and lipolysis, inhibited lipogenesis, and hence improved hepatic lipid metabolism in the NAFLD animal model (Wei et al., 2018).

Alcoholic liver disease (ALD)

Excessive alcohol consumption has become one of the global healthcare problems, and approximately 3.8% of all global deaths are attributing to alcohol use and alcohol disorders (Rehm et al., 2009). Over time, excessive alcohol intake can lead to the development of various chronic diseases and other life-threatening problems such as hypertension, cardiovascular disease, stroke, liver disease, and cancers. However, as the liver is the primary site for ethanol metabolism, it withstands the earliest and the highest degree of tissue injury (Lieber, 2000). Excessive alcohol consumption develops a wide spectrum of liver injury including steatosis, hepatitis, fibrosis/cirrhosis, and carcinoma (Osna et al., 2017). Still, until now, there are no FDA-approved therapies for ALD even though some therapies are being used for optimal ALD management including abstinence from alcohol, natural and artificial steroids, nutritional supplements, and liver transplantation, which is the utmost treatment for ALD patients. Kim et al. (2016) reported an increasing trend in using natural products with hepatoprotective effects.

Pollen is a natural material rich in nutrients and has been reported to possess a hepatoprotective effect (Eraslan et al., 2009a). However, until now, limited data are available on pollen and its treatment effects on ALD. Only one study was found to be reported on pollen and its different extracts and its hepatoprotective effects against liver damage caused by alcohol (Liping et al., 2008).

The use of bee bread in treating ALD has been reported by Čeksterytė et al. (2012), using a mixture of bee bread and honey in alcohol abuse patients who have chronic hepatitis. The results showed decreased levels of serum CRP, liver enzymes AST and ALT, and disappearance of yellowing eye sclera and skin pigmentation, and reduced liver size.

Moreover, previous studies have also pointed out the use of pollen and bee bread active components and compositions in the treatment of ALD. Intake of caffeic acid has remarkedly decreased serum and liver phospholipids, TG, FFA, cholesterol (Karthikesan and Pari, 2008), and the levels of liver oxidative stress and inflammation, meanwhile increased the liver antioxidants activities (Karthikesan and Pari, 2007) in alcohol-induced rats. In addition, previous studies have also reported similar outcomes and improved liver injury after administrations of naringin (Zhou et al., 2019), quercetin, rutin (Lee et al., 2019), gallic acid (Kartkaya et al., 2013), naringenin (Jayaraman et al., 2012), ascorbic acid (Abhilash et al., 2014), and β-carotene (Lin et al., 2009) in the ALD management.

Cirrhosis

Some of the drugs like paracetamol and hydrocortisone exerted cirrhotic changes in the liver if administered in overdose. Pollen applied as nutritional treatment in a rat model of liver cirrhosis induced by CCl_4 resulted in improved liver damage and fibrosis and ameliorated liver degeneration as evidenced by significant reductions in the AST, ALT, transforming growth factor-β (TGF-β), and hepatocyte growth factor (HGF) levels. The authors also found out that the albumin level was significantly increased in cirrhotic rats treated with Noveliver, a supplement composed of pollen, albumin, phospholipids, and yeast rich in selenium. It was suggested that it might be attributed to the higher content of amino acids in the pollen used in the study (Shi et al., 2010). Cirrhosis is a pathological consequence of chronic liver disease involving a self-healing response of fibrosis mediated by various cytokines and the formation of an extracellular matrix. It has also been shown that pollen collected from Masson pine has antifibrotic activity in a rat model of hepatic fibrosis. Its effect was observed in reducing the mRNA expression of TGF-β and α-smooth muscle actin (SMA), and increasing the matrix-degrading enzymes compared to the hepatic fibrosis rat group, hence, reduced the progression of liver fibrosis. Additionally, it also reduced the collagen hyperplasia at the hepatic portal area as compared to hepatic fibrosis rats (Cong et al., 2015). Up to date, no study is conducted on the effects of bee bread on changes in a cirrhotic animal model in a preclinical state.

Flavonoid and phenolic compounds, which are reported to be found in pollen and bee bread, could serve as therapeutic supplements in alleviating hepatofibrotic changes in the cirrhotic liver. In a study involving CCl_4-induced hepatotoxicity rats, the expressions of profibrotic mediators such as TGF-β and collagen type 1 alpha 1 chain (COL1A1) are downregulated in the presence of polyphenolic-rich extracts containing gallic, vanillic, caffeic, syringic, *p*-coumaric, ferulic, ellagic, benzoic, *o*-coumaric, and salicylic compounds. This has further restored the hepatic function by preventing fibrogenesis and the death of hepatic cells (Abu-Serie and Habashy, 2020). Resveratrol administration for 2 weeks demonstrated prevention of deleterious effect on liver cirrhosis by reducing the expressions of TGF-β, α-SMA, collagen, and liver fibrosis in a cirrhotic rat model (Di Pascoli et al., 2013). Improvement in fibrotic scarring and ballooning of the hepatocytes is also observed in CCl_4-induced hepatotoxicity rats administered with extract rich in polyphenolic compounds containing caffeic acid and epicatechin (Chowdhury et al., 2015).

Hepatocellular carcinoma

Although there is a lack of literature data reporting the relationship between pollen intake and hepatocellular carcinoma, pollen might have the potential to be one of the therapeutic tools in ameliorating the hepatocellular carcinoma progression as supported by established antiproliferative and antiapoptotic activities demonstrated in colon cancer cell lines in mice (Margaoan et al., 2016). The observed antiproliferative property is not only confined to colon cancer, but there are also few inconsistent data showing the antiproliferative effect of pollen on HepG2 (human liver carcinoma) cells (Chu et al., 2013). However, another study found the less cytotoxic activity of cancer cell-related to HepG2 cells after 48-h exposure to pollen extract (Kustiawan et al., 2014). Nevertheless, the available data on how pollen might exhibit antiproliferative effect is inconclusive and need to be elucidated in the future study.

In a recent study, the mechanical approach of antitumor activity of pollen might be related to its regulation on PI3K/AKT pathway, a critical pathway in human cancer. The inhibition, therefore, prevents abnormal cell proliferation and massive growth of tumor cells (Ran et al., 2020). Besides, pollen has been added as one of the natural remedies in Chinese medicine to treat patients with primary liver diseases (Dai et al., 2016). Besides its beneficial role as antiproliferative activity, pollen may also be helpful in cancer patients with anorexia and cachexia due to its role as an appetite stimulant apart from improving muscle performance (Moruzzi et al., 2019). On the other hand, bee bread from Portugal has been shown to exhibit inhibition against HepG2 cell line growth which indicated its antitumor property.

Additionally, bee bread has shown no toxicity effect toward normal cell lines in this study, indicating a good candidate for future antitumor agents (Sobral et al., 2017). In a recent finding, Egyptian citrus bee bread appeared to produce high inhibition activity against HepG2 cell lines in a dose-dependent manner. The cytotoxicity activity of bee bread might be related to its nutritional value and polyphenolic compounds present in the bee bread sample (Elsayed et al., 2021).

Powerful phenolic constituents such as caffeic and *p*-coumaric acid could be the most cytotoxic phenolic class toward inhibition of growth in the tumor cells (Falcão et al., 2019), probably due to its actions in regulating apoptosis, cell cycle arrestment, and inhibition of cell metastasis (Watanabe et al., 2011). Gallic acid was demonstrated to be a potent antiproliferative agent in diethylnitrosamine-induced hepatocellular carcinoma rats by improving the proliferating cell nuclear antigen (PCNA)-positive stained cells (Jagan et al., 2008). A nutritional diet rich in resveratrol could be a promising anticancer agent. In a recent study, resveratrol inhibited the viability and proliferation of hepatocellular cells and induced hepatocellular cancer cells apoptosis, and downregulated the apoptotic markers through the regulation of PI3K/AKT signaling pathway, hence leading to suppression of tumor growth (Chai et al., 2017).

Safety considerations of pollen and bee bread

Apart from bioactive compounds, pollen may also contain contaminants such as heavy metals, pesticides, herbicides, mycotoxins, bacteria, and antibiotics (Sattler et al., 2016), as it is exposed to different contaminating sources in its environment. For that reason, it should also be analyzed for monitoring the potential contaminants.

A study found that there are few problems encountered by the researchers concerning the presence of contaminants such as bacterial, herbicides, pesticides, heavy metals, and mycotoxin, which may impact an individual's health. Among these, higher contaminants are related to the presence of bacterial and mycotoxin problems. Fungi-related mycotoxin production of Aspergillus species has been shown to cause DNA damage and reduce the antioxidant component in the pollen due to their higher ochratoxin A level

(Cavin et al., 2007). Another study reported an allergic reaction, including systemic anaphylaxis, in one person who ingested pollen (Makris et al., 2010). Anemophilous pollen grains have been shown to cause an allergic reaction in few patients with underlying atopy disease (Choi et al., 2015). Other than that, allergen sources from Asteraceae plants possibly induce an adverse allergic reaction in children after pollen ingestion (Martín-Muñoz et al., 2010). Nevertheless, the allergies related to ingested bee bread are infrequent.

Pollen is a naturally occurring product that involves environmental exposures to the soil, water, and anthropogenic chemical pollutants, creating serious risks to human health. These environmental factors are also caused by accumulations of heavy metals from the soil, which highly depend on environmental activities. Few studies have reported the contamination with various heavy metal elements, including arsenic, cadmium, mercury, and plumbum, in fresh pollen samples from different regions (Dinkov and Stratev, 2016; Roman, 2009). Besides, certain microorganisms are quickly grown in pollen and bee products during the collection, transportation, and storage processes, because of the nutrient-rich factor and high humidity conditions, which further attract fungi, including yeast and mold (Estevinho et al., 2012).

Future pollen and bee bread research

Pollen and bee bread are good candidates to be commercialized as nutrient-rich food products in the future. It is suggested to harvest bee products daily and immediately store them in the freezer to prevent spoilage and maintain the quality to the maximum. The frozen bee products samples should be processed as soon as possible once they are thawed. Longer storage duration without proper maintenance in cold storage could affect the scavenging activity of pollen (Campos et al., 2008). Additionally, other storage techniques could also be implemented instead of doing it in the mainstream method. For example, applying liquid nitrogen and then freezing step preserves the biological and nutritive values optimally, essential for better intestinal digestion (Wollgast and Anklam, 2000). The main difficulty is due to variation in pollen and bee bread compositions, thus might contribute to differences in biological activities. Preparation of pollen and bee bread may vary between the manufacturers and product development and might also different between batches from one manufacturer to another. Hence, standardization is strongly needed for the efficacy and reproducibility of their beneficial biological effects and further clinical studies. For these reasons, future study is needed to obtain a high nutritional value of these bee's products concerning their developments in the production methods. Although significant attention has been focusing on the storage and production of pollen and bee bread, these bee products have a great potential to be accepted for their integration with modern medicine.

Based on the available literature data, the specific compounds of pollen and bee bread as curative drugs have not yet been implemented. An authorized manufacturer may play its role in cases related to the bioactive compounds' toxicity, hence increasing the pharmaceutical manufacturing and industrial economy. Other than that, future studies to define the association between metabolic factors and biochemical functions at the hepatomolecular level are also required.

Furthermore, before the products are ready for human usage, microbiological and mycotoxicological tests are recommended in pollen and bee bread analysis. A production with standard quality control should be implemented to prevent further contamination involving dehydrating process, cold chain process, and avoidance of UV exposure. Based on available data, pollen and bee bread have a significant amount of chemicals and bioactive constituents hence gaining attention as nutritional food for human usage. Another obstacle in applying pollen and bee bread in modern medicine is the variation in species related to bioactive composition. Apart from the varieties in their bioactive and chemical compositions and biological activities, the diversity of natural compounds in pollen and bee bread is relatively unknown. Nevertheless, pollen and bee bread have shown promising potential as NAFLD, ALD, and antitumor therapeutic agents as they contain substantial chemical and bioactive compounds harboring synergistic activity against the growth of hepatic steatosis, fibrosis, cirrhosis, and tumor cells which require more future studies to elucidate their potential mechanism of actions.

Conclusions

There is now sizeable evidence that pollen and bee bread are a natural antioxidant, antiinflammatory, antiobesity, hepatoprotective, and anticancer agents and their bioactive components and nutrients. The action mechanisms of compositions further support the beneficial role of pollen and bee bread for liver health.

References

Abhilash, P.A., Harikrishnan, R., Indira, M., 2014. Ascorbic acid suppresses endotoxemia and NF-κB signaling cascade in alcoholic liver fibrosis in guinea pigs: a mechanistic approach. Toxicol. Appl. Pharmacol. 274 (2), 215–224.

Abouda, Z., Zerdani, I., Kalalou, I., Faid, M., Ahami, M.T., 2011. The antibacterial activity of Moroccan bee bread and bee-pollen (fresh and dried) against pathogenic bacteria. Res. J. Microbiol. 6 (4), 376–384.

Abu-Serie, M.M., Habashy, N.H., 2020. Vitis vinifera polyphenols from seedless black fruit act synergistically to suppress hepatotoxicity by targeting necroptosis and pro-fibrotic mediators. Sci. Rep. 10, 2452. https://doi.org/10.1038/s41598-020-59489-z.

Acar, A., Yalçin, E., Çavuşoğlu, K., 2018. Protective effects of β-carotene against ammonium sulfate toxicity: biochemical and histopathological approach in mice model. J. Med. Food 21 (11), 1145–1149.

Ahmad, S., Zeb, A., 2019. Effects of phenolic compounds from aqueous extract of *Trifolium repens* against acetaminophen-induced hepatotoxicity in mice. J. Food Biochem. 43 (9), e12963.

Al-Salem, H.S., Al-Yousef, H.M., Ashour, A.E., Ahmed, A.F., Amina, M., Issa, I.S., Bhat, R.S., 2020. Antioxidant and hepatorenal protective effects of bee pollen fractions against propionic acid-induced autistic feature in rats. Food Sci. Nutr. 8 (9), 5114–5127.

Anđelković, B., Jevtić, G., Marković, J., Mladenović, M., Peševa, V., 2014. Quality of honey bee bread collected in spring. J. Hyg. Eng. Des. 1, 450–454. http://arhiva.nara.ac.rs/handle/123456789/699. Available from: [19 April 2021].

Anderson, K.E., Carroll, M.J., Sheehan, T., Mott, B.M., Maes, P., Corby-Harris, V., 2014. Hive-stored pollen of honey bees: many lines of evidence are consistent with pollen preservation, not nutrient conversion. Mol. Ecol. 23 (23), 5904–5917.

Ares, A.M., Valverde, S., Bernal, J.L., Nozal, M.J., Bernal, J., 2018. Extraction and determination of bioactive compounds from bee pollen. J. Pharm. Biomed. Anal. 147, 110–124. https://doi.org/10.1016/j.jpba.2017.08.009.

Attia, Y.A., Al-Hanoun, A., Tag El-Din, A.E., Bovera, F., Shewika, Y.E., 2011. Effect of bee pollen levels on productive, reproductive and blood traits of NZW rabbits. J. Anim. Physiol. Anim. Nutr. 95 (3), 294–303.

Attia, Y.A., Abd Al-Hamid, A.E., Ibrahim, M.S., Al-Harthi, M.A., Bovera, F., Elnaggar, A.S., 2014. Productive performance, biochemical and hematological traits of broiler chickens supplemented with propolis, bee pollen, and mannan oligosaccharides continuously or intermittently. Livest. Sci. 164, 87–95. https://doi.org/10.1016/j.livsci.2014.03.005.

Attia, Y.A., Bovera, F., El-Tahawy, W.S., El-Hanoun, A.M., Al-Harthi, M.A., Habiba, H.I., 2015. Productive and reproductive performance of rabbits does as affected by bee pollen and/or propolis, inulin and/or mannan-oligosaccharides. World Rabbit Sci. 23 (4), 273–282.

Bakour, M., Al-Waili, N.S., El Menyiy, N., Imtara, H., Figuira, A.C., Al-Waili, T., Lyoussi, B., 2017. Antioxidant activity and protective effect of bee bread (honey and pollen) in aluminium-induced anemia, elevation of inflammatory markers and hepato-renal toxicity. J. Food Sci. Technol. 54 (13), 4205–4212.

Bakour, M., El Menyiy, N., El Ghouizi, A., Lyoussi, B., 2021. Hypoglycemic, hypolipidemic and hepato-protective effect of bee bread in streptozotocin-induced diabetic rats. Avicenna J. Phytomedicine 11 (4), 343–352.

Baltrušaityte, V., Venskutonis, P.R., Čeksteryte, V., 2007. Radical scavenging activity of different floral origin honey and beebread phenolic extracts. Food Chem. 101 (2), 502–514.

Barajas, J., Cortes-Rodriguez, M., Rodríguez-Sandoval, E., 2012. Effect of temperature on the drying process of bee pollen from two zones of Colombia. J. Food Process Eng. 35 (1), 134–148.

Barene, I., Daberte, I., Siksna, S., 2015. Investigation of bee bread and development of its dosage forms. Medicinos Teorija Ir Praktika 21 (1), 16–22.

Basim, E., Basim, H., Özcan, M., 2006. Antibacterial activities of Turkish pollen and propolis extracts against plant bacterial pathogens. J. Food Eng. 77 (4), 992–996.

Bayram, N.E., Gercek, Y.C., Çelik, S., Mayda, N., Kostić, A.Ž., Dramićanin, A.M., Özkök, A., 2021. Phenolic and free amino acid profiles of bee bread and bee pollen with the same botanical origin—similarities and differences. Arab. J. Chem. 14 (3), 1–7.

Bevzo, V.V., Grygor'eva, N.P., 1997. Effect of bee pollen extract on glutathione system activity in mice liver under X-ray irradiation. Ukr. Biokhim. Zh. (1978) 69 (4), 115–117. http://www.ncbi.nlm.nih.gov/pubmed/9583135. Available from: (12 April 2021).

BinMowyna, M.N., AlFaris, N.A., 2021. Kaempferol suppresses acetaminophen-induced liver damage by upregulation/activation of SIRT1. Pharm. Biol. 59 (1), 146–156.

Bugianesi, E., Leone, N., Vanni, E., Marchesini, G., Brunello, F., Carucci, P., Musso, A., De Paolis, P., Capussotti, L., Salizzoni, M., 2002. Expanding the natural history of non-alcoholic steatohepatitis: from cryptogenic cirrhosis to hepatocellular carcinoma. Gastroenterology 123 (1), 134–140.

Caglayan, C., Kandemir, F.M., Darendelioğlu, E., Yıldırım, S., Kucukler, S., Dortbudak, M.B., 2019. Rutin ameliorates mercuric chloride-induced hepatotoxicity in rats via interfering with oxidative stress, inflammation and apoptosis. J. Trace Elem. Med. Biol. 56, 60–68. https://doi.org/10.1016/j.jtemb.2019.07.011.

Campos, M.G.R., Bogdanov, S., de Almeida-Muradian, L.B., Szczesna, T., Mancebo, Y., Frigerio, C., Ferreira, F., 2008. Pollen composition and standardisation of analytical methods. J. Apic. Res. 47 (2), 154–161.

Carpes, S.T., de Alencar, S.M., Cabral, I.S.R., Oldoni, T.L.C., Mourão, G.B., Haminiuk, C.W.I., da Luz, C.F.P., Masson, M.L., 2013. Polyphenols and palynological origin of bee pollen of Apis mellifera L. from Brazil. Characterization of polyphenols of bee pollen. CyTA J. Food 11 (2), 150–161.

Casas-Grajales, S., Muriel, P., 2015. Antioxidants in liver health. World J. Gastrointest. Pharmacol. Ther. 6 (3), 59–72.

Cavin, C., Delatour, T., Marin-Kuan, M., Holzhauser, D., Higgins, L., Bezencon, C., 2007. Reduction in antioxidant defences may contribute to ochratoxin a toxicity and carcinogenicity. Toxicol. Sci. 96 (1), 30–39.

Čeksterytė, V., Kazlauskas, S., Račys, J., 2006. Composition of flavonoids in Lithuanian honey and beebread. Biologija 2 (2), 28–33.

Čeksterytė, V., Balžekas, J., Baltuškevičius, A., Jurgevičius, E., 2012. The use of beebread-honey mixture in the treatment of liver diseases in alcohol-dependent patients. Chem. Technol. 60 (2), 62–66.

Chai, R., Fu, H., Zheng, Z., Liu, T., Ji, S., Li, G., 2017. Resveratrol inhibits proliferation and migration through SIRT1 mediated post-translational modification of PI3K/AKT signaling in hepatocellular carcinoma cells. Mol. Med. Rep. 16 (6), 8037–8044.

Chang, C.J., Tzeng, T.F., Liou, S.S., Chang, Y.S., Liu, I.M., 2011. Kaempferol regulates the lipid-profile in high-fat diet-fed rats through an increase in hepatic PPAR levels. Planta Med. 77 (17), 1876–1882.

Chao, J., Huo, T.L., Cheng, H.Y., Tsai, J.C., Liao, J.W., Lee, M.S., Qin, X.M., Hsieh, M.T., Pao, L.H., Peng, W.H., 2014. Gallic acid ameliorated impaired glucose and lipid homeostasis in high fat diet-induced NAFLD mice. PLoS One 9 (6), e96969.

Cheng, N., Ren, N., Gao, H., Lei, X., Zheng, J., Cao, W., 2013. Antioxidant and hepatoprotective effects of Schisandra chinensis pollen extract on CCl_4-induced acute liver damage in mice. Food Chem. Toxicol. 55, 234–240. https://doi.org/10.1016/j.fct.2012.11.022.

Cheng, N., Chen, S., Liu, X., Zhao, H., Cao, W., 2019. Impact of Schisandrachinensis bee pollen on non-alcoholic fatty liver disease and gut microbiota in highfat diet induced obese mice. Nutrients 11 (2), 346. https://doi.org/10.3390/nu11020346.

Choi, J., Jang, Y., Oh, J., Kim, C., Hyun, I., 2015. Bee pollen-induced anaphylaxis: a case report and literature review. Allergy Asthma Immunol. Res. 7 (5), 513–517.

Chowdhury, M.R.H., Sagor, M.A.T., Tabassum, N., Potol, M.A., Hossain, H., Alam, M.A., 2015. Supplementation of Citrus maxima peel powder prevented oxidative stress, fibrosis, and hepatic damage in carbon tetrachloride (CCl_4) treated rats. Evid. Based Complement. Alternat. Med. 2015, 1–11. https://doi.org/10.1155/2015/598179.

Chu, H.L., Mao, H., Feng, W., Liu, J.W., Geng, Y., 2013. Effects of sulfated polysaccharide from Masson pine (Pinus massoniana) pollen on the proliferation and cell cycle of HepG2 cells. Int. J. Biol. Macromol. 55, 104–108. https://doi.org/10.1016/j.ijbiomac.2012.12.025.

Cong, T., Jin, X.Y., Zhao, L., Ma, L., Li, R.S., Zhao, P., Guo, C.J., 2015. Anti-fibrotic effects of the Masson pine pollen aqueous extract on hepatic fibrosis rat model. Int. J. Clin. Exp. Pathol. 8 (5), 4651–4661.

Conte, G., Benelli, G., Serra, A., Signorini, F., Bientinesi, M., Nicolella, C., Mele, M., Canale, A., 2017. Lipid characterization of chestnut and willow honeybeecollected pollen: impact of freeze-drying and microwave-assisted drying. J. Food Compos. Anal. 55, 12–19. https://doi.org/10.1016/j.jfca.2016.11.001.

Dai, M., Yang, Y.W., Guo, W.H., Wang, F.L., Xiao, G.M., Li, Y.M., Yang, H.Z., 2016. Addition and subtraction theory of TCM using Xiao-Chaihu-decoction and naturopathy in predicting survival outcomes of primary liver cancer patients: a prospective cohort study. Evid. Based Complement. Alternat. Med. 2016, 1–9. https://doi.org/10.1155/2016/4723530.

Das, S., Choudhuri, D., 2020. Calcium supplementation shows a hepatoprotective effect against high-fat diet by regulating oxidative-induced inflammatory response and lipogenesis activity in male rats. J. Tradit. Complement. Med. 10 (5), 511–519.

de Melo, I.L.P., de Freitas, A.S.D., Barth, O.M., de Almeida-Muradian, L.B.D., 2009. Correlation between nutritional composition and floral origin of dried bee pollen. Rev. Inst. Adolfo Lutz 68 (3), 346–353.

Denisow, B., Denisow-Pietrzyk, M., 2016. Biological and therapeutic properties of bee pollen: a review. J. Sci. Food Agric. 96 (13), 4303–4309.

Di Pascoli, M., Diví, M., Rodríguez-Vilarrupla, A., Rosado, E., Gracia-Sancho, J., Vilaseca, M., Bosch, J., García-Pagán, J.C., 2013. Resveratrol improves intrahepatic endothelial dysfunction and reduces hepatic fibrosis and portal pressure in cirrhotic rats. J. Hepatol. 58 (5), 904–910.

Dinkov, D., Stratev, D., 2016. The content of two toxic heavy metals in Bulgarian bee pollen. Int. Food Res. J. 23 (3), 1343–1345.

Donnelly, K.L., Smith, C.I., Schwarzenberg, S.J., Jessurun, J., Boldt, M.D., Parks, E.J., 2005. Sources of fatty acids stored in liver and secreted via lipoproteins in patients with nonalcoholic fatty liver disease. J. Clin. Invest. 115 (5), 1343–1351.

El Shater, A.E.R.A., Ali, R.A., Ahlam, F.R., 2019. Effect of selenium and bee pollen against immunotoxicity and hepatotoxicity induced by cadmium in male albino rats. Egypt Acad. J. Biol. Sci. 11 (2), 1–19.

Eleazu, C., Suleiman, J.B., Othman, Z.A., Zakaria, Z., Nna, V.U., Hussain, N.H.N., Mohamed, M., 2020. Bee bread attenuates high fat diet induced renal pathology in obese rats via modulation of oxidative stress, downregulation of NF-kB mediated inflammation and Bax signalling. Arch. Physiol. Biochem. https://www.tandfonline.com/doi/full/10.1080/13813455.2020.1752258. Available from: [21 Jun 2021].

El-Maadawy, W.H., El-Din, S.H.S., Ezzat, S.M., Hammam, O.A., Safar, M.M., Saleh, S., El-Lakkany, N.M., 2021. Rutin ameliorates hepatic fibrosis via targeting hepatic stellate cells' activation, proliferation and apoptosis. Int. J. Geogr. Inf. Syst. 27 (3), 322–341.

Elsayed, N., El-Din, H.S., Altemimi, A.B., Ahmed, H.Y., Pratap-Singh, A., Abedelmaksoud, T.G., 2021. In vitro antimicrobial, antioxidant and anticancer activities of Egyptian citrus beebread. Molecules 26 (9), 1–13.

Eraslan, G., Kanbur, M., Silici, S., 2009a. Effect of carbaryl on some biochemical changes in rats: the ameliorative effect of bee pollen. Food Chem. Toxicol. 47 (1), 86–91.

Eraslan, G., Kanbur, M., Silici, S., Liman, B.C., Altinordulu, Ş., Sarica, Z.S., 2009b. Evaluation of protective effect of bee pollen against propoxur toxicity in rat. Ecotoxicol. Environ. Saf. 72 (3), 931–937.

Estevinho, L.M., Rodrigues, S., Pereira, A.P., Feás, X., 2012. Portuguese bee pollen: palynological study, nutritional and microbiological evaluation. Int. J. Food Sci. Technol. 47 (2), 429–435.

Falcão, S.I., Calhelha, R.C., Touzani, S., Lyoussi, B., Ferreira, I.C.F.R., Vilas-Boas, M., 2019. In vitro interactions of Moroccan propolis phytochemical's on human tumor cell lines and anti-inflammatory properties. Biomol. Ther. 9 (315), 1–11.

Fanaei, H., Mard, S.A., Sarkaki, A., Goudarzi, G., Khorsandi, L., 2021. Gallic acid treats dust-induced NAFLD in rats by improving the liver's anti-oxidant capacity and inhibiting ROS/NFκβ/TNFα inflammatory pathway. Iran. J. Basic Med. Sci. 24 (2), 240–247.

Farag, M.E., Helmy, A.A., El-Nimer, A.M.M., 2021. Effect of using bee bread extract as feed additives on digestibility and productive performance of maternal goats during suckling period. Egypt. J. Appl. Sci. 36 (2), 1–20.

Feás, X., Vázquez-Tato, M.P., Estevinho, L., Seijas, J.A., Iglesias, A., 2012. Organic bee pollen: botanical origin, nutritional value, bioactive compounds, antioxidant activity and microbiological quality. Molecules 17 (7), 8359–8377.

Freire, K.R.L., Lins, A.C.S., Dorea, M.C., Santos, F.A.R., Camara, C.A., Silva, T.M.S., 2012. Palynological origin, phenolic content, and antioxidant properties of honeybee-collected pollen from Bahia, Brazil. Molecules 17 (2), 1652–1664.

Fukunishi, S., Sujishi, T., Takeshita, A., Ohama, H., Tsuchimoto, Y., Asai, A., Tsuda, Y., Higuchi, K., 2014. Lipopolysaccharides accelerate hepatic steatosis in the development of nonalcoholic fatty liver disease in Zucker rats. J. Clin. Biochem. Nutr. 54 (1), 39–44.

Furusawa, E., Chou, S.C., Hirazumi, A., Melera, A., 1995. Antitumour potential of pollen extract on Lewis lung carcinoma implanted intraperitoneally in syngeneic mice. Phytother. Res. 9 (4), 255–259.

Habryka, C., Kruczek, M., Drygaś, B., 2016. Bee products used in apitherapy. World Sci. News 48, 254–258. https://www.infona.pl//resource/bwmeta1.element.psjd-09a97842-03de-4aa6-8cac-f1bc068a7d8c. Available from: [15 April 2021].

Hoang, M.H., Jia, Y., Lee, J.H., Kim, Y., Lee, S.J., 2019. Kaempferol reduces hepatic triglyceride accumulation by inhibiting Akt. J. Food Biochem. 43 (11), 1–7.

Hsu, C.L., Yen, G.C., 2007. Effect of gallic acid on high fat diet-induced dyslipidaemia, hepatosteatosis and oxidative stress in rats. Br. J. Nutr. 98 (4), 727–735.

Isidorov, V.A., Isidorova, A.G., Sczczepaniak, L., 2009. Gas chromatographic–mass spectrometric investigation of the chemical composition of beebread. J. Food Chem. 115 (3), 1056–1063.

Ivanišová, E., Kačániová, M., Frančáková, H., Petrová, J., Hutková, J., Brovarskyi, V., Velychko, S., Adamchuk, L., Schubertová, Z., Musilová, J., 2015. Bee bread—perspective source of bioactive compounds for future. Potravinarstvo 91 (1), 592–598.

Jagan, S., Ramakrishnan, G., Anandakumar, P., Kamaraj, S., Devaki, T., 2008. Antiproliferative potential of gallic acid against diethylnitrosamine-induced rat hepatocellular carcinoma. Mol. Cell. Biochem. 319 (51), 51–59.

Jayaraman, J., Jesudoss, V.A.S., Menon, V.P., Namasivayam, N., 2012. Anti-inflammatory role of naringenin in rats with ethanol induced liver injury. Toxicol. Mech. Methods 22 (7), 568–576.

Jin, X., Cong, T., Zhao, L., Ma, L., Li, R., Zhao, P., Guo, C., 2015. The protective effects of Masson pine pollen aqueous extract on CCl_4-induced oxidative damage of human hepatic cells. Int. J. Clin. Exp. Med. 8 (10), 17773–17780.

Kalantari, H., Foruozandeh, H., Khodayar, M.J., Siahpoosh, A., Saki, N., Kheradmand, P., 2018. Antioxidant and hepatoprotective effects of Capparis spinosa L. fractions and quercetin on tert-butyl hydroperoxide-induced acute liver damage in mice. J. Tradit. Complement. Med. 8 (1), 120–127.

Kaplan, M., Karaoglu, Ö., Eroglu, N., Silici, S., 2016. Fatty acid and proximate composition of bee bread. Food Technol. Biotechnol. 54 (4), 497–504.
Karandish, M., Tamimi, M., Shayesteh, A.A., Haghighizadeh, M.H., Jalali, M.T., 2013. The effect of magnesium supplementation and weight loss on liver enzymes in patients with non-alcoholic fatty liver disease. J. Res. Med. Sci. 18 (7), 573–579.
Karthikesan, K., Pari, L., 2007. Beneficial effect of caffeic acid on alcohol-induced alterations in lipid peroxidation and antioxidant defense in rats. Toxicol. Mech. Methods 17 (9), 527–534.
Karthikesan, K., Pari, L., 2008. Caffeic acid alleviates the increased lipid levels of serum and tissues in alcohol-induced rats. Fundam. Clin. Pharmacol. 22 (5), 523–527.
Kartkaya, K., Oğlakçi, A., Şentürk, H., Bayramoğlu, G., Canbek, M., Kanbak, G., 2013. Investigation of the possible protective role of gallic acid on paraoxanase and arylesterase activities in livers of rats with acute alcohol intoxication. Cell Biochem. Funct. 31 (3), 208–213.
Kaškonienė, V., Ruočkuvienė, G., Kaškonas, P., Akuneca, I., Maruška, A., 2015. Chemometric analysis of bee pollen based on volatile and phenolic compound compositions and antioxidant properties. Food Anal. Methods 8 (5), 1150–1163.
Khalifa, S.A.M., Elashal, M., Kieliszek, M., Ghazala, N.E., Farag, M.A., Saeed, A., Xiao, J., Zou, X., Khatib, A., Göransson, U., El-Seedi, H.R., 2020. Recent insights into chemical and pharmacological studies of bee bread. Trends Food Sci. Technol. 97, 300–316. https://www.sciencedirect.com/science/article/pii/S0924224419301657.
Kieliszek, M., Piwowarek, K., Kot, A.M., Błażejak, S., Chlebowska-Śmigiel, A., Wolska, I., 2018. Pollen and bee bread as new health-oriented products: a review. Trends Food Sci. Technol. 71 (January 2018), 170–180. https://www.sciencedirect.com/science/article/pii/S0924224417301681.
Kim, H.P., Son, K.H., Chang, H.W., Kang, S., 2004. Anti-inflammatory plant flavonoids and cellular action mechanisms. J. Pharmacol. Sci. 96, 229–245. Available from: https://www.jstage.jst.go.jp/article/jphs/advpub/0/advpub_0_0411110005/_pdf.
Kim, M.-S., Ong, M., Qu, X., 2016. Optimal management for alcoholic liver disease: conventional medications, natural therapy or combination? World J. Gastroenterol. 22 (1), 8–23.
Kim, H.M., Kim, Y., Lee, E.S., Huh, J.H., Chung, C.H., 2018. Caffeic acid ameliorates hepatic steatosis and decreased endoplasmic reticulum stress in high-fat diet-induced obese mice by regulating autophagy. Nutrition 55, 63–70. Available from: https://yonsei.pure.elsevier.com/en/publications/caffeic-acid-ameliorates-hepatic-steatosis-and-decreased-endoplas.
Kim, A., Wu, X., Allende, D.S., Nagy, L.E., 2021. Gene deconvolution reveals aberrant liver regeneration and immune cell infiltration in alcohol-associated hepatitis. Hepatology. Available from: https://aasldpubs.onlinelibrary.wiley.com/doi/epdf/10.1002/hep.31759.
Kocot, J., Kiełczykowska, M., Luchowska-Kocot, D., Kurzepa, J., Musik, I., 2018. Antioxidant potential of propolis, bee pollen, and royal jelly: possible medical application. Oxid. Med. Cell. Longev. 2018, 1–29. Article ID 7074209.
Komosinska-Vassev, K., Olczyk, P., Kaźmierczak, J., Mencner, L., Olczyk, K., 2015. Bee pollen: chemical composition and therapeutic application. Evid. Based Complement. Alternat. Med. 2015, 1–6. Article ID 297425.
Kustiawan, P.M., Puthong, S., Arung, E.T., Chanchao, C., 2014. In vitro cytotoxicity of Indonesian stingless bee products against human cancer cell lines. Asian Pac. J. Trop. Biomed. 4 (7), 549–556.
Lee, H., Ahn, J., Shin, S.S., Yoon, M., 2018. Ascorbic acid inhibits visceral obesity and nonalcoholic fatty liver disease by activating peroxisome proliferator-activated receptor α in high-fat-diet-fed C57BL/6J mice. Int. J. Obes. (Lond) 43 (8), 1620–1630.
Lee, S., Lee, J., Lee, H., Sung, J., 2019. Relative protective activities of quercetin, quercetin-3-glucoside, and rutin in alcohol-induced liver injury. J. Food Biochem. 43 (11), 1–9.
Li, X., Wang, R., Zhou, N., Wang, X., Liu, Q., Bai, Y., Bai, Y., Liu, Z., Yang, H., Zou, J., Wang, H., Shi, T., 2013. Quercetin improves insulin resistance and hepatic lipid accumulation in vitro in a NAFLD cell model. Biomed. Rep. 1 (1), 71–76.

Li, Y., Qi, Y., Ritho, J., Zhang, Y., Zheng, X., Zhou, J., Sun, L., 2015. Characterization of flavonoid glycosides from rapeseed bee pollen using a combination of chromatography, spectrometry and nuclear magnetic resonance with a step-wise separation strategy. Nat. Prod. Res. 30 (2), 228–231.

Li, X., Gong, H., Yang, S., Yang, L., Fan, Y., Zhou, Y., 2017. Pectic bee pollen polysaccharide from *Rosa rugosa* alleviates diet-induced hepatic steatosis and insulin resistance via induction of AMPK/mTOR-mediated autophagy. Molecules 22 (5), 1–13.

Li, Q.Q., Wang, K., Marcucci, M.C., Sawaya, A.C.H.F., Hu, L., Xue, X.F., Wu, L.M., Hu, F.L., 2018. Nutrient-rich bee pollen: a treasure trove of active natural metabolites. J. Funct. Foods 49, 472–484. Available from: https://www.sciencedirect.com/science/article/pii/S1756464618304845.

Lieber, C.S., 2000. Alcoholic liver disease: new insights in pathogenesis lead to new treatments. J. Hepatol. 32 (1), 113–128.

Lin, W.T., Huang, C.C., Lin, T.J., Chen, J.R., Shieh, M.J., Peng, H.C., Yang, S.C., Huang, C.Y., 2009. Effects of β-carotene on antioxidant status in rats with chronic alcohol consumption. Cell Biochem. Funct. 27 (6), 344–350.

Linskens, H.F., Jorde, W., 1997. Pollen as food and medicine—a review. Econ. Bot. 51 (1), 78–86.

Liping, S., Daqian, W., Lei, L., Wenjun, P., 2008. Study on preventive and curative effects of rape bee pollen extracts on alcoholic liver disease of rats. Food Sci. 29 (10), 543–545.

Liu, Q., Liu, S., Chen, L., Zhao, Z., Du, S., Dong, Q., Xin, Y., Xuan, S., 2019. Role and effective therapeutic target of gut microbiota in NAFLD/NASH. Exp. Ther. Med. 18 (3), 1935–1944.

Lv, Y., Gao, X., Luo, Y., Fan, W., Shen, T., Ding, C., Yao, M., Song, S., Yan, L., 2019. Apigenin ameliorates HFD-induced NAFLD through regulation of the XO/NLRP3 pathways. J. Nutr. Biochem. 71, 110–121. no. Sept 2019, Available from: https://www.sciencedirect.com/science/article/pii/S0955286318312798.

Makris, M., Koulouris, S., Koti, I., Aggelides, X., Sideri, K., Chliva, C., Vassilatou, E., Kalogeromitros, D., 2010. Temporal relationship of allergic rhinitis with asthma and other co-morbidities in a Mediterranean country: a retrospective study in a tertiary reference allergy clinic. Allergol. Immunopathol. 38 (5), 246–253.

Malhi, H., Kaufman, R.J., 2011. Endoplasmic reticulum stress in liver disease. J. Hepatol. 54 (4), 795–809.

Margaoan, R., Zahan, M., Marghitas, L.A., Dezmirean, D., Erler, S., Bobis, O., 2016. Antiproliferative activity and apoptotic effects of *Filipendula Ulmaria* pollen against C26 mice colon tumour cells. J. Apic. Sci. 60 (1), 135–145.

Markiewicz-Żukowska, R., Naliwajko, S.K., Bartosiuk, E., Moskwa, J., Isidorov, V., Soroczynka, J., Borawska, M.H., 2013. Chemical composition and antioxidant activity of beebread, and its influence on the glioblastoma cell line (U87MG). J. Apic. Sci. 57 (2), 147–157.

Marrocco, I., Altieri, F., Peluso, I., 2017. Measurement and clinical significance of biomarkers of oxidative stress in humans. Oxid. Med. Cell. Longev. 2017, 1–32. Article ID6501046, Available from: https://www.hindawi.com/journals/omcl/2017/6501046/.

Martín-Muñoz, M.F., Bartolome, B., Caminoa, M., Bobolea, I., Garcia Ara, M.C., Quirce, S., 2010. Bee pollen: a dangerous food for allergic children. Identification of responsible allergens. Allergol. Immunopathol. 38 (5), 263–265.

Maruyama, H., Sakamoto, T., Araki, Y., Hara, H., 2010. Anti-inflammatory effect of bee pollen ethanol extract from *Cistus* sp. of Spanish on carrageenan-induced rat hind paw edema. BMC Complement. Altern. Med. 10 (1), 1–11.

Mirza, M.S., 2011. Obesity, visceral fat, and NAFLD: querying the role of adipokines in the progression of non-alcoholic fatty liver disease. ISRN Gastroenterol 2011, 1–11. Article ID 592404, Available from: https://downloads.hindawi.com/archive/2011/592404.pdf.

Mohdaly, A.A.A., Mahmoud, A.A., Roby, M.H.H., Smetanska, I., Ramadan, M.F., 2015. Phenolic extract from propolis and bee pollen: composition, antioxidant and antibacterial activities. Food Biochem. 39 (5), 538–547.

Mokdad, A.A., Lopez, A.D., Shahraz, S., Lozano, R., Mokdad, A.H., Stanaway, J., Murray, C.J.L., Naghavi, M., 2014. Liver cirrhosis mortality in 187 countries between 1980 and 2010: a systematic analysis. BMC Med. 12 (1), 1–24.

Morais, M., Moreira, L., Feás, X., Estevinho, L.M., 2011. Honeybee-collected pollen from five Portuguese natural parks: palynological origin, phenolic content, antioxidant properties and antimicrobial activity. Food Chem. Toxicol. 49 (5), 1096–1101.

Moruzzi, G., Modica, D., Digangi, G., Saraceno, A., 2019. Adjuvant treatment for cancer pre-cachexia with vitamins, royal jelly, pollen and cod liver oil: a retrospective, non-comparative clinical study. Farmacia 18 (2), 1–7.

Moscatiello, S., Manini, R., Marchesini, G., 2007. Diabetes and liver disease: an ominous association. Nutr. Metab. Cardiovasc. Dis. 17 (1), 63–70.

Mu, H.N., Zhou, Q., Yang, R.Y., Tang, W.Q., Li, H.X., Wang, S.M., Li, J., Chen, W.X., Dong, J., 2021. Caffeic acid prevents non-alcoholic fatty liver disease induced by a high-fat diet through gut microbiota modulation in mice. Food Res. Int. 143, 1–13. no. May 2021, Available from: https://www.sciencedirect.com/science/article/pii/S0963996921001393.

Nagai, T., Nagashima, T., Myoda, T., Inoue, R., 2004. Preparation and functional properties of extracts from bee bread. Nahrung/Food 48 (3), 226–229.

Naranjo, P.R.D., Sánchez, S.J., Paramás, G.A.M., Gonzalo, R.J.C., 2004. Liquid chromatographic–mass spectrometric analysis of anthocyanin composition of dark blue bee pollen from *Echium plantagineum*. J. Chromatogr. A 1054 (1–2), 205–210.

Nayik, G.A., Shah, T.R., Muzaffar, K., Wani, S.A., Gull, A., Majid, I., Bhat, F.M., 2014. Honey: its history and religous significance: a review. Univ. J. Pharm. 3 (1), 5–8.

Oliveira, C.P.M.S., Da Costa Gayotto, L.C., Tatai, C., Della Nina, B.I., Lima, E.S., Abdalla, D.S.P., Lopasso, F.P., Laurindo, F.R.M., Carrilho, F.J., 2003. Vitamin C and vitamin E in prevention of non-alcoholic fatty liver disease (NAFLD) in choline deficient diet fed rats. Nutr. J. 2 (1), 1–5.

Osna, N.A., Donohue, T.M., Kharbanda, K.K., 2017. Alcoholic liver disease: pathogenesis and current management. Alcohol Res. 38 (2), 147–161.

Othman, Z.A., Ghazali, W.S.W., Noordin, L., Yusof, N.A.M., Mohamed, M., 2020. Phenolic compounds and the anti-atherogenic effect of bee bread in high-fat diet-induced obese rats. Antioxidants 9 (1), 1–12.

Owayss, A.A., Rady, M.M., Gadallah, F., 2004. Pigmentation of some honeybee, *Apis mellifera* L., products. Fayoum J. Agric. Res. Dev. 18 (2), 121–132.

Özcan, M., 2004. Inhibition of *aspergillus parasiticus* NRRL 2999 by pollen and propolis extracts. J. Med. Food 7 (1), 114–116.

Perumpail, B., Li, A., John, N., Sallam, S., Shah, N., Kwong, W., Cholankeril, G., Kim, D., Ahmed, A., 2018. The role of vitamin E in the treatment of NAFLD. Diseases 6 (4), 86.

Pinto, B., Caciagli, F., Riccio, E., Reali, D., Arić, A., Balog, T., Likić, S., Scarpato, R., 2010. Antiestrogenic and antigenotoxic activity of bee pollen from *Cystus incanus* and *Salix alba* as evaluated by the yeast estrogen screen and the micronucleus assay in human lymphocytes. Eur. J. Med. Chem. 45 (9), 4122–4128.

Porras, D., Nistal, E., Martínez-Flórez, S., Pisonero-Vaquero, S., Olcoz, J.L., Jover, R., González-Gallego, J., García-Mediavilla, M.V., Sánchez-Campos, S., 2017. Protective effect of quercetin on high-fat diet-induced non-alcoholic fatty liver disease in mice is mediated by modulating intestinal microbiota imbalance and related gut-liver axis activation. Free Radic. Biol. Med. 102 (2017), 188–202.

Prahastuti, S., Ladi, J.E., Dewi, K., Albertina, F., Imam, M.K., 2020. The effect of bee pollen on SGOT, SGPT levels and liver histopathological images of male rats Wistar induced by high fat diet. J. Med. Health 2 (5), 51–60.

Rahman, M.M., Alam, M.N., Ulla, A., Sumi, F.A., Subhan, N., Khan, T., Sikder, B., Hossain, H., Reza, H.M., Alam, M.A., 2017. Cardamom powder supplementation prevents obesity, improves glucose intolerance, inflammation and oxidative stress in liver of high carbohydrate high fat diet induced obese rats. Lipids Health Dis. 16 (1), 1–12.

Ran, L., Chen, F., Zhang, J., Mi, J., Lu, L., Yan, Y., Cao, Y., 2020. Antitumor effects of pollen polysaccharides from Chinese wolfberry on DU145 cells via the PI3K/AKT pathway in vitro and in vivo. Int. J. Biol. Macromol. 152, 1164–1173. no. June 2020.

Rehm, J., Mathers, C., Popova, S., Thavorncharoensap, M., Teerawattananon, Y., Patra, J., 2009. Global burden of disease and injury and economic cost attributable to alcohol use and alcohol-use disorders. Lancet 373 (9682), 2223–2233.

Reports of the Scientific Committee for Food, 2010. Scientific Panel on Dietetic Products Nutrition and Allergies. Tolerable upper intake levels for vitamins and minerals. EFSA J.

Roman, A., 2009. Concentration of chosen trace elements of toxic properties in bee pollen loads. Pol. J. Environ. Stud. 18 (2), 265–272.

Rzepecka-Stojko, A., Kabała-Dzik, A., Kubina, R., Jasik, K., Kajor, M., Wrześniok, D., Stojko, J., 2018. Protective effect of polyphenol-rich extract from bee pollen in a high-fat diet. Molecules 23 (805), 1–18.

Sanjeev, S., Bidanchi, R.M., Murthy, M.K., Gurusubramanian, G., Roy, V.K., 2019. Influence of ferulic acid consumption in ameliorating the cadmium-induced liver and renal oxidative damage in rats. Environ. Sci. Pollut. Res. 26 (20), 20631–20653.

Saral, O., Yildiz, O., Canpolat, S., Ozturk, F., 2016. Apitherapy products enhances the recovery of CCL_4-induced hepatic apitherapy products enhances the recovery of CCL_4-induced hepatic damages in rats. Turk. J. Med. Sci. 46, 194–202. Available from: https://journals.tubitak.gov.tr/medical/issues/sag-16-46-1/sag-46-1-30-1411-35.pdf.

Sattler, J.A.G., de Melo, A.A.M., do Nascimento, K.S., de Melo, I.L.P., Mancini-Filho, J., Sattler, A., de Almeida-Muradian, L.B., 2016. Essential minerals and inorganic contaminants (barium, cadmium, lithium, lead and vanadium) in dried bee pollen produced in Rio Grande do Sul State, Brazil. Food Sci. Technol. 36 (3), 505–509.

Serra Bonvehí, J., Soliva Torrentó, M., Centelles Lorente, E., 2001. Evaluation of polyphenolic and flavonoid compounds in honeybee-collected pollen produced in Spain. J. Agric. Food Chem. 49 (4), 1848–1853.

Shi, H.B., Kong, M., Chen, G., Zhao, J., Shi, H.L., Chen, Y., Rowan, F.G., 2010. Compound pollen protein nutrient increases serum albumin in cirrhotic rats. Gastroenterol. Res. 3 (6), 253–261.

Silva, T.M., Camara, C.A., Lins, A.C., Agra, M., De, F., Silva, E.M., Reis, I.T., Freitas, B.M., 2009. Chemical composition, botanical evaluation and screening of radical scavenging activity of collected pollen by the stingless bees *Melipona rufiventris* (Uruçu-amarela). An. Acad. Bras. Cienc. 81 (2), 173–178.

Singh, R., Kaushik, S., Wang, Y., Xiang, Y., Novak, I., Komatsu, M., Tanaka, K., Cuervo, A.M., Czaja, M.J., 2009. Autophagy regulates lipid metabolism. Nature 458 (7242), 1131–1135.

Sobral, F., Calhelha, R.C., Barros, L., Dueñas, M., Tomás, A., Santos-Buelga, C., Vilas-Boas, M., Ferreira, I.C.F.R., 2017. Flavonoid composition and antitumor activity of bee bread collected in Northeast Portugal. Molecules 22 (2), 248.

Sousa, J.N., Paraíso, A.F., Andrade, J.M.O., Lelis, D.F., Santos, E.M., Lima, J.P., Monteiro-Junior, R.S., D'Angelo, M.F.S.V., de Paula, A.M.B., Guimarães, A.L.S., Santos, S.H.S., 2020. Oral gallic acid improve liver steatosis and metabolism modulating hepatic lipogenic markers in obese mice. Exp. Gerontol. 134, 1–9. 110881.

Spencer, J.P.E., Chowrimootoo, G., Choudhury, R., Debnam, E.S., Srai, S.K., Rice-Evans, C., 1999. The small intestine can both absorb and glucuronidate luminal flavonoids. FEBS Lett. 458, 224–230. Available from: https://febs.onlinelibrary.wiley.com/doi/pdf/10.1016/S0014-5793(99)01160-6.

Spencer, J.P., Abd El Mohsen, M.M., Rice-Evans, C., 2004. Cellular uptake and metabolism of flavonoids and their metabolites: implications for their bioactivity. Arch. Biochem. Biophys. 423 (1), 148–161.

Stephen, A., 2014. Pollen—a microscopic wonder of plant kingdom. Int. J. Adv. Res. Biol. Sci. 1 (9), 45–62.

Suleiman, J.B., Nna, V.U., Zakaria, Z., Othman, Z.A., Eleazu, C.O., Abu Bakar, A.B., Ahmad, A., Usman, U.Z., Abdul Rahman, W.F.W., Mohamed, M., 2020. Protective effects of bee bread on testicular oxidative stress, NF-κB-mediated inflammation, apoptosis and lactate transport decline in obese male rats. Biomed. Pharmacother. 131, 1–13. November 2020, Available from: https://www.sciencedirect.com/science/article/pii/S0753332220309744.

Sun, L., Guo, Y., Zhuang, Y., 2017. Antioxidant and anti-tyrosinase activities of phenolic extracts from rape bee pollen and inhibitory melanogenesis by CAMP/MITF/TYR pathway in B16 mouse melanoma cells. Front. Pharmacol. 8 (104), 1–9.

Tanaka, M., Sato, A., Kishimoto, Y., Mabashi-Asazuma, H., Kondo, K., Lida, K., 2020. Gallic acid inhibits lipid accumulation via AMPK pathway and suppresses apoptosis and macrophage-mediated inflammation in hepatocytes. Nutrients 12 (1479), 1–14.

Tarko, T., Duda-Chodak, A., Zajac, N., 2013. Digestion and absorption of phenolic compounds assessed by in vitro simulation methods. A review. Ann. Natl. Inst. Hyg. 64 (2), 79–84.

Tavdidishvili, D., Khutsidze, T., Pkhakadze, M., Vanidze, M., Kalandia, A., 2014. Flavonoids in Georgian bee bread and bee pollen. J. Chem. Chem. Eng. 8, 676–681 (Available from: ProQuest).

Thakur, M., Nanda, V., 2020. Composition and functionality of bee pollen: a review. Trends Food Sci. Technol. 98, 82–106. April 2020, Available from: https://www.sciencedirect.com/science/article/pii/S0924224419310210.

Tokoro, M., Gotoh, K., Kudo, Y., Hirashita, Y., Iwao, M., Arakawa, M., Endo, M., Oribe, J., Masaki, T., Honda, K., Kakuma, T., Seike, M., Murakami, K., Shibata, H., 2021. α-Tocopherol suppresses hepatic steatosis by increasing CPT-1 expression in a mouse model of diet-induced non-alcoholic fatty liver disease. Obes. Sci. Pract. 7 (1), 91–99.

Tomás-Barberán, F.A., Tomás-Lorente, F., Ferreres, F., Garcia-Viegueira, C., 1989. Flavonoids as biochemical markers of the plant origin of bee pollen. J. Sci. Food Agric. 47 (3), 337–340.

Ulusoy, E., Kolayli, S., 2014. Phenolic composition and antioxidant properties of Anzer bee pollen. J. Food Biochem. 38 (2014), 73–82. Available from: https://onlinelibrary.wiley.com/doi/full/10.1111/jfbc.12027.

Urcan, A., Al Marghitas, L., Dezmirean, D.S., Bobiş, O., Bonta, V., Mureşan, C.I., Mărgăoan, R., 2017. Chemical composition and biological activities of beebread—review. Bull. UASVM Anim. Sci. Biotechnol. 74 (1), 1843. 536.

Van Hoof, V.O., Deng, J.T., De Broe, M.E., 1997. How do plasma membranes reach the circulation? Clin. Chim. Acta 266 (1), 23–31.

Vásquez, A., Olofsson, T.C., 2009. The lactic acid bacteria involved in the production of bee pollen and bee bread. J. Apic. Res. Bee World 48 (3), 189–195.

Vinothiya, K., Ashokkumar, N., 2017. Modulatory effect of vanillic acid on antioxidant status in high fat diet-induced changes in diabetic hypertensive rats. Biomed. Pharmacother. 87, 640–652. no. March 2017, Available from https://www.sciencedirect.com/science/article/pii/S0753332216318376.

Wang, O., Liu, J., Cheng, Q., Guo, X., Wang, Y., Zhao, L., Zhou, F., Ji, B., 2015. Effects of ferulic acid and γ-Oryzanol on high-fat and high-fructose diet-induced metabolic syndrome in rats. PLoS One 10 (2), 1–14.

Watanabe, M.A.E., Amarante, M.K., Conti, B.J., Sforcin, J.M., 2011. Cytotoxic constituents of propolis inducing anticancer effects: a review. J. Pharm. Pharmacol. 63 (11), 1378–1386.

Wei, J., Lei, G.H., Fu, L., Zeng, C., Yang, T., Peng, S.F., 2016. Association between dietary vitamin C intake and non-alcoholic fatty liver disease: a cross-sectional study among middle-aged and older adults. PLoS One 11 (1), 1–10.

Wei, C.C., Luo, Z., Hogstrand, C., Xu, Y.H., Wu, L.X., Chen, G.H., Pan, Y.X., Song, Y.F., 2018. Zinc reduces hepatic lipid deposition and activates lipophagy via $Zn2^{+}$/MTF-1/PPARa and $Ca2^{+}$/CaMKKb/AMPK pathways. FASEB J. 32 (12), 6666–6680.

Williamson, G., 2017. The role of polyphenols in modern nutrition. Nutr. Bull. 42 (3), 226–235.

Wollgast, J., Anklam, E., 2000. Review on polyphenols in Theobroma cacao: changes in composition during the manufacture of chocolate and methodology for identification and quantification. Food Res. Int. 33 (6), 423–447.

Yan, S., Wang, K., Wang, X., Ou, A., Wang, F., Wu, L., Xue, X., 2021. Effect of fermented bee pollen on metabolic syndrome in high-fat diet-induced mice. Food Sci. Human Wellness 10 (3), 345–355.

Yıldız, O., Can, Z., Saral, Ö., Yuluğ, E., Öztürk, F., Aliyazıcıoğlu, R., Canpolat, S., Kolaylı, S., 2013. Hepatoprotective potential of chestnut bee pollen on carbon tetrachloride-induced hepatic damages in rats. Evid. Based Complement. Alternat. Med. 2013, 1–9. Article ID 461478. Available from: https://www.hindawi.com/journals/ecam/2013/461478/.

Yue, S., Xue, N., Li, H., Huang, B., Chen, Z., Wang, X., 2020. Hepatoprotective effect of apigenin against liver injury via the non-canonical nf-κb pathway *in vivo* and *in vitro*. Inflammation 43, 1634–1648. Available from: https://link.springer.com/article/10.1007%2Fs10753-020-01238-5.

Zhang, Y., Yang, F., Jamali, M.A., Peng, Z., 2016. Antioxidant enzyme activities and lipid oxidation in rape (*Brassica campestris* L.) bee pollen added to salami during processing. Molecules 21 (1439), 1–13.

Zhang, Y., Qu, X., Gao, H., Zhai, J., Tao, L., Sun, J., Song, Y., Zhang, J., 2020. Quercetin attenuates NLRP3 inflammasome activation and apoptosis to protect INH-induced liver injury via regulating SIRT1 pathway'. Int. Immunopharmacol. 85, 1–9. August 2020, Available from: https://www.sciencedirect.com/science/article/pii/S1567576920302836.

Zhang, Z., Liu, S., Qi, Y., Aluo, Z., Zhang, L., Yu, L., Li, Q., Luo, Z., Sun, Z., Zhou, L., Li, Y., 2021. Calcium supplementation relieves high-fat diet-induced liver steatosis by reducing energy metabolism and promoting lipolysis'. J. Nutr. Biochem. 94, 1–11. August 2021, Available from: https://www.sciencedirect.com/science/article/pii/S0955286321000656.

Zhou, C., Lai, Y., Huang, P., Xie, L., Lin, H., Zhou, Z., Mo, C., Deng, G., Yan, W., Gao, Z., Huang, S., Chen, Y., Sun, X., Lv, Z., Gao, L., 2019. Naringin attenuates alcoholic liver injury by reducing lipid accumulation and oxidative stress. Life Sci. 216, 305–312. January 2018, Available from: https://www.sciencedirect.com/science/article/pii/S0024320518304089.

Zhu, X., Xiong, T., Liu, P., Guo, X., Xiao, L., Zhou, F., Tang, Y., Yao, P., 2018. Quercetin ameliorates HFD-induced NAFLD by promoting hepatic VLDL assembly and lipophagy via the IRE1a/XBP1s pathway. Food Chem. Toxicol. 114, 52–60. April 2018, Available from: https://www.sciencedirect.com/science/article/pii/S027869151830084X.

Zuluaga, C.M., Serrato, J.C., Quicazan, M.C., 2015. Chemical, nutritional and bioactive characterization of Colombian bee-bread. Chem. Eng. Trans. 43, 175–180.

CHAPTER 13

Bee bread and gut microbiota

Carlos M. Zuluaga-Dominguez[a] and Carlos A. Fuenmayor[b]
[a]Universidad Nacional de Colombia, Sede Bogotá, Facultad de Ciencias Agrarias, Departamento de Desarrollo Rural y Agroalimentario, Bogotá, Colombia
[b]Universidad Nacional de Colombia, Sede Bogotá, Instituto de Ciencia y Tecnología de Alimentos, Bogotá, Colombia

Contents

Abbreviations

GABA gamma-aminobutyric acid
GAE gallic acid equivalents
FLAB fructophilic lactic acid bacteria
LAB lactic acid bacteria
LPS lipopolysaccharide
PUFA polyunsaturated fatty acids
QE quercetin equivalent

Introduction

Bee bread can be defined as a mixture of pollen, nectar/honey, and living microorganisms, including lactic acid bacteria, obtained by an anaerobic fermentative process that takes place in the comb cells of the hive (Anderson

Bee Products and Their Applications in the Food and Pharmaceutical Industries
https://doi.org/10.1016/B978-0-323-85400-9.00010-1

et al., 2014). Corbicular pollen, i.e., pollen prior to becoming bee bread, is collected by beekeepers with pollen traps located at the entrance of hives. Due to its rich concentration of nutrients and bioactive compounds, it is the second beekeeping product with the greatest economic potential, after honey (Krell, 1996). As a food product for human consumption, bee pollen is a good source of proteins, carbohydrates, dietary fiber, lipids with a prevalence of polyunsaturated fatty acids, several vitamins, and minerals, as well as phenolic acids, flavonoids, carotenoids, phytosterols, and other potentially bioactive compounds (Denisow and Denisow-Pietrzyk, 2016; Li et al., 2018; Thakur and Nanda, 2020).

Despite its concentration of nutrients, the digestibility and bioavailability of corbicular pollen appear to be relatively limited for monogastric in part due to the resistant microstructure that covers pollen grains, known as exine animals (Zhou et al., 2018; Zuluaga Domínguez et al., 2014; Wu et al., 2020). In contrast, it has been proposed that, in bee bread, the action of anaerobic lactic acid bacteria and yeasts, the natural pH reduction, as well as the presence of various enzymes, would make its nutrients more digestible compared to non-fermented pollen (Di Cagno et al., 2019; Khalifa et al., 2020; Kieliszek et al., 2018). Indeed, it has been found that, rather than nutrient conversion, due to the presence of honey, nectar, bee secretions, and the properties of pollen itself, bee bread allows for a more microbiologically safe food product (Anderson et al., 2013, 2014; Disayathanoowat et al., 2020), and even a potential source of bacteria with human probiotic activity (Moreno Galarza, 2012).

In sum, bee bread is a valuable source of nutrients for human consumption, including protein, polyunsaturated fats, carbohydrates, and phenolic compounds, featuring the presence of potentially probiotic microorganisms, and therefore, it could be beneficial for the gut microbiota (Bakour et al., 2019). This chapter reviews and discusses the nutritional composition of bee bread, the main biochemical principles, lactic bacteria and yeast involved in its production in the beehive, the evidence of possible probiotic and bio-functional potential, and the current developments on bioprocessing using solid-state fermentation.

Chemical composition of bee bread

The chemical composition of corbicular pollen varies according to its botanical/geographical origin. Moreover, due to the addition of nectar, honey, wax, and microbial and enzymatic actions, the biochemical compositions of bee bread differ (Khalifa et al., 2020). The constituents of bee bread include water (moisture), carbohydrates, especially simple sugars, proteins, including active enzymes, lipids, polysaccharides, minerals, vitamins, and a large amount of other bioactive molecules in low or trace concentrations. Besides, bee bread has a greater content of lactic acid, up to 3.2% (Barene et al., 2015), which is lower or negligible in fresh pollen, from the anaerobic fermentation (Herbert and Shimanuki, 1978; Human and Nicolson, 2006). Table 13.1 presents the contents of moisture, protein, lipids, and ash of bee bread.

Moisture

The moisture content appears to vary in a wide range, as presented in Table 13.1, from 9% to more than 30%. Corbicular pollen has also a wide variation range attributed to the diversity of pollen composition (Donkersley et al., 2017), as well as to the altitude and different climatic conditions of the hive environment (Kaplan et al., 2016). There are disagreeing reports regarding the effect of pollen storage on the moisture content differences between corbicular pollen and bee bread. Human and Nicolson (2006) reported found that for South African bee bread samples from *Aloe greatheadii* var. *davyana* the pollen storage entailed a significant increase in water content, arguably due to the addition of nectar and glandular secretions by the worker bees. On the other hand, Herbert and Shimanuki (1978) did not find significant differences between corbicular pollen and bee bread. However, Zuluaga et al. (2015a) reported that moisture content of bee bread was noticeably lower than that of corbicular pollens from the same origins.

Proteins and amino acids

Among the macronutrients present in bee bread, a prominent place certainly belongs to proteins. The protein content is highly variable depending on the studies: on a dry basis, the percentage varies from 14% to 37.3%. Many amino acids have been found in bee bread proteins at variable concentrations, including proline, alanine, arginine, aspartic acid, cysteine, glutamic acid, glycine, histidine, isoleucine, leucine, lysine, methionine, phenylalanine, serine, threonine, tryptophan, tyrosine, and valine (Salazar-González and Díaz-Moreno, 2016). In bee bread samples from the United Kingdom, Donkersley et al. (2017) found glycine to be the amino acid with the higher mass fraction of the total amino acids (0.17 ± 0.13); other dominant amino acids included valine (0.14 ± 0.04) and methionine (0.08 ± 0.01).

Table 13.1 Moisture, protein, lipids, and ash content of bee bread from different origins.

Country of origin	Moisture[a]	Total protein	Lipid (fat)	Ash	References
Latvia	7.7%–11.1%	20.3%–21.7%	0.7%–1.6%	–	Barene et al. (2015)
United Kingdom	29% ± 18%	629 ± 290 mg/g	38 ± 2 mg/g	–	Donkersley et al. (2017)
United States	15.8%–21.9%[b]	23.3%[b]	5.4%[b]	2.8%[b]	Herbert and Shimanuki (1978)
Turkey	12.3%–15.9%	14.8%–24.3%	5.9%–11.5%	1.9%–2.5%	Kaplan et al. (2016)
Portugal	9%–19%	14.1%–20.9%			Tomás et al. (2017)
Colombia	7.8%–19.1%	23.1 ± 2.9%[b]	1.7%–5.5% (3.4 ± 1.1%)[b]	2.5 ± 0.2[b]	Zuluaga et al. (2015a)
South Africa	21.0%	–	–	3.6 ± 0.2[b]	Human and Nicolson (2006)
Morocco	–	20.0%[b]	1.9%[b]	3.32%[b]	Bakour et al. (2019)
Malaysia	–	18.4%	4.5%[b]	–	Othman et al. (2019)

a Data originally reported in dry basis, transformed to wet basis for comparison.
[b] Dry basis.

According to the results of Standifer et al. (1980), there were slight differences in the content of total protein of bee bread and corbicular pollen: it was very similar when stored between 7 and 21 days (22%), but slightly dropped in 42 days for bee bread (18.9%); the amino acid profile did not change significantly during the storage time, but losses as high as 20% were observed for several individual amino acids in bee bread concerning corbicular pollen, arguably due to the microbiota metabolism. Similarly, other studies have indicated that protein concentration in bee bread is slightly lower than in corbicular pollen (DeGrandi-Hoffman et al., 2013; Herbert and Shimanuki, 1978).

In Colombian samples, in vitro digestibility of bee bread was found to be significantly higher (45.7–94.7 g protein digested/100 g total protein) compared to that of corbicular pollen (38.7–85.3 g protein digested/100 g total protein) of the same origin (Zuluaga et al., 2015a), whereas in Italian samples, the digestibility of corbicular pollen (62.1%) was improved when the same pollen was fermented with a lactic acid bacteria starter (75.1%) (Di Cagno et al., 2019). More in vitro and in vivo studies are needed to quantify the actual protein digestibility of bee bread for human consumption, considering the wide variations between pollens from different botanical origins.

Lipids

The lipid content of bee bread quantified by various authors ranges from less than 1% to nearly 13%; in general, total lipids are found to be lower in bee bread than in fresh pollen, and it is certainly affected by the botanical origin. Fatty acid profiles are quite complex and variable as well; however, several studies have shown that the lipid fraction is characterized by higher levels of polyunsaturated fatty acids, featuring the presence of several omega 3 (ω-3) fats, which seems to be linked to bee feed preferences. The analysis carried out by Bakour et al. (2019) in Morocco indicated a total of 25 fatty acids, including 14 saturated and 11 unsaturated. The levels PUFA, essential fatty acids for nutrition that are not synthesized by the human body, were particularly high ($64.7 \pm 0.4\%$), with α-linoleic ($25 \pm 1\%$), arachidonic ($23.2 \pm 0.5\%$), and linoleic ($14.2 \pm 0.1\%$) as the main ones.

A study with bee bread samples from Lithuania (Urcan et al., 2017) identified 22 fatty acids, with the highest content (27%–43.8%) of octadecatrienoic acid (ω-3). Kaplan et al. (2016) found a similar but more complex profile in samples from Turkey, with 37 fatty acids, of which the predominant unsaturated ones were (9*Z*,12*Z*,15*Z*)-octadeca-9,12,15-trienoic,

(9*Z*,12*Z*)-octadeca-9,12-dienoic, (*Z*)-octadeca-9-enoic, and the main saturated ones were hexadecenoic and octadecanoic acids. They reported that the ratio of unsaturated to saturated fats was always higher than 1 (1.4–2.4) and that cotton bee bread was the richest in ω-3. Given the fatty acid profile reported from several origins, those with higher total fat concentrations of these samples might be a dietary significant source of essential polyunsaturated fats.

Ash content and minerals

The ash (inorganic matter) content ranges around 2.4%–3.6% of the dry matter, which is a rather high value for a plant-origin food (Table 13.1) that highlights the concentration of minerals in this product. Bee bread is the source of minerals consumed by nurse bees in the conversion of royal jelly (Anderson et al., 2014). Table 13.2 shows the content of several mineral elements. According to Pohl et al. (2020), element composition has been widely reported for bee pollen, and more rarely for bee bread; however, minerals in pollen and bee bread have similar profiles, thus the most abundant macro elements in these products are potassium followed by phosphorous, sulfur, calcium, and magnesium; also sodium, iron, manganese, aluminum, and zinc are found. Among microelements are copper, barium, and strontium, and trace elements and contaminants that have been found in some bee bread or pollen samples include barium, cadmium, cobalt, chromium, mercury, lithium, nickel, lead, astatine, antimony, and vanadium. Loper et al. (1980) found potassium to be the predominant mineral in almond pollen from the United States, followed by calcium, magnesium, and zinc. Similarly, Bakour et al. (2019) reported potassium, followed by phosphorous, calcium, and magnesium as the main minerals in samples from Morocco. Therefore, bee pollen and bee bread are valuable alternatives for avoiding mineral deficiency.

Carbohydrates

The carbohydrates, in particular, the monosaccharides fructose and glucose, constitute a major part of bee bread, and the levels found to vary according to the geographical and botanical origin. However, there are fewer reports focused on the concentration and profiles of carbohydrates compared to the other macronutrients. Table 13.3 summarizes the main reports available on the carbohydrate concentration of bee bread. Nevertheless, it is important to note that the proximate analysis approach to estimate the total carbohydrate content can be very inexact for chemically complex matrices.

Table 13.2 Minerals reported in bee bread from different origins.

Country of origin	Mineral content (mg/kg, dry basis)							References
	Na	K	Mg	Ca	Zn	Fe	Others	
United States	–	5050	649	910	81	–	Pb: 6	Loper et al. (1980)
Morocco	142	3380	610	1980	33.1	273	P: 2510 Mn: 26 Cu: 700 Pb: 7 Cd: 5.4	Bakour et al. (2019)
Serbia	–	7700	2100	5500	32.2	115.8	P: 7200 Mn: 33.1	Andjelkovic et al. (2015)

Table 13.3 Reported content of total carbohydrates (by difference), sugars and starch of bee bread from different origins.

Country of origin	Reported content of carbohydrates	Reported content of sugars	Reported content of starch	References
Morocco	74.8%[a]	18%[a]	N.D.	Bakour et al. (2019)
Poland	19.5 ± 5%	N.D.	N.D.	Markiewicz-Żukowska et al. (2013)
Malaysia	32.7%–59.6%[a]	N.D.	N.D.	Othman et al. (2019)
Portugal	58%–78%	9.9%–27.2%[a] fructose 6.6%–17.5%[a] glucose 0%–4.5% arabinose		Tomás et al. (2017)
United States	N.D.	27.9%[a] reducing sugars 2.5%[a] nonreducing sugars	<D.L.	Herbert and Shimanuki (1978)
United Kingdom	N.D.	130 mg/g ± 63 reducing sugars 119 mg/g ± 85 nonreducing sugars	13 mg/g ± 8	Donkersley et al. (2017)

N.D., no data reported; <D.L., under detection limit.
[a] Dry basis.

Bee bread contains more reducing sugars than corbicular pollen, arguably due to the addition of honey or nectar and the hydrolysis of polysaccharides. Herbert and Shimanuki (1978) found, on a dry basis, reducing sugar contents between 14.0% and 37.0% (average of 27.9%) and nonreducing sugars from 1.5% to 5.7% (average of 2.5%). A review of Khalifa et al. (2020) report percentages of total carbohydrates between 24% and 35%, with a predominance of monosaccharides: fructose (18.9%) and glucose (11.5%), as well as oligosaccharides including turanose, maltose, trehalose, and erlose; however, the sources of these data are unclear.

Bee pollen contains variable amounts of polysaccharides, including starch or starch-like compounds, as well as digestion-resistant polysaccharides and other molecules classifiable as dietary fiber, whose contents have been reported from less than 1% to over 20% (Salazar-González and Díaz-Moreno, 2016). Some studies have indicated a potential prebiotic effect of raw and fermented bee pollen in monogastric animals, arguably linked to the dietary fiber content and the presence of phenolic compounds (Aabed et al., 2019; Kačániová et al., 2013). However, there is very little knowledge regarding how the fermentation of bee bread affects the pollen polysaccharides chemistry, and there are few reports of the actual concentration of nondigestible, or poorly digestible, carbohydrates in bee bread classifiable as dietary fiber. Herbert and Shimanuki (1978) reported a mean content of crude fiber from the United States of 10.6% (5.8%–18.5%) of the dry matter, 0.5% (0.2%–0.7%) of pectins; they also found that, unlike corbicular pollen, bee bread presented starch content below the detection limit, indicating the enzymatic/microbial breakdown of digestible polysaccharides during fermentation. In samples of the United Kingdom, Donkersley et al. (2017) found starch contents of $1.3 \pm 0.8\%$.

Phenolic compounds

Corbicular pollen contains variable amounts of phenolic compounds with antioxidant activity (Salazar-González and Díaz-Moreno, 2016), and the fermentation of bee bread with lactic acid bacteria might favor the antioxidant activity of their phenolic-rich extracts (Di Cagno et al., 2019). Since phenolic compounds are secondary metabolites of plants, the phenolic profile is very specific to the botanical and geographical origins and collecting period (Markiewicz-Żukowska et al., 2013).

In Colombian samples, the total phenolic compounds varied between 2.5 and 13.7 mg GAE/g (mg GAE/g) (Zuluaga et al., 2015b). Stanciu et al. (2007) reported total phenolics as high as 22.7 mg GAE/g in samples from

Romania. Among phenolic compounds, flavonoids seem to be one of the major classes. In Colombian samples, the total flavonoid concentration ranged between 1.9 and 4.5 mg QE/g (Zuluaga et al., 2015a). Ivanišová et al. (2015) reported the contents of flavonoids of 13.6–18.2 μg QE/g (dry basis). Borycka et al. (2015) found 3.9, 24.6, and 16.3 mg QE/g for extracts with water, ethanol, and methanol, respectively.

As expected, a great variety of specific phenolic compounds has been identified in bee bread of different origins. Bakour et al. (2019) found 13 compounds, mainly flavonols glycoside derivatives, in particular quercetin, kaempferol, isorhamnetin, and methylherbacetin, from Morocco. Sobral et al. (2017) reported 32 compounds, in samples from Portugal, with flavonol derivatives, mainly quercetin, kaempferol, myricetin, isorhamnetin, and herbacetin glycoside derivatives, as the predominant phenolic compounds. Isidorov et al. (2009) identified naringenin, kaempferol, apigenin, isorhamnetin, quercetin, many of which probably are present in glycoside forms from the Baltic region. Markiewicz-Żukowska et al. (2013) reported kaempferol and apigenin as the main flavonoids in Polish samples.

Compounds related to aroma and flavor

Corbicular pollen and bee bread are rich in aroma- and taste-active compounds. Furthermore, lactic fermentation is expected to improve the sensory characteristics highly demanded by consumers (Mărgăoan et al., 2020). Sugars, such as fructose and glucose are related to sweetness, whereas lactic acid and other organic acids are responsible for sourness. Glutamic acid has been reported to increase because of pollen fermentation with lactic bacteria (Di Cagno et al., 2019). This, as well as the presence of several cations (K, Na), suggests that bee bread could trigger umami receptors in the tongue; however, there is no experimental evidence of the perception of this basic taste in bee bread consumption. Moreover, flavonoids are important flavor molecules, characterized as astringent, sour, fruity, and sweet (Mărgăoan et al., 2020).

Regarding volatile odor compounds, Kaškonienė et al. (2018) found that bee bread from Lithuania contains 32 compounds, including unsaturated aliphatic acids, alcohols, aldehydes, terpene derivatives, ketones, nitriles, and furfural, and the profiles are linked to the floral source of pollen. Dimethyl sulfide, 1-heptadecene, acetic acid, nonane, and furfural, were the most concentrated compounds of the volatile fraction of the samples analyzed.

Vitamins

Bee bread contains a wide variety of vitamins including both water-soluble and fat-soluble vitamers. Water-soluble vitamins C, B1, B2, nicotinic acid, folic acid, ascorbic acid, pantothenic acid, niacin, inositol, and pyridoxine have been found (Andjelkovic et al., 2015; Khalifa et al., 2020; Kieliszek et al., 2018). Vitamins of the fat-soluble group E, K, and provitamin A (carotenoids, such as β-carotene) have been also reported at variable concentrations (Barene et al., 2015; Khalifa et al., 2020). Interestingly, vitamin K is higher in bee bread than corbicular pollen, apparently due to the metabolism of fermenting microorganisms (Haydak and Vivino, 1950). Further research is needed to establish the nutritional significance for human consumption of vitamins at the concentrations found in bee bread.

Biochemical principles related to bee bread production in the hives

The processes of pollen transformation into bee bread are accompanied by a series of biochemical reactions resulting from the metabolism of microorganisms (bacteria and yeast) and bee digestive enzymes on the wide variety of chemical substrates, at the honeycomb conditions (anaerobicity, temperature, and water activity) during the storage time. Therefore, bee bread production inside the beehive has been assimilated to an ensiling/solid-state fermentation bioprocess, that prevents spoilage, and possibly improves palatability and digestibility (Gilliam, 1979a; Human and Nicolson, 2006; Di Cagno et al., 2019).

Worker bees (foragers) collect the floral pollen and transport it to the hive using their corbiculae. In this process, bees add regurgitated nectar to enhance the pellet formation. Once in the beehive, younger worker bees pack the pollen in the cells of the brood comb and cover it with a thin layer of honey (Gilliam, 1979b). Ultimately, the regurgitated nectar, the enzyme-rich glandular secretions, and the inoculated microorganisms (which will be discussed below) certainly change pollen microbiologically and biochemically (Gilliam et al., 1989; Loper et al., 1980; Standifer et al., 1980). Due to the environment of the covered pollen cells in the brood combs, the biochemical changes that lead to bee bread production are mainly attributed to the microbial action of anaerobic bacteria and yeasts. There is evidence that bees would prefer freshly stored pollen (<72 h) (Anderson et al., 2014); however, changes in the microbial population, and therefore the biochemistry of bee bread, have been documented for much longer periods of 6 weeks or more (Gilliam, 1997).

The main biochemical changes observed are: (i) the net increase in acidity/pH reduction that indicates active fermentation, (ii) the increase in lipid content concerning floral pollen, and (iii) the improved availability of amino acids in the pollen protein (Gilliam, 1997; Standifer et al., 1980). There are fewer reports regarding the effect of bee bread production on carbohydrate biochemistry; however, digestible polysaccharide contents (starch) were reported as lower than those of corbicular pollen (Herbert and Shimanuki, 1978), which agrees with the amylase activity of microorganisms isolated from bee gut and bee bread (García et al., 2006). The glucose resulting from starch breakdown forms gluconic acid and oxygen peroxide by the action of glucose oxidase, especially at the more external layers, contributing to the stabilization of microbial populations (Anderson et al., 2014). Other reported biochemical effects of bee bread productions are an increase of antimicrobial activity, related to the presence of certain potentially beneficial/probiotic microorganisms, such as *Lactobacillus kunkeei* (Anderson et al., 2014; Di Cagno et al., 2019), an increase in antioxidant activity, concerning corbicular pollen (Zuluaga et al., 2015b), and an increase in vitamin K content (Haydak and Vivino, 1950).

Ultimately, the biochemical modifications that pollen grains undergo during bee bread production are largely a consequence of the metabolism of the active bacteria and fungi, which produce enzymes, vitamins, antimicrobial substances including organic acid (e.g., lactic acid). For instance, *Bacillus* spp. from pollen and bee bread have carbohydrases and proteolytic enzymes (Gilliam, 1979a), most of the yeast species isolated are enzymatically able to ferment glucose and sucrose (Gilliam, 1979b), *Acinetobacter* release proteases (Disayathanoowat et al., 2020), and mold species produced enzymes to metabolize proteins, lipids, and carbohydrates (Gilliam et al., 1989), among which *Cladosporium* might promote further pH decrease reducing contaminant *Enterobacteria* (Disayathanoowat et al., 2020). Interestingly, from the perspective of gut biochemistry and the microbiome of bees, the consumption of bee bread appears to have an important role. Intestinal trypsin, chymotrypsin, myristate lipase, acid phosphatase, and cystine aminopeptidase appear to depend on bee bread consumption (Gilliam et al., 1988), as well as the levels of total amino acids and most individual amino acids, which increase with bee bread consumption (Gilliam, 1997).

The microbial succession, the complexity of enzymatic activities, the occurrence of fermentation by yeasts and lactic acid bacteria, and the enhancement of stability by the preservative effect of the organic acids produced are all analogous to the biochemistry of the plant foods ensiling bioprocesses (Gilliam, 1997).

Microorganisms involved in bee bread production

Different microorganisms involved in the production of bee bread have been reported, grouped in bacteria, fungi, and yeasts. Such organisms come mainly from the pollen itself or the gastrointestinal tract of bees. Some studies have shown that the microorganisms present in bee bread have antimicrobial activity against different pathogens, and, as a consequence of their metabolism, they can increase the levels of antioxidant activity and the content of available phenols. This has led to the postulate that there is a synergy between LAB and bees to defend each other against possible pathogens (Vásquez et al., 2012). In addition, the wide flora involved promotes different metabolic routes for fermentation; however, fructophilic lactic acid bacteria are predominant. Other common microorganisms during bee bread making processes are *Streptococcus* and *Lactobacillus* (Xu et al., 2019), as well as *Pediococcus*, *Leuconostoc*, and *Weissella* (Kim et al., 2017; Keshri et al., 2019; Soundharrajan et al., 2019).

The fermentation process from pollen to bee bread involves four stages, an initial one of 12 h where occurs growth of all kinds of microorganisms including yeasts, a second anaerobic phase where the LAB colonize the substrate, consume the nutrients metabolized by the yeasts, and reduce the pH by producing lactic acid. The third stage is characterized by the disappearance of *Streptococcus* and the growth of *Lactobacillus*. Finally, the fourth stage, which usually starts on the seventh day, occurs when the lactic acid concentration makes the presence of LAB unfeasible (Khalifa et al., 2020). The main metabolic activity of microorganisms involves an anaerobic process from glycolysis that converts pyruvate molecules to lactic acid, and eventually, from lactic acid to acetic acid, ethanol, or CO_2. In parallel, the LAB can carry out other types of metabolic activities that lead to the production of volatile molecules that add flavor to bee bread (De Filippis et al., 2020). Vásquez and Olofsson (2009) affirm that during the process of obtaining bee bread, the pH drops to levels of 3.8, being a very acidic medium so that the LAB could remain viable in times greater than 2 months. However, this time is sufficient for LAB to also act as regulators of colonization and proliferation of spoilage microorganisms (Carpes et al., 2007; Vásquez et al., 2012), thanks to the production of hydrogen peroxide that acts as an inhibitor of the growth of pathogens (Bang et al., 2003). The abundance of LAB in bee products depends on the season, the health of bees, the available flower resources, and contact with older bees inside the colony (Lamei, 2018; Mathialagan et al., 2018).

In the particular case of LAB, Gilliam (1979a) found microorganisms of the genus *Bacillus* as those predominant in pollen and bee bread, specifically *Bacillus subtilis*. In general, *Bacillus* can withstand adverse environments, especially at low pH as occurs in bee bread, as well as the ability to ferment glucose and, to a lesser extent, xylose, mannitol, arabinose, and trehalose. Vásquez et al. (2012) found more than 50 different LAB in different species of bees from the Apini and Meliponini tribes, from Malaysia, Thailand, Sweden, Kenya, and Mexico, with a microbiota dominated by *Lactobacillus kunkeii*. A similar result was found by McFrederick et al. (2012) and Neveling et al. (2012), while Endo and Salminen (2013) found a common presence of *Fructobacillus fructosus* in bee bread. Disayathanoowat et al. (2020) found *Acinetobacter* and *L. kunkeii* in different simples of bee bread. Filannino et al. (2016a) reported *L. kunkeii*, *F. fructosus*, *Lactobacillus plantarum*, and *L. Lactobacillus fermentum* in pollen and bee bread samples. It has been claimed that given the frequency with which these microorganisms have been found, there may be coevolution and mutualistic interactions with the hosts (Endo and Salminen, 2013).

Perhaps the most recognized microorganism in bee bread is *L. kunkeii*, which has typically been described as a food-altering microorganism, particularly wine (Edwards et al., 1998; Huang et al., 1996). This is a fructophilic microorganism, with preferential growth under aerobic conditions and incapable of using glucose as a substrate in anaerobiosis, but rather fructose and only grows adequately when there are external electron acceptors (Endo and Salminen, 2013; Endo et al., 2012; Neveling et al., 2012). The observation over other insects that base their diet on flower nectar, where fructose is the majority carbohydrate, has shown that FLAB have a greater capacity to colonize the gastrointestinal tract compared to other types of microorganisms (Endo and Salminen, 2013).

Recently, it has been mentioned that *L. kunkeii* has a protective function against the proliferation of pathogens such as *Penicillium larvae*, which mainly affects larvae, being a natural biocontrol system (Bisson et al., 2017). This action is mainly due to its ability to rapidly produce acetic and lactic acid, which, in combination with the formic acid produced by some *Bifidobacterium*, creates an environment of high acidity or low pH to inhibit the growth of different species of yeasts, affecting fermentation for the production of alcoholic beverages (Bisson et al., 2017). The presence of *L. kunkeii* in conjunction with other LAB species is believed to contribute to the flavor, aroma, and texture of bee bread, in part, through the secretion of metabolites (Tajabadi et al., 2011).

Historically, some microorganisms other than LAB have been suggested as responsible for the production of bee bread, among them, *Escherichia*, *Streptococcus* (Chevtchik, 1950), *Pseudomonas*, *Saccharomyces* (Pain and Maugenet, 1966), *Bacillus*, and *Torulopsis* (Gilliam, 1979a, 1979b). In that sense, Gilliam (1979b) identified *Torulopsis magnoliae*, *Candida guilliermondii* var. *Guilliermondii*, and *Cryptococcus albidus* as the most abundant yeasts in pollen and bee bread. Although there have been contradictory reports, it is agreed that these yeasts can ferment glucose, sucrose, and maltose, as well as exhibit a low capacity to use lactose as a substrate (Egorova and Babeva, 1967; Wilson and Marvin, 1929). The importance of the production of bee bread by yeasts is given thanks to their ability to produce complex B vitamins, which are later used as nutrients for LAB (Egorova, 1971).

Gilliam et al. (1989) identified different types of fungi in pollen stored for up to 6 weeks, mainly finding *Penicillium*, *Mucorales*, and *Aspergillus*. During the storage time of pollen, the fungal profile varies in *Penicillium corylophilum* and *Penicillium crustosum* for the first week, *Penicillium cyclopium*, and *P. corylophilum* in week two, to finally find in weeks 3 and 6 *P. corylophilum* and *Aspergillus niger*, respectively. It is then suggested that the presence of fungi and molds contributes to the production of antibiotics, organic acids, and enzymes that improve the availability of nutrients in bee bread and provide elements for the metabolism of LAB, in addition to providing antimicrobial capacity to the product (Gilliam et al., 1989).

Čadež et al. (2015) reported different species of yeasts in bee bread, particularly of the genus *Zygosaccharomyces* (*Z. rouxii*, *Z. mellis*, and *Z. siamensis*). Meanwhile, Sinpoo et al. (2017) frequently found the species *Cladosporium* sp. in samples of bee bread, of which Keller et al. (2014) had previously found, along with *Aspergillus*. Other reported fungi have been *Aspergilli*, *Mucorales*, *Penicillia*, and *Peyronelia* sp. (Gilliam and Prest, 1977, 1987). Table 13.4 presents a summary of the main species of microorganisms reported in bee bread.

Functional potential of bee bread on gut microbiota

Evidence shows that there is a link between the activity of the human intestinal microbiota and some aspects of health, mainly in the cardiovascular and neural system, as well as action on immune and metabolic activity. Some bacteria have shown a greater beneficial effect in the construction of such a microbiota, for example, different species of *Lactobacillus* improve the intestinal barrier function against pathogenic bacteria (Karczewski et al., 2010;

Table 13.4 Some microorganisms found in pollen and bee bread.

Authors	Microorganisms
Vásquez et al. (2012)	*Lactobacillus* y *Bifidobacterium*. Mainly, *Lactobacillus kunkeii*
Gilliam (1979b)	*Torulopsis magnoliae, Candida guilliermondii* var. *guilliermondii, Cryptococcus albidus, C. guilliermondii* var. *guilliermondii, Cryptococcus albidus* var. *albidus, Cryptococcus albidus* var. *diffluens, Cryptococcus laurentii* var. *laurentii, Cryptococcus laurentii* var. *magnus, Kloeckera apiculata, Metschnikowia pulcherrima, Rhodotorula pallida, Rhodotorula pilimanae, Torulopsis norvegica*
Gilliam (1979a)	*Bacillus megaterium, B. licheniformis, B. pumilus, B. circulaus, B. subtilis*
Endo and Salminen (2013)	*L. kunkeei* and *F. fructosus*
Gilliam et al. (1989)	*P. corylophilum, P. crustosum, P. cyclopium, P. corylophilum*, and *Aspergillus niger*
Janashia et al. (2018)	Bacteria: *Fructobacillus tropaeola, L. kunkeii, Bifidobacterium asteroides, Streptomyces* sp. Fungi: *Candida floricola, Lecythophora mutabilis, Zygosaccharomyces rouxii, Rhodotorula slooffiae. Cladosporium cladosporioides, Alternaria* sp. *Alternaria infectoria, Aspergillus niger, Penicillium oxalicum, Aureobasidium pullulans, Nigrospora sphaerica*

Xue et al., 2017). Likewise, *Lactobacillus* has shown antiinflammatory action by suppressing the secretion of different cytokines (Chen et al., 2015). In the case of *Bifidobacterium*, they can inhibit pathogenic bacteria and suppress the secretion of pro-inflammatory substances (Xue et al., 2017), and control different intestinal diseases such as cancer, irritable bowel syndrome, or allergies (Arseneault-Bréard et al., 2012; Fu et al., 2017; Persson et al., 2013). Both *Lactobacillus* and *Bifidobacterium* have shown anticancer effects, reduce lactose intolerance (Almeida et al., 2012), serve as antiallergic (O'Sullivan et al., 1992), strengthen the immune system, support drug treatments, improve concentration, memory, and the functioning of the digestive system, as well as regulate cholesterol content and lipid levels in the blood (Kieliszek et al., 2018).

Some benefits have been reported in the host health, such as an improvement in the digestion of lactose by the production of β-galactosidase, as well as an increase in the levels of conjugated linoleic acid, bioactive peptides, and phenolic compounds (Marco et al., 2017). Other effects are the production of exopolysaccharides with cholesterol-lowering, antidiabetic, antioxidant, and immunomodulatory properties (Nampoothiri et al., 2017). Recently, it has been indicated that such exopolysaccharides modify the rheological characteristics of fermented products (Hassan, 2008; Behare et al., 2009). The production of GABA has also been reported; this molecule promotes beneficial actions such as lowering blood pressure, muscle relaxation, and physiological stress (Pessione and Cirrincione, 2016).

The symbiotic activities between *Lactobacillus* and *Bifidobacterium* have also shown a favorable effect related to the reduction of acute stress and depression (Arseneault-Bréard et al., 2012; Savignac et al., 2015). On the other hand, *Enterococcus* have shown antitumor and anticancer activity, as well as modulation of the immune system (Azad et al., 2018). It has been established that *Lactobacillus* is commonly found as the dominant microorganisms from the duodenum to the ileum (Reuter, 2001), while they are present in a reduced proportion in the colon (Nishiyama et al., 2016). Meanwhile, LAB have been reported to promote the inhibition of Gram-positive (*Listeria* spp., *Bacillus* spp., *Micrococcus* spp.) and Gram-negative bacteria (*Escherichia coli*, *Salmonella* spp., *Helicobacter pylori*, and *Pseudomonas aeruginosa*) (Belhadj et al., 2014), due to the production of organic acids, hydrogen peroxide, and bacteriocins.

In the case of yeasts, *Saccharomyces* have probiotic activity and, it has been shown that within their beneficial effects on the body is the increase of bacteroidetes in the intestinal microbiota, which some researchers associate with the prevention of obesity (Indiani et al., 2018; Magne et al., 2020); however, so far the results for such an assertion are not yet conclusive. Additionally, yeasts can reduce inflammatory processes and increase the production of short chain fatty acids (Brun et al., 2017; Everard et al., 2014; Li et al., 2015; Yu et al., 2017).

The fermentation of plant matrices, such as pollen, relies on the ability of LAB to adapt to food and metabolize available nutrients. In particular, such adaptation processes have been commonly reported for *L. plantarum*, as it is the most frequent lactic microorganism in natural and induced fermentations of products of plant origin (Filannino et al., 2014a, 2016a; Golomb and Marco, 2015). Such mechanisms promote an improvement in the bioavailability and bioactivity of the secondary metabolites, as well as an increase in the availability of postbiotics with the consequent benefit for human health. *Lactobacillus* has been reported to hydrate polyunsaturated fatty acids thanks to their metabolism, producing hydroxyl fatty acids, which are compounds with antioxidant and immunomodulatory activity (Bergamo et al., 2014; Fernández Murga et al., 1999). LAB are also efficient for the synthesis of polyols (e.g., xylitol, erythritol, mannitol, among others) with shown health-promoting effects (reduction in calories, reduction in the glycemic index, anticarcinogenic, prebiotic), given their redox balance that promotes the oxidation of NADH that accumulate in the catabolic processes of carbohydrates (Monedero et al., 2010).

Phenolic compounds participate in most of the metabolic pathways of LAB, as has been reported for *L. plantarum*, *Weisella* sp., or *L. mesenteroides* (Filannino et al., 2014b). However, the concentration or type of phenolic compounds affects the growth and viability of LAB, particularly at high concentrations where adverse effects have been reported on the cell wall of microorganisms, as well as on their ability to catabolize carbohydrates (Filannino et al., 2018). However, LAB typically reported in plant matrices, such as *L. plantarum*, *L. curvatus*, *Weissella* spp., *L. brevis*, *L. fermentum*, and *L. mesenteroides* have shown tolerance to high levels of phenolics (Filannino et al., 2014a). The products resulting from the metabolism of phenolic compounds show important physiological implications in human health, where they convert to biologically active compounds thanks to the function of different enzymes such as glycosyl hydrolase, phenolic acid decarboxylase, phenolic acid reductase, and esterase (Filannino et al., 2018). Some LAB species convert flavonoids into aglycones, which have greater bioactivity in humans and act as adjuvants to restore homeostasis under conditions of cellular stress (Landete et al., 2014, 2015; Curiel et al., 2015; Lee et al., 2015; Filannino et al., 2016b; Francesca et al., 2016).

Regarding the production of bioactive peptides, although it is true that the human enzyme cocktail, generated mainly by proteases and peptidases, can release bioactive peptides bound to food and then absorbed in the intestine, LAB have a greater capacity to achieve their release (Pessione and Cirrincione, 2016). In some species, such as *Lactobacillus delbrueckii* subsp. *Lactis*, bioactive peptides shown antihypertensive and metal-chelating activities (Pessione and Cirrincione, 2016). Additionally, some peptides have shown antimicrobial activity both in Gram-positive bacteria when binding with teichoic acids, as well as in Gram-negative bacteria when binding with LPS, to later interact with negatively charged bacterial membranes (Vorland, 1999); the subsequent mechanism is postulated as the depolarization action of the molecules that make up the junction between the peptide and the cell membrane (Lohner and Blondelle, 2005) or removal/chelation of the iron-bound membrane that affects cell viability (Pessione and Cirrincione, 2016).

These peptides can also act as antioxidant compounds that can control the peroxidation of fatty acids (Hartmann and Meisel, 2007) and act as radical scavengers to eliminate free radicals that affect cells or tissues; therefore, antioxidants are essential to minimize the prevalence of degenerative diseases (Pessione and Cirrincione, 2016). Antioxidant properties have been found for the amino acids Tyr, Trp, His, and Lys (Hernández-Ledesma et al., 2008). However, more studies are still required to demonstrate how, and to what extent, food proteolysis phenomena occur at the intestinal level.

Another characteristic derived from protein metabolism is related to the production of some bioactive amines. For example, beta-phenylethylamine, derived from phenylalanine, has been shown to have beneficial effects on the body, such as the regulation of appetite/satiety balance (Shimazu and Miklya, 2004). In addition, it has already been shown the production of melatonin for some LAB is commonly found in foods of plant origin (*L. bulgaricus*, *L. acidophilus*, *L. casei*, and *L. plantarum*), which affects the regulation of sleep and behavior, and the control of immunity, inflammation, and carcinogenesis (Tan et al., 2014). The development in the study of bioactive peptide production and its availability remains promising and still has many unresolved concerns.

Doğanyiğit et al. (2020) observed an increase in leptin levels in a group of people who were given bee bread. Leptin is a hormone associated with the regulation of appetite and the feeling of satiety, so the consumption of bee bread may be associated with a favorable effect on the reduction of obesity. Nechaeva (2009) has made different claims of the continuous use of pollen, among which stands out the improvement in brain functions, such as memory, learning, and concentration, as well as an increase in the ability of the immune system to combat infectious agents. Yakusheva (2010) also reported benefits in the cardiovascular and renal systems, bladder, and urethra.

The antioxidant activity of bee bread has been suggested as an aid in the prevention of obesity since mainly phenolic compounds intervene in different metabolic processes inducing: (i) a lower food intake, (ii) reduction in lipogenesis, (iii) increased lipolysis, (iv) fatty acid oxidation, (v) inhibition of adipocyte differentiation and growth, and (vi) attenuation of inflammatory responses and suppression of oxidative stress, where results in chickens have shown reductions in triglyceride and cholesterol levels when bee bread is consumed periodically (Arafa et al., 2016).

It is important to note that intestinal microbiota is referred to as different types of microorganisms, including bacteria, archaea, and fungi, that live in a close relationship with the human host (Belkaid and Hand, 2014).

In particular, it should be clarified that the microorganisms that are usually ingested with food represent a transient microbiome, whose possibility of colonization in the intestine depends on factors such as the type of exposure and frequency of consumption, this means that the fraction is unknown. of microorganisms that come from food that remains effectively in the intestine (De Filippis et al., 2020). In particular, there are no conclusive studies that refer to the microbiota present in the small intestine, due to the difficulty of accessing this part of the human body, differently from invasive surgical procedures (Derrien and van Hylckama Vlieg, 2015; El Aidy et al., 2015; Stolaki et al., 2019; Van den Bogert et al., 2014).

Approximately the human gastrointestinal tract is home to around 10^{14} microorganisms from hundreds of species, which constitutes the intestinal microbiota (Bäckhed et al., 2005). Bacterial divisions include Firmicutes, Bacteroidetes, proteobacteria, actinobacteria, fusobacteria, verrumicrobia, cyanobacteria, and spirochaetes (Vrieze et al., 2010). Of such divisions, bacteroidetes and firmicutes represent up to 90% of the population (Salazar et al., 2016). The distribution of such microorganisms is variable between the stomach and the small intestine but highly concentrated in the colon, where populations of up to 10^{12} bacteria/g are found. The distribution and quantity of microbiota vary according to several factors, such as age, genetic factors of each individual, diet, use of medication, among others (De Filippo et al., 2010; Dethlefsen and Relman, 2011).

In particular, different factors affect the viability of LAB during their passage through the gastrointestinal tract. Among them, the structure of the food can be numbered, considering that the difficulty of disintegrating it in the stomach affects the ability of microorganisms to be attached to the gastrointestinal tract, where denser and more viscous products better protect LAB but prevent an effective release in the GIT (Gomand et al., 2019). Other factors include pH, water activity, buffer capacity, and nutritional content. A pH close to neutrality (5–6) is defined as optimal for growth and to facilitate adhesion to intestinal cells. Water activity should remain low during storage to promote protection and stability. Likewise, a high buffer capacity protects LAB during storage and reduces stress due to mechanical, physicochemical, and thermal modifications, which could increase the pH of the stomach, improving the protection of LAB during digestion. Solid matrices are recommended to improve the growth of LAB since it reduces the decay in the characteristics of the food during conservation, as well as preventing LAB from being eliminated in the stomach as they are embedded in the structure of the food. Regarding the composition, a high fat content has an equally positive effect on the buffer capacity, while oxygen must be limited to avoid the death of LAB (Gomand et al., 2019).

Simulation of production of bee bread from bee pollen

For several years, attempts have been made to develop processes that emulate the fermentation process of bee bread, trying to obtaining a product with a standard composition and in enough quantities to maintain regular marketing, and in response to concerns related to the difficulty of extract bee bread without destroying the integrity of the hives and maintaining the safety of the food, whose contamination typically occurs during storage time in the traps, or in the harvesting and handling stages done by the beekeeper (De-Melo et al., 2016; Deveza et al., 2015; Kieliszek et al., 2018).

The first developments reported in the literature mention that Pain and Maugenet (1966) inoculated *Lactobacillus* in pollen; however, the result was noticeably different in sensory terms than traditional bee bread. Vamanu et al. (2006) carried out in vitro processes to obtain bee bread by inoculating *L. acidophilus* in pollen with or without the inclusion of water at different pH and temperature conditions. The ideal parameters for the fermentation process were pH 4.5–6.5 and 30–42 °C. Later, Vamanu et al. (2008) used *L. acidophilus*, *L. plantarum*, and a mixture of the two of them to obtain a product with probiotic characteristics. The results showed a proportional relationship between the pollen concentration and the viability of *L. acidophilus*, while there was a balance in the viability of both microorganisms at low pollen concentrations.

Fuenmayor (2009) carried out studies related to the fermentation of Colombian pollen with *L. acidophilus*, *L. paracasei*, and two commercial cultures typically used in milk fermentation, finding that the best conditions were when inoculation of 10^8 CFU/mL of *L. acidophilus* was performed at 25 °C, using a sterile pollen-water mixture at a 2:1 ratio. Vamanu et al. (2010) compared the effectiveness to ferment pollen of six different microorganisms: *Lactococcus fermentum*, *L. plantarum* 1 and 2, *L. paracasei*, and *Bifidobacterium bifidum* 1 and 2, these latter, in junction with inulin and lactulose as prebiotics, were the best conditions for fermentation. Salazar-González and Díaz-Moreno (2014) evaluated the use of LAB and yeasts for the development of artificial bee bread, using *L. acidophilus*, *L. plantarum*, *Saccharomyces cerevisiae*, and a mixture of *L. plantarum* and *S. cerevisiae*, finding that the best process conditions were under a pollen-water ratio 1:1 and incubation at 35 °C when the LAB and yeast consortium was used, in addition to showing a significant increase in the content of phenolic compounds.

Duan et al. (2015) carried out different pollen fermentation tests using *L. plantarum* MRS3 and *B. subtilis* natto ATCC15245, finding a product with characteristics similar to natural bee bread. Additionally, the incorporation of such microorganisms induced the production of nattokinase, a useful component in the development of drugs with thrombolytic function. Salazar-González and Díaz-Moreno (2016) reported low viability in the growth of LAB in pollen due to adverse conditions for their growth and use of nutrients. In that sense, Misiewicz et al. (2015) stated that, in general, bacteria are not very tolerant of media with high osmotic pressure, such as bee bread.

Uțoiu et al. (2018) carried out the fermentation of pollen with a symbiotic culture of bacteria and yeast known as Kombucha/SCOBY, finding a structurally modified product compared to bee bread. The authors reported the destruction of the pollen wall from the outside to the inside of the grain caused by the bacteria present in the culture, mainly *Lactobacillus* sp. or *Pediococcus* sp., promoting greater availability of nutrients and bioactive compounds. Kaškonienė et al. (2018) carried out pollen fermentation tests with *Lactobacillus rhamnosus* GG, finding a noticeable increase in the content of flavonoids compared to unprocessed pollen, between 1.6 and 2.4 times.

Di Cagno et al. (2019) developed a protocol for the elaboration of artificial bee bread from the inoculation of *L. kunkeii* and *Hanseniaspora uvarum*. The pectic enzymes secreted by *H. uvarum* were useful in breaking the pollen wall and the subsequent release of nutrients, thus demonstrating the relationship between bacteria and other microorganisms to achieve the modification of pollen. Zuluaga-Dominguez and Quicazan (2019) evaluated the structural modification of pollen grains when subjected to fermentation using *S. cerevisiae*, *L. plantarum*, and commercial Choozit® culture. The results showed that the joint use of yeast and LAB increased the level of bioactive compounds such as phenols and flavonoids by 31% and 39%, respectively. The authors also demonstrated a variation in pollen structure after fermentation from the evaluation of photomicrographs by Scanning Electron Microscopy and results from Differential Scanning Calorimetry. The results obtained through the use of commercial culture, in terms of practicality and economy, is a much more accessible strategy for beekeepers interested in producing bee bread in culture environment.

Yan et al. (2019) fermented pollen using *L. bulgaricus*, *S. thermophilus*, and dry yeast, observing an increase of 9%, 69%, and 18% in phenolic compounds, oligopeptides, and fructose, compared to fresh pollen, as well as an increase in the level of some B-complex vitamins and amino acids. Shirsat et al. (2019) subjected pollen to fermentation with *L. lactis*, finding process conditions of 35 °C for the first 96 h and then 20 °C for the next 72 h, under a moisture content between 35% and 40%. The fermented product showed an increase of 1.5% in protein and 2% in free amino acids; in contrast, the polyphenol content decreased by 1.8%.

In summary, it is observed that fermentation of pollen in culture environment has shown a series of benefits regarding the improvement of nutritional and bioactive indicators that present it as an attractive option for food transformation. However, further research is needed to know with certainty the effect that the inclusion of microorganisms has on the structural modification of pollen, as well as the possibility of establishing a standard fermentation process.

Conclusion

There is a growing trend among people toward the increasingly frequent consumption of minimally processed foods, with traceable links to the origin and a close relationship to nature. Bee bread, despite being considered a relatively new food, has been described as a promising food thanks to its nutritional and functional characteristics. Although bee bread composition is dependent on the botanical and geographical origin, it has proven to be a food that goes beyond being just a nutritional source, and the studies carried out agree that its frequent consumption results in a general benefit for consumers, even indicated as a potential preventive in the appearance and development of certain diseases. The research carried out so far leads to the consideration that bee bread promotes a series of beneficial effects for intestinal microbiota preservation and its consequent improvement in health. The biochemical processes that take place during the conversion of pollen to bee bread, which in addition to improving the bioavailability of nutrients, favors the obtaining of different bioactive compounds that serve as a nutrient for gut microbiota helping to maintain their population and preventing colonization by other potentially pathogenic microorganisms. Despite the progress reached so far, a deeper investigation is required at a clinical level to find direct links between the consumption of bee bread and the potential pharmacological benefit of its consumption. Likewise, it is necessary to find alternatives for the inclusion of bee bread in the consumers' diet that promote its massive consumption, which can be achieved through the design of new foods with added value related to health and well-being, both physical and mental.

References

Aabed, K., et al., 2019. Ameliorative effect of probiotics (Lactobacillus paracaseii and Protexin®) and prebiotics (propolis and bee pollen) on clindamycin and propionic acid-induced oxidative stress and altered gut microbiota in a rodent model of autism. Cell. Mol. Biol. 65 (1), 1–7.

Almeida, C.C., et al., 2012. Beneficial effects of long-term consumption of a probiotic combination of Lactobacillus casei Shirota and Bifidobacterium breve Yakult may persist after suspension of therapy in lactose-intolerant patients. Nutr. Clin. Pract. 27 (2), 247–251. Wiley Online Library.

Anderson, K.E., et al., 2013. Microbial ecology of the hive and pollination landscape: bacterial associates from floral nectar, the alimentary tract and stored food of honey bees (Apis mellifera). PloS One 8 (12), e83125. Public Library of Science.

Anderson, K.E., et al., 2014. Hive-stored pollen of honey bees: many lines of evidence are consistent with pollen preservation, not nutrient conversion. Mol. Ecol. 23 (23), 5904–5917. Wiley Online Library.

Andjelkovic, B., et al., 2015. Quality of honey bee bread collected in spring. International Symposium on Animal Science. Institute for Zootechnics of the Faculty of Agriculture University of Belgrade and Department of Animal Sciences of the Faculty of Agriculture University of Novi Sad, pp. 450–454.

Arafa, S.A., et al., 2016. Effect of adding bee-bread and bee-pollen as antioxidants on productive performance and functional properties of HY-line hens strain. Egypt. Poult. Sci. J. 36 (4).

Arseneault-Bréard, J., et al., 2012. Combination of Lactobacillus helveticus R0052 and Bifidobacterium longum R0175 reduces post-myocardial infarction depression symptoms and restores intestinal permeability in a rat model. Br. J. Nutr. 107 (12), 1793–1799. Cambridge University Press.

Azad, M., et al., 2018. Probiotic species in the modulation of gut microbiota: an overview. Biomed. Res. Int. 2018. Hindawi.

Bäckhed, F., et al., 2005. Host-Bacterial Mutualism in the Human Intestine. Science 307 (5717), 1915–1920. https://doi.org/10.1126/science.1104816.

Bakour, M., et al., 2019. Bee bread as a functional product: chemical composition and bioactive properties. LWT 109, 276–282. Elsevier.

Bang, L.M., Buntting, C., Molan, P., 2003. The effect of dilution on the rate of hydrogen peroxide production in honey and its implications for wound healing. J. Altern. Complement. Med. 9 (2), 267–273. Mary Ann Liebert, Inc.

Barene, I., Daberte, I., Siksna, S., 2015. Investigation of bee bread and development of its dosage forms. Proteins 24, 20–30.

Behare, P., et al., 2009. Exopolysaccharides of lactic acid bacteria: a review. J. Food Sci. Technol. 46 (1), 1–11.

Belhadj, H., et al., 2014. Phenotypic and genotypic characterization of some lactic acid bacteria isolated from bee pollen: a preliminary study. Biosci. Microbiota, Food Health 33 (1), 11–23. BMFH Press.

Belkaid, Y., Hand, T.W., 2014. Role of the microbiota in immunity and inflammation. Cell 157 (1), 121–141. https://doi.org/10.1016/j.cell.2014.03.011.

Bergamo, P., et al., 2014. Immunomodulatory activity of a gut microbial metabolite of dietary linoleic acid, 10-hydroxy-cis-12-octadecenoic acid, associated with improved antioxidant/detoxifying defences. J. Funct. Foods 11, 192–202. https://doi.org/10.1016/j.jff.2014.10.007.

Bisson, L.F., et al., 2017. The two faces of Lactobacillus kunkeei: wine spoilage agent and bee probiotic. Catalyst: Discov. Pract. 1 (1), 1–11.

Borycka, K., Grabek-Lejko, D., Kasprzyk, I., 2015. Antioxidant and antibacterial properties of commercial bee pollen products. J. Apic. Res. 54 (5), 491–502. Taylor & Francis.

Brun, P., et al., 2017. Saccharomyces boulardii CNCM I-745 supplementation reduces gastrointestinal dysfunction in an animal model of IBS. PLoS One 12 (7), e0181863. Public Library of Science San Francisco, CA USA.

Čadež, N., et al., 2015. Zygosaccharomyces favi sp. nov., an obligate osmophilic yeast species from bee bread and honey. Antonie Van Leeuwenhoek 107 (3), 645–654. Springer.

Carpes, S., et al., 2007. Study of preparations of bee pollen extracts, antioxidant and antibacterial activity. Cienc. Agrotecnol. 31 (6), 1818–1825.

Chen, L., et al., 2015. Lactobacillus acidophilus suppresses colitis-associated activation of the IL-23/Th17 axis. J. Immunol. Res. 2015. Hindawi.

Chevtchik, V., 1950. Mikrobiologie Pyloveho Kvaseni. vol. 323 Publication from Faculty of Science, Universty of Masaryk, Czech Republic, pp. 103–130.

Curiel, J.A., et al., 2015. Lactic acid fermentation as a tool to enhance the antioxidant properties of Myrtus communis berries. Microb. Cell Fact. 14 (1), 67. https://doi.org/10.1186/s12934-015-0250-4.

De Filippis, F., Pasolli, E., Ercolini, D., 2020. The food-gut axis: lactic acid bacteria and their link to food, the gut microbiome and human health. FEMS Microbiol. Rev. 44 (4), 454–489. Oxford University Press.

De Filippo, C., et al., 2010. Impact of diet in shaping gut microbiota revealed by a comparative study in children from Europe and rural Africa. Proc. Natl. Acad. Sci. 107 (33), 14691–14696. https://doi.org/10.1073/pnas.1005963107.

DeGrandi-Hoffman, G., Eckholm, B., Huang, M., 2013. A comparison of bee bread made by Africanized and European honey bees (Apis mellifera) and its effects on hemolymph protein titers. Apidologie 44 (52–63).

De-Melo, A.A.M., et al., 2016. Effect of processing conditions on characteristics of dehydrated bee-pollen and correlation between quality parameters. LWT Food Sci. Technol. 65, 808–815. https://doi.org/10.1016/j.lwt.2015.09.014.

Denisow, B., Denisow-Pietrzyk, M., 2016. Biological and therapeutic properties of bee pollen: a review. J. Sci. Food Agric. 96 (13), 4303–4309. Wiley Online Library.

Derrien, M., van Hylckama Vlieg, J.E.T., 2015. Fate, activity, and impact of ingested bacteria within the human gut microbiota. Trends Microbiol. 23 (6), 354–366. https://doi.org/10.1016/j.tim.2015.03.002.

Dethlefsen, L., Relman, D.A., 2011. Incomplete recovery and individualized responses of the human distal gut microbiota to repeated antibiotic perturbation. Proc. Natl. Acad. Sci. 108 (supplement 1), 4554–4561. https://doi.org/10.1073/pnas.1000087107.

Deveza, M.V., et al., 2015. Mycotoxicological and palynological profiles of commercial brands of dried bee pollen. Braz. J. Microbiol. 46 (4), 1171–1176. https://doi.org/10.1590/S1517-838246420140316.

Di Cagno, R., et al., 2019. Novel solid-state fermentation of bee-collected pollen emulating the natural fermentation process of bee bread. Food Microbiol. 82, 218–230. Elsevier.

Disayathanoowat, T., et al., 2020. Different dynamics of Bacterial and fungal communities in hive-stored bee bread and their possible roles: a case study from two commercial honey bees in China. Microorganisms 8 (2), 264. Multidisciplinary Digital Publishing Institute.

Doğanyiğit, Z., et al., 2020. Histological, Immunohistochemical and Biochemica L Effects of Bee Bread on Stomach Tissue of Obese Rats. Comenius University.

Donkersley, P., et al., 2017. Nutritional composition of honey bee food stores vary with floral composition. Oecologia 185 (4), 749–761. Springer.

Duan, C., et al., 2015. Optimization of fermentation condition of man-made bee-bread by response surface methodology. In: Advances in Applied Biotechnology. Springer, pp. 353–363.

Edwards, C.G., et al., 1998. Lactobacillus kunkeei sp. nov.: a spoilage organism associated with grape juice fermentations. J. Appl. Microbiol. 84 (5), 698–702. Wiley Online Library.

Egorova, A.I., 1971. Preservative microflora in stored pollen. Veterinariya 8, 40–41.

Egorova, A., Babeva, I., 1967. Yeast flora of the honey bee (Apis mellifica L.). Izvestiya Akademii Nauk Armyanskoi SSR, Nauki o Zemle 2, 127–132.

El Aidy, S., van den Bogert, B., Kleerebezem, M., 2015. The small intestine microbiota, nutritional modulation and relevance for health. Curr. Opin. Biotechnol. 32, 14–20. https://doi.org/10.1016/j.copbio.2014.09.005.

Endo, A., Salminen, S., 2013. Honeybees and beehives are rich sources for fructophilic lactic acid bacteria. Syst. Appl. Microbiol. 36 (6), 444–448. https://doi.org/10.1016/j.syapm.2013.06.002.

Endo, A., et al., 2012. Characterization and emended description of Lactobacillus kunkeei as a fructophilic lactic acid bacterium. Int. J. Syst. Evol. Microbiol. 62 (3), 500–504. Microbiology Society.

Everard, A., et al., 2014. Saccharomyces boulardii administration changes gut microbiota and reduces hepatic steatosis, low-grade inflammation, and fat mass in obese and type 2 diabetic db/db mice. mBio 5 (3).

Fernández Murga, M.L., et al., 1999. Permeability and stability properties of membranes formed by lipids extracted from Lactobacillus acidophilus grown at different temperatures. Arch. Biochem. Biophys. 364 (1), 115–121. https://doi.org/10.1006/abbi.1998.1093.

Filannino, P., Gobbetti, M., et al., 2014a. Hydroxycinnamic acids used as external acceptors of electrons: an energetic advantage for strictly heterofermentative lactic acid bacteria. Appl. Environ. Microbiol. 80 (24), 7574–7582. https://doi.org/10.1128/AEM.02413-14. Edited by C. A. Elkins. American Society for Microbiology Journals.

Filannino, P., Cardinali, G., et al., 2014b. Metabolic responses of Lactobacillus plantarum strains during fermentation and storage of vegetable and fruit juices. Appl. Environ. Microbiol. 80 (7), 2206–2215. https://doi.org/10.1128/AEM.03885-13. Edited by M. J. Pettinari. American Society for Microbiology Journals.

Filannino, P., Cavoski, I., et al., 2016a. Lactic acid fermentation of cactus cladodes (Opuntia ficus-indica L.) generates flavonoid derivatives with antioxidant and anti-inflammatory properties. PLoS One 11 (3), e0152575. https://doi.org/10.1371/journal.pone.0152575. Public Library of Science.

Filannino, P., Di Cagno, R., et al., 2016b. Transcriptional reprogramming and phenotypic switching associated with the adaptation of Lactobacillus plantarum C2 to plant niches. Sci. Rep. 6 (1), 27392. https://doi.org/10.1038/srep27392.

Filannino, P., Di Cagno, R., Gobbetti, M., 2018. Metabolic and functional paths of lactic acid bacteria in plant foods: get out of the labyrinth. Curr. Opin. Biotechnol. 49, 64–72. Elsevier.

Francesca, N., et al., 2016. Optimised method for the analysis of phenolic compounds from caper (Capparis spinosa L.) berries and monitoring of their changes during fermentation. Food Chem. 196, 1172–1179. https://doi.org/10.1016/j.foodchem.2015.10.045.

Fu, L., et al., 2017. Bifidobacterium infantis potentially alleviates shrimp tropomyosin-induced allergy by tolerogenic dendritic cell-dependent induction of regulatory T cells and alterations in gut microbiota. Front. Immunol. 8, 1536. Frontiers.

Fuenmayor, C., 2009. Aplicación de bioprocesos en polen de abejas para el desarrollo de un suplemento nutricional protéico. Universidad Nacional de Colombia.

García, D.G., Mogollón, M.A.R., Nieves, J.S., 2006. Cultured microbiological content of the intestinal tract and stored pollen of Apis mellifera (Hymenoptera: Apidae). Acta Biológica Colombiana 11 (1), 123–129. Universidad Nacional de Colombia.

Gilliam, M., 1979a. Microbiology of pollen and bee bread: the genus Bacillus. Apidologie 10 (3), 269–274.

Gilliam, M., 1979b. Microbiology of pollen and bee bread: the yeasts. Apidologie 10 (1), 43–53.

Gilliam, M., 1997. Identification and roles of non-pathogenic microflora associated with honey bees. FEMS Microbiol. Lett. 155 (1), 1–10. Blackwell Publishing Ltd Oxford, UK.

Gilliam, M., Prest, D.B., 1977. The mycoflora of selected organs of queen honey bees, Apis mellifera. J. Invertebr. Pathol. 29 (2), 235–237. Elsevier.

Gilliam, M., Prest, D.B., 1987. Microbiology of feces of the larval honey bee, Apis mellifera. J. Invertebr. Pathol. 49 (1), 70–75. Elsevier.

Gilliam, M., et al., 1988. Factors affecting development of chalkbrood disease in colonies of honey bees, Apis mellifera, fed pollen contaminated with Ascosphaera apis. J. Invertebr. Pathol. 52 (2), 314–325. Elsevier.

Gilliam, M., Prest, D.B., Lorenz, B.J., 1989. Microbiology of pollen and bee bread: taxonomy and enzymology of molds. Apidologie 20 (1), 53–68. EDP Sciences.

Golomb, B.L., Marco, M.L., 2015. Lactococcus lactis metabolism and gene expression during growth on plant tissues. J. Bacteriol. 197 (2), 371–381. https://doi.org/10.1128/JB.02193-14. Edited by W.W. Metcalf. American Society for Microbiology Journals.

Gomand, F., et al., 2019. Food matrix design for effective lactic acid bacteria delivery. Annu. Rev. Food Sci. Technol. 10, 285–310. Annual Reviews.

Hartmann, R., Meisel, H., 2007. Food-derived peptides with biological activity: from research to food applications. Curr. Opin. Biotechnol. 18 (2), 163–169. https://doi.org/10.1016/j.copbio.2007.01.013.

Hassan, A.N., 2008. ADSA foundation scholar award: possibilities and challenges of exopolysaccharide-producing lactic cultures in dairy foods. J. Dairy Sci. 91 (4), 1282–1298. https://doi.org/10.3168/jds.2007-0558.

Haydak, M.H., Vivino, A.E., 1950. The changes in the thiamine, riboflavin, niacin and pantothenic acid content in the food of female honeybees during growth with a note on the vitamin K activity of royal jelly and beebread. Ann. Entomol. Soc. Am. 43 (3), 361–367.

Herbert, E., Shimanuki, H., 1978. Chemical composition and nutritive value of bee collected and bee stored pollen. Apidologie 9 (1), 33–40.

Hernández-Ledesma, B., Recio, I., Amigo, L., 2008. β-Lactoglobulin as source of bioactive peptides. Amino Acids 35 (2), 257–265. https://doi.org/10.1007/s00726-007-0585-1.

Huang, Y.-C., et al., 1996. Relationship between sluggish fermentations and the antagonism of yeast by lactic acid bacteria. Am. J. Enol. Vitic. 47 (1), 1–10. Am Soc Enol Viticulture.

Human, H., Nicolson, S.W., 2006. Nutritional content of fresh, bee-collected and stored pollen of Aloe greatheadii var. davyana (Asphodelaceae). Phytochemistry 67 (14), 1486–1492.

Indiani, C.M.D.S.P., et al., 2018. Childhood obesity and firmicutes/bacteroidetes ratio in the gut microbiota: a systematic review. Childhood Obesity 14 (8), 501–509. https://doi.org/10.1089/chi.2018.0040. Mary Ann Liebert, Inc. Publishers.

Isidorov, V.A., et al., 2009. Gas chromatographic–mass spectrometric investigation of the chemical composition of beebread. Food Chem. 115 (3), 1056–1063. Elsevier.

Ivanišová, E., et al., 2015. Bee bread-perspective source of bioactive compounds for future. Potravinarstvo Slovak J. Food Sci. 9 (1), 592–598.

Janashia, I., et al., 2018. Beneficial protective role of endogenous lactic acid bacteria against mycotic contamination of honeybee beebread. Probiotics Antimicrob. Proteins 10 (4), 638–646. Springer.

Kačániová, M., et al., 2013. The effects of bee pollen extracts on the broiler chicken's gastrointestinal microflora. Res. Vet. Sci. 95 (1), 34–37. Elsevier.

Kaplan, M., et al., 2016. Fatty acid and proximate composition of bee bread. Food Technol. Biotechnol. 54 (4), 497–504. Prehrambeno-biotehnološki fakultet, Sveučilište u Zagrebu.

Karczewski, J., et al., 2010. Regulation of human epithelial tight junction proteins by Lactobacillus plantarum in vivo and protective effects on the epithelial barrier. Am. J. Physiol. Gastrointest. Liver Physiol. 298 (6), G851–G859. American Physiological Society Bethesda, MD.

Kaškonienė, V., et al., 2018. The impact of solid-state fermentation on bee pollen phenolic compounds and radical scavenging capacity. Chem. Pap. 72 (8), 2115–2120. Springer.

Keller, K.M., et al., 2014. Fungi infection in honeybee hives in regions affected by Brazilian sac brood. Arquivo Brasileiro de Medicina Veterinária e Zootecnia 66 (5), 1471–1478. SciELO Brasil.

Keshri, J., et al., 2019. Bacterial dynamics of wheat silage. Front. Microbiol. 10, 1532. Frontiers.

Khalifa, S.A.M., et al., 2020. Recent insights into chemical and pharmacological studies of bee bread. Trends Food Sci. Technol. 97, 300–316. https://doi.org/10.1016/j.tifs.2019.08.021.

Kieliszek, M., et al., 2018. Pollen and bee bread as new health-oriented products: a review. Trends Food Sci. Technol. 71, 170–180. Elsevier.

Kim, H.S., Han, O.K., Kwak, Y.-S., 2017. Complete genome sequence and functional study of the fibrinolytic enzyme-producing bacterium Leuconostoc holzapfelii 5H4, a silage probiotic. J. Genomics 5, 32. Ivyspring International Publisher.

Krell, R., 1996. Value-Added Products From Beekeeping. Food and Agriculture Organization.

Lamei, S., 2018. The Effect of Honeybee-Specific Lactic Acid Bacteria on American Foulbrood Disease of Honeybees. Lund University.

Landete, J.M., et al., 2014. Aryl glycosidases from Lactobacillus plantarum increase antioxidant activity of phenolic compounds. J. Funct. Foods 7, 322–329. https://doi.org/10.1016/j.jff.2014.01.028.

Landete, J., et al., 2015. Effect of soaking and fermentation on content of phenolic compounds of soybean (Glycine max cv. Merit) and mung beans (Vigna radiata [L] Wilczek). Int. J. Food Sci. Nutr. 66 (2), 203–209. https://doi.org/10.3109/09637486.2014.986068. Taylor & Francis.

Lee, N.K., et al., 2015. Conversion of flavonols kaempferol and quercetin in mulberry (Morus alba L.) leaf using plant-fermenting Lactobacillus plantarum. J. Food Biochem. 39 (6), 765–770. https://doi.org/10.1111/jfbc.12176. John Wiley & Sons, Ltd.

Li, M., et al., 2015. Oral administration of Saccharomyces boulardii ameliorates carbon tetrachloride-induced liver fibrosis in rats via reducing intestinal permeability and modulating gut microbial composition. Inflammation 38 (1), 170–179. Springer.

Li, Q.-Q., et al., 2018. Nutrient-rich bee pollen: a treasure trove of active natural metabolites. J. Funct. Foods 49, 472–484. Elsevier.

Lohner, K., Blondelle, S.E., 2005. Molecular mechanisms of membrane perturbation by antimicrobial peptides and the use of biophysical studies in the design of novel peptide antibiotics. In: Combinatorial Chemistry & High Throughput Screening, pp. 241–256. Available at: https://www.ingentaconnect.com/content/ben/cchts/2005/00000008/00000003/art00005.

Loper, G., et al., 1980. Biochemistry and microbiology of bee-collected almond (prunus dulcis) pollen and bee bread. I. Fatty acids, sterols, vitamins and minerals. Apidologie 11 (1), 63–73.

Magne, F., et al., 2020. The firmicutes/bacteroidetes ratio: a relevant marker of gut dysbiosis in obese patients? Nutrients 12, 1474–1490.

Marco, M.L., et al., 2017. Health benefits of fermented foods: microbiota and beyond. Curr. Opin. Biotechnol. 44, 94–102. https://doi.org/10.1016/j.copbio.2016.11.010.

Mărgăoan, R., et al., 2020. Impact of fermentation processes on the bioactive profile and health-promoting properties of bee bread, mead and honey vinegar. Processes 8 (9), 1081. Multidisciplinary Digital Publishing Institute.

Markiewicz-Żukowska, R., et al., 2013. Chemical composition and antioxidant activity of beebread, and its influence on the glioblastoma cell line (U87MG). J. Apic. Sci. 57 (2), 147–157. Sciendo.

Mathialagan, M., Johnson, Y., Thangaraj, E., 2018. Isolation, characterization and identification of probiotic lactic acid bacteria (LAB) from honey bees. Int. J. Curr. Microbiol. App. Sci. 7, 849–906.

McFrederick, Q.S., et al., 2012. 'Environment or kin: whence do bees obtain acidophilic bacteria? Mol. Ecol. 21 (7), 1754–1768. Wiley Online Library.

Misiewicz, A., Kieliszek, M., Czarniak, K., 2015. New strain of *Lactobacillus delbrueckii* bacteria and its use for the production of bee pollen. Patent: WO2015093997.

Monedero, V., Pérez-Martínez, G., Yebra, M.J., 2010. Perspectives of engineering lactic acid bacteria for biotechnological polyol production. Appl. Microbiol. Biotechnol. 86 (4), 1003–1015. https://doi.org/10.1007/s00253-010-2494-6.

Moreno Galarza, L.J., 2012. Aislamiento y Selección de Lactobacillus sp con potencial probiótico a partir de pan de abejas. Facultad de Ciencias. Universidad Nacional de Colombia, Bogotá D.C., Colombia. 109 pp.

Nampoothiri, K.M., et al., 2017. Chapter 3—health benefits of exopolysaccharides in fermented foods. In: Frias, J., Martinez-Villaluenga, C., Peñas, E. (Eds.), Fermented Foods in Health and Disease Prevention. Academic Press, Boston, pp. 49–62.

Nechaeva, N., 2009. Changes of Functinal and Sport Medicine After Intake of Bee Products. Ryazan State Medical University, Ryazan, Russia.

Neveling, D.P., Endo, A., Dicks, L.M.T., 2012. Fructophilic Lactobacillus kunkeei and Lactobacillus brevis isolated from fresh flowers, bees and bee-hives. Curr. Microbiol. 65 (5), 507–515. Springer.

Nishiyama, K., Sugiyama, M., Mukai, T., 2016. Adhesion properties of lactic acid bacteria on intestinal mucin. Microorganisms 4 (3), 34. Multidisciplinary Digital Publishing Institute.

O'Sullivan, M.G., et al., 1992. Probiotic bacteria: myth or reality? Trends Food Sci. Technol. 3, 309–314. https://doi.org/10.1016/S0924-2244(10)80018-4.

Othman, Z.A., et al., 2019. Nutritional, phytochemical and antioxidant analysis of bee bread from different regions of Malaysia. Indian J. Pharm. Sci. 81 (5), 955–960. OMICS International.

Pain, J., Maugenet, J., 1966. Recherches biochimiques et physiologiques sur le pollen emmagasiné par les abeilles. Annales de l'abeille 9, 209–236.

Persson, E.K., et al., 2013. Dendritic cell subsets in the intestinal lamina propria: ontogeny and function. Eur. J. Immunol. 43 (12), 3098–3107. Wiley Online Library.

Pessione, E., Cirrincione, S., 2016. Bioactive molecules released in food by lactic acid bacteria: encrypted peptides and biogenic amines. Front. Microbiol. 7, 876. Frontiers.

Pohl, P., et al., 2020. Element analysis of bee-collected pollen and bee bread by atomic and mass spectrometry–methodological development in addition to environmental and nutritional aspects. Trends Anal. Chem., 115922. Elsevier.

Reuter, G., 2001. The Lactobacillus and Bifidobacterium microflora of the human intestine: composition and succession. Curr. Issues Intest. Microbiol. 2 (2), 43–53.

Salazar, N., et al., 2016. Exopolysaccharides produced by lactic acid bacteria and bifidobacteria as fermentable substrates by the intestinal microbiota. Crit. Rev. Food Sci. Nutr. 56 (9), 1440–1453. Taylor & Francis.

Salazar-González, C.Y., Díaz-Moreno, C., 2014. Características bioactivas y sensoriales en procesos de fermentación en fase sólida de polen apícola. Encuentro Nacional de Investigación y Desarrollo - ENID.

Salazar-González, C., Díaz-Moreno, C., 2016. The nutritional and bioactive aptitude of bee pollen for a solid-state fermentation process. J. Apic. Res. 55 (2), 161–175. Taylor & Francis.

Savignac, H.M., et al., 2015. Bifidobacteria modulate cognitive processes in an anxious mouse strain. Behav. Brain Res. 287, 59–72. Elsevier.

Shimazu, S., Miklya, I., 2004. Pharmacological studies with endogenous enhancer substances: β-phenylethylamine, tryptamine, and their synthetic derivatives. Prog. Neuro-Psychopharmacol. Biol. Psychiatry 28 (3), 421–427. https://doi.org/10.1016/j.pnpbp.2003.11.016.

Shirsat, D.V., Kad, S.K., Wakhle, D.M., 2019. Solid state fermentation of bee-collected pollen. Int. J. Curr. Microbiol. App. Sci. 8 (5), 1557–1563.

Sinpoo, C., Williams, G.R., Chantawannakul, P., 2017. Dynamics of fungal communities in corbicular pollen and bee bread. Chiang Mai J. Sci. 44, 1244–1256.

Sobral, F., et al., 2017. Flavonoid composition and antitumor activity of bee bread collected in Northeast Portugal. Molecules 22 (2), 248. Multidisciplinary Digital Publishing Institute.
Soundharrajan, I., et al., 2019. Probiotic and triticale silage fermentation potential of Pediococcus pentosaceus and Lactobacillus brevis and their impacts on pathogenic Bacteria. Microorganisms 7 (9), 318. Multidisciplinary Digital Publishing Institute.
Stanciu, O., Marghitas, L.A., Dezmirean, D., 2007. Examination of antioxidant capacity of beebread extracts by different complementary assays. Bull. Univ. Agric. Sci. Vet. Med. Cluj-Napoca 64 (1–2). Animal Science and Biotechnologies.
Standifer, L.N., et al., 1980. Biochemistry and microbiology of pollen collected by honey bees (Apis mellifera L.) from almond, Prunus dulcis. II. Protein, amino acids and enzymes. Apidologie 11 (2), 163–171. EDP Sciences.
Stolaki, M., et al., 2019. Microbial communities in a dynamic in vitro model for the human ileum resemble the human ileal microbiota. FEMS Microbiol. Ecol. 95 (8). https://doi.org/10.1093/femsec/fiz096.
Tajabadi, N., et al., 2011. Detection and identification of Lactobacillus bacteria found in the honey stomach of the giant honeybee Apis dorsata. Apidologie 42 (5), 642. Springer.
Tan, D.-X., et al., 2014. Fundamental issues related to the origin of melatonin and melatonin isomers during evolution: relation to their biological functions. Int. J. Mol. Sci. 15, 15858–15890.
Thakur, M., Nanda, V., 2020. Composition and functionality of bee pollen: a review. Trends Food Sci. Technol. 98, 82–106. Elsevier.
Tomás, A., et al., 2017. Potentialities of beebread as a food supplement and source of nutraceuticals: botanical origin, nutritional composition and antioxidant activity. J. Apic. Res. 56 (3), 219–230. Taylor & Francis.
Urcan, A.C., et al., 2017. Chemical composition and biological activities of beebread–review. Bull. Univ. Agric. Sci. Vet. Med. Cluj-Napoca 74 (1), 6–14. Animal Science and Biotechnologies.
Uţoiu, E., et al., 2018. Bee collected pollen with enhanced health benefits, produced by fermentation with a Kombucha consortium. Nutrients 10 (10), 1365. Multidisciplinary Digital Publishing Institute.
Vamanu, A., et al., 2006. Identification of a lactic bacterium strain used for obtaining a pollen-based probiotic product. Turk. J. Biol. 30, 75–80.
Vamanu, A., et al., 2008. Obtaining of a symbiotic product based on lactic acid bacteria, pollen and honey. Pak. J. Biol. Sci. 11 (4), 613–617.
Vamanu, E., et al., 2010. The antioxidant effect of a functional product based on probiotic biomass, pollen and honey. Sci. Pap. Anim. Sci. Biotechnol. 43 (1), 331–336.
Van den Bogert, B., et al., 2014. Comparative genomics analysis of Streptococcus isolates from the human small intestine reveals their adaptation to a highly dynamic ecosystem. PLoS ONE 8 (12), e83418. https://doi.org/10.1371/journal.pone.0083418. Public Library of Science.
Vásquez, A., Olofsson, T.C., 2009. The lactic acid bacteria involved in the production of bee pollen and bee bread. J. Apic. Res. 48 (3), 189–195. https://doi.org/10.3896/IBRA.1.48.3.07.
Vásquez, A., et al., 2012. Symbionts as major modulators of insect health: lactic acid bacteria and honeybee. PLoS ONE 7 (3), 133138.
Vorland, L.H., 1999. Lactoferrin: a multifunctional glycoprotein. APMIS 107 (7–12), 971–981. https://doi.org/10.1111/j.1699-0463.1999.tb01499.x. John Wiley & Sons, Ltd.
Vrieze, A., et al., 2010. The environment within: how gut microbiota may influence metabolism and body composition. Diabetologia 53 (4), 606–613. https://doi.org/10.1007/s00125-010-1662-7.
Wilson, H.F., Marvin, G.E., 1929. On the occurrence of the yeasts which may cause the spoilage of honey. J Econ. Entomol. 22 (3), 513–517. Oxford University Press Oxford, UK.

Wu, W., et al., 2020. In vitro and in vivo digestion comparison of bee pollen with or without wall-disruption. J. Sci. Food Agric. 101 (7), 2744–2755 (Wiley Online Library).

Xu, D., et al., 2019. Modulation of metabolome and bacterial community in whole crop corn silage by inoculating homofermentative Lactobacillus plantarum and heterofermentative Lactobacillus buchneri. Front. Microbiol. 9, 3299. Frontiers.

Xue, L., et al., 2017. Probiotics may delay the progression of nonalcoholic fatty liver disease by restoring the gut microbiota structure and improving intestinal endotoxemia. Sci. Rep. 7, 45176. Nature Publishing Group.

Yakusheva, E., 2010. Pollen and bee bread: physico-chemical properties. Biological and pharmacological effects. Use in medical practice. In: Theoretical and Practical Basics of Apitherapy, pp. 84–97.

Yan, S., et al., 2019. Analysis of improved nutritional composition of bee pollen (*Brassica campestris* L.) after different fermentation treatments. Int. J. Food Sci. Technol. 54 (6), 2169–2181. Wiley Online Library.

Yu, L., et al., 2017. Saccharomyces boulardii administration changes gut microbiota and attenuates D-galactosamine-induced liver injury. Sci. Rep. 7 (1), 1–7. Nature Publishing Group.

Zhou, W., et al., 2018. Simulated digestion and fermentation in vitro by human gut microbiota of polysaccharides from bee collected pollen of Chinese wolfberry. J. Agric. Food Chem. 66 (4), 898–907. ACS Publications.

Zuluaga Domínguez, C.M., Serrato Bermudez, J.C., Quicazán De Cuenca, M.C., 2014. Valorization alternatives of Colombian bee-pollen for its use as food resource—a structured review. Vitae 21 (3).

Zuluaga, C.M., Serrato, J.C., Quicazan, M.C., 2015a. Bee-pollen structure modification by physical and biotechnological processing: influence on the availability of nutrients and bioactive compounds. Chem. Eng. Trans. 43. https://doi.org/10.3303/CET1543014.

Zuluaga, C.M., Serrato, J.C., Quicazan, M.C., 2015b. Chemical, nutritional and bioactive characterization of Colombian bee-bread. Chem. Eng. Trans. 43. https://doi.org/10.3303/CET1543030.

Zuluaga-Dominguez, C.M., Quicazan, M., 2019. Effect of fermentation on structural characteristics and bioactive compounds of bee-pollen based food. J. Apic. Sci. 63 (2), 209–222. https://doi.org/10.2478/jas-2019-0016.

CHAPTER 14

Therapeutic application of bee venom for rheumatoid arthritis

Lin Ang[a], Eunhye Song[b], Hye Won Lee[c], and Myeong Soo Lee[a]
[a]KM Science Research Division, Korea Institute of Oriental Medicine, Daejeon, Korea
[b]Global Cooperation Center, Korea Institute of Oriental Medicine, Daejeon, Korea
[c]KM Convergence Research Division, Korea Institute of Oriental Medicine, Daejeon, Korea

Contents

Introduction

Historical context

Bee venom has been widely studied for its therapeutic properties. The use of bee products, including bee venom, honey, pollen, propolis, and royal jelly, dates back to ancient Egypt, China, and Greece, and their medicinal properties are cited in religious text such as Vedas, Bible, and Quran (Fratellone et al., 2016; Wehbe et al., 2019). The practice of bee venom therapy in traditional medicine can also be traced back to 5000 years ago (Rose, 1994; Zhang et al., 2018). In modern medicine, the history of the use of bee venoms goes back to ancient Greece, where the father of modern medicine, Hippocrates, also referred to the medical uses of bee stings in his practice (Kim, 2013). Since the 16th century, bee venom is used to treat other

Bee Products and Their Applications in the Food and Pharmaceutical Industries
https://doi.org/10.1016/B978-0-323-85400-9.00001-0

illnesses such as gout, baldness, urinary disorders, and cancer. The potential of bee venom therapy has been widely studied to treat illness and progressed with publications of scientific studies during the 19th century (Kim, 2013).

In 1859, the first scientific report on the usage of bee stings for rheumatic conditions and skin cancer was published by a French physician Dr. Desjardins in *Abeille Medical* (Table 14.1). Decades after, an Austrian physician Dr. Philip Terc (the Father of Modern Apitherapy) published the first report, "Report about a peculiar connection between the bee stings and rheumatism," in *Vienna Medical Press* and became the foremost pioneer who systematically applied bee venom for rheumatism. His report showed that beekeepers who are often stung by bees rarely suffered from arthritis and other musculoskeletal diseases, indicating the connection between the bee venom and rheumatism (Hellner et al., 2008; Zhang et al., 2018). In the 1900s, Viennese physician Dr. Rudolph Tertsch published research studies on the application of bee venom in the treatment of rheumatic fever, showing huge progress in the medical understanding of bee venom therapy. One of his significant findings were 82% out of 660 rheumatic arthritis patients were fully recovered, 15% showed improvements, and only 3% were unsuccessful in showing signs of recovery.

In 1935, American physician Dr. Bodog Beck coined the term "bee venom therapy" and published his findings in *Bee venom therapy: Bee venom, its nature, and its effect on arthritic and rheumatoid conditions*. He also initiated the practice of bee venom therapy in the United States based on his expertise in treating arthritic and rheumatoid conditions using bee venom. The Master of Apitherapy Charles Mraz, who followed Beck, had been highly credited for his remarkable results in using bee venom therapy to treat various illnesses and was the founding member of the American Apitherapy Society

Table 14.1 Brief history of bee venom therapy.

Year	Significant events
1859	First scientific report on the usage of bee stings for rheumatic conditions by Dr. Desjardins
1888	The connection between bee stings and rheumatism first found and reported by Dr. Philip Terc
1912	Scientific research studies on the successful treatment of rheumatic fever using bee venom by Dr. Rudolph Tertsch
1935	The term "Bee venom therapy" was introduced by Dr. Bodog Beck
1989	The American Apitherapy Society was formed with Charles Mraz as cofounder
2003	The first standardized form of injectable honeybee venom by Dr. Christopher Kim

(https://apitherapy.org/). He also initiated clinical research of bee venom therapy at the Sloan-Kettering Institute and the Walter Reed Army Institute and established the purity standard for dried whole venom. In 2003, American-Korean physician Dr. Christopher Kim filed a patent for the standardized injectable form of honeybee venom (purified *Apis mellifera* venom), which was brand-named *Apitoxin* and certified as a bio-therapeutic drug in South Korea.

Type of application method of bee venom therapy

In the distant past, bee venom therapy was applied by putting honey on the skin of the treatment sites to attract the bees, and a cloth was then put over the bees to provoke them to sting. In modern times, bee venom therapy can be applied through three approaches: live bee sting, bee venom injection, and bee venom acupuncture (Zhang et al., 2018). A live bee sting is a traditional approach of bee venom therapy used in Asian countries such as China and South Korea. The patient is directly stung by live bees where the stinger is being inserted into the skin, followed by the separation of venom sac from the bee's abdomen as it pulls away. The autonomous repetitive contraction of the inserted stinger then continues to inject the venom into the muscles (Chen and Lariviere, 2010). The bee venom injection approach is developed more recently where pure bee venom is extracted using methods such as electric stimulus and being injected into the treatment site using a syringe with accurate placement and dosing of bee venom (Kim et al., 2005b; Zhang et al., 2018). Bee venom acupuncture is similar to the first two approaches, live bee sting and bee venom injection, except that meridian system acupoints or ashi points, are selected as treatment sites (Lee et al., 2014b, c). It can also be applied indirectly using acupuncture needles by applying bee venom to the needle tip before inserting into the acupoint (Abd El-Wahed et al., 2019; Cherniack and Govorushko, 2018). This approach utilizes both pharmacological actions of bee venom and simulation effects of acupuncture which is considered more efficient than the other two approaches (Kwon et al., 2001). On the other hand, there are also several noninvasive applications of bee venom therapy, such as inhalation, ointments, and iontophoresis (Kim, 2013).

Pharmaceutical properties and medical uses of bee venom

Bioactive compounds in bee venom

Bee venom is a clear liquid consist of 88% of water and other molecule contents. It contains many active components, including peptides, enzymes, and amines, with various pharmacological effects (Table 14.2).

Table 14.2 Major compounds of bee venoms and their biological activities.

Compounds (% by weight in dry bee venom)	Biological activity
Peptides	
Melittin (40–60)	Antiapoptotic[a], antiarrhythmic[b], antiarthritis[c], antiatherosclerotic[d], anticancer[e], antidiabetic[f], antifibrotic[g], antiinflammatory[h], antinociceptive[i], antiangiogenic[j], antibacterial[k], antifungal[l], antitumor[m], antiviral[n], analgesic[o], pro-apoptotic[p]
Apamin (1–3)	Antiatherosclerotic[q], antifibrotic[r], antiinflammatory[s], antibacterial[t], antifungal[u], neuroprotection[v]
Mast cell degranulating peptide (1–3)	Antiallergic[w], antiinflammatory[x], neuroprotection[y]
Secapin (1–2)	Antielastolytic[z], antifibrinolytic[z], antibacterial[z]
Adolapin (0.1–0.8)	Antiinflammatory[aa], analgesic[aa], antipyretic[ab]
Enzymes	
Phospholipase A2 (10 – 12)	Antiarthritis[ac], antibacterial[ad], anticancer[ae], antiparasitic[af], antiviral[ag], allergenicity[ah], antiinflammatory[ai], neuroprotection[aj]
Hyaluronidase (1.5–3)	Antiallergic[ak], hydrolyzing hyaluronic acid activation[al]

[a] Lee et al. (2011b).
[b] Yalcin et al. (2009).
[c] Jeong et al. (2015).
[d] Kim et al. (2011a).
[e] Lim et al. (2019), Rady et al. (2017).
[f] Hossen et al. (2017).
[g] Li et al. (2020).
[h] Lin et al. (2017).
[i] Son et al. (2007).
[j] Shin et al. (2013).
[k] Marques Pereira et al. (2020).
[l] Memariani and Memariani (2020).
[m] Park et al. (2010).
[n] Memariani et al. (2020).
[o] Choi et al. (2019).
[p] Kong et al. (2016).
[q] Kim et al. (2012).
[r] Kim et al. (2017).
[s] Lee et al. (2020).
[t] El-Seedi et al. (2020).
[u] Shin et al. (2017).
[v] Mohammadi-Rad et al. (2019).
[w] Buku et al. (2001).
[x] Banks et al. (1990).
[y] Kondo et al. (1992).
[z] Lee et al. (2016b).
[aa] Shkenderov and Koburova (1982).
[ab] Koburova (1984).
[ac] Duchez et al. (2019).
[ad] Leandro et al. (2015).
[ae] Ho et al. (2010).
[af] Boutrin et al. (2008).
[ag] Fenard et al. (2001).
[ah] Corthésy et al. (2016).
[ai] Landucci et al. (2000).
[aj] Ham et al. (2019), Kim et al. (2019).
[ak] Marković-Housley et al. (2000).
[al] Csoka et al. (2001).

Melittin is the main constituent of bee venom, comprising 40%–60% of dry bee venom by weight, followed by apamin. Compared to other components of bee venom, both melittin and apamin are of higher importance for concentration and their broad spectrum of pharmacological activities. Mast cell degranulating peptide constituting 1%–3% of dry bee venom is considered a classic releaser of histamine and has shown to have neuroprotective properties in recent years. Other peptides such as secapin and adolapin also demonstrate various pharmacological properties even though they have a lower concentration in bee venom. Phospholipase A2 is also relatively one of the main constituents of bee venom as this enzyme constitutes 10%–20% of dry bee venom. Phospholipase A2 is shown to have various interactions with melittin and exhibits many pharmacological properties. Hyaluronidase, known since the 1940s, constitutes 1.5%–3% of dry bee venom and plays a role in breaking down hyaluronic acid in tissues.

Therapeutic applications

Bee venom has been historically used for therapeutic purposes in various conditions such as rheumatism, antiinflammatory disorders, and skin disease. In recent years, many clinical and experimental studies have also shown its application and potential effects on immunological disorders, neurological conditions, infectious diseases, and many more (Table 14.3).

Bee venom therapy for rheumatoid arthritis

Rheumatoid arthritis is a chronic autoimmune disorder with unknown etiology that affects the joints in the hands, wrists, and knees. RA symptoms include warm and swollen joints, stiff joints (worse after inactivity), and painful joints that eventually result in bone erosion and joint deformity. Bee venom therapy has been used to treat RA for millennia, and its effectiveness has been extensively researched.

In vivo studies for rheumatoid arthritis

In the etiology and pathogenesis study of rheumatoid arthritis, animal models have been widely used, especially rodent models. These models include induced arthritis models (e.g., collagen-induced arthritis, adjuvant-induced arthritis, streptococcal cell wall-induced arthritis, collagen-antibody-induced arthritis, pristane-induced arthritis, proteoglycan-induced arthritis, zymosan-induced arthritis, and antigen-induced arthritis) and genetically manipulated spontaneous arthritis models (e.g., tumor necrosis factor

Table 14.3 Various health claims of bee venom.

Therapeutic applications	
Musculoskeletal conditions	Osteoarthritis[a] Gout[b] Adhesive capsulitis[c] Low back pain[d] Temporomandibular disorder[e]
Autoimmune conditions	Rheumatoid arthritis[f] Multiple sclerosis[g] Allergic rhinitis[h] Systemic lupus erythematosus [i] Autoimmune encephalitis[j]
Neurological conditions	Parkinson's disease[k] Alzheimer's disease[l] Amyotrophic lateral sclerosis [m] Stroke[n]
Cardiovascular conditions	Hypertension[o] Arrhythmias[p] Atherosclerosis[q]
Infectious conditions (viral, bacterial, and fungal)	Herpes simplex virus[r] *Mycobacterium Chelonae*[s]
Endocrine disorders	Diabetes mellitus[t] Polycystic ovarian syndrome[v]
Tumors and cancers[u]	–
Skin conditions	Atopic dermatitis[w] Psoriasis[x] Acne[y] Wrinkles[z] Varicose veins[aa]
Psychological disorders	Depression[ab]
Other conditions	Hepatic fibrosis[ac] Pelvic inflammatory disease[ad]

The table above only listed the major health claims of bee venom therapy and does not imply the effectiveness of bee venom or that these conditions are cured.

[a] Conrad et al. (2019).
[b] Lee et al. (2020).
[c] Koh et al. (2013).
[d] Seo et al. (2017).
[e] Nitecka-Buchta et al. (2014).
[f] Lee et al. (2014b, c).
[g] Karimi et al. (2012).
[h] Shin et al. (2014).
[i] Lee et al. (2011a).
[j] Lee et al. (2016a).
[k] Hartmann et al. (2016).
[l] Ham et al. (2019).
[m] Yang et al. (2010).
[n] Sun et al. (2018).
[o] Yalcin et al. (2009).
[p] Kang et al. (2020).
[q] Uddin et al. (2016).
[r] Cho et al. (2013).
[s] Hassan et al. (2019).
[t] Yasin et al. (2018).
[u] Jo et al. (2012), Park et al. (2011).
[v] You et al. (2016).
[w] Eltaher et al. (2015).
[x] Han et al. (2016).
[y] Han et al. (2015).
[z] el Amraoui et al. (2020).
[aa] Journals et al. (2015).
[ab] Lee et al. (2014a).
[ac] Mohamed and Ewida (2016).
[ad] Kim and Han (2020).

(TNF)-α transgenic, K/BxN serum transfer arthritis (STA), SKG monogenic model, and severe combined immunodeficient (SCID) model). The collagen-induced arthritis model is considered the gold standard for rheumatoid arthritis (Asquith et al., 2009; Choudhary et al., 2018).

In a study using murine type II collagen-induced arthritis model, bee venom administered via bee venom acupuncture showed that the production of proinflammatory cytokines TNF-α was suppressed, and the mean arthritis index and the number of arthritic limbs were lower in the treated mice compared to the control mice (Lee et al., 2004). The effect of bee venom acupuncture on type II collagen-induced arthritis rats was assessed in another study which reported that bee venom significantly increased the activity of proteolytic enzymes and lowered the free radical reactive oxygen species level in synovial fluid (Suh et al., 2006). Higher doses of bee venom injection (15 and 20 μL/100 g) were also reported to be more effective in decreasing the incidence of arthritis (Kim et al., 2008). Other studies using similar animal models also claimed that bee venom acupuncture significantly lowered the arthritis index in treated mice compared to control mice (Xian et al., 2016) and could mediate α2-adrenergic receptors for antinociceptive effect (Baek et al., 2006). In a recent study, phospholipase A2 of bee venom was also found to induce Foxp3 + regulatory Treg cells polarization and alleviated arthritic behaviors (Choi et al., 2021).

Previous studies on the therapeutic effect of bee venom using complete Freund's adjuvant arthritis animal model suggested that the antinociceptive effect was greater when administered in acupoint than nonacupoint (Kwon et al., 2001). The water-soluble fraction of bee venom could be the constituent responsible for this effect (Kwon et al., 2002), suggesting that the effect of bee venom may depend on the administration site and constituent. In another animal study, melittin of whole bee venom presented antiarthritis properties by reducing arthritis-induced nociceptive behaviors (Li et al., 2010). A study that examined the therapeutic properties of bee venom also reported that bee venom therapy in adjunct with methotrexate improved the arthritis indexes and analgesic effect by reducing both NF-κB and TNF-α expression, as compared to methotrexate alone in the complete Freund's adjuvant-induced arthritic rats (Darwish et al., 2013).

Several studies also reported that the antiinflammatory effect of bee venom was mediated by central muscarinic type 2 receptors through adrenal medullary activity and involved activation of the coeruleospinal pathway and sympathetic preganglionic neurons (Kwon et al., 2003; Yoon et al., 2005, 2007). It was also reported that endogenous glucocorticoids

(Saad Rached et al., 2010) and hydrophile peptide fragment of melittin contributed to the antiinflammatory effect of bee venom (Yu et al., 2005).

In vitro studies for rheumatoid arthritis

On the other hand, there are also ex vivo models (e.g., tissue explants) and in vitro models (e.g., two-dimensional culture or three-dimensional coculture) (Damerau and Gaber, 2020). Numerous in vitro studies have attempted to examine the antiarthritis and antiinflammatory effects of bee venom. In a rheumatoid synovial fibroblasts study, it was reported that bee venom-induced apoptosis and inhibited cell proliferation by increasing caspase-3 expression (Hong et al., 2005). Another study demonstrated that bee venom also stimulated aromatase activation in FLG19.1 and hOB cells in vitro, which could be beneficial for bone metabolism and favorable for rheumatoid arthritis conditions (Kim et al., 2005a). Both bee venom and melittin were found to suppress IκB kinase and NF-κB activities and the generation of inflammatory mediators in raw 264.7 cells (Park et al., 2007). They also inactivated the JNK pathway, contributing to the antiarthritis and antiinflammatory effect (Park et al., 2008). Another study reported that melittin induced apoptosis in apoptosis-resistant synoviocytes by inhibiting the activation of transcription factors and expression of antiapoptotic genes (Kim et al., 2011b). Melittin also produced antiarthritic effects by reducing L-1β secretion and promoting apoptosis, as well as inhibited MMP3 production and NF-κB activities (Nah et al., 2008). Table 14.4 summarized the major publications of in vivo and in vitro studies of rheumatoid arthritis.

Clinical studies for rheumatoid arthritis

Due to the complex nature of rheumatoid arthritis as a systemic autoimmune disease, the treatment of rheumatoid arthritis entails a variety of therapeutics (Lee et al., 2014b, c). RA affects about 350 million people worldwide (Qamar and Mukherjee, 2021), 0.5%–1% of the population in the industrialized countries (Ebel and O'Dell, 2021), and the incidence was reported to increase with age, with the highest incidence in the '60s and '70s (Eriksson et al., 2013; Helmick et al., 2008). Rheumatoid arthritis was also reported to affect 1% or 2% of the worldwide population aged 20–50 years (Akram et al., 2021).

The diagnosis of rheumatoid arthritis is comprehensively based on clinical symptoms, physical assessments, laboratory data, and imaging (Ebel and O'Dell, 2021). As the evidence demonstrates early intervention as the key factor for the treatment of rheumatoid arthritis, the classification criteria

Table 14.4 In vivo and in vitro studies for rheumatoid arthritis.

Model or samples	BV and/or its compounds	Administration or dose	References
In vivo			
Type II CIA mice	Whole BV	BV acupuncture (bilateral ST36 acupoint)	Lee et al. (2004)
Type II CIA mice	Phospholipase A2	BV injection (intraperitoneally injection)	Choi et al. (2021)
Type II CIA rats	Whole BV	BV acupuncture (bilateral BL23 acupoint)	Suh et al. (2006)
Type II CIA rats	Whole BV	BV acupuncture (bilateral ST36 acupoint)	Baek et al. (2006)
Type II CIA rats	Whole BV	BV injection (intraperitoneally injection)	Kim et al. (2008)
Type II CIA rats	Whole BV	BV acupuncture (bilateral ST36 acupoint)	Xian et al. (2016)
CFA-induced arthritis rats	Whole BV	BV acupuncture (bilateral ST36 acupoint)	Kwon et al. (2001)
CFA-induced arthritis rats	Water soluble fraction of BV	BV acupuncture (bilateral ST36 acupoint)	Kwon et al. (2002)
CFA-induced arthritis rats	Whole BV and melittin	BV acupuncture (bilateral ST36 acupoint)	Li et al. (2010)
CFA-induced arthritis rats	Whole BV	BV injection (subcutaneous injection)	Darwish et al. (2013)
Air pouch model in mice	Whole BV	BV injection (subcutaneous injection)	Kwon et al. (2003)
Air pouch model in mice	Whole BV	BV injection (subcutaneous injection)	Yoon et al. (2005)
Air pouch model in mice	Whole BV	BV injection (subcutaneous injection)	Yoon et al. (2007)

Continued

Table 14.4 In vivo and in vitro studies for rheumatoid arthritis—cont'd

Model or samples	BV and/or its compounds	Administration or dose	References
OIA rabbits	Melittin	BV injection (intramuscular injection)	Yu et al. (2005)
AIA rabbits	Whole BV	BV injection (subcutaneous injection)	Saad Rached et al. (2010)
In vitro			
FLS of RA patients	Melittin	0.001–10 μg/mL	Nah et al. (2008)
FLS of RA patients	Melittin	1 μg/mL and 2 μg/mL	Kim et al. (2011b)
FLS of RA patients	Melittin	–	He et al. (2020)
SF of RA patients	Whole BV	10 μg/mL	Hong et al. (2005)
Human leukemic cell line FLG 29.1 and the primary first-passage HOb	Whole BV	5.0 μL/mL	Kim et al. (2005a)
Raw 264.7 cells, THP-1 cells, and synoviocytes of RA patients	Whole BV and melittin	5 or 10 μg/mL, and 0.5, 1 and 5 μg/mL, respectively	Park et al. (2007)
Raw 264.7 cells, THP-1 cells, and synoviocytes of RA patients	Whole BV and melittin	0.5–5 μg/mL, and 5 or 10 μg/mL, respectively	Park et al. (2008)

AIA, antigen-induced arthritis; *BV*, bee venom; *CFA*, complete Freund's adjuvant; *CIA*, collagen-induced arthritis; *FLS*, fibroblast-like synoviocytes; *Hob*, human osteoblasts; *OIA*, ovalbumin-induced arthritis; *RA*, rheumatoid arthritis; *SF*, synovial fibroblasts.

(Table 14.5) by the American College of Rheumatology and European League Against Rheumatism are regularly used for diagnosis, and the classification criteria are intended for promoting prompt treatments (Ebel and O'Dell, 2021). From the classification criteria, a total score of 6 or more is considered rheumatoid arthritis (Cush, 2021). In addition, the patient with (1) at least one joint with definite clinical synovitis (swelling) and (2) synovitis not better explained by another disease should only be considered.

Table 14.5 The 2010 American College of Rheumatology/European League Against Rheumatism classification criteria for rheumatoid arthritis.

Criteria	Score
Joint distribution	
1 large joint	0
2 – 10 large joints	1
1 – 3 small joints (large joints excluded)	2
4 – 10 small joints (large joints excluded)	3
> 10 joints (at least 1 small joint)	5
Serology	
Negative RF and negative ACPA	0
Low-positive RF or low-positive ACPA (≤3 times of ULN)	2
High-positive RF or high-positive ACPA (>3 times of ULN)	3
Acute-phase reactants	
Normal CRP and normal ESR	0
Abnormal CRP or abnormal ESR	1
Symptom duration	
< 6 weeks	0
≥6 weeks	1

ACPA, anticyclic citrullinated protein antibody; *CRP*, C-reactive protein; *ESR*, erythrocyte sedimentation rate; *RF*, rheumatoid factor; *ULN*, upper limit of normal.
Data from Cush, J.J., 2021. Rheumatoid arthritis: early diagnosis and treatment. Med. Clin. N. Am. 105, 355–365; with permission.

The treatment for rheumatoid arthritis involves a comprehensive approach to relieve pain, decrease inflammation, restore joint activities (Akram et al., 2021). According to Akram et al. (2021), there still is no cure for rheumatoid arthritis. Still, treatments to manage rheumatoid arthritis have enhanced substantially, with approximately 60%–90% of patients with arthritis using complementary and alternative medicines (Akram et al., 2021). The high usage of an integrative and alternative approach to managing arthritis-related pain could imply its efficacy on arthritis. Especially when the treatment begins in the early stage, joint injury and disability could be prevented or remain in remission (Park, 2009).

In a recent Chinese randomized controlled trial, 120 rheumatoid arthritis patients were included in the trial, with 60 patients in the control group and 60 patients in the bee venom group (Chen et al., 2018). The treatment group received 5–15 bee stings of appropriate acupoints specific to their conditions for about 5 min every other day. In contrast, the control group received oral administration of methotrexate (10 mg, once a week) and celecoxib (0.2 g, once a day). The treatment lasted for 8 weeks. Chen et al. (2018) concluded that bee venom therapy is safe and effective for rheumatoid arthritis without any significant adverse effects. However, the trial data showed no significant difference between the two groups in the effective rate ($P>0.05$).

Another randomized controlled trial involved 100 rheumatoid arthritis patients, where 50 patients in the bee venom group received bee sting of acupoints and oral administration of methotrexate, sulfasalazine, meloxicam. In comparison, the 50 patients in the control group received the medication mentioned above only (Liu et al., 2008). The treatment lasted for 3 months, and the treatment of BV with conventional drugs significantly improved in terms of joint swelling, pain, and morning stiffness duration. The relapse rate was 12% in the bee venom group, while it was 32% in the control group ($P<0.05$). An integrative intervention involving bee venom therapy and medication was more effective for treating rheumatoid arthritis with dropped relapse rate. In addition, when supplemented by bee venom therapy, the doses of medication could be reduced.

In a Korean randomized controlled trial involving 80 rheumatoid arthritis patients for 3 months, the patient in the bee venom acupuncture group had significantly lower tender joint counts ($P=0.002$), swollen joint counts ($P=0.001$), and morning stiffness duration ($P=0.009$) (Lee et al., 2003).

Previously, the clinical evidence of bee venom acupuncture for rheumatoid arthritis has been systematically analyzed and reviewed (Lee et al., 2014b, c). However, the number of trials, their quality, and the total sample size were too low to draw firm conclusions that the safety and efficacy of bee venom acupuncture are still not confirmed.

Safety of bee venom therapy

Different forms of bee venom usage in medical treatments include administering live bee stings, injections of bee venom, and purified and diluted bee venom acupuncture into acupoints (Lee et al., 2014b, c). While the use of bee venom for treatment is common in some Asian countries, incidences of allergic reactions during treatments are a serious concern for both practitioner and patient (Ahn et al., 2016).

There have been numerous reports on adverse events of bee venom acupuncture, and safety has been a crucial part and an important concern in using bee venom acupuncture. In a study evaluating 130 cases of bee venom treatment, 37 cases (28.46%) experienced localized itching, and 41 cases (31.54%) experienced localized pain (Jung et al., 2013). Park et al. evaluated the adverse events associated with bee venom therapy from 145 studies. They found bee venom therapy associated with adverse events such as systematic reactions, large local reactions, local reactions, skin problems, and nonspecific reactions (Park et al., 2015). Large local reactions included swelling of ≥ 10 cm in diameter for more than a day. Local reactions included local pruritus, edema, or erythema. BV demonstrated a higher relative risk for the adverse events with a relative risk of 3.61 (95% CI, 2.10–6.20) than the normal saline injection in randomized controlled trials (Park et al., 2015). Some of the adverse events are minor, but some are serious and even lethal.

Other adverse events of bee venom acupuncture reported in previous studies severe ulnar nerve injury (Park et al., 2017), systemic immediate hypersensitive reactions (Jo and Roh, 2015), acute lung injury (Chae et al., 2015), Guillain–Barre syndrome (Lee et al., 2015), immune thrombocytopenia (Abdulsalam et al., 2016), and *Mycobacterium chelonae* infection (Cho et al., 2014). In addition, infectious skin disease and dermatoses associated with bee venom acupuncture were observed (Park et al., 2016). Some of the dermatologic problems which may be associated with bee venom acupuncture include abscess, pyoderma, lipoatrophy, and hypertrophic scar (Park et al., 2016), chronic folliculitis (Lee et al., 2013), and giant dermatofibroma (Rhee et al., 2009).

Serious adverse events of bee venom treatment include anaphylaxis and disseminated intravascular coagulation. Cherniack and Govorushko published a review on the safety of bee venom (Cherniack and Govorushko, 2018), and in the review, the authors discussed two fatal cases. An anaphylaxis case of a woman receiving bee venom acupuncture every month for 2 years was identified. This case was rare and surprising because the woman had no medical history of reactions or allergies to bee stings, yet she experienced an acute shock during the treatment and died. Another case was also a woman who received bee venom acupuncture treatment earlier on without an adverse event but experienced disseminated intravascular coagulation during the second treatment and died of hypovolemic shock.

Although bee venom acupuncture is being used in clinical practice, the safety and adverse events of bee venom acupuncture in clinical practice have not been evaluated. The systematic review evaluating the safety and adverse events of bee venom in clinical studies failed to identify the possible risk factors for adverse events of bee venom therapy, such as bee venom dose, frequency of treatment, administration site, or administration form (live bee, acupuncture, etc.). Therefore, firm conclusions on the safety of bee venom acupuncture cannot be drawn from the current evidence in the absence of further investigations of bee venom acupuncture in clinical practice (Cherniack and Govorushko, 2018).

Since no concrete evidence is available on risk factors of adverse events related to bee venom acupuncture, practitioners handling bee venom acupuncture should be alert to the incidence and occurrence of adverse events, follow clinical guidelines for monitoring adverse events, and receive relevant trainings (Park et al., 2015). Anaphylaxis can be caused by bee venom, and in some cases, the patients who successfully received the bee venom therapy earlier experienced the adverse event. The number of published reports of toxicity is small, while the incidence of adverse events is not fully known. Refining bee venom to remove harmful substances may potentially limit its toxicity (Cherniack and Govorushko, 2018).

Conclusion

Many experimental and clinical studies have shown that bee venom has promising pharmaceutical properties. However, the effectiveness of bee venom for rheumatoid arthritis or other health conditions remains premature and inconclusive. Bee venom therapy can yet be considered a proven treatment option as issues relating to standardization, safety assurance, mechanism of action, and metabolic pathway still need to be addressed. More effects should be directed to mechanism studies and clinical trials with adequate designs to comprehend better and define bee venom's role in health applications.

References

Abd El-Wahed, A.A., Khalifa, S.A.M., Sheikh, B.Y., Farag, M.A., Saeed, A., Larik, F.A., Koca-Caliskan, U., AlAjmi, M.F., Hassan, M., Wahabi, H.A., Hegazy, M.-E.F., Algethami, A.F., Büttner, S., El-Seedi, H.R., 2019. Bee venom composition: from chemistry to biological activity. In: Rahman, A. (Ed.), Studies in Natural Products Chemistry. vol. 60. Elsevier, pp. 459–484, https://doi.org/10.1016/B978-0-444-64181-6.00013-9 (Chapter 13).

Abdulsalam, M.A., Ebrahim, B.E., Abdulsalam, A.J., 2016. Immune thrombocytopenia after bee venom therapy: a case report. BMC Complement. Altern. Med. 16, 107. https://doi.org/10.1186/s12906-016-1091-3.

Ahn, Y.-J., Shin, J.-S., Lee, J., Lee, Y.J., Kim, M.-R., Shin, Y.-S., Park, K.B., Kim, E.J., Kim, M.-J., Lee, J.-W., Lee, H.D., Lee, Y., Kim, S., Chung, H.-J., Ha, I.-H., 2016. Safety of essential bee venom pharmacopuncture as assessed in a randomized controlled double-blind trial. J. Ethnopharmacol. 194, 774–780. https://doi.org/10.1016/j.jep.2016.11.012.

Akram, M., Daniyal, M., Sultana, S., Owais, A., Akhtar, N., Zahid, R., Said, F., Bouyahya, A., Ponomarev, E., Ali Shariat, M., Thiruvengadam, M., 2021. Traditional and modern management strategies for rheumatoid arthritis. Clin. Chim. Acta 512, 142–155. https://doi.org/10.1016/j.cca.2020.11.003.

Asquith, D.L., Miller, A.M., McInnes, I.B., Liew, F.Y., 2009. Animal models of rheumatoid arthritis. Eur. J. Immunol. 39 (8), 2040–2044. https://doi.org/10.1002/eji.200939578.

Baek, Y.H., Huh, J.E., Lee, J.D., Choi, D.Y., Park, D.S., 2006. Antinociceptive effect and the mechanism of bee venom acupuncture (apipuncture) on inflammatory pain in the rat model of collagen-induced arthritis: mediation by alpha2-adrenoceptors. Brain Res. 1073-1074, 305–310. https://doi.org/10.1016/j.brainres.2005.12.086.

Banks, B.E., Dempsey, C.E., Vernon, C.A., Warner, J.A., Yamey, J., 1990. Anti-inflammatory activity of bee venom peptide 401 (mast cell degranulating peptide) and compound 48/80 results from mast cell degranulation in vivo. Br. J. Pharmacol. 99 (2), 350–354. https://doi.org/10.1111/j.1476-5381.1990.tb14707.x.

Boutrin, M.C.F., Foster, H.A., Pentreath, V.W., 2008. The effects of bee (*Apis mellifera*) venom phospholipase A2 on Trypanosoma brucei brucei and enterobacteria. Exp. Parasitol. 119 (2), 246–251. https://doi.org/10.1016/j.exppara.2008.02.002.

Buku, A., Price, J.A., Mendlowitz, M., Masur, S., 2001. Mast cell degranulating peptide binds to RBL-2H3 mast cell receptors and inhibits IgE binding. Peptides 22 (12), 1993–1998. https://doi.org/10.1016/S0196-9781(01)00542-3.

Chae, W., Kim, S.H., Lee, Y.H., Lee, B.H., Lee, J.H., Woo, J.J., 2015. Acute lung injury after bee sting acupuncture. Allergy Asthma Respir. Dis. 3 (2), 4.

Chen, J., Lariviere, W.R., 2010. The nociceptive and anti-nociceptive effects of bee venom injection and therapy: a double-edged sword. Prog. Neurobiol. 92 (2), 151–183. https://doi.org/10.1016/j.pneurobio.2010.06.006.

Chen, S.Y., Zhou, P., Qin, Y., 2018. Treatment of rheumatoid arthritis by bee-venom acupuncture. Zhen Ci Yan Jiu 43 (4), 251–254. https://doi.org/10.13702/j.1000-0607.170506.

Cherniack, E.P., Govorushko, S., 2018. To bee or not to bee: the potential efficacy and safety of bee venom acupuncture in humans. Toxicon 154, 74–78. https://doi.org/10.1016/j.toxicon.2018.09.013.

Cho, S.Y., Peck, K.R., Kim, J., Ha, Y.E., Kang, C.-I., Chung, D.R., Lee, N.Y., Song, J.-H., 2013. Mycobacterium Chelonae infections associated with bee venom acupuncture. Clin. Infect. Dis. 58 (5), e110–e113. https://doi.org/10.1093/cid/cit753.

Cho, S.Y., Peck, K.R., Kim, J., Ha, Y.E., Kang, C.-I., Chung, D.R., Lee, N.Y., Song, J.-H., 2014. Mycobacterium Chelonae infections associated with bee venom acupuncture. Clin. Infect. Dis. 58 (5), e110–e113. https://doi.org/10.1093/cid/cit753.

Choi, S., Chae, H.K., Heo, H., Hahm, D.-H., Kim, W., Kim, S.K., 2019. Analgesic effect of melittin on oxaliplatin-induced peripheral neuropathy in rats. Toxins (Basel) 11 (7), 396. https://www.mdpi.com/2072-6651/11/7/396.

Choi, G.M., Lee, B., Hong, R., Park, S.Y., Cho, D.E., Yeom, M., Park, H.J., Bae, H., Hahm, D.H., 2021. Bee venom phospholipase A2 alleviates collagen-induced polyarthritis by inducing Foxp3(+) regulatory T cell polarization in mice. Sci. Rep. 11 (1), 3511. https://doi.org/10.1038/s41598-021-82298-x.

Choudhary, N., Bhatt, L.K., Prabhavalkar, K.S., 2018. Experimental animal models for rheumatoid arthritis. Immunopharmacol. Immunotoxicol. 40 (3), 193–200. https://doi.org/10.1080/08923973.2018.1434793.

Conrad, V.J., Hazan, L.L., Latorre, A.J., Jakubowska, A., Kim, C.M.H., 2019. Efficacy and safety of honey bee venom (*Apis mellifera*) dermal injections to treat osteoarthritis knee pain and physical disability: a randomized controlled trial. J. Altern. Complement. Med. 25 (8), 845–855. https://doi.org/10.1089/acm.2019.0121.

Corthésy, B., Lassus, A., Terrettaz, J., Tranquart, F., Bioley, G., 2016. Efficacy of a therapeutic treatment using gas-filled microbubble-associated phospholipase A2 in a mouse model of honeybee venom allergy. Allergy 71 (7), 957–966. https://doi.org/10.1111/all.12859.

Csoka, A.B., Frost, G.I., Stern, R., 2001. The six hyaluronidase-like genes in the human and mouse genomes. Matrix Biol. 20 (8), 499–508. https://doi.org/10.1016/S0945-053X(01)00172-X.

Cush, J.J., 2021. Rheumatoid arthritis: early diagnosis and treatment. Med. Clin. N. Am. 105 (2), 355–365. https://doi.org/10.1016/j.mcna.2020.10.006.

Damerau, A., Gaber, T., 2020. Modeling rheumatoid arthritis in vitro: from experimental feasibility to physiological proximity. Int. J. Mol. Sci. 21 (21). https://doi.org/10.3390/ijms21217916.

Darwish, S.F., El-Bakly, W.M., Arafa, H.M., El-Demerdash, E., 2013. Targeting TNF-α and NF-κB activation by bee venom: role in suppressing adjuvant induced arthritis and methotrexate hepatotoxicity in rats. PLoS One 8 (11). https://doi.org/10.1371/journal.pone.0079284, e79284.

Duchez, A.C., Boudreau, L.H., Naika, G.S., Rousseau, M., Cloutier, N., Levesque, T., Gelb, M.H., Boilard, E., 2019. Respective contribution of cytosolic phospholipase A2α and secreted phospholipase A(2) IIA to inflammation and eicosanoid production in arthritis. Prostaglandins Other Lipid Mediat. 143. https://doi.org/10.1016/j.prostaglandins.2019.106340, 106340.

Ebel, A.V., O'Dell, J.R., 2021. Clinical features, diagnosis, and treatment of rheumatoid arthritis. Phys. Assist. Clin. 6 (1), 41–60. https://doi.org/10.1016/j.cpha.2020.08.004.

el Amraoui, M., Frikh, R., Hjira, N., Mohammed, B., 2020. Bee venom: a case of effectiveness on skin varicosities veins with review of its dermatological benefits. Ann. Dermatol. Res. 4, 6–8. https://doi.org/10.29328/journal.adr.1001010.

El-Seedi, H., Abd El-Wahed, A., Yosri, N., Musharraf, S.G., Chen, L., Moustafa, M., Zou, X., Al-Mousawi, S., Guo, Z., Khatib, A., Khalifa, S., 2020. Antimicrobial properties of *Apis mellifera*'s bee venom. Toxins 12 (7), 451. https://doi.org/10.3390/toxins12070451.

Eltaher, S., Mohammed, G.F., Younes, S., Elakhras, A., 2015. Efficacy of the apitherapy in the treatment of recalcitrant localized plaque psoriasis and evaluation of tumor necrosis factor-alpha (TNF-α) serum level: a double-blind randomized clinical trial. J. Dermatol Treat. 26 (4), 335–339. https://doi.org/10.3109/09546634.2014.990411.

Eriksson, J.K., Neovius, M., Ernestam, S., Lindblad, S., Simard, J.F., Askling, J., 2013. Incidence of rheumatoid arthritis in Sweden: a nationwide population-based assessment of incidence, its determinants, and treatment penetration. Arthritis Care Res. 65 (6), 870–878. https://doi.org/10.1002/acr.21900.

Fenard, D., Lambeau, G., Maurin, T., Lefebvre, J.-C., Doglio, A., 2001. A peptide derived from bee venom-secreted phospholipase A_2 inhibits replication of T-cell tropic HIV-1 strains via interaction with the CXCR4 chemokine receptor. Mol. Pharmacol. 60 (2), 341. https://doi.org/10.1124/mol.60.2.341.

Fratellone, P.M., Tsimis, F., Fratellone, G., 2016. Apitherapy products for medicinal use. J. Altern. Complement. Med. 22 (12), 1020–1022. https://doi.org/10.1089/acm.2015.0346.

Ham, H.J., Han, J.H., Lee, Y.S., Kim, K.C., Yun, J., Kang, S.K., Park, Y., Kim, S.H., Hong, J.T., 2019. Bee venom soluble phospholipase A2 exerts neuroprotective effects in a lipopolysaccharide-induced mouse model of Alzheimer's disease via inhibition of nuclear factor-kappa B [original research]. Front. Aging Neurosci. 11 (287). https://doi.org/10.3389/fnagi.2019.00287.

Han, S.M., Hong, I.P., Woo, S.O., Chun, S.N., Park, K.K., Nicholls, Y.M., Pak, S.C., 2015. The beneficial effects of honeybee-venom serum on facial wrinkles in humans. Clin. Interv. Aging 10, 1587–1592. https://doi.org/10.2147/cia.S84940.

Han, S.M., Pak, S.C., Nicholls, Y.M., Macfarlane, N., 2016. Evaluation of anti-acne property of purified bee venom serum in humans. J. Cosmet. Dermatol. 15 (4), 324–329. https://doi.org/10.1111/jocd.12227.

Hartmann, A., Müllner, J., Meier, N., Hesekamp, H., van Meerbeeck, P., Habert, M.O., Kas, A., Tanguy, M.L., Mazmanian, M., Oya, H., Abuaf, N., Gaouar, H., Salhi, S., Charbonnier-Beaupel, F., Fievet, M.H., Galanaud, D., Arguillere, S., Roze, E., Degos, B., Grabli, D., Lacomblez, L., Hubsch, C., Vidailhet, M., Bonnet, A.M., Corvol, J.C., Schüpbach, M., 2016. Bee venom for the treatment of Parkinson disease – a randomized controlled clinical trial. PLoS One 11 (7). https://doi.org/10.1371/journal.pone.0158235, e0158235.

Hassan, A.K., El-kotby, D.A., Tawfik, M.M., Badr, R.E., Bahgat, I.M., 2019. Antidiabetic effect of the Egyptian honey bee (*Apis mellifera*) venom in alloxan-induced diabetic rats. J. Basic Appl. Zool. 80 (1), 58. https://doi.org/10.1186/s41936-019-0127-x.

He, S.D., Tan, N., Sun, C.X., Liao, K.H., Zhu, H.J., Luo, X.G., Zhang, J.Y., Li, D.Y., Huang, S.G., 2020. Treatment with Melittin induces apoptosis and autophagy of fibroblast like synoviocytes in patients with rheumatoid arthritis. Curr. Pharm. Biotechnol. 21 (8), 734–740. https://doi.org/10.2174/1389201021666191210110826.

Hellner, M., Winter, D., von Georgi, R., Münstedt, K., 2008. Apitherapy: usage and experience in German beekeepers. Evid. Based Complement. Alternat. Med. 5. https://doi.org/10.1093/ecam/nem052, 827582.

Helmick, C.G., Felson, D.T., Lawrence, R.C., Gabriel, S., Hirsch, R., Kwoh, C.K., Liang, M.H., Kremers, H.M., Mayes, M.D., Merkel, P.A., Pillemer, S.R., Reveille, J.D., Stone, J.H., National Arthritis Data, W, 2008. Estimates of the prevalence of arthritis and other rheumatic conditions in the United States: part I. Arthritis Rheum. 58 (1), 15–25. https://doi.org/10.1002/art.23177.

Ho, J.N., Lee, S.B., Lee, S.S., Yoon, S.H., Kang, G.Y., Hwang, S.G., Um, H.D., 2010. Phospholipase A2 activity of peroxiredoxin 6 promotes invasion and metastasis of lung cancer cells. Mol. Cancer Ther. 9 (4), 825–832. https://doi.org/10.1158/1535-7163.Mct-09-0904.

Hong, S.J., Rim, G.S., Yang, H.I., Yin, C.S., Koh, H.G., Jang, M.H., Kim, C.J., Choe, B.K., Chung, J.H., 2005. Bee venom induces apoptosis through caspase-3 activation in synovial fibroblasts of patients with rheumatoid arthritis. Toxicon 46 (1), 39–45. https://doi.org/10.1016/j.toxicon.2005.03.015.

Hossen, M.S., Gan, S.H., Khalil, M.I., 2017. Melittin, a potential natural toxin of crude bee venom: probable future arsenal in the treatment of diabetes mellitus. J. Chem. 2017, 4035626. https://doi.org/10.1155/2017/4035626.

Jeong, Y.-J., Shin, J.-M., Bae, Y.-S., Cho, H.-J., Park, K.-K., Choe, J.-Y., Han, S.-M., Moon, S.-K., Kim, W.-J., Choi, Y.H., Kim, C.-H., Chang, H.-W., Chang, Y.-C., 2015. Melittin has a chondroprotective effect by inhibiting MMP-1 and MMP-8 expressions via blocking NF-κB and AP-1 signaling pathway in chondrocytes. Int. Immunopharmacol. 25 (2), 400–405. https://doi.org/10.1016/j.intimp.2015.02.021.

Jo, N., Roh, J., 2015. Systemic immediate hypersensitive reactions after treatment with sweet bee venom: a case report. J. Pharmacopuncture 18 (4), 59–62. https://doi.org/10.3831/KPI.2015.18.039.

Jo, M., Park, M.H., Kollipara, P.S., An, B.J., Song, H.S., Han, S.B., Kim, J.H., Song, M.J., Hong, J.T., 2012. Anti-cancer effect of bee venom toxin and melittin in ovarian cancer cells through induction of death receptors and inhibition of JAK2/STAT3 pathway. Toxicol. Appl. Pharmacol. 258 (1), 72–81. https://doi.org/10.1016/j.taap.2011.10.009.

Journals, I., Wahab, S.D.A.E., Eita, L.H., 2015. The Effectiveness of Live Bee Sting Acupuncture on Depression.

Jung, D.J., Lee, H.G., Choi, Y.M., Song, B.Y., Yook, T.H., Kim, J.U., 2013. The clinical study on 130 cases with sweet bee venom treatment. J. Korean Acupunct. Moxib. Soc. 30 (5), 211–217. https://doi.org/10.13045/acupunct.2013060.

Kang, G.H., Lee, S., Choi, D.B., Shin, D., Kim, J., Yang, H., Bae, H., 2020. Bee venom phospholipase A2 ameliorates atherosclerosis by modulating regulatory T cells. Toxins (Basel) 12 (10). https://doi.org/10.3390/toxins12100609.

Karimi, A., Ahmadi, F., Parivar, K., Nabiuni, M., Haghighi, S., Imani, S., Afrouzi, H., 2012. Effect of honey bee venom on Lewis rats with experimental allergic encephalomyelitis, a model for multiple sclerosis. Iran. J. Pharm. Res. 11 (2), 671–678.

Kim, C.M.H., 2013. Apitherapy – bee venom therapy. In: Grassberger, M., Sherman, R.A., Gileva, O.S., Kim, C.M.H., Mumcuoglu, K.Y. (Eds.), Biotherapy – History, Principles and Practice: A Practical Guide to the Diagnosis and Treatment of Disease Using Living Organisms. Springer, Netherlands, pp. 77–112, https://doi.org/10.1007/978-94-007-6585-6_4.

Kim, M., Han, C.H., 2020. Pharmacopuncture for stroke survivors: a systematic review of randomized controlled trials in South Korea. Complement. Ther. Clin. Pract. 40. https://doi.org/10.1016/j.ctcp.2020.101179, 101179.

Kim, K.-S., Choi, U.S., Lee, S.-D., Kim, K.-H., Chung, K.-H., Chang, Y.-C., Park, K.-K., Lee, Y.-C., Kim, C.-H., 2005a. Effect of bee venom on aromatase expression and activity in leukaemic FLG 29.1 and primary osteoblastic cells. J. Ethnopharmacol. 99 (2), 245–252. https://doi.org/10.1016/j.jep.2005.02.025.

Kim, Y.-S., Jun, H., Chae, Y., Park, H.-J., Kim, B.H., Chang, I.-M., Kang, S.-K., Lee, H.-J., 2005b. The practice of Korean medicine: an overview of clinical trials in acupuncture. Evid. Based Complement. Alternat. Med. 2. https://doi.org/10.1093/ecam/neh102, 345063.

Kim, K.W., Shin, Y.S., Kim, K.S., Chang, Y.C., Park, K.K., Park, J.B., Choe, J.Y., Lee, K.G., Kang, M.S., Park, Y.G., Kim, C.H., 2008. Suppressive effects of bee venom on the immune responses in collagen-induced arthritis in rats. Phytomedicine 15 (12), 1099–1107. https://doi.org/10.1016/j.phymed.2008.02.015.

Kim, S.-J., Park, J.-H., Kim, K.-H., Lee, W.-R., Kim, K.-S., Park, K.-K., 2011a. Melittin inhibits atherosclerosis in LPS/high-fat treated mice through atheroprotective actions. J. Atheroscler. Thromb. 18 (12), 1117–1126. https://doi.org/10.5551/jat.8474.

Kim, S.K., Park, K.Y., Yoon, W.C., Park, S.H., Park, K.K., Yoo, D.H., Choe, J.Y., 2011b. Melittin enhances apoptosis through suppression of IL-6/sIL-6R complex-induced NF-κB and STAT3 activation and Bcl-2 expression for human fibroblast-like synoviocytes in rheumatoid arthritis. Joint Bone Spine 78 (5), 471–477. https://doi.org/10.1016/j.jbspin.2011.01.004.

Kim, S.-J., Park, J.-H., Kim, K.-H., Lee, W.-R., Pak, S.C., Han, S.-M., Park, K.-K., 2012. The protective effect of Apamin on LPS/fat-induced atherosclerotic mice. Evid. Based Complement. Alternat. Med. 2012. https://doi.org/10.1155/2012/305454, 305454.

Kim, J.Y., An, H.J., Kim, W.H., Park, Y.Y., Park, K.D., Park, K.K., 2017. Apamin suppresses biliary fibrosis and activation of hepatic stellate cells. Int. J. Mol. Med. 39 (5), 1188–1194. https://doi.org/10.3892/ijmm.2017.2922.

Kim, K.H., Lee, S.Y., Shin, J., Hwang, J.T., Jeon, H.N., Bae, H., 2019. Dose-dependent neuroprotective effect of standardized bee venom phospholipase A(2) against MPTP-induced Parkinson's disease in mice. Front. Aging Neurosci. 11, 80. https://doi.org/10.3389/fnagi.2019.00080.

Koburova, K., 1984. Antipyretic effect of a polypeptide from bee venom-adolapin. Eksp. Med. Morfol. 23, 143–148. https://ci.nii.ac.jp/naid/10007595898/en/.

Koh, P.S., Seo, B.K., Cho, N.S., Park, H.S., Park, D.S., Baek, Y.H., 2013. Clinical effectiveness of bee venom acupuncture and physiotherapy in the treatment of adhesive capsulitis: a randomized controlled trial. J. Shoulder Elb. Surg. 22 (8), 1053–1062. https://doi.org/10.1016/j.jse.2012.10.045.

Kondo, T., Ikenaka, K., Fujimoto, I., Aimoto, S., Kato, H., Ito, K.-I., Taguchi, T., Morita, T., Kasai, M., Mikoshiba, K., 1992. K+ channel involvement in induction of synaptic enhancement by mast cell degranulating (MCD) peptide. Neurosci. Res. 13 (3), 207–216. https://doi.org/10.1016/0168-0102(92)90060-P.

Kong, G.M., Tao, W.H., Diao, Y.L., Fang, P.H., Wang, J.J., Bo, P., Qian, F., 2016. Melittin induces human gastric cancer cell apoptosis via activation of mitochondrial pathway. World J. Gastroenterol. 22 (11), 3186–3195. https://doi.org/10.3748/wjg.v22.i11.3186.

Kwon, Y.-B., Lee, J.-D., Lee, H.-J., Han, H.-J., Mar, W.-C., Kang, S.-K., Beitz, A.J., Lee, J.-H., 2001. Bee venom injection into an acupuncture point reduces arthritis associated edema and nociceptive responses. Pain 90 (3), 271–280. https://doi.org/10.1016/S0304-3959(00)00412-7.

Kwon, Y.B., Lee, H.J., Han, H.J., Mar, W.C., Kang, S.K., Yoon, O.B., Beitz, A.J., Lee, J.H., 2002. The water-soluble fraction of bee venom produces antinociceptive and anti-inflammatory

effects on rheumatoid arthritis in rats. Life Sci. 71 (2), 191–204. https://doi.org/10.1016/s0024-3205(02)01617-x.
Kwon, Y.-B., Kim, H.-W., Ham, T.-W., Yoon, S.-Y., Roh, D.-H., Han, H.-J., Beitz, A.-J., Yang, I.-S., Lee, J.-H., 2003. The anti-inflammatory effect of bee venom stimulation in a mouse air pouch model is mediated by adrenal medullary activity. J. Neuroendocrinol. 15 (1), 93–96. https://doi.org/10.1046/j.1365-2826.2003.00951.x.
Landucci, E.C.T., Toyama, M., Marangoni, S., Oliveira, B., Cirino, G., Antunes, E., de Nucci, G., 2000. Effect of crotapotin and heparin on the rat paw oedema induced by different secretory phospholipases A2. Toxicon 38 (2), 199–208. https://doi.org/10.1016/S0041-0101(99)00143-9.
Leandro, L.F., Mendes, C.A., Casemiro, L.A., Vinholis, A.H.C., Cunha, W.R., Almeida, R.D., Martins, C.H.G., 2015. Antimicrobial activity of apitoxin, melittin and phospholipase A2 of honey bee (Apis mellifera) venom against oral pathogens. An. Acad. Bras. Cienc. 87, 147–155. http://www.scielo.br/scielo.php?script=sci_arttext&pid=S0001-37652015000100147&nrm=iso.
Lee, S.H., Lee, H.J., Baek, Y.H., Kim, S.Y., Park, J.K., Hong, S.J., Yang, H.I., Kim, K.S., Lee, J.D., Choi, D.Y., Lee, D.I., Lee, Y.H., 2003. Effects of bee venom on the pain, edema, and acute inflammatory reactant of rheumatoid arthritis patients. J. Acupunct. Res. 20 (2), 77–84. http://www.e-jar.org/journal/view.php?number=967.
Lee, J.D., Kim, S.Y., Kim, T.W., Lee, S.H., Yang, H.I., Lee, D.I., Lee, Y.H., 2004. Anti-inflammatory effect of bee venom on type II collagen-induced arthritis. Am. J. Chin. Med. 32 (3), 361–367. https://doi.org/10.1142/s0192415x04002016.
Lee, H., Lee, E.J., Kim, H., Lee, G., Um, E.J., Kim, Y., Lee, B.Y., Bae, H., 2011a. Bee venom-associated Th1/Th2 immunoglobulin class switching results in immune tolerance of NZB/W F1 murine lupus nephritis. Am. J. Nephrol. 34 (2), 163–172. https://doi.org/10.1159/000329731.
Lee, W.-R., Park, J.-H., Kim, K.-H., Park, Y.-Y., Han, S.-M., Park, K.-K., 2011b. Protective effects of melittin on transforming growth factor-β1 injury to hepatocytes via anti-apoptotic mechanism. Toxicol. Appl. Pharmacol. 256 (2), 209–215. https://doi.org/10.1016/j.taap.2011.08.012.
Lee, N.R., Lee, S.-Y., Lee, W.-S., 2013. Granulomatous inflammation with chronic folliculitis as a complication of bee sting acupuncture. Indian J. Dermatol. Venereol. Leprol. 79, 554.
Lee, W.R., Kim, K.H., An, H.J., Kim, J.Y., Lee, S.J., Han, S.M., Pak, S.C., Park, K.K., 2014a. Apamin inhibits hepatic fibrosis through suppression of transforming growth factor β1-induced hepatocyte epithelial-mesenchymal transition. Biochem. Biophys. Res. Commun. 450 (1), 195–201. https://doi.org/10.1016/j.bbrc.2014.05.089.
Lee, J.A., Son, M.J., Choi, J., Jun, J.H., Lee, M.S., 2014b. Bee venom acupuncture for rheumatoid arthritis: a systematic review protocol. BMJ Open 4 (4). https://doi.org/10.1136/bmjopen-2013-004602, e004602.
Lee, J.A., Son, M.J., Choi, J., Jun, J.H., Kim, J.I., Lee, M.S., 2014c. Bee venom acupuncture for rheumatoid arthritis: a systematic review of randomised clinical trials. BMJ Open 4 (11). https://doi.org/10.1136/bmjopen-2014-006140, e006140.
Lee, H.J., Park, I.S., Lee, J.-I., Kim, J.-S., 2015. Guillain-Barre syndrome following bee venom acupuncture. Intern. Med. 54 (8), 975–978. https://doi.org/10.2169/internalmedicine.54.2238.
Lee, M.J., Jang, M., Choi, J., Lee, G., Min, H.J., Chung, W.S., Kim, J.I., Jee, Y., Chae, Y., Kim, S.H., Lee, S.J., Cho, I.H., 2016a. Bee venom acupuncture alleviates experimental autoimmune encephalomyelitis by upregulating regulatory T cells and suppressing Th1 and Th17 responses. Mol. Neurobiol. 53 (3), 1419–1445. https://doi.org/10.1007/s12035-014-9012-2.
Lee, K.S., Kim, B.Y., Yoon, H.J., Choi, Y.S., Jin, B.R., 2016b. Secapin, a bee venom peptide, exhibits anti-fibrinolytic, anti-elastolytic, and anti-microbial activities. Dev. Comp. Immunol. 63, 27–35. https://doi.org/10.1016/j.dci.2016.05.011.

Lee, Y.M., Cho, S.-N., Son, E., Song, C.-H., Kim, D.-S., 2020. Apamin from bee venom suppresses inflammation in a murine model of gouty arthritis. J. Ethnopharmacol. 257. https://doi.org/10.1016/j.jep.2020.112860, 112860.

Li, J., Ke, T., He, C., Cao, W., Wei, M., Zhang, L., Zhang, J.X., Wang, W., Ma, J., Wang, Z.R., Shao, Z.J., 2010. The anti-arthritic effects of synthetic melittin on the complete Freund's adjuvant-induced rheumatoid arthritis model in rats. Am. J. Chin. Med. 38 (6), 1039–1049. https://doi.org/10.1142/s0192415x10008457.

Li, L., Zhang, S., Wei, L., Wang, Z., Ma, W., Liu, F., Shen, Y., Zhang, S., Zhang, X., Hang, Y., Qian, Y., 2020. Anti-fibrotic effect of melittin on TRIM47 expression in human embryonic lung fibroblast through regulating TRIM47 pathway. Life Sci. 256. https://doi.org/10.1016/j.lfs.2020.117893, 117893.

Lim, H.N., Baek, S.B., Jung, H.J., 2019. Bee venom and its peptide component melittin suppress growth and migration of melanoma cells via inhibition of PI3K/AKT/mTOR and MAPK pathways. Molecules 24 (5), 929. https://www.mdpi.com/1420-3049/24/5/929.

Lin, L., Zhu, B.-P., Cai, L., 2017. Therapeutic effect of melittin on a rat model of chronic prostatitis induced by complete Freund's adjuvant. Biomed. Pharmacother. 90, 921–927. https://doi.org/10.1016/j.biopha.2017.04.055.

Liu, X.D., Zhang, J.L., Zheng, H.G., Liu, F.Y., Chen, Y., 2008. Clinical randomized study of bee-sting therapy for rheumatoid arthritis. Zhen Ci Yan Jiu 33 (3), 197–200.

Marković-Housley, Z., Miglierini, G., Soldatova, L., Rizkallah, P.J., Müller, U., Schirmer, T., 2000. Crystal structure of hyaluronidase, a major allergen of bee venom. Structure 8 (10), 1025–1035. https://doi.org/10.1016/S0969-2126(00)00511-6.

Marques Pereira, A.F., Albano, M., Bérgamo Alves, F.C., Murbach Teles Andrade, B.F., Furlanetto, A., Mores Rall, V.L., Delazari dos Santos, L., de Oliveira Orsi, R., Fernandes Júnior, A., 2020. Influence of apitoxin and melittin from *Apis mellifera* bee on *Staphylococcus aureus* strains. Microb. Pathog. 141. https://doi.org/10.1016/j.micpath.2020.104011, 104011.

Memariani, H., Memariani, M., 2020. Anti-fungal properties and mechanisms of melittin. Appl. Microbiol. Biotechnol. 104 (15), 6513–6526. https://doi.org/10.1007/s00253-020-10701-0.

Memariani, H., Memariani, M., Moravvej, H., Shahidi-Dadras, M., 2020. Melittin: a venom-derived peptide with promising anti-viral properties. Eur. J. Clin. Microbiol. Infect. Dis. 39 (1), 5–17. https://doi.org/10.1007/s10096-019-03674-0.

Mohamed, E., Ewida, M., 2016. Efficacy of bee venom phonphoresis in treatment of chronic pelvic inflammatory diseases. Int. J. PharmTech Res. 9, 66–71.

Mohammadi-Rad, M., Ghasemi, N., Aliomrani, M., 2019. Evaluation of apamin effects on myelination process in C57BL/6 mice model of multiple sclerosis. Res. Pharm. Sci. 14 (5), 424–431. https://doi.org/10.4103/1735-5362.268203.

Nah, S.S., Ha, E., Mun, S.H., Won, H.J., Chung, J.H., 2008. Effects of melittin on the production of matrix metalloproteinase-1 and -3 in rheumatoid arthritic fibroblast-like synoviocytes. J. Pharmacol. Sci. 106 (1), 162–166. https://doi.org/10.1254/jphs.sc0070215.

Nitecka-Buchta, A., Buchta, P., Tabeńska-Bosakowska, E., Walczyńska-Dragoń, K., Baron, S., 2014. Myorelaxant effect of bee venom topical skin application in patients with RDC/TMD Ia and RDC/TMD Ib: a randomized, double blinded study. Biomed. Res. Int. 2014. https://doi.org/10.1155/2014/296053, 296053.

Park, Y.-B., 2009. Recent trends and guidelines for the management of rheumatoid arthritis. Korean J. Med. 76 (1), 18–24. http://www.ekjm.org/journal/view.php?number=19679.

Park, H.J., Son, D.J., Lee, C.W., Choi, M.S., Lee, U.S., Song, H.S., Lee, J.M., Hong, J.T., 2007. Melittin inhibits inflammatory target gene expression and mediator generation via interaction with IκB kinase. Biochem. Pharmacol. 73 (2), 237–247. https://doi.org/10.1016/j.bcp.2006.09.023.

Park, H.J., Lee, H.J., Choi, M.S., Son, D.J., Song, H.S., Song, M.J., Lee, J.M., Han, S.B., Kim, Y., Hong, J.T., 2008. JNK pathway is involved in the inhibition of inflammatory target gene expression and NF-kappaB activation by melittin. J. Inflamm. 5 (1), 7. https://doi.org/10.1186/1476-9255-5-7.

Park, J.H., Jeong, Y.-J., Park, K.-K., Cho, H.-J., Chung, I.-K., Min, K.-S., Kim, M., Lee, K.-G., Yeo, J.-H., Park, K.-K., Chang, Y.-C., 2010. Melittin suppresses PMA-induced tumor cell invasion by inhibiting NF-κB and AP-1-dependent MMP-9 expression. Mol. Cell 29 (2), 209–215. https://doi.org/10.1007/s10059-010-0028-9.

Park, M.H., Choi, M.S., Kwak, D.H., Oh, K.W., Yoon, D.Y., Han, S.B., Song, H.S., Song, M.J., Hong, J.T., 2011. Anti-cancer effect of bee venom in prostate cancer cells through activation of caspase pathway via inactivation of NF-κB. Prostate 71 (8), 801–812. https://doi.org/10.1002/pros.21296.

Park, J.H., Yim, B.K., Lee, J.-H., Lee, S., Kim, T.-H., 2015. Risk associated with bee venom therapy: a systematic review and meta-analysis. PLoS One 10 (5). https://doi.org/10.1371/journal.pone.0126971, e0126971.

Park, S.-M., Kim, W.-J., Mun, J.-H., Kim, H.-S., Ko, H.-C., Kim, B.-S., Kim, M.-B., Song, M., 2016. Adverse events associated with acupuncture: a clinicopathologic review. Int. J. Dermatol. 55 (7), 757–763. https://doi.org/10.1111/ijd.12914.

Park, J.S., Park, Y.G., Jang, C.H., Cho, Y.N., Park, J.H., 2017. Severe ulnar nerve injury after bee venom acupuncture at a traditional Korean medicine clinic: a case report. Ann. Rehabil. Med. 41 (3), 483–487. https://doi.org/10.5535/arm.2017.41.3.483.

Qamar, T., Mukherjee, S., 2021. Genetic approaches for the diagnosis and treatment of rheumatoid arthritis through personalized medicine. Gene Rep. 23. https://doi.org/10.1016/j.genrep.2021.101173, 101173.

Rady, I., Siddiqui, I.A., Rady, M., Mukhtar, H., 2017. Melittin, a major peptide component of bee venom, and its conjugates in cancer therapy. Cancer Lett. 402, 16–31. https://doi.org/10.1016/j.canlet.2017.05.010.

Rhee, D.Y., Lee, H.W., Chung, W.K., Chang, S.E., Lee, M.W., Choi, J.H., Moon, K.C., Koh, J.K., 2009. Giant dermatofibroma with granular cell changes: side-effect of bee-venom acupuncture? Clin. Exp. Dermatol. 34 (5), e18–e20. https://doi.org/10.1111/j.1365-2230.2008.03149.x.

Rose, A., 1994. Bee in Balance: A Guide to Healing the Whole Person with Honeybees, Oriental Medicine and Commonsense. Starpoint Enterprises Ltd.

Saad Rached, I.C.F., Castro, F.M., Guzzo, M.L., de Mello, S.B.V., 2010. Anti-inflammatory effect of bee venom on antigen-induced arthritis in rabbits: influence of endogenous glucocorticoids. J. Ethnopharmacol. 130 (1), 175–178. https://doi.org/10.1016/j.jep.2010.04.015.

Seo, B.K., Han, K., Kwon, O., Jo, D.J., Lee, J.H., 2017. Efficacy of bee venom acupuncture for chronic low back pain: a randomized, double-blinded, sham-controlled trial. Toxins (Basel) 9 (11). https://doi.org/10.3390/toxins9110361.

Shin, J.-M., Jeong, Y.-J., Cho, H.-J., Park, K.-K., Chung, I.-K., Lee, I.-K., Kwak, J.-Y., Chang, H.-W., Kim, C.-H., Moon, S.-K., Kim, W.-J., Choi, Y.-H., Chang, Y.-C., 2013. Melittin suppresses HIF-1α/VEGF expression through inhibition of ERK and mTOR/p70S6K pathway in human cervical carcinoma cells. PLoS One 8 (7). https://doi.org/10.1371/journal.pone.0069380, e69380.

Shin, S.H., Kim, Y.H., Kim, J.K., Park, K.K., 2014. Anti-allergic effect of bee venom in an allergic rhinitis mouse model. Biol. Pharm. Bull. 37 (8), 1295–1300. https://doi.org/10.1248/bpb.b14-00102.

Shin, S.H., Ye, M.K., Choi, S.Y., Park, K.K., 2017. The effects of melittin and apamin on airborne fungi-induced chemical mediator and extracellular matrix production from nasal polyp fibroblasts. Toxins (Basel) 9 (11). https://doi.org/10.3390/toxins9110348.

Shkenderov, S., Koburova, K., 1982. Adolapin – a newly isolated analgetic and anti-inflammatory polypeptide from bee venom. Toxicon 20 (1), 317–321. https://doi.org/10.1016/0041-0101(82)90234-3.

Son, D.J., Lee, J.W., Lee, Y.H., Song, H.S., Lee, C.K., Hong, J.T., 2007. Therapeutic application of anti-arthritis, pain-releasing, and anti-cancer effects of bee venom and its constituent compounds. Pharmacol. Ther. 115 (2), 246–270. https://doi.org/10.1016/j.pharmthera.2007.04.004.

Suh, S.-J., Kim, K.-S., Kim, M.-J., Chang, Y.-C., Lee, S.-D., Kim, M.-S., Kwon, D.Y., Kim, C.-H., 2006. Effects of bee venom on protease activities and free radical damages in synovial fluid from type II collagen-induced rheumatoid arthritis rats. Toxicol. in Vitro 20 (8), 1465–1471. https://doi.org/10.1016/j.tiv.2006.06.016.

Sun, Y., Han, M., Shen, Z., Huang, H., Miao, X., 2018. Anti-hypertensive and cardioprotective effects of a novel apitherapy formulation via upregulation of peroxisome proliferator-activated receptor-α and -γ in spontaneous hypertensive rats. Saudi J. Biol. Sci. 25 (2), 213–219. https://doi.org/10.1016/j.sjbs.2017.10.010.

Uddin, M.B., Lee, B.H., Nikapitiya, C., Kim, J.H., Kim, T.H., Lee, H.C., Kim, C.G., Lee, J.S., Kim, C.J., 2016. Inhibitory effects of bee venom and its components against viruses in vitro and in vivo. J. Microbiol. 54 (12), 853–866. https://doi.org/10.1007/s12275-016-6376-1.

Wehbe, R., Frangieh, J., Rima, M., El Obeid, D., Sabatier, J.-M., Fajloun, Z., 2019. Bee venom: overview of main compounds and bioactivities for therapeutic interests. Molecules 24 (16), 2997. https://doi.org/10.3390/molecules24162997.

Xian, P.F., Chen, Y., Yang, L., Liu, G.T., Peng, P., Wang, S.X., 2016. Effect of bee venom injection on TrkA and TRPV1 expression in the dorsal root ganglion of rats with collagen-induced arthritis. Nan Fang Yi Ke Da Xue Xue Bao 36 (6), 838–841.

Yalcin, M., Aydin, C., Savci, V., 2009. Cardiovascular effect of peripheral injected melittin in normotensive conscious rats: mediation of the central cholinergic system. Prostaglandins Leukot. Essent. Fat. Acids 81 (5), 341–347. https://doi.org/10.1016/j.plefa.2009.10.001.

Yang, E.J., Jiang, J.H., Lee, S.M., Yang, S.C., Hwang, H.S., Lee, M.S., Choi, S.M., 2010. Bee venom attenuates neuroinflammatory events and extends survival in amyotrophic lateral sclerosis models. J. Neuroinflammation 7, 69. https://doi.org/10.1186/1742-2094-7-69.

Yasin, M., Elhosary, E., Hamada, H., Yousef, A., Shahin, M., 2018. Effect of bee venom phonophoresis in obese polycystic ovarian women: a single blind randomized controlled trial. J. Appl. Pharm. Sci. 8 (1), 159–164.

Yoon, S.-Y., Kim, H.-W., Roh, D.-H., Kwon, Y.-B., Jeong, T.-O., Han, H.-J., Lee, H.-J., Choi, S.-M., Ryu, Y.-H., Beitz, A.J., Lee, J.-H., 2005. The anti-inflammatory effect of peripheral bee venom stimulation is mediated by central muscarinic type 2 receptors and activation of sympathetic preganglionic neurons. Brain Res. 1049 (2), 210–216. https://doi.org/10.1016/j.brainres.2005.05.020.

Yoon, S.-Y., Kwon, Y.-B., Kim, H.-W., Roh, D.-H., Seo, H.-S., Han, H.-J., Lee, H.-J., Beitz, A.J., Hwang, S.-W., Lee, J.-H., 2007. Peripheral bee venom's anti-inflammatory effect involves activation of the coeruleospinal pathway and sympathetic preganglionic neurons. Neurosci. Res. 59 (1), 51–59. https://doi.org/10.1016/j.neures.2007.05.008.

You, C.E., Moon, S.H., Lee, K.H., Kim, K.H., Park, C.W., Seo, S.J., Cho, S.H., 2016. Effects of emollient containing bee venom on atopic dermatitis: a double-blinded, randomized, base-controlled, multicenter study of 136 patients. Ann. Dermatol. 28 (5), 593–599. https://doi.org/10.5021/ad.2016.28.5.593.

Yu, M., Wang, Y., Li, X., Fang, S., Xue, J., Song, J., 2005. Synthesis of peptide fragment of melittin and the function of rheumatoid arthritis cure. Sheng Wu Yi Xue Gong Cheng Xue Za Zhi 22 (5), 1031–1035.

Zhang, S., Liu, Y., Ye, Y., Wang, X.R., Lin, L.T., Xiao, L.Y., Zhou, P., Shi, G.X., Liu, C.Z., 2018. Bee venom therapy: potential mechanisms and therapeutic applications. Toxicon 148, 64–73. https://doi.org/10.1016/j.toxicon.2018.04.012.

CHAPTER 15

Bee venom: Antitumor activity and its therapeutic applications

Mamdouh Ibrahim Nassar
Faculty of Science, Cairo University, Cairo, Egypt

Contents

Introduction

Insects make up the largest and most biodiversity group of organisms on the planet. Likewise, the magnitude of the chemical diversity they produce and utilize is also one of the most impressive in the living world. Among the many species of insects, only very few can defend themselves with a sting and venom injection during stinging. All insects that can sting are members of the order Hymenoptera, including ants, wasps, and bees. The sting is always at or near the abdominal end rather than the head. Therefore, the pain inflicted by a honeybee defending its colony is not caused by a bite, as is frequently said, but by a sting.

Bee venom (BV) can be of benefit in treating a large number of ailments. Honeybee venom is produced by two glands associated with the sting apparatus of worker bees. When a bee stings, it does not typically inject all of the 0.15–0.3 mg of venom held in a full venom sac (Crane, 1999; Schumacher

Bee Products and Their Applications in the Food and Pharmaceutical Industries
https://doi.org/10.1016/B978-0-323-85400-9.00014-9

et al., 1990). This therapeutic value was already known to many ancient civilizations. Today, the only uses of bee venom are for human and veterinary medicine. The single most promising bee venom product involving bee venom deals with its use as antiarthritis and antitumor effects (Son et al., 2007). In addition, the antimicrobial properties of bee venom melittin and mastoparan have been the most studied and developed among the components of wasp and bee venoms.

Considering the venoms produced by arthropods, bee venom is the most studied type regarding its anticancer activities, due mainly to two substances that have been isolated and characterized: melittin and phospholipase A2 (PLA2). Although these proteins are the two major components in the venom of the species *Apis mellifera* and many studies have been published describing their antitumoral effects (Ownby et al., 1997), BV is a very complex mixture of components that may cause other physiological effects. Bee venom comprises a complex mixture of active peptides, enzymes, and amines (Dotimas et al., 1987; Habermann, 1972). Besides melittin and PLA2, other important components are histamines, catecholamines, and polyamines. These BV components are promising for the future therapeutic applications of human diseases treatment as potential new molecules and systems that are coming to light.

This chapter discusses the biologically active components of honeybee venom by focusing on its enzymes and peptides and their therapeutic effects considering antiinflammatory, antimicrobial, and antitumor activities. In addition, the antimicrobial activity of bee venom was examined, particularly for its effects on dermal microorganisms to ensure healthy skin. The neuroprotective potential of bee venom for Parkinson's disease and Alzheimer's was also evaluated in improving cognitive functions.

The physical and chemical attributes of bee venom

Honeybee venom is a clear, odorless, watery liquid, and it irritates when coming into contact with mucous membranes or eyes. It is a transparent liquid that dries up quickly even at room temperature with basic pH (4.5–5.5) used by bees proteins for defense (Somwongin et al., 2018). Dried venom takes on a light yellow and brown color due to the oxidation of some of the venom proteins.

Bee venom contains a number of very volatile compounds which are easily lost during collection; it is considered a rich source of enzymes, peptides, and biogenic amines. Bee venom is soluble in water before dry in air and insoluble in alcohol and ammonium sulfate. When it comes in contact with air, it forms grayish-white crystals.

A large number of studies have been carried out on the composition of honeybee venom. Piek (1986) covers the venom apparatus's morphology, venom collection, the pharmacological effects of bee venom, and allergies to the Hymenoptera venom of bees, wasps, and ants in a book. Bee venom is a liquid of biologically active substances comprising proteins, peptides, enzymes, and other small molecules (Table 15.1). About 88% of venom is water. The glucose, fructose, and phospholipid contents of venom are similar to those in a bee's blood (Crane, 1990). At least 18 pharmacologically active components have been described, including various enzymes, peptides, and amines (Gauldie et al., 1976).

The identification of the biologically active components of bee venom against *Sitophilus granarius* (Coleoptera-Curculionidae) has been analyzed by column chromatography, infrared (IR), ultraviolet (UV), and mass spectrophotometer (MS) (Nassar, 2013). The elementary analysis of the active components revealed the percentage of Carbon (C) (5.47%), Hydrogen (H) (8%), Nitrogen (N) (68%), and Oxygen (O) (17.96%). The spectrophotometric analysis of bee venom suggested that its molecular weight (MW) of Mass Spectrum is 2847.7 MW, and charts of UV, IR, and elementary analysis showed the empirical chemical formula of the active components of bee venom is $C_{31}H_{228}N_{38}O_{32}$ (Nassar, 2013). The IR maxima (wavenumber) with its transition is given in Table 15.2. Its formula may suggest the active component was contained in the form of an aromatic structure. This structure was attached to some functional groups such as NH, CO, CH, and CL (Nassar, 2013).

Table 15.1 Components of honeybee venom.

Enzymes	Phospholipase A_2, hyaluronidase, acid phosphomonoesterase, lysophospholipase, glucosidase
Proteins and peptides	Melittin, apamine, mast cell degranulating peptide, secapin, procamine, adolapin, protease inhibitor, tertiapin, small peptides
Physiologically active amines	Histamine, dopamine, noradrenaline
Amino acids	Aminobutyric acid, amino acids
Sugars	Glucose, fructose

Table 15.2 Infrared (IR) maxima (wavenumber) with its transition (Nassar, 2013).

Peak no.	Maxima (cm^{-1})	% Transition	Peak no.	Maxima (cm^{-1})	% Transition
1	3338.18	53.7303	7	1053.91	64.2488
2	2931.27	78.5415	8	920.843	94.4015
3	1641.13	88.5232	9	865.882	96.5969
4	1413.57	84.1679	10	819.598	95.1972
5	1353.78	86.0324	11	775.244	92.9522
6	1253.5	88.1592	12	628.68	88.9303

Biological active components of bee venom

Enzymes

Enzymes in honeybee venom include phospholipase A2 (PLA2), phospholipase B (PLB), hyaluronidase, which show cytotoxicity, and other nontoxic hydrolytic enzymes such as phosphatase and α-glucosidase.

Phospholipase A2

Phospholipase A2 (PLA2) plays an important role in diverse cellular processes due to metabolizing phospholipids and producing fatty acids and lysophospholipids. According to a review by Dennis et al. (2011), PLA2 has been classified into three broad classes based on the cellular distribution: secreted (sPLA2), cytosolic (cPLA2), and Ca2+-independent PLA2 (iPLA2). sPLA2 isolated initially from human synovial fluid of arthritic knee joints was involved in diverse inflammatory processes such as arthritis (Boilard et al., 2010). Thus, bee venom PLA2 conferred protective immunity by inducing a T helper cell type 2 response and additional IgE antibodies response (Palm et al., 2013). Furthermore, bee venom PLA2 was able to modulate regulatory T cells to exert an antiinflammatory effect (Kim et al., 2014, 2015), ameliorate neuropathic pain (Lee and Bae, 2016), and antiparasitic amoeboid on chitosan-nanoparticles (Saber et al., 2017).

Phospholipase B

Phospholipase B (PLB), first characterized as extremely thermostable (Doery and Pearson, 1964), is present from bacteria to mammals. PLB catalyzes deacylation of both sn-1 and sn-2 ester linkage of diacyl phospholipids, remodeling the fatty acids. With a relatively large molecular mass, PLB is known to have broader roles than PLA2 in vivo (Saito, 2014).

Hyaluronidase

Hyaluronidase enzyme induces the cleavage of hyaluronan, the major glycosaminoglycan of connective tissue, and facilitates the diffusion of venom components into the tissues and blood circulation, causing systemic envenomation (Wiezel et al., 2015).

Phosphatase

Phosphatase enzyme, the prominent hydrolase in the dephosphorylation process, removes a phosphate group from an organic compound by hydrolysis. Thus, acid phosphatase containing bee venom inhibited tumor angiogenesis and metastasis by blocking phosphorylation of a particular growth factor such as vascular endothelial growth factor (Huh et al., 2010). Bee venom acid phosphatase is known to be a potent releaser of histamine from sensitized human basophils (Grunwald et al., 2006).

α-Glucosidase

α-Glucosidase is an exoenzyme hydrolyzing terminal glycosidic bonds and releasing α-glucose from the nonreducing end of the substrate chain. These enzymes are widely distributed in microorganisms, plants, and animal tissues. α-Glucosidase possesses starch-hydrolyzing activity by releasing D-glucose from the nonreducing end side of the substrate (Chiba, 1997). For the entomopathogenic bacterium *Lysinibacillus sphaericus* to elicit high larvicidal activity, subunits of binary protoxin have to bind to α-glucosidases on the surface of the midgut microvilli of larvae to perform larvicidal activity against some major vectors of several human pathogens (Ferreira et al., 2014).

Peptides

In addition to the two major peptides, melittin and apamin, bee venom contains many other peptides such as mast cell degranulating (MCD) peptide, adolapin, tertiapin, secapin, melittin F, and cardiopep. Mast cell degranulating (MCD) peptide is a cationic peptide with 22 amino acid residues similar to apamin, being cross-linked by two disulfide bonds (Dotimas et al., 1987). This peptide is a potent antiinflammatory agent; however, at a low concentration, it is a strong mediator of mast cell degranulation and histamine release from mast cells, which are present in the blood supply and all tissues perfused by blood (Ziai et al., 1990). Some are cell-penetrating peptides and elicit cell lysis, while others are neurotoxins affecting the nervous system.

Melittin

Melittin is the principal toxic component in the bee venom and is a cationic, hemolytic peptide. It is a small linear peptide composed of 26 amino acid residues in which the amino-terminal region is predominantly hydrophobic. In contrast, the carboxy-terminal region is hydrophilic due to the presence of a stretch of positively charged amino acids. This amphiphilic property of melittin has resulted in melittin being used as a suitable model peptide for monitoring lipid-protein interactions in membranes. The sequence of melittin is Gly-Ile-Gly-Ala-Val-Leu-Lys-Val-Leu-Thr-Thr-Gly-Leu-Pro-Ala-Leu-Ile-Ser-Trp-Ile-Lys-Arg-Lys-Arg-Gln-Gln (Gevod and Birdi, 1984). A large-scale fractionation of crude bee venom has resulted in the isolation of melittin, a 19-amino-acid residue peptide which is a fragment of melittin consisting of residues of 8–26 (Gauldie et al., 1976). In the crystalline state, each polypeptide chain of melittin is composed of two α-helices aligned approximately 120 degree to each other, and its overall shape is that of a bent rod (Terwilliger and Eisenberg, 1982).

Apamin

Apamin is a peptide neurotoxin comprising 18 amino acid residues tightly cross-linked by the presence of two disulfide bonds that connect position 1 with 11 and position 3 with 15 (Pease and Wemmer, 1988). Apamin selectively blocks the small conductance of Ca2+-dependent K+ channels (SK channels) expressed in the central nervous system (CNS). Apamin is an octadecapeptide having a molecular weight of 2039 (Habermann and Reiz, 1965) with 18 peptides of which their sequence elucidated by von Haux, Sowerthal, and Habermann cited in Gauldie et al. (1976). Octadecapeptide, apamin, is a centrally acting neurotoxin that produces motor hyperexcitability by crossing the blood–brain barrier (Strong, 1990). This high penetration and specific distribution to the CNS make the design of an apamin-based drug delivery system feasible. Recently, researchers constructed an efficient CNS-targeting drug delivery system specific for spinal cord injury using apamin as a targeting ligand which showed high efficacy in tissue repair following spinal cord injury (Wu et al., 2014). Apamin is a remarkably stable molecule connected with two intermolecular disulfide bridges and an amidated C-terminal carboxyl group (Strong, 1990). Apamin possesses an α-helical core with arginine-13 located at the center of the helical region and adjacent arginine-14, and these two residues are essential for the maintenance of neurotoxic activity (Hider and Ragnarsson, 1980). Apamin has long been known as a selective inhibitor of small conductance calcium-activated potassium (SK) channels. It inhibits a few cloned SK channel subtypes exclusively with different affinities. The cloning of the SK channels opened the avenue for the molecular knowledge of the interaction of apamin with the pore region, which gives some insight into the development of subunit-specific inhibitors of apamin (Nolting et al., 2007).

Mast cell degranulating (MCD) peptide

Mast cell degranulating peptide (MCD peptide) contains 22 residues and occurs in the bee venom of the common (Gauldie et al., 1976). It has potent antiinflammatory activity. According to the high-resolution two-dimensional nuclear magnetic resonance (NMR), MCD peptide has a conformation with an N-terminal β-turn and a C-terminal α-helix which are connected by two tight turns (Kumar et al., 1988). Since bee venom contains only a small amount of histamine, MCD peptide contributes additional histamine by releasing it from mast cell granules by fusion of the granule membranes with the mast cell membrane and exocytosis of the granule contents without lysis of mast cells (Karalliedde, 1995).

Adolapin

Adolapin is a peptide that was first isolated from bee venom in the 1980s. It exerts a potent analgesic effect and antiinflammatory activity in rats, blocking prostaglandin isolated and named by Shkenderov and Koburova (1982). The effects of adolapin are thought to be due to its ability to inhibit the prostaglandin (PG) synthesis system via proinflammatory cyclooxygenase (COX) enzyme inhibition. The antiinflammatory activity of adolapin was most marked concerning carrageenin, prostaglandin, adjuvant rat hind paw edemas, and adjuvant polyarthritis. Likely, central mechanisms are also involved in the analgetic action of adolapin. Koburova et al. (1985) revealed that adolapin inhibited the activity of PLA2 and lipoxygenase from human platelets. In addition, adolapin displayed an antipyretic effect by inhibiting the mean temperature rise similar to other nonsteroid analgetics.

Tertiapin

Tertiapin prevents dose-dependently ACh-induced AV blocks in mammalian hearts by inhibiting IKACh. The solution structure of tertiapin, a 21-residue bee venom peptide, has been characterized by circular dichroism (CD), two-dimensional nuclear magnetic resonance (NMR) spectroscopy, and distance geometry (Xu and Nelson, 1993). It consists of type 4 reverse turn and an α-helix which are connected by a loop formed by an extended β sheet. Four cysteines within the polypeptide chain form two disulfide bonds (Xu and Nelson, 1993). It was found that tertiapin blocked the inward-rectifier potassium channels (Jin and Lu, 1998). In this respect, tertiapin prevented acetylcholine-induced complete heart block by inhibiting inward-rectifier potassium channels (Drici et al., 2000). These papers proved the role of tertiapin as a powerful ligand for purifying functional channels and screening pharmaceutical agents against these channels.

Secapin

Secapin peptide from the venom of the Asiatic honeybee (*Apis cerana*) exhibits antifibrinolytic, antielastolytic, and antimicrobial activities can be used as an agent for biomedical applications (Kim et al., 2017). Secapin may possess a similar function to melittin, which lyses cancer cells when presented with phospholipase A2 (PLA2), which activates and induces caspase and metalloproteinase that can damage tumor cells (Gajski and Garaj-Vrhovac, 2013). Secapin has a 24-residue peptide containing a large proportion of proline and one disulfide bridge (Gauldie et al., 1976).

Cardiopep

Cardiopep is an acronym for a cardioactive polypeptide isolated from bee venom with an estimated molecular weight of 1940. Cardiopep comprises about 0.7% of the bee venom when isolated on a chromatographic column (Vick et al., 1974). Cardiopep, isolated from bee venom, has beta adrenergic and antiarrhythmic effects (regulate heat rhythm). Cardiopep is nontoxic (LD_{50} = 15 mg/kg) when compared to whole bee venom (LD_{50} = 35 mg/kg). It is a potent beta-adrenergic-like stimulant not entirely blocked by propranolol, possessing definite antiarrhythmic properties.

General medicinal properties of bee venom

Honeybee's venom has long been used in traditional medicine for the treatment of chronic inflammatory diseases. A number of studies have shown that bee venom exhibits various physiological activities, such as antiarthritic, antiinflammatory, and anticancer actions (Son et al., 2007). Bee venom, as mentioned above, is a complex mixture of proteins and peptides, enzymes, amines, amino acids, sugars, phospholipids, and volatile compounds (Hanson et al., 1974).

In human liver diseases, bee venom was able to inhibit the growth of SMMC-7721 human hepatoma cells by inducing apoptotic cell death via the reduction of the Ki67, a protein after injected to the BALB/c nude mice, lead to delay the tumor cell growth and the induction of apoptosis by DNA fragmentation (Wang et al., 2008). Other evidence is the induction of apoptosis by bee venom on MCF7 breast cancer cells because of morphological changes, inhibition of proliferation, induction of the production of reactive oxygen species (ROS), and dysfunction of the mitochondrial membrane potential (Ip et al., 2008).

Bee venom melittin is by far the peptide with the greatest antitumor activity isolated from BV, acting in different ways upon the physiology of cancer cells. Melittin is a small and amphiphilic peptide containing 26 amino acid residues and is the principal toxin derived from the venom. The sequence of melittin is Gly-Ile-Gly-Ala-Val-Leu-Lys-Val-Leu-Thr-Thr-Gly-Leu-Pro-Ala-Leu-Ile-Ser-Trp-Ile-Lys-Arg-Lys-Arg-Gln-Gln (Gevod and Birdi, 1984). Melittin exhibits antimicrobial activities and proinflammatory effects (Sumikura et al., 2003), inducing perturbations in the cell membrane and damage to enzyme systems (Habermann, 1972). Several cancer cells, including leukemia, renal, lung, liver, prostate, bladder, and mammary cancer cells, can be targets of bee venom melittin (Son et al., 2007). Chueng (1982) has shown that melittin can bind calmodulin, which has a role in cellular proliferation. Most recent studies have shown that melittin kills tumor cells by apoptosis through several cancer cell death mechanisms, including the activation of caspase and matrix metalloproteinases (MMP) (Holle et al., 2003).

Currently, bee venom has been given to nanotechnology advances in the fight against parasitic diseases by using nanoparticles as antiparasitic amoeboids on chitosan-nanoparticles (Saber et al., 2017). Furthermore, bee venom is applied as an antitumor, either as a chemotherapy-delivery agent to induce apoptosis. Bee venom melittin is one of the most powerful inhibitors of calmodulin activity and, as such, is an inhibitor of cell growth, clonogenicity of human and murine leukemic cells (Hait et al., 1983, 1985; Lee and Hait, 1985).

Antitumor activity of bee venom

Cancer is the major public burden in all developed and developing countries accounting for nearly 10 million deaths in 2020 (Ferlay et al., 2020). In all types of cancer, genetic alterations give rise to changes in expression, activation, or localization of regulatory proteins in the cells, affecting the signaling pathways that alter their response to regulatory stimuli and allow unrestricted cell growth. Various therapies have been used for treating cancer, such as chemotherapy, radiotherapy, immunotherapy, and gene therapy. However, still, there is an urgent need to find a better natural, safe way to treat cancer with little effect on normal cells. Anticancer drug developments from natural biological resources are ventured throughout the world.

Bee products have long been used in traditional medicine. Their raw materials, crude extracts, and purified active compounds have exhibited interesting bioactivities, including antimicrobial, antiinflammatory, and antioxidant activities. In addition, they have been widely used in the treatment of many immune-related diseases and in recent times in treating tumors (Premratanachai and Chanchao, 2014). Bee venom has been found to have anticancer activities in different cancer cell lines involving the breast, liver, and prostate. Its mechanism action causes cleavage of genomic DNA and inhibition of cell migration, indicating induction of apoptosis (Nassar, 2019). Immunohistochemistry studies demonstrated that BV decreased the expression of Bcl-2 and P16. Bee venom melittin links to some tumor cells at a higher affinity than healthy cells, antimetastatic, antiinvasive, and anti-angiogenesis impact (Nassar, 2019). Bee venom has traditionally been used as a nonsteroidal antiinflammatory drug to relieve pain and treat chronic inflammatory diseases, such as rheumatoid arthritis and multiple sclerosis, as well as in the treatment of tumors (Son et al., 2007). Also, the pegylated (polyethylene glycol-coated) immunoliposome coupled to a humanized antihepatocarcinoma single-chain antibody fragment and loaded with a bee venom peptide fraction displayed high antitumor activity and resulted in a significant reduction in tumor size compared to nontargeted liposomes (Hu et al., 2006).

Melittin, a major peptide component of bee venom, is an attractive candidate for cancer therapy. This agent has shown a variety of anticancer effects in preclinical cell culture and animal model systems. Despite convincing efficacy data against a variety of cancers, it has a hemolytic activity. Several optimization approaches include the utilization of nanoparticle-based melittin delivery, which inhibits tumor cell growth and induces apoptosis, thereby indicating a potential application of this venom peptide as an alternative or complementary medicine for the treatment of human cancers. Melittin's action involved membrane pore formation (Laine et al., 1988), resulting in the disruption of the tumor cell membranes (Benachir and Lafleur, 1995). Cell death was found to be associated with phospholipase A2 (PLA2) activation (Shier, 1979). Saini et al. (1999) demonstrated that melittin caused cell lysis in human monocytic leukemia. The hypertrophy was reported to be due to a change in membrane permeability to ions (Tosteson and Tosteson, 1981).

Melittin has a cytotoxic effect against several cancer cell lines, suggesting that it plays a role in mediated membrane cytolysis (Sarin et al., 1993). Regarding procytotoxic melittin systems, a design was based on the mixture of melittin with the anionic detergent sodium dodecyl sulfate formulated into poly(D,L-lactide-*co*-glycolide acid) nanoparticles by an emulsion solvent diffusion method. The inhibitory in vitro effects of these 130 nm-diameter melittin-loaded nanoparticles on breast cancer MCF-7 cells were claimed to be promising (Yang et al., 2009).

Bee venom has a prime role in defense because it contains biochemically active compounds such as enzymes and peptides. Antimicrobial peptides in bee venom are considered a phospholipase activating protein; they inhibit superoxide production. Also, it can induce degranulation of neutrophils; therefore, it is used as a potent antiinflammatory effect and protective medicine for many diseases (Mousavizadeh and Ghasemi, 2021).

Liver fibrosis is the excessive accumulation of extracellular matrix proteins, including collagen, in most types of chronic liver diseases. Advanced liver fibrosis results in cirrhosis, liver failure, and portal hypertension and often requires liver transplantation. In industrialized countries, the leading causes of liver fibrosis include chronic HCV infection, alcohol abuse, and nonalcoholic steatohepatitis (NASH). The accumulation of proteins distorts the hepatic architecture by forming a fibrous scar, and the subsequent development of nodules of regenerating hepatocytes defines cirrhosis. Park et al. (2010) reported that transforming growth factor β1, TGF-β1, decreased cell viabilities and induced hepatocyte apoptosis. However, adding the 10 ng/mL of bee venom significantly increased the viability of the TGF-β1-treated hepatocyte. Kim et al. (2014) demonstrated that bee venom phospholipase PLA2 protects against hepatic dysfunction and induces antiinflammatory cytokine production in acetaminophen-injected mice. The researchers suggested that PLA2 may have therapeutic potential in preventing acetaminophen-induced hepatotoxicity (Kim et al., 2014).

In addition, melittin exerts antiapoptotic effects against transforming growth factor β1, TGF-β1-induced injury to hepatocytes via the mitochondrial pathway (Lee et al., 2011). As such, these found that an optimal dose of bee venom and melittin can serve to protect cells against TGF-β1-mediated injury. Melittin attenuates liver injury in thioacetamide-treated mice through modulating inflammation and fibrogenesis (Park et al., 2010). The study revealed that the melittin mechanism was suppressed of necrosis factor-κB (NF-κB) promoter sequence. Hepatic stellate cells (HSCs) are perisinusoidal cells that adopt a myofibroblast-like phenotype and represent the cornerstone of the fibrotic response in the liver (Park et al., 2014; Sarem et al., 2006).

Acute hepatic failure is characterized by hepatic encephalopathy, severe coagulopathy, jaundice, and hydroperitoneum (Sun et al., 2011). Upon stimulation with D-galactosamine and lipopolysaccharide (LPS), secretion of various proinflammatory cytokines and hepatic necrosis occurs, leading to decreased antioxidant enzymes (Gong et al., 2010). This effect is associated with significant increases in alanine aminotransferase (ALT) activity and tumor necrosis factor-α (TNF-α) level in liver serum, ultimately leading to extremely high lethality (Yamada et al., 2008). Park et al. (2012) found that melittin prevents D-galactosamine/LPS-induced liver failure by suppressing apoptosis and the inflammatory response in the mouse liver. Melittin decreased the high lethality rate, alleviated hepatic pathological injury, attenuated hepatic inflammatory responses, and inhibited hepatocyte apoptosis. This study provides evidence that melittin offers an alternative for the prevention of acute hepatic failure.

On the other hand, bee venom apamin acts antifibrosis or antiepithelial–mesenchymal transition (EMT) mechanism by examining the effect on transforming growth factor β1, TGF-β1-treated hepatocytes, or CCl4-injected animal model. This action demonstrated that administration of apamin significantly increased the expression of epithelial marker E-cadherin and decreased mesenchymal marker vimentin in the TGF-β1-treated hepatocytes. In particular, apamin suppressed the expression of Smad-independent and Smad-dependent signaling pathways in hepatocytes. These results demonstrated the potential of apamin for the prevention of antiepithelial–mesenchymal transition (EMT) progression induced by TGF-β1 in vitro (Lee et al., 2014).

Alzheimer's disease

Alzheimer's disease (AD) is a severe neurodegenerative disorder characterized by the accumulation of two hallmark proteins, amyloid-β (Aβ) peptides and neurofibrillary tangles, that play critical roles in neuroinflammation, including the production of proinflammatory cytokines and activation of microglial cells and complements. Thus, the Aβ pathway has become a significant target for developing disease-modifying drugs that stabilize or slow the AD pathological process. A clinical Aβ vaccination trial (AN1792) was initiated, and modest beneficial effects were observed in a cohort of Aβ1–42-immunized AD patients (Gilman et al., 2005). As an alternative strategy,

Ye et al. (2016) showed that bvPLA2 could be used as a treatment to block the progression of Alzheimer's disease in transgenic mice. A better understanding of Tregs may therefore facilitate the development of new alternative strategies for modulating the inflammatory responses involved in AD. Adoptive transfer of Tregs improved cognitive deficits and reduced cortical and hippocampal Aβ plaque deposition in a mouse AD model (3xTg-AD). In addition, depletion of Tregs resulted in marked aggravation of the spatial learning deficits and a decrease in glucose metabolism.

Bee venom phospholipase A2 (bvPLA2) treatment exerted neuroprotective effects in a mouse model of Alzheimer's disease, AD via microglial inactivation, and reduced $CD4^+$ T-cell infiltration. Baek et al. (2018) reported that demonstrate that bee venom, bvPLA2-mediated expansion of Tregs together with coimmunization of Aβ42 vaccine may have distinct effects on immunosuppressive responses, facilitating a balance in homeostasis in the context of neurodegenerative diseases, including Alzheimer's disease, AD. BvPLA2-mediated induction of Tregs improved cognitive function, reduced Aβ deposition and the production of inflammatory cytokines, and increased glucose uptake in Aβ-vaccinated 3xTg-AD mice. Also, bvPLA2-induced Tregs significantly reduced inflammatory infiltration in the brain and spinal cord following Aβ immunization and injection of pertussis toxin (PT) as a model for neuroinflammation. These findings suggest that Treg induction by bee venom, bvPLA2 treatment of Alzheimer's disease may improve the safety of Aβ-based vaccines against central nervous system (CNS) inflammation (Baek et al., 2018).

Parkinson's disease

Parkinson's disease (PD) is a degenerative movement disorder that leads to progressive disability in patients. Generally, PD caused loss of dopaminergic neurons as significant neuropathological markers. The cardinal motor symptoms of Parkinson's disease (PD), akinesia, rigidity, and rest tremor are due to degeneration of dopaminergic neurons in the substantia nigra pars compacta (SNpc) with subsequent depletion of dopamine in the striatum. Parkinson's disease (PD) is the second most common neurodegenerative disorder of aging and the most common movement disorder (Mhyre et al., 2012).

Abnormal microglial activation is also a pathological sign in different neurodegenerative diseases, including PD (Iakovakis et al., 2018). Many researchers investigated the effect of BV on the migration of leukocytes or microglial activation in animal and cellular models. Other tests evaluated the neuroprotective potential of BV acupuncture therapy (BVA) against rotenone-induced oxidative stress, neuroinflammation, and apoptosis in PD mice models (Khalil et al., 2015). Bee venom can prevent dopamine depletion after administering rotenone, a pesticide in PD mice models and BV experiments. The treatment with BV effectively repressed DNA damage. Also, it inhibited the expression of the apoptotic Bax, Bcl-2, and caspase-3 genes in the brain (Wehbe et al., 2019). Bee venom in this study proved its ability to prevent dopamine depletion after the administration of rotenone. In addition, the locomotor activity was reestablished after treating PD mice models with BV. The treatment effectively repressed DNA damage and inhibited the expression of the apoptotic Bax, Bcl-2, and caspase-3 genes in the brain of PD mice. It has also been shown that BV can protect dopaminergic neurons from degeneration (Wehbe et al., 2019).

Apamin accounts for about 2%–3% dry weight of bee venom and is a peptide neurotoxin containing amino acid residues tightly cross-linked by two disulfide bonds. Apamin is well known for its pharmacological functions, which irreversibly block Ca2+-activated K+ (SK) channels (Gu et al., 2020). Apamin has long been known as a specifically selective blocker of small conductance Ca2+-activated K+ (SK) channels (Mourre et al., 1997). These channels play a pivotal role in various pathophysiological responses, such as atherosclerosis, Parkinson's disease. Apamin regulates gene expression in various signal transduction pathways involved in cell development. After intraperitoneal injection in animals, apamin locates not only in the brain's gray matter but also in the liver and the adrenal cortex (Deschaux and Bizot, 1997). Unlike melittin, apamin is a potent and selective antagonist of the SK channel. This SK channel subtype mediates slow hyperpolarization after a train of action potentials, regulating neuronal excitability (Van der Staay et al., 1999). This binding specificity provides apamin electrical properties, which have been exploited in biomedical research. The use of apamin was patented against the disadvantages of the drugs used in treating Parkinson's disease as a single dose of injection in the range of 1–10 μg of apamin (Hartman et al. cited in Moreno and Giralt, 2015). This patent claims that apamin protects undamaged neurons and restores the function of silent neurons. Another patent proposes a pharmaceutical composition comprising apamin as an active ingredient to treat Parkinson's Disease and related Parkinsonian disorders (Thomas and Justin cited in Cornara et al., 2017).

Antimicrobial properties of bee venom

Various peptides derived from bee venom, including mastoparan, melittin, apamin, and secapin, melectin, have been reported to exhibit potent antimicrobial activity (Ko et al., 2020). Many researchers isolated and characterized remarkable antimicrobial peptides (AMPs) from the bee venom. These peptides display high antibacterial activity, including phospholipase A2, hyaluronidase, acid phosphomonoesterase, α-D-glucosidase, and lysophospholipase (Banks and Shipolini, 1986). Among them, melittin, a water-soluble cationic amphipathic 26 amino acid α-helical peptide, is a very nonspecific cytolytic peptide that attacks all microbial lipid membranes leading to significant toxicity (Pan et al., 2011). Antibacterial activity of bee venom was observed against both Gram-negative and Gram-positive bacteria (Perumal Samy et al., 2007). Nakatsuji et al. (2009) reported that bee venom controlled the growth of *Staphylococcus aureus* bacteria. Moreover, bee venom also exhibited antibacterial activities against skin bacteria such as *Propionibacterium acnes*, *Staphylococcus epidermidis*, and *Streptococcus pyogenes* (Han et al., 2010). Due to the observed antibacterial properties, bee venom is suggested as an alternative to antibiotic therapy (Saini et al., 1997). The antibacterial actions of phospholipase A2 and melittin were also assessed on oral pathogens like *Streptococcus salivarius*, *S. sobrinus*, *S. mutans*, *S. mitis*, *S. sanguinis*, *Lactobacillus casei*, and *Enterococcus faecalis*. Due to the observed good antibacterial activity, the scientists proposed that apitoxin and its component melittin could be a potential agent against oral pathogens and positively impact tooth decays (Leandro et al., 2015).

On the other hand, bee venom has been exhibited prominent antifungal activities against *Trichophyton mentagrophytes* and *Trichophyton rubrum* fungus, which is much stronger than that of fluconazole (Yu et al., 2012). Also, the antiviral properties of bee venom have been reported since it significantly inhibited as a noncytotoxic effect by the replication of enveloped viruses such as Influenza A virus (PR8), Vesicular Stomatitis Virus (VSV), Respiratory Syncytial Virus (RSV), and Herpes Simplex Virus (HSV) (Uddin et al., 2016). A recent epidemiological study in China showed that local beekeepers have a level of immunity against SARS-CoV-2 with and without previous exposure to the virus. Bee venom's antiinflammatory properties are associated with melittin and phospholipase A2 (PLA2), which show activity against enveloped and nonenveloped viruses (Amer et al., 2021; Kasozi et al., 2020).

Furthermore, in parallel to antimicrobial peptides for therapeutic use in humans, these bee venom peptides can be applied to fight economically significant plant pathogenic bacteria, which are currently one of the major factors limiting crop production worldwide (Han et al., 2010).

As the largest organ in the human body, the skin undergoes an aging process like all the other organs. Both intrinsic and extrinsic factors contribute to skin aging via collagen alteration. While chronic sun exposure, especially to ultraviolet (UV) rays, is the dominant extrinsic aging cause, the passage of time is also a significant intrinsic aging factor. In addition, *P. acnes* is a chronic dermatologic problem with multiple factors involved in its pathogenesis. Bee venom has been proposed to be a promising treatment for many pathogenic organisms that contribute to the development of healthcare-associated infections (HAI) due to its antibacterial activity against skin pathogens. Bee venom has been demonstrated to inhibit and kill gram-negative bacteria through an in vitro experiment in addition to the inhibition of the growth of *Escherichia coli*, *Acinetobacter baumannii*, and *Pseudomonas aeruginosa* (Nassar et al., 2018). These findings strongly suggest that whole bee venom could be an effective antimicrobial agent for skin bacteria. In addition, bee venom was topically applied to investigate the efficacy of cosmetics containing skin bacteria, acne vulgaris. Many ingredients have been added to these cosmeceuticals in addition to the bee venom. Bee-venom serum was developed as a cosmeceutical for reducing the wrinkle formation associated with the aging process and act as antiwrinkle effects of bee venom-containing cosmetics in humans (Han et al., 2015).

Conclusion

The therapeutic interests of crude bee venom and its main compounds, particularly melittin, have been discussed for their antiinflammatory, antimicrobial, and antitumor activities. Bee venom affects inflammation signaling pathways and inducing the inhibition of proinflammatory genes expression. Bee venom also possesses a neuroprotective potential in neurodegenerative diseases such as Parkinson's disease, Alzheimer's, and amyotrophic lateral sclerosis (ALS) by significantly blocking their progression and improving cognitive functioning. In terms of antitumor activity, both melittin and BV have a cytotoxic effect on cancer cells and a significant antimetastatic activity. Optimization approaches focus on the possible use of nanoparticle-based delivery of bee venom or melittin to avoid their nonspecific cytotoxic effect. The antiviral activity of BV is also promising since BV and melittin have notable toxic effects against a broad spectrum of enveloped viruses, including the challenging HIV and few nonenveloped viruses. Finally, the clinical application of BV therapy is still a long way ahead. Still, researchers believe that the ongoing work on this topic will eventually allow BV and its compounds to be considered definitive candidates in various therapies in upcoming years.

References

Amer, A.E., Essa, E.E., Rady, M.H., Al-Sayed, A.K., Mahmoud, D.M., 2021. Antibacterial activity and mechanism of action of the wasp Vespa Orientalis venom peptides. Egypt. J. Zool. 75 (75), 60–69.

Baek, H., Lee, C.J., Choi, D.B., Kim, N.S., Kim, Y.S., Ye, Y.J., Kim, Y.S., Kim, J.S., Shim, I., Bae, H., 2018. Bee venom phospholipase A2 ameliorates Alzheimer's disease pathology in Aβ vaccination treatment without inducing neuro-inflammation in a 3xTg-AD mouse model. Sci. Rep. 8 (1), 1–11.

Banks, B.E.C., Shipolini, R.A., 1986. Chemistry and pharmacology of honeybee venom. In: Piek, T. (Ed.), Venoms of the Hymenoptera. Academic Press, London, UK, pp. 329–416.

Benachir, T., Lafleur, M., 1995. Study of vesicle leakage induced by melittin. Biochim. Biophys. Acta Biomembr. 1235 (2), 452–460.

Boilard, E., Lai, Y., Larabee, K., Balestrieri, B., Ghomashchi, F., Fujioka, D., Gobezie, R., Coblyn, J.S., Weinblatt, M.E., Massarotti, E.M., Thornhill, T.S., 2010. A novel anti-inflammatory role for secretory phospholipase A2 in immune complex-mediated arthritis. EMBO Mol. Med. 2 (5), 172–187.

Chiba, S., 1997. Molecular mechanism in α-glucosidase and glucoamylase. Biosci. Biotechnol. Biochem. 61 (8), 1233–1239.

Chueng, W.Y., 1982. Calmodulin: an overview. Fed. Proc. 41, 2253–2257.

Cornara, L., Biagi, M., Xiao, J., Burlando, B., 2017. Therapeutic properties of bioactive compounds from different honeybee products. Front. Pharmacol. 8, 412.

Crane, E., 1990. Bees and Beekeeping: Science Practice and World Resources. Comstock Publishing, Ithaca, NY, USA, p. 593.

Crane, E., 1999. The World History of Beekeeping and Honey Hunting. Routledge, New York, NY.

Dennis, E.A., Cao, J., Hsu, Y.H., Magrioti, V., Kokotos, G., 2011. Phospholipase A2 enzymes: physical structure, biological function, disease implication, chemical inhibition, and therapeutic intervention. Chem. Rev. 111, 6130–6185.

Deschaux, O., Bizot, J.C., 1997. Effect of apamin, a selective blocker of Ca2+-activated k+-channel, on habituation and passive avoidance responses in rats. Neurosci. Lett. 227, 57–60. https://doi.org/10.1016/S0304-3940(97)00301-7.

Doery, H.M., Pearson, J.E., 1964. Phospholipase B in snake venoms and bee venom. Biochem. J. 92, 599–602.

Dotimas, E.M., Hamid, K.R., Hider, R.C., Ragnarsson, U., 1987. Isolation and structure analysis of bee venom mast cell degranulating peptide. Biochim. Biophys. Acta 911, 285–293.

Drici, M.D., Diochot, S., Terrenoire, C., Romey, G., Lazdunski, M., 2000. The bee venom peptide tertiapin underlines the role of I(KACh) in acetylcholine-induced atrioventricular blocks. Br. J. Pharmacol. 131, 569–577.

Ferlay, J., Ervik, M., Lam, F., Colombet, M., Mery, L., Piñeros, M., et al., 2020. Global Cancer Observatory: Cancer Today. International Agency for Research on Cancer, Lyon. Available from: https://gco.iarc.fr/today. (Accessed 29 August 2021).

Ferreira, L.M., Romão, T.P., Nascimento, N.A., Costa Mda, C., Rezende, A.M., de Melo Neto, O.P., Silva-Filha, M.H., 2014. Non conserved residues between Cqm1 and Aam1 mosquito α-glucosidases are critical for the capacity of Cqm1 to bind the binary toxin from *Lysinibacillus sphaericus*. Insect Biochem. Mol. Biol. 50, 34–42.

Gajski, G., Garaj-Vrhovac, V., 2013. Melittin: a lytic peptide with anticancer properties. Environ. Toxicol. Pharmacol. 2, 697–705.

Gauldie, J., Hanson, J.M., Rumjanek, F.D., Shipolini, R.A., Vernon, C.A., 1976. The structures of some peptides from bee venom. Eur. J. Biochem. 61, 369–376.

Gevod, V.S., Birdi, K.S., 1984. Melittin and the 8-26 fragment. Differences in ionophoric properties as measured by monolayer method. Biophys. J. 45, 1079–1083.

Gilman, S., Koller, M., Black, R.S., Jenkins, L., Griffith, S.G., Fox, N.C., Eisner, L., Kirby, L., Rovira, M.B., Forette, F., Orgogozo, J.M., 2005. Clinical effects of Aβ immunization (AN1792) in patients with AD in an interrupted trial. Neurology 64 (9), 1553–1562.

Gong, X., Luo, F.L., Zhang, L., Li, H.Z., Wu, M.J., Li, X.H., Wang, B., Hu, N., Wang, C.D., Yang, J.Q., 2010. Tetrandrine attenuates lipopolysaccharide-induced fulminant hepatic failure in D-galactosamine-sensitized mice. Int. Immunopharmacol. 10, 357–363.

Grunwald, T., Bockisch, B., Spillner, E., Ring, J., Bredehorst, R., Ollert, M.W., 2006. Molecular cloning and expression in insect cells of honeybee venom allergen acid phosphatase (Api m 3). J. Allergy Clin. Immunol. 117, 848–854.

Gu, H., Han, S.M., Park, K.K., 2020. Therapeutic effects of apamin as a bee venom component for non-neoplastic disease. Toxins 12 (3), 195.

Habermann, E., 1972. Bee and wasp venoms. Science 177, 314–322.

Habermann, E., Reiz, K.G., 1965. On the biochemistry of bee venom peptides, melittin and apamin. Biochem. Z. 343 (2), 192–203.

Hait, W.N., Cadman, E., Benz, C., Cole, J., Weiss, B., 1983. Inhibition of growth of L1210 cyclic leukemic cells by inhibitors of nucleotide phosphodiesterase and calmodulin. Cancer Res. 2, 5–9.

Hait, W.N., Grais, L., Benz, C., Cadman, E.C., 1985. Inhibition of growth of leukemic cells by inhibitors of calmodulin: phenothiazines and melittin. Cancer Chemother. Pharmacol. 14, 202–205.

Han, S., Lee, K., Yeo, J., Baek, H., Park, K., 2010. Antibacterial and anti-inflammatory effects of honeybee (*Apis mellifera*) venom against acne-inducing bacteria. J. Med. Plants Res. 4 (6), 459–464.

Han, S.M., Hong, I.P., Woo, S.O., Chun, S.N., Park, K.K., Nicholls, Y.M., Pak, S.C., 2015. The beneficial effects of honeybee-venom serum on facial wrinkles in humans. Clin. Interv. Aging 10, 1587–1592.

Hanson, J.M., Morley, J., Soria-Herrera, C., 1974. Anti-inflammatory property of 401 (MCD-peptide), a peptide from the venom of the bee *Apis mellifera* (L.). Br. J. Pharmacol. 50 (3), 383.

Hider, R.C., Ragnarsson, U., 1980. A proposal for the structure of apamin. FEBS Lett. 111, 189–193.

Holle, L., Song, W., Holle, E., Wei, Y., Wagner, T., Yu, X., 2003. A matrix metalloproteinase 2 cleavable melittin/avidin conjugate specifically targets tumor cells in vitro and in vivo. Int. J. Oncol. 22, 93–98.

Hu, H., Chen, D., Liu, Y., Deng, Y., Yang, S., Qiao, M., Zhao, J., Zhao, X., 2006. Target ability and therapy efficacy of immunoliposomes using a humanized antihepatoma disulfide-stabilized Fv fragment on tumor cells. J. Pharm. Sci. 95 (1), 192–199.

Huh, J.E., Baek, Y.H., Lee, M.H., Choi, D.Y., Park, D.S., Lee, J.D., 2010. Bee venom inhibits tumor angiogenesis and metastasis by inhibiting tyrosine phosphorylation of VEGFR-2 in LLC-tumor-bearing mice. Cancer Lett. 292 (1), 98–110.

Iakovakis, D., Hadjidimitriou, S., Charisis, V., Bostantzopoulou, S., Katsarou, Z., Hadjileontiadis, L.J., 2018. Touchscreen typing-pattern analysis for detecting fine motor skills decline in early-stage Parkinson's disease. Sci. Rep. 8 (1), 1–13.

Ip, S.W., Liao, S.S., Lin, S.Y., Lin, J.P., Yang, J.S., Lin, M.L., Chen, G.W., Lu, H.F., Lin, M.W., Han, S.M., Chung, J.G., 2008. The role of mitochondria in bee venom-induced apoptosis in human breast cancer MCF7 cells. In Vivo 22, 237–245.

Jin, W., Lu, Z., 1998. A novel high-affinity inhibitor for inward-rectifier K+ channels. Biochemistry 37 (38), 13291–13299.

Karalliedde, L., 1995. Animal toxins. Br. J. Anaesth. 74 (3), 319–327.

Kasozi, K.I., Niedbała, G., Alqarni, M., Zirintunda, G., Ssempijja, F., Musinguzi, S.P., Usman, I.M., Matama, K., Hetta, H.F., Mbiydzenyuy, N.E., Batiha, G.E.S., 2020. Bee venom—a potential complementary medicine candidate for SARS-CoV-2 infections. Front. Public Health 8, 755.

Khalil, W.K., Assaf, N., ElShebiney, S.A., Salem, N.A., 2015. Neuroprotective effects of bee venom acupuncture therapy against rotenone-induced oxidative stress and apoptosis. Neurochem. Int. 80, 79–86.

Kim, H., Keum, D.J., Kwak, J.W., Chung, H.S., Bae, H., 2014. Bee venom phospholipase A2 protects against acetaminophen-induced acute liver injury by modulating regulatory T cells and IL-10 in mice. PLoS One 9 (12), e114726.

Kim, H., Lee, H., Lee, G., Jang, H., Kim, S.S., Yoon, H., Kang, G.H., Hwang, D.S., Kim, S.K., Chung, H.S., Bae, H., 2015. Phospholipase A2 inhibits cisplatin-induced acute kidney injury by modulating regulatory T cells by the CD206 mannose receptor. Kidney Int. 88 (3), 550–559.

Kim, B.Y., Lee, K.S., Ok, M., Jin, B.R., 2017. Synthetic secapin bee venom peptide exerts an anti-microbial effect but not a cytotoxic or inflammatory response. J. Asia Pac. Entomol. 20 (1), 151–155.

Ko, S.J., Park, E., Asandei, A., Choi, J.Y., Lee, S.C., Seo, C.H., Luchian, T., Park, Y., 2020. Bee venom-derived antimicrobial peptide melectin has broad-spectrum potency, cell selectivity, and salt-resistant properties. Sci. Rep. 10 (1), 1–12.

Koburova, K.L., Michailova, S.G., Shkenderov, S.V., 1985. Further investigation on the anti-inflammatory properties of adolapin—bee venom polypeptide. Acta Physiol. Pharmacol. Bulg. 11 (2), 50–55.

Kumar, N.V., Wemmer, D.E., Kallenbach, N.R., 1988. Structure of P401 (mast cell degranulating peptide) in solution. Biophys. Chem. 31 (1–2), 113–119.

Laine, R.O., Morgan, B.P., Esser, A.F., 1988. Comparison between complement and melittin hemolysis: anti-melittin antibodies inhibit complement lysis. Biochemistry 27 (14), 5308–5314.

Leandro, L.F., Mendes, C.A., Casemiro, L.A., Vinholis, A.H., Cunha, W.R., Almeida, R.D., Martins, C.H., 2015. Antimicrobial activity of apitoxin, melittin and phospholipase A2 of honey bee (*Apis mellifera*) venom against oral pathogens. An. Acad. Bras. Cienc. 87, 147–155.

Lee, G., Bae, H., 2016. Bee venom phospholipase A2: yesterday's enemy becomes today's friend. Toxins 8 (2), 48.

Lee, G.L., Hait, W.N., 1985. Inhibition of growth of C6 astrocytoma cells by inhibitors of calmodulin. Life Sci. 36 (4), 347–354.

Lee, W.R., Park, J.H., Kim, K.H., Park, Y.Y., Han, S.M., Park, K.K., 2011. Protective effects of melittin on transforming growth factor-β1 injury to hepatocytes via anti-apoptotic mechanism. Toxicol. Appl. Pharmacol. 256 (2), 209–215.

Lee, W.R., Kim, K.H., An, H.J., Kim, J.Y., Lee, S.J., Han, S.M., Pak, S.C., Park, K.K., 2014. Apamin inhibits hepatic fibrosis through suppression of transforming growth factor β1-induced hepatocyte epithelial–mesenchymal transition. Biochem. Biophys. Res. Commun. 450 (1), 195–201.

Mhyre, T.R., Boyd, J.T., Hamill, R.W., Maguire-Zeiss, K.A., 2012. Parkinson's disease. Subcell. Biochem. 65, 389–455. https://doi.org/10.1007/978-94-007-5416-4_16.

Moreno, M., Giralt, E., 2015. Three valuable peptides from bee and wasp venoms for therapeutic and biotechnological use: melittin, apamin and mastoparan. Toxins 7 (4), 1126–1150.

Mourre, C., Fournier, C., Soumireu-Mourat, B., 1997. Apamin, a blocker of the, induces neurodegeneration of Purkinje cells exclusively. Brain Res. 778 (2), 405–408.

Mousavizadeh, L., Ghasemi, S., 2021. Genotype and phenotype of COVID-19: their roles in pathogenesis. J. Microbiol. Immunol. Infect. 54 (2), 159–163.

Nakatsuji, T., Kao, M.C., Fang, J.Y., Zouboulis, C.C., Zhang, L., Gallo, R.L., Huang, C.M., 2009. Antimicrobial property of lauric acid against Propionibacterium acnes: its therapeutic potential for inflammatory acne vulgaris. J. Investig. Dermatol. 129 (10), 2480–2488.

Nassar, M.I., 2013. The potential of natural venom of *Apis mellifera* for the control of grains weevil adults (Sitophilus granarius-coleopter-curculionidae). Int. J. Entomol. Res. 1 (1), 25–31.

Nassar, M.I., 2019. The cancer diseases and potentiality of bee venom toward therapeutic tools antitumor. CPQ Med. 7 (3), 01–17.

Nassar, M.I., Mawgoud, Y.M.A., Dawoud, M.E., Helal, S.F., Kamel, H.M.M., 2018. Effect of honey bee venom against intensive care units' environmental pathogens. Egypt. J. Exp. Biol. 14 (2), 237–244. https://doi.org/10.5455/egyjebb.20180809061247.

Nolting, A., Ferraro, T., D'hoedt, D., Stocker, M., 2007. An amino acid outside the pore region influences apamin sensitivity in small conductance Ca2+-activated K+ channels. J. Biol. Chem. 282 (6), 3478–3486. https://doi.org/10.1074/jbc.M607213200.

Ownby, C.L., Powell, J.R., Jiang, M.S., Fletcher, J.E., 1997. Melittin and phospholipase A2 from bee (*Apis mellifera*) venom cause necrosis of murine skeletal muscle in vivo. Toxicon 35 (1), 67–80.

Palm, N.W., Rosenstein, R.K., Yu, S., Schenten, D.D., Florsheim, E., Medzhitov, R., 2013. Bee venom phospholipase A2 induces a primary type 2 response that is dependent on the receptor ST2 and confers protective immunity. Immunity 39 (5), 976–985.

Pan, H., Soman, N.R., Schlesinger, P.H., Lanza, G.M., Wickline, S.A., 2011. Cytolytic peptide nanoparticles ('NanoBees') for cancer therapy. Wiley Interdiscip. Rev. Nanomed. Nanobiotechnol. 3 (3), 318–327.

Park, J.H., Jeong, Y.J., Park, K.K., Cho, H.J., Chung, I.K., Min, K.S., Kim, M., Lee, K.G., Yeo, J.H., Park, K.K., Chang, Y.C., 2010. Melittin suppresses PMA-induced tumor cell invasion by inhibiting NF-κB and AP-1-dependent MMP-9 expression. Mol. Cell 29 (2), 209–215.

Park, J.H., Kim, K.H., Lee, W.R., Han, S.M., Park, K.K., 2012. Protective effect of melittin on inflammation and apoptosis in acute liver failure. Apoptosis 17 (1), 61–69.

Park, J.H., Lee, W.R., Kim, H.S., Han, S.M., Chang, Y.C., Park, K.K., 2014. Protective effects of melittin on tumor necrosis factor-α induced hepatic damage through suppression of apoptotic pathway and nuclear factor-kappa B activation. Exp. Biol. Med. 239 (12), 1705–1714.

Pease, J.H., Wemmer, D.E., 1988. Solution structure of apamin determined by nuclear magnetic resonance and distance geometry. Biochemistry 27, 8491–8498. https://doi.org/10.1021/bi00422a029.

Perumal Samy, R., Gopalakrishnakone, P., Thwin, M.M., Chow, T.K.V., Bow, H., Yap, E.H., Thong, T.W.J., 2007. Antibacterial activity of snake, scorpion and bee venoms: a comparison with purified venom phospholipase A2 enzymes. J. Appl. Microbiol. 102 (3), 650–659.

Piek, T. (Ed.), 1986. Venoms of the Hymenoptera: Biochemical, Pharmacological and Behavioural Aspects. Academic Press, Inc., London, p. 547.

Premratanachai, P., Chanchao, C., 2014. Review of the anticancer activities of bee products. Asian Pac. J. Trop. Biomed. 4 (5), 337–344.

Saber, A.E.S., Abdelwahab, A.K., El Amir, A.M., Nassar, M.I., Zohdi, H.F., 2017. Bee venom loaded chitosan nanoparticles as treatment for amoebiasis in mice. J. Egypt. Soc. Parasitol. 47 (2), 443–458.
Saini, S.S., Peterson, J.W., Chopra, A.K., 1997. Melittin binds to secretory phospholipase A2 and inhibits its enzymatic activity. Biochem. Biophys. Res. Commun. 238 (2), 436–442.
Saini, S.S., Chopra, A.K., Peterson, J.W., 1999. Melittin activates endogenous phospholipase D during cytolysis of human monocytic leukemia cells. Toxicon 37, 1605–1619.
Saito, K., 2014. Reminiscence of phospholipase B in Penicillium notatum. Proc. Jpn. Acad. Ser. B Phys. Biol. Sci. 90 (9), 333–346.
Sarem, M., Znaidak, R., Macías, M., Rey, R., 2006. Las células estrelladas del hígado: su importancia en condiciones normales y patológicas. Gastroenterol. Hepatol. 29 (2), 93–101. https://doi.org/10.1157/13083906.
Sarin, A., Adams, D.H., Henkart, P.A., 1993. Protease inhibitors selectively block T cell receptor-triggered programmed cell death in a murine T cell hybridoma and activated peripheral T cells. J. Exp. Med. 178 (5), 1693–1700.
Schumacher, M.J., Schmidt, J.O., Egen, N.B., Lowry, J.E., 1990. Quantity, analysis, and lethality of European and Africanized honey bee venoms. Am. J. Trop. Med. Hyg. 43 (1), 79–86.
Shier, W.T., 1979. Activation of high levels of endogenous phospholipase A2 in cultured cells. Proc. Natl. Acad. Sci. 76 (1), 195–199.
Shkenderov, S., Koburova, K., 1982. Adolapin—a newly isolated analgetic and anti-inflammatory polypeptide from bee venom. Toxicon 20 (1), 317–321.
Somwongin, S., Chantawannakul, P., Chaiyana, W., 2018. Antioxidant activity and irritation property of venoms from Apis species. Toxicon 145, 32–39.
Son, D.J., Lee, J.W., Lee, Y.H., Song, H.S., Lee, C.K., Hong, J.T., 2007. Therapeutic application of anti-arthritis, pain-releasing, and anti-cancer effects of bee venom and its constituent compounds. Pharmacol. Ther. 115 (2), 246–270.
Strong, P.N., 1990. Potassium channel toxins. Pharmacol. Ther. 46 (1), 137–162.
Sumikura, H., Andersen, O.K., Drewes, A.M., Arendt-Nielsen, L., 2003. A comparison of hyperalgesia and neurogenic inflammation induced by melittin and capsaicin in humans. Neurosci. Lett. 337 (3), 147–150.
Sun, H., Chen, L., Zhou, W., Hu, L., Li, L., Tu, Q., Chang, Y., Liu, Q., Sun, X., Wu, M., Wang, H., 2011. The protective role of hydrogen-rich saline in experimental liver injury in mice. J. Hepatol. 54 (3), 471–480.
Terwilliger, T.C., Eisenberg, D., 1982. The structure of melittin. II. Interpretation of the structure. J. Biol. Chem. 257 (11), 6016–6022.
Tosteson, M.T., Tosteson, D.C., 1981. The sting: melittin forms channels in lipid bilayers. Biophys. J. 36 (1), 109–116.
Uddin, M.B., Lee, B.H., Nikapitiya, C., Kim, J.H., Kim, T.H., Lee, H.C., Kim, C.G., Lee, J.S., Kim, C.J., 2016. Inhibitory effects of bee venom and its components against viruses in vitro and in vivo. J. Microbiol. 54 (12), 853–866.
Van der Staay, F.J., Fanelli, R.J., Blokland, A., Schmidt, B.H., 1999. Behavioral effects of apamin, a selective inhibitor of the SKCa-channel, in mice and rats. Neurosci. Biobehav. Rev. 23 (8), 1087–1110.
Vick, J.A., Shipman, W.H., Brooks Jr., R., 1974. Beta adrenergic and anti-arrhythmic effects of cardiopep, a newly isolated substance from whole bee venom. Toxicon 12 (2), 139–142.
Wang, C., Chen, P., Jin, H., Yan, X., Gan, L., Li, Y., Zhou, S., Chang, J., Wang, Y., Yang, G., He, G., 2008. Nidus vespae protein inhibiting proliferation of HepG2 hepatoma cells through extracellular signal-regulated kinase signaling pathways and inducing G1 cell cycle arrest. Acta Biochim. Biophys. Sin. 40 (11), 970–978.

Wehbe, R., Frangieh, J., Rima, M., El Obeid, D., Sabatier, J.M., Fajloun, Z., 2019. Bee venom: overview of main compounds and bioactivities for therapeutic interests. Molecules 24 (16), 2997.

Wiezel, G.A., dos Santos, P.K., Cordeiro, F.A., Bordon, K.C., Selistre-de-Araújo, H.S., Ueberheide, B., Arantes, E.C., 2015. Identification of hyaluronidase and phospholipase B in Lachesis muta rhombeata venom. Toxicon 107, 359–368.

Wu, J., Jiang, H., Bi, Q., Luo, Q., Li, J., Zhang, Y., Chen, Z., Li, C., 2014. Apamin-mediated actively targeted drug delivery for treatment of spinal cord injury: more than just a concept. Mol. Pharm. 11 (9), 3210–3222.

Xu, X., Nelson, J.W., 1993. Solution structure of tertiapin determined using nuclear magnetic resonance and distance geometry. Proteins 17 (2), 124–137.

Yamada, I., Goto, T., Takeuchi, S., Ohshima, S., Yoneyama, K., Shibuya, T., Kataoka, E., Segawa, D., Sato, W., Dohmen, T., Anezaki, Y., 2008. Mao (*Ephedra sinica* Stapf) protects against D-galactosamine and lipopolysaccharide-induced hepatic failure. Cytokine 41 (3), 293–301.

Yang, L., Cui, F., Shi, K., Cun, D., Wang, R., 2009. Design of high payload PLGA nanoparticles containing melittin/sodium dodecyl sulfate complex by the hydrophobic ion-pairing technique. Drug Dev. Ind. Pharm. 35 (8), 959–968.

Ye, M., Chung, H.S., Lee, C., Yoon, M.S., Yu, A.R., Kim, J.S., Hwang, D.S., Shim, I., Bae, H., 2016. Neuroprotective effects of bee venom phospholipase A2 in the 3xTg AD mouse model of Alzheimer's disease. J. Neuroinflammation 13 (1), 1–12.

Yu, A.R., Kim, J.J., Park, G.S., Oh, S.M., Han, C.S., Lee, M.Y., 2012. The antifungal activity of bee venom against dermatophytes. J. Appl. Biol. Chem. 55 (1), 7–11.

Ziai, M.R., Russek, S., Wang, H.C., Beer, B., Blume, A.J., 1990. Mast cell degranulating peptide: a multi functional neurotoxin. J. Pharm. Pharmacol. 42 (7), 457–461

CHAPTER 16

An overview of risks and benefits of bee products

Elena Azzini
CREA—Research Centre for Food and Nutrition, Via Ardeatina, Rome, Italy

Contents

Introduction

The most recent "International Standard on Honey" revised by the Codex Alimentarius (2000) Commission (FAO 2018. Food and Agriculture Organization of the United Nations; World Health Organization (WHO), 2018): "*Honey is the food product that honeybees produce from the nectar of flowers or from the secretions coming from living parts of plants or that they are found on the same, which they forage, transform, combine with their own specific substances and leave to mature in the honeycombs of the hive.*" This definition states what honey is and what it is not: It is a sweet substance elaborated by honeybees, from the nectar of flowers or honeydew, and neither from other sweet

Bee Products and Their Applications in the Food and Pharmaceutical Industries
https://doi.org/10.1016/B978-0-323-85400-9.00009-5

products nor insect, and any substance can be added or subtracted from the bee product, and anything else could be defined as honey. To date, about 300 different types of honey are recognized, depending on the nectar of different flowers foraged by bees. It consists mainly of carbohydrates, but it has other compounds such as proteins, vitamins, amino acids, minerals, and organic acids. Furthermore, pure raw honey also contains polyphenols, flavonoids, alkaloids, glycosides, and volatile compounds (El-Soud, 2012). From the perspective of honey as a food for human beings, it is a natural sweetener and a valuable source of components able to exert antioxidant, antimicrobial, antiinflammatory, antiproliferative, anticancer, and antimetastatic effects (Fig. 16.1) as reported in various studies (Alvarez-Suarez et al., 2013; Cianciosi et al., 2018; Samarghandian et al., 2017). But honey is not just food, and bees are not just honey-maker: the close relationship between these insects and the planet must also be taken into account. Nevertheless, on the one hand, bees are bio-indicators of the healthy environment, preserving and protecting the status of the habitats through the pollination process; on the other hand, the food security (i.e., the variety and the quantity of food available) depends on their every activity.

Bees pollinate 80% of plant species, safeguarding food diversity and some of the most critical monocultures, which would not exist otherwise. Einstein assumed: "*If the bee disappeared off the face of the Earth, man would only have few years left to live.*" This statement places these small insects in the center of our ecosystem and our survival. Referring to bees, it must also

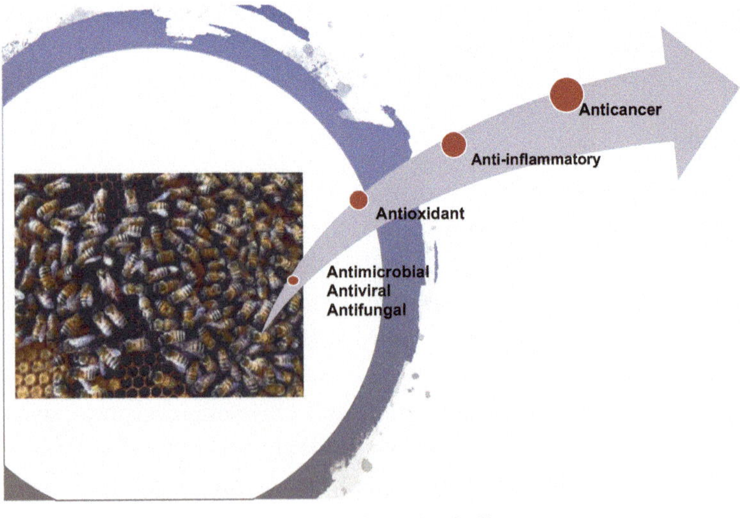

Fig. 16.1 Honey activities to exert its beneficial effects.

be considered the entire process, which is much broader and involves many sectors, including agriculture, breeding, production and consumption, and territory. During the last decades, "apitherapy" was developed, an alternative medicine branch based on honey, due to its properties and health benefits. This chapter aims to summarize the current literature highlighting the healing effects and possible risks of bee product consumption.

Health benefits of bee products

Honey

The overall composition, including 36% of glucose and 41% of fructose associated with organic acids, mineral salts, enzymes, aromas, and many other substances, makes honey unique and exceptional. As shown in Table 16.1, its proximate composition confers to honey a high energy power level (Marletta and Camilli, 2019). Even if the water content (18%) gives it a lower caloric value in percentage than sugar, it provides 304 cal/100 g against it the 400 of sucrose or cooking sugar. Its composition, mainly simple sugars, allows it to be easily digested. In fact, in sugar, glucose and fructose are bonded together to form sucrose (Özcan et al., 2006). It, therefore, offers an immediate energy supply that makes it a suitable component of nutrition, mainly in the diet of athletes, the elderly, and school-aged children, mainly when energy needs are high and fast. The sweetening power of honey is higher than regular cooking sugar, putting sucrose's sweetening power at 100, that of fructose at 173, and that of glucose at 74. Honey with a high fructose content (e.g., acacia) is mostly sweeter than high glucose concentration (e.g., rape) (Bogdanov et al., 2008). Therefore, on a dietary level, the consumption of honey allows realizing a small calorie saving, even if its nutritional profile is not substantially different from sucrose.

Table 16.1 Proximate composition of honey (Marletta and Camilli, 2019).

Edible part (%)	100	Sodium (mg)	11
Energy (kcal)	304	Potassium (mg)	51
Water (g)	18.0	Calcium (mg)	5
Protein (g)	0.6	Phosphorus (mg)	6
Lipid (g)	0	Thiamin (mg)	Trace
Carbohydrates (g)	80.3	Riboflavin (mg)	0.04
Starch (g)	0	Niacin (mg)	0.30
Soluble sugars (g)	80.3	Vitamin A (ret. eq.) (μg)	0
Fiber (g)	0	Vitamin C (mg)	1
Iron (mg)	0.5	Vitamin E (mg)	0

The flavor can vary depending on the plants from which it is produced (acacia, chestnut, wildflower, and so on); what makes it unique as a sweet substance is its natural origin, from bees and plants enriches its flavor and aroma before reaching human consumption. In addition to consumer's preference, the sensory evaluation plays a vital role in defining product standards and in the related controls to distinguish the botanical origin of honey as well as to identify and quantify certain defects, including fermentation, impurities, off-odors, and flavors (Piana et al., 2004; Oddo and Bogdanov, 2004). For its caloric supply, the recommended daily consumption is the same as monosaccharides and disaccharides naturally present and/or added to foods (WHO, 2015). It has no other nutritional value, and its excessive intake should be avoided.

Natural honey contains antioxidants, phytochemicals, and other bioactive substances that interact synergistically to benefit humans. As well known to exert its healing effect, the bioactive compounds should be bioavailable, and currently, only a few studies on bioavailability in humans have been performed. Schramm et al. (2003) reported an increased total phenolic content and total antioxidant and reducing capacities in plasma upon consumption of 1.5 g of honey/kg body, indicating a potential improvement in counteracting oxidative stress. However, no polyphenols substances or their metabolites were found in plasma, probably due to the unclear metabolism process of these compounds. The overall healing properties attributed to honey over time are numerous, producing beneficial effects in almost all human body organs. Honey would act favorably on various disorders of the respiratory, circulatory, and digestive systems, on the liver, and in the treatment of dental and oral diseases/conditions (Fig. 16.2).

Yaghoobi et al. (2008) investigated the effect of 70 g of natural honey daily consumption for 30 days on cardiovascular risk factors in overweight or obese patients. The results showed that honey caused a mild reduction in body weight, body fat, and total cholesterol; a significant decrease in LDL-C, triacylglycerol, and CRP; and increased HDL-C (3.3%) level in subjects with typical values. In contrast, in patients with elevated variables, honey caused a reduction in total cholesterol by 3.3%, LDL-C by 4.3%, triacylglycerol by 19%, and CRP by 3.3% ($P < 0.05$).

Recently, in a cross-sectional study of adults aged 20–90 years affected by nonalcoholic fatty liver disease (NAFLD), the honey consumption 2–6 times/week was inversely associated with NAFLD, whereas consuming honey more than one time/day had no association with NAFLD (Zhang et al., 2021). Furthermore, research by Atwa et al. (2014) reported that eating raw honey does not give rise to gingivitis and periodontal disease. The noncarcinogenic

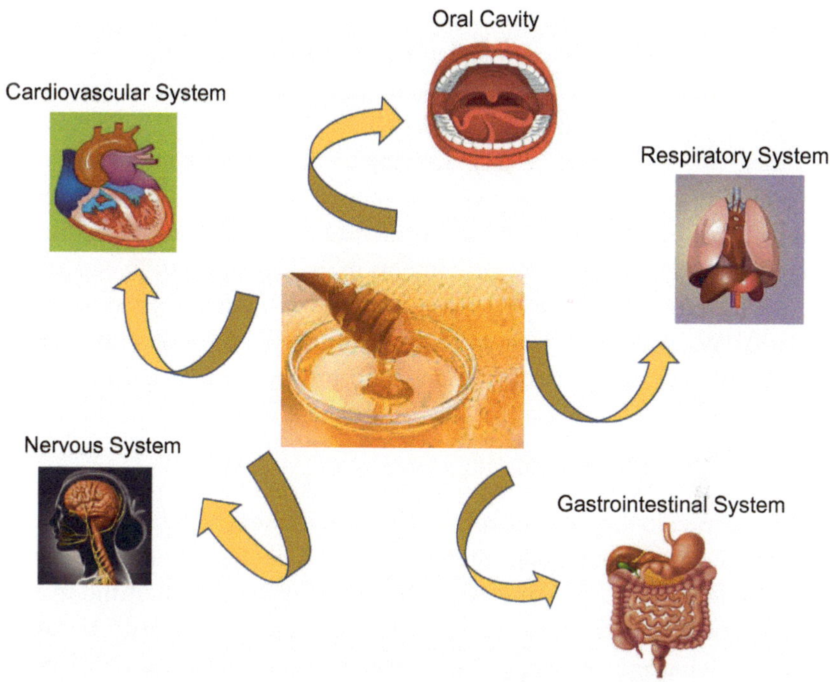

Fig. 16.2 Human systems affected by honey consumption.

potential of honey could be due to the protective role of synergistic honey constituents. The high sugar content of honey makes it a very hyperosmolar preparation, making it advantageous as a topical preparation for skin burns and other superficial infections, acting as an antibacterial agent. For example, local application of honey was helpful in the treatment of postsurgical wounds in neonates that failed to respond to conventional systemic and local antibiotic treatment (Vardi et al., 1998). Recently, the potential therapeutic use of honey as a remedy for cough in the pediatric age group was reviewed and confirmed by Toorani (2019). By Rossi and Marrazzo (2021), honey and other honeybee products have been suggested for implants based on biocompatible materials and biomaterial-based dressings for protecting target tissues from inflammation and infection threat. Moreover, target tissues can suffer injuring inflammation. The authors reviewed the available and most recent literature reporting the use of these natural products combined with different polymeric scaffolds to provide original insights into wound healing and other tissue regenerative approaches.

In vitro antiinflammatory models support the dual role of honey as an immune-modulatory by antioxidant properties and the ability to inhibit proinflammatory enzymes (such as COX, LOX, and iNOS). Honey seems to act on pro-inflammatory mediators, including nitric oxide, cytokines, and chemokines, as well as by downregulating the inflammatory transcription factors (NF-κB and MAPK) (Ranneh et al., 2021; Schieber and Chandel, 2014; Silva et al., 2021). In a regular diet supplemented with daily consumption of 1.2 g/kg body weight honey dissolved in 250 mL of water during a 2-week test period, honey increased antioxidant agents, serum iron and blood indices, and trace elements in healthy subjects. In addition, it decreased immunoglobulin E, liver and muscle enzymes, and fasting blood sugar (Al-Waili, 2003). Based on current evidence, some of these actions have occasionally been verified; there is no proper clinical trial to support each claim because, above all, honey is never uniform, never identical, and never stable.

On the other hand, there is no in-depth research on the composition of the honey used in the experiments carried out. In the same way, the pharmacological activities attribute to unifloral kinds of honey coming from plants are not scientifically confirmed due to the high complexity of the matrix. On the other hand, antibacterial activity has been verified, both in honey and dilute solutions, due to the sugar concentration and its acid pH. This last activity, attributed for a long time to a substance of unknown nature defined with the generic name of "inhibin," is due to the action of the enzyme glucose oxidase, which, under dilution conditions, produces hydrogen peroxide and gluconic acid starting from glucose (Albaridi, 2019; Aliyazicioglu and Boukraâ, 2015). To confer antibiotic activity to the honey solutions would be the accumulation of hydrogen peroxide (which is subsequently destroyed); the biological significance should probably be protecting the honey in formation from microbial attack (when the inhibition system due to the high sugar concentration is still not efficient). This mechanism would be based on the antibacterial activity carried out by honey on wounds and could also explain some other traditionally referred activities in this product. Other substances present in honey (polyphenols) also seem to have an antibiotic-like activity. Probably most of the traditionally recognized benefits of honey can be traced back to a generic trophic action and to the emollient, mildly laxative, hepatoprotective, and detoxifying effect of fructose.

Some recent research has shown that honey contains numerous compounds (mainly flavonoids and phenolic substances) with antioxidant potential. However, the type and quantity of these antioxidants are highly variable depending on the botanical origin of honey, and it is also necessary to evaluate their actual bioavailability. However, it is better to consider honey as a food rather than as a drug: even if it is necessary to remember that it is not a complete food due to the negligible content of proteins, lipids, and vitamins, its nutritional value, as well as its pleasantness, is undoubtedly more sustainable with valid arguments.

Bee pollen

Bee pollen is defined as the agglutination of flower pollen, nectar, and honeybee salivary substances. The vegetal origin can vary greatly, depending on the landscape and the season (Lau et al., 2019). In addition, while pausing in various other places, the bee contaminates this pollen mixture with molds, spores, and bacteria. Therefore, it is impossible to produce a stable, consistent, and clean product from this contaminated source. Thus, the major problem with bee pollen is not the pollen itself but the bee's collection, storage, and processing methods (Ulbricht et al., 2009). Due to its composition including proteins, carbohydrates, lipids, minerals, enzymes, and coenzymes as well as vitamins (vitamins C, E, and β-carotene, as provitamin A, B-group, and D vitamins) and phenolic compounds, bee pollen demonstrates a wide range of activities including antifungal, antimicrobial, antiviral, antiinflammatory, hepatoprotective, immunostimulant, and local analgesic (Feás et al., 2012; Kroyer and Hegedus, 2001).

Kassyanenko et al. (2010) evaluated the efficacy of treatment with beekeeping products (honey, pollen, bee bread) on lipidemic profile in patients with pathologic lipid metabolism. In 157 patients (64 males and 93 females), aged 39–72 (mean age 61.7 + 8.5 years), a significant hypolipidemic effect was registered in patients taking honey in combination with pollen (total cholesterol decreased by 18.3%, LDL cholesterol by 23.9%) in overweight (body mass index—BMI 25–30) and obese (BMI over 30) patients, only in loss of body mass. Probably its hypoglycemic role could be attributed to the presence of unsaturated fatty acids (e.g., omega-3 and α-ALA), phospholipids, and phytosterols through regulating glycerophospholipid and sphingolipid metabolisms, potentially involved in building permeability barriers and alleviating intestinal oxidative stress and then could control the gastrointestinal fat absorption (Li et al., 2020). In a comprehensive review, in patients receiving tamoxifen and aromatase inhibitors/inactivators, Münstedt et al. (2015) compared a pollen–honey mixture with pure honey (placebo) alleviating menopausal symptoms on a total of 49 breast cancer patients in a randomized crossover trial. Approximately 70.9% of patients reported an improvement in their symptoms (~25%) with respect placebo effect using the Menopause Rating Scale (MRS) assessment.

Royal jelly

Royal jelly (RJ) composition has been reported to vary with seasonal and regional conditions (Bincoletto et al., 2005; Fontana et al., 2004). It contains sugars, proteins, fatty acids, amino acids, vitamins, salts, and trace elements. The biologically active components of royal jelly include various proteins and peptides, the fatty acid 10-hydroxy-2-decenoic acid (10-HDA), and unidentified components with estrogenic activity. About 90% of proteins in royal jelly largely belong to a family called MRJP (apalbumins). As its name highlights, RJ is essential for the sexual differentiation and longevity of the queen bee. RJ is a creamy, whitish, strongly acidic secretion from the mandibular and hypopharyngeal glands of *Apis mellifera* nurse bees. As reported, RJ improves heart health by maintaining a balanced level of cholesterol and lipids in the body and other claimed effects related to immune system, metabolism, vascular function, skin health, and so on.

In a randomized placebo-controlled, double-blind trial, the effects of long-term royal jelly ingestion on healthy humans were studied. Upon 3000 mg of RJ per day for 6 months, subjects improved erythropoiesis, glucose tolerance, and mental health. The authors suggested that RJ induced an acceleration of conversion from DHEA-S to T by activation of 3β-HSD2 and/or 17β-HSD3 through antioxidant enzyme potential of RJ (Morita et al., 2012). In a randomized, placebo-controlled study, Chiu et al. (2017) examined the hypolipidemic beneficial effect of royal jelly through evaluating anthropometric measurements, lipid profile, and various hormone levels in mildly hypercholesterolemic participants. Upon consumption of nine capsules (350 mg/capsule) of royal jelly or placebo/day, respectively, no significant changes in anthropometric parameters, triglyceride, and high-density lipoprotein cholesterol levels were observed for 3 months. No considerable changes were noted in hepatic and renal markers. Royal jelly consumption significantly reduced cholesterol and low-density lipoprotein cholesterol levels and significantly ameliorated ($P < 0.05$) the concentration of sex hormones.

The hypocholesterolemic property of RJ could be due to the presence of two major RJ proteins, specifically MRJP1 and MRJP2. MRJP1 could interact with bile acids increasing fecal bile acid excretion and fecal cholesterol excretion and enhance the hepatic cholesterol catabolism. Thus, RJ supplementation could reduce the risk of cardiovascular disease. The same picture was observed by Guo et al. (2007) upon 6 g/day of royal jelly in healthy individuals for 4 weeks. A significant reduction ($P < 0.05$) in TC and LDL levels was observed compared to the controls. No significant differences in serum high-density lipoprotein (HDL) or triglyceride concentrations were recorded. The RJ protein could increase small VLDL uptake by active hepatic lipoprotein receptors or suppress hepatic sterols synthesis. In addition to a probable regulation of lipid metabolism by RJ unsaturated fatty acids composition. Furthermore, a systematic review and metaanalysis of randomized clinical trials by Mahboobi et al. (2019), RJ supplementation does not improve markers of glycemia: RJ supplementation did not beneficially affect markers of glycemia such as fasting plasma glucose (FPG) and hemoglobin A1C (HbA1C).

Propolis

As recently reviewed by Anjum et al. (2019), propolis is a beehive product with a complex chemical composition including 30% wax, 50% resin and vegetable balsam, 10% essential/aromatic oils, 5% pollen, and 5% other substances. Its chemical composition and biological properties may vary according to geographical location and different plant sources, and the extraction procedure used (Sforcin, 2007). Propolis provides different biological activities through various distinct preparations, extracts, and isolated components from various sources. The high modifiable concentrations of over 300 presumed biologically active constituents, mainly flavones, and flavonols (Kumazawa et al., 2004), can compromise propolis's standardization and universal chemical characterization (Bankova, 2005).

Steinberg et al. (1996), in a pilot study (10 volunteers), evaluated the short-term antibacterial effect of propolis solution and honey on salivary's total bacteria and *Streptococcus mutans*, commonly found in the human oral cavity and a significant contributor to tooth decay. Propolis demonstrated an antibacterial effect on salivary bacterial counts. Honey decreased salivary counts of total bacteria and *S. mutans* for 1 h after its application. The antibacterial effect of propolis and honey tested was attributed to its osmolarity effect, which prevents bacterial growth, thereby mitigating the pathogenesis of dental caries. Recently, propolis was found to be as efficient as chlorhexidine in reducing plaque, gingivitis, and dental caries pathogens in a randomized controlled trial conducted by Bapat et al. (2021). On 120 participants randomized to 4 mouth rinse study groups twice a day for 3 months by hot ethanolic propolis extract, cold ethanolic propolis extract, chlorhexidine, and distilled water, a decline in the concentration of *S. mutans* was observed in samples collected after the use of mouth rinse ($P < 0.05$), as well as a decrease in the *L. acidophilus* and *S. mutans* count concerning baseline.

Mujica et al. (2017) evaluated the antioxidant effect of propolis in humans after oral administration (twice daily, 15 drops each time, 90 days) of commercially available propolis solution on the oxidative status and lipid profile in a Chilean population. Propolis supplementation decreased lipid peroxidation derivative products (67% by TBARS), increased the reduced glutathione (175% GSH) level compared to the baseline, and increased the HDL concentration. The authors concluded that propolis supplementation could reduce the risk of cardiovascular events.

The therapeutic benefits of propolis in metabolic syndrome and its chronic disease have been reviewed by Zulhendri et al. (2021). The molecular and biochemical mechanisms of action of propolis include the inhibition of the expression and interaction of advanced glycation end products (AGEs) and their receptors (RAGEs), the inhibition of the proinflammatory signaling cascades, and the promotion of the cellular antioxidant systems. The most investigated compound of propolis, a hydrophobic bioactive polyphenolic ester, caffeic acid phenethyl ester (CAPE), demonstrated a wide range of pharmacological activities through several mechanisms of action in protecting the central nervous system in numerous in vitro and preclinical studies (Balaha et al., 2021).

Risks associated with consumption of bee products

"*All things are poison, and nothing is without poison; the dosage alone makes it so a thing is not a poison*," as Paracelsus said a substance could produce the harmful effect associated with its toxic properties only if it reaches a susceptible biological system within the body in a high enough concentration. Moreover, the appropriate dose of a compound depends on several factors, such as the consumer's age, health, and several other conditions. It should always be kept in mind that natural products are not always automatically safe, and dosages can be crucial due to the higher interindividual variability and insufficient scientific information in the supplemented composition of honey and its products. In fact, with natural products, recommended/optimal doses are challenging to balance efficacy and safety, and there is currently insufficient available dosing from well-designed clinical trials regarding honey and its products. Moreover, their preparation may vary from different manufacturers and from batch to batch within one manufacturer. In addition, the active components of a product are not the same, the standardization is quite impossible, and the clinical effects of different supplementation cannot be comparable.

Environmental contaminants

People are constantly increasing attention to the risks associated with pollution of the environment and food products' leading to a growing demand for safer products. Such a trend creates new market opportunities, new niches that attract an increasing number of producers. About a third of our food reserves would disappear without the work of domestic and wild bees and other pollinators. Due to some choices of modern agriculture such as monoculture, the elimination of hedges, the use of pesticides, and the alteration and fragmentation of natural areas, the environment has become inhospitable for most pollinating insects. Therefore, the sustainable use of plant protection products must represent a concrete response to the conservation and protection of bees and other pollinators. The European Directive that introduced a specific regulation on the use of plant protection products, downstream of the process of authorization and marketing of the same, is 2009/128/EC of the European Parliament and of the Council of October 21, 2009, establishing a framework for Community action to achieve sustainable use of pesticides. Pesticides are toxic substances whose residue limits in food are constantly updated. A detailed list is given in Regulation (EC) 396/2005, and where not specified, a generic residual limit of active substances not exceeding 10 μg/kg limit also extended to imported products.

In the European Union countries, Regulation 470/2009 prohibits the administration of antibiotics even for therapeutic purposes, even if accidental contamination may occur. However, antibiotics in other countries such as South America, where a large part of the global honey supply is produced, are legal. Therefore, residue analysis is extremely important, especially in imported honey. An action limit of 5.0 μg/kg has been set for antibiotics such as tetracyclines, streptomycin, sulfonamides, and tylosin, above which the sample cannot be considered compliant (Al-Waili et al., 2012). The lack of a defined maximum value must be understood as an obligation of the absence of the residue itself. Among the drug residues, the limits for Amitraz and its metabolites are 200 μg/kg while for Coumaphos, the limit is 100 μg/kg, these substances are so-called dual use; for veterinary treatments (Regulation (EU) No. 37/2010) to protect honeybees against Varroa mites; for plant protection products. Unless specified, the maximum residue limit of 10 μg/kg applies as set out in EC Regulation 396/2005. No limits are reported for natural products useful against Varroa, such as oxalic acid, menthol, eucalyptol, thymol, etc.

Tosi et al. (2018), in a 3-year survey of Italian honeybee pollen, found widespread and prolonged contamination by multiple insecticides and fungicides, highlighting as surrounding agricultural, industrial, and other anthropogenic activities represent a source of contamination. In addition, the bee-collected pollen is a valuable tool for environmental monitoring and the detection of illegal uses of pesticides. Still, about residue control in animals and derivative products, it should be emphasized that from January 01, 2000, it is no longer possible to place veterinary drugs on the market for which the fate in the finished food product is not known. Soil, polluted water used for irrigation, vehicular emissions, industrial activities, and adaptation of bad agricultural practices in the surrounding agricultural and industrial areas can cause heavy metals accumulation in the plant's tissues. Regarding heavy metals, including zinc, iron, cobalt, manganese, and chromium, the limit of 0.1 mg/kg has been set for the lead. In contrast, no limits have been set for cadmium and any other contaminants.

Microbial contamination and mycotoxins

The presence of microorganisms in honey affects both the sensory quality and the healthiness of the product. The microorganisms in honey are bacteria, yeasts, and molds. These residues come from bees, nectar, and external sources. Due to their specific chemical characteristics, most bacteria and microbes do not reproduce or grow in honey. In addition, some unprocessed or raw honey products, including raw honey, comb honey, wild honey, etc., are gaining popularity in recent years. Some consumers associate raw honey (i.e., unfiltered, unheated/unpasteurized honey) or wild honey (i.e., from bees that live in the wild) with specific health benefits. However, it is important to note that these products can carry some risks.

Medina et al. (2004) tested the toxigenic capability of bee pollen for the production of ochratoxin A by a strain of *Aspergillus ochraceus*. The ochratoxin A accumulation was significantly higher in liquid cultures supplemented with 5% bee pollen than in pollen-free ones. Thus, bee pollen may constitute a significant risk factor concerning consumers' diet. On a total of 90 ready-to-eat bee pollen samples, González et al. (2005) found contamination with several potential mycotoxin-producing species, including *Penicillium verrucosum*, *Aspergillus niger* aggregate, *Aspergillus carbonarius*, *Aspergillus ochraceus*, *Aspergillus flavus*, *Aspergillus parasiticus*, and *Alternaria* spp.

Allergic reactions

Raw honey may contain allergens; during its processing, honey is filtered to the extent that all or most of the fine particles and pollen grains typically found in suspension will be removed. Unlike processed honey, raw honey may contain small amounts of pollen grains. According to the literature, there were cases of allergic reactions after ingestion of raw honey that contained pollen. Symptoms could range from itching in the oral mucosa to potentially life-threatening anaphylaxis, depending on the type of pollen (Greenberger and Flais, 2001). People allergic to pollen or with severe seasonal allergies, such as hay fever, should be aware of the potential risk of eating raw honey. With a good understanding of plants in the environment around the hives by beekeepers and blending of honey from different sources by packers to achieve a dilution effect if toxic substances are present, honey can be enjoyed without issues. In addition, bee pollen consumption might result in an increased chance of bruising or bleeding in people on warfarin (Coumadin) therapy. Other allergic symptoms can include itching, swelling, shortness of breath, light-headedness, and severe whole-body reactions, while a hypersensitivity reaction includes symptoms such as gastrointestinal upset, rash, erythema, asthma, hay fever, nausea, abdominal cramps, diarrhea, and vomiting.

Chemicals occurring during processing and storage

Among the various components present in the honey pathways, the hydroxymethylfurfural (HMF) aromatic aldehyde is formed by sugar degradation in an acidic environment. This molecule is practically absent in fresh honey. Still, it is produced during storage: for this reason, although it is a natural component of honey, it is considered an indicator of freshness, and the maximum acceptable values are defined by law. The European legislation has set the maximum limit of HMF in honey at 40 mg/kg since these levels are effective under the climatic conditions of Europe, for kinds of honey with declared origin from regions with a tropical climate and their blends with the same types, no more than 80 mg/kg. Among the minor constituents, it is also possible to find pollen, wax, and some undesirable compounds with pharmacological activity (Sabatini, 2007). However, contaminants of a chemical and biological nature arising from industrialization, from the treatments of beekeepers for the treatment of diseases, and the environment have been found in honey. They are the object of attention by the organizations' control (De-Melo et al., 2017).

Adulteration of honey

Food adulteration could compromise not only the quality of food products but also the consumer's health resulting from their consumption. According to the Codex Alimentarius regulation, consumers have the right to receive truthful information about the food they will consume. Recently, Fakhlaei et al. (2020) reviewed the different types of adulteration, common sugar adulterants, and detection methods and suggested a clear perspective toward the impact of honey adulteration on human health. Sugars can be used for direct adulteration and indirect adulteration. During direct adulteration, a certain ratio of syrups is added to harvested honey to increase its sweet taste (Guler et al., 2014; Soares et al., 2017), while in indirect adulteration, the bees are overfed with sugar syrups to increase the honey yield in hives (Chen et al., 2014; Soares et al., 2017). Another honey adulteration procedure is blending by mixing pure and high-quality honey with cheap and low-quality honey; however, recently, synthetic honey addition to pure one is becoming a more frequent type of adulteration (Cordella et al., 2005; Jaafar et al., 2020).

The high fructose corn syrup, corn sugar syrup, inverted sugar syrup, and cane sugar syrup are reported as the most frequent sugar syrups for honey adulteration, followed by palm sugar, rice syrup, and inulin syrup with a high preference toward high fructose corn syrup from simple isomerization of corn sugar syrup. These sugar adulterants can have health disadvantages toward human health. Thus, glucose, which is present in combination with fructose in honey, enhances the absorption of fructose and promotes its hepatic actions through its enhanced delivery to the liver. The liver is the most common organ affected by honey adulterants, followed by the kidney, heart, and brain. Adulteration increases the consumer's blood sugar, which can cause diabetes, abdominal weight gain, and obesity, raise the level of blood lipids, and can cause high blood pressure (Awasthi et al., 2014; Bogdanov et al., 2008).

Poisonous honey

Some types of honey were once known for their harmful properties due to the species they had foraged. The nectar of these plants may not be harmful to bees but harmful to human health. Examples of poisoning derive from relatively ancient times from a species of azalea (*Azalea pontica*), although it is believed that perhaps the Greeks with honey also consumed the brood and the pollen supplies. Some honey varieties are considered undesirable, such as privet and ailanthus, because a very small amount of nectar can give a strong and unpleasant taste to a whole harvest. Some toxic kinds of honey have been reported in Spain, mainly deriving from a type of herb, *Coriaria myrtifolia* L., and from the ferruginous rhododendron (*Rhododendron ferrugineum* L.), which abounds in the central Spanish Pyrenees. Some species can sometimes give toxic honey in other regions, including cherry laurel, aconite, belladonna, hemlock, datura, foxglove, Hyoscyamus, etc. Still, these species are not at all, or very little, visited by bees who find plenty of other plants, and the risk is therefore nonexistent. The symptoms of poisoning due to honey consumption depend on the types and levels of the toxins. Rhododendrons produce a natural toxin known as grayanotoxins, causing symptoms like dizziness, weakness, excessive perspiration, excessive salivation, numbing of limbs, nausea to develop quickly after consumption. To dilute the level of toxins, the small-scale beekeepers blend the rhododendron honey with nontoxic ones. There are over 500 different pyrrolizidine alkaloids, a large group of natural toxins found in many plants worldwide and could exert chronic effects on the liver (EFSA Panel on Contaminants in the Food Chain (CONTAM), 2011; Wiedenfeld, 2011).

Honey and infant botulism

Although the high sugar content makes the honey composition relatively safe from a microbiological perspective by inhibiting bacterial growth, some bacteria manage to survive in its matrix. Raw honey can contain the genera Bacillus and Clostridium, both capable of producing spores (Dembek et al., 2007). These spores can germinate in the colon of children less than 1 year of age due to the still immature intestinal flora at this age. This results in producing a toxin that causes the classic manifestations of infant botulism, a disease of newborns that can be fatal if not properly recognized and treated (Cagan et al., 2010). These phenomena do not occur in older children and adults because the fully developed intestinal flora can block the life cycle of the spores. Therefore, under 12 months of life, all types of honey are not recommended, even pasteurized ones, because pasteurization cannot heal the product by eliminating the spores.

Conclusion

For several years now, the EU has been providing support to the beekeeping sector through national apiculture programs and rural development measures considering environment for a sustainable agriculture in protecting bee health and improve biodiversity. As the European Food Safety Authority (EFSA, 2021) stated: "*Bees are critically important in the environment, sustaining biodiversity by providing essential pollination for a wide range of crops and wild plants. They contribute to human health and wellbeing directly through the production of honey and other food and feed supplies such as: pollen, wax for food processing, propolis in food technology, and royal jelly as a dietary supplement and ingredient in food. Beyond the essential value of pollination to maintaining biodiversity, the global annual monetary value of pollination has been estimated at hundreds of billions of euros. In view of the important ecological and economic value of bees, there is a need to monitor and maintain healthy bee stocks, not just locally or nationally but globally.*" Moreover, due to the increasingly globalized food supply, strengthening food safety systems among countries is becoming more important. Nowadays, food safety is the consumer's concern. Simple rules could control the disorders resulting from the consumption of honey and its products contaminated by microbial pathogens, toxic chemicals, or allergens, including avoiding crude honey consumption for susceptible people, buying honey from a reliable source, and avoiding feeding honey to infants less than 1-year-old.

Even if human studies provide outcomes related to the health effects, the origin and composition of bee products used were not specified as well as it was not possible to characterize them nor the specific components of bee products mediating the functions for which the claims were made consequently it's difficult to establish the cause-and-effect relationship between healing effects and the consumption of honey and its products. The interesting healing properties highlighted in vitro and in vivo should be better understood and performed in more and well design randomized controlled trials to evaluating the effectiveness of honey and its products as therapeutic agents.

References

Albaridi, N.A., 2019. Antibacterial potency of honey. Int. J. Microbiol. 2019, 2464507. https://doi.org/10.1155/2019/2464507. 31281362. PMC6589292.

Aliyazicioglu, R., Boukraâ, L., 2015. Honey: the natural "inhibine". Anti-Infect. Agents 13, 42–49.

Alvarez-Suarez, J.M., Giampieri, F., Battino, M., 2013. Honey as a source of dietary antioxidants: structures, bioavailability and evidence of protective effects against human chronic diseases. Curr. Med. Chem. 20 (5), 621–638. https://doi.org/10.2174/092986713804999358. 23298140.

Al-Waili, N.S., 2003. Effects of daily consumption of honey solution on hematological indices and blood levels of minerals and enzymes in normal individuals. J. Med. Food 6 (2), 135–140. https://doi.org/10.1089/109662003322233549. 12935325.

Al-Waili, N., Salom, K., Al-Ghamdi, A., Ansari, M.J., 2012. Antibiotic, pesticide, and microbial contaminants of honey: human health hazards. Sci. World J. 2012. https://doi.org/10.1100/2012/930849, 930849. Epub 2012 Oct 14 23097637. PMC3477659.

Anjum, S.I., Ullah, A., Khan, K.A., Attaullah, M., Khan, H., Ali, H., Bashir, M.A., Tahir, M., Ansari, M.J., Ghramh, H.A., Adgaba, N., 2019. Composition and functional properties of propolis (bee glue): a review. Saudi J. Biol. Sci. 26 (7), 1695–1703.

Atwa, A.A., AbuShahba, R.Y., Mostafa, M., Mohamed, I.H., 2014. Effect of honey in preventing gingivitis and dental caries in patients undergoing orthodontic treatment. Saudi Dent J. 26 (3), 108–114. https://doi.org/10.1016/j.sdentj.2014.03.001.

Awasthi, S., Jain, K., Das, A., Alam, R., Surti, G., Kishan, N., 2014. Analysis of food quality and food adulterants from different departmental & local grocery stores by qualitative analysis for food safety. IOSR J. Environ. Sci. Toxicol. Food Technol. 8, 22–26.

Balaha, M., De Filippis, B., Cataldi, A., di Giacomo, V., 2021. CAPE and neuroprotection: a review. Biomolecules 11 (2), 176. https://doi.org/10.3390/biom11020176. 33525407. PMC7911454.

Bankova, V., 2005. Chemical diversity of propolis and the problem of standardization. J. Ethnopharmacol. 100 (1–2), 114–117.

Bapat, S., Nagarajappa, R., Ramesh, G., Bapat, K., 2021. Effect of propolis mouth rinse on oral microorganisms—a randomized controlled trial. Clin. Oral Investig. 25, 1–8.

Bincoletto, C., Eberlin, S., Figueiredo, C.A., Luengo, M.B., Queiroz, M.L., 2005. Effects produced by Royal Jelly on haematopoiesis: relation with host resistance against Ehrlich ascites tumour challenge. Int. Immunopharmacol. 5 (4), 679–688.

Bogdanov, S., Jurendic, T., Sieber, R., Gallmann, P., 2008. Honey for nutrition and health: a review. J. Am. Coll. Nutr. 27 (6), 677–689. https://doi.org/10.1080/07315724.2008.10719745. 19155427.

Cagan, E., Peker, E., Dogan, M., Caksen, H., 2010. Infant botulism. Eurasian J. Med. 42, 92–94.

Chen, Q., Qi, S., Li, H., Han, X., Ouyang, Q., Zhao, J., 2014. Determination of rice syrup adulterant concentration in honey using three-dimensional fluorescence spectra and multivariate calibrations. Spectrochim. Acta A Mol. Biomol. Spectrosc. 131, 177–182.

Chiu, H.F., Chen, B.K., Lu, Y.Y., Han, Y.C., Shen, Y.C., Venkatakrishnan, K., Golovinskaia, O., Wang, C.K., 2017. Hypocholesterolemic efficacy of royal jelly in healthy mild hypercholesterolemic adults. Pharm. Biol. 55 (1), 497–502.

Cianciosi, D., Forbes-Hernández, T.Y., Afrin, S., Gasparrini, M., Reboredo-Rodriguez, P., Manna, P.P., Zhang, J., Bravo Lamas, L., Martínez Flórez, S., Agudo Toyos, P., Quiles, J.L., Giampieri, F., Battino, M., 2018. Phenolic compounds in honey and their associated health benefits: a review. Molecules 23 (9), 2322. https://doi.org/10.3390/molecules23092322. 30208664. PMC6225430.

Codex Alimentarius, 2000. Codex Alimentarius, Alinorm 01/25. 2000. Drafty revised standard for honey at step 8 of the codex procedure; EU directive/1/110/2001 of 02/12/2001(L10/47).

Cordella, C., Militao, J.S., Clément, M.-C., Drajnudel, P., Cabrol-Bass, D., 2005. Detection and quantification of honey adulteration via direct incorporation of sugar syrups or bee-feeding: preliminary study using high-performance anion exchange chromatography with pulsed amperometric detection (HPAEC-PAD) and chemometrics. Anal. Chim. Acta 531, 239–248.

Dembek, Z.F., Smith, L.A., Rusnak, J.M., 2007. Botulism: cause, effects, diagnosis, clinical and laboratory identification, and treatment modalities. Disaster Med. Public Health Prep. 1, 122–134.

De-Melo, A.A.M., de Almeida-Muradian, L.B., Sancho, M.T., Pascual-Maté, A., 2017. Composition and properties of Apis melliferahoney: a review. J. Apic. Res. https://doi.org/10.1080/00218839.2017.1338444.

EFSA, 2021. European Food Safety Authority. Bee Health. Available from: https://www.efsa.europa.eu/en/topics/topic/bee-health. (Accessed 7 June 2021).

EFSA Panel on Contaminants in the Food Chain (CONTAM), 2011. Scientific opinion on pyrrolizidine alkaloids in food and feed. EFSA J. 9 (11), 134. https://doi.org/10.2903/j.efsa.2011.2406. 2406. Available from: www.efsa.europa.eu/efsajournal.

El-Soud, N.H.A., 2012. Honey between traditional uses and recent medicine. Maced J. Med. Sci. https://doi.org/10.3889/MJMS.1957-5773.2012.0213.

Fakhlaei, R., Selamat, J., Khatib, A., Razis, A.F.A., Sukor, R., Ahmad, S., Babadi, A.A., 2020. The toxic impact of honey adulteration: a review. Foods 9 (11), 1538.

FAO 2018. Food and Agriculture Organization of the United Nations; World Health Organization (WHO), 2018. In: Codex Alimentarius Commission (Ed.), General Standard for Food Additives, CODEX STAN 192–1995. FAO/WHO, Rome, Italy/Geneva, Switzerland. Available from: http://www.fao.org/gsfaonline/docs/CXS_192e.pdf.

Feás, X., Vázquez-Tato, M.P., Estevinho, L., Seijas, J.A., Iglesias, A., 2012. Organic bee pollen: botanical origin, nutritional value, bioactive compounds, antioxidant activity and microbiological quality. Molecules 17 (7), 8359–8377.

Fontana, R., Mendes, M.A., De Souza, B.M., Konno, K., César, L.M.M., Malaspina, O., Palma, M.S., 2004. Jelleines: a family of antimicrobial peptides from the royal jelly of honeybees (*Apis mellifera*). Peptides 25 (6), 919–928.

González, G., Hinojo, M.J., Mateo, R., Medina, A., Jiménez, M., 2005. Occurrence of mycotoxin producing fungi in bee pollen. Int. J. Food Microbiol. 105 (1), 1–9.

Greenberger, P.A., Flais, M.J., 2001. Bee pollen-induced anaphylactic reaction in an unknowingly sensitized subject. Ann. Allergy Asthma Immunol. 86 (2), 239–242.

Guler, A., Kocaokutgen, H., Garipoglu, A.V., Onder, H., Ekinci, D., Biyik, S., 2014. Detection of adulterated honey produced by honeybee (*Apis mellifera* L.) colonies fed with different levels of commercial industrial sugar (C3 and C4 plants) syrups by the carbon isotope ratio analysis. Food Chem. 155, 155–160.

Guo, H., Saiga, A., Sato, M., Miyazawa, I., Shibata, M., Takahata, Y., Morimatsu, F., 2007. Royal jelly supplementation improves lipoprotein metabolism in humans. J. Nutr. Sci. Vitaminol. 53 (4), 345–348.

Jaafar, M., Othman, M., Yaacob, M., Talip, B., Ilyas, M., Ngajikin, N., Fauzi, N., 2020. A review on honey adulteration and the available detection approaches. Int. J. Integr. Eng. 12, 125–131.

Kassyanenko, V., Komisarenko, I., Dubtsova, E., 2010. Influence of honey, pollen and bee bread on serum cholesterin of patients with pathological lipid metabolism. In: Beekeeping, Apitherapy and Life Quality. International Industrial Academy, pp. 81–82.

Kroyer, G., Hegedus, N., 2001. Evaluation of bioactive properties of pollen extracts as functional dietary food supplement. Innovative Food Sci. Emerg. Technol. 2 (3), 171–174.

Kumazawa, S., Hamasaka, T., Nakayama, T., 2004. Antioxidant activity of propolis of various geographic origins. Food Chem. 84 (3), 329–339.

Lau, P., Bryant, V., Ellis, J.D., Huang, Z.Y., Sullivan, J., Schmehl, D.R., Cabrera, A.R., Rangel, J., 2019. Seasonal variation of pollen collected by honey bees (*Apis mellifera*) in developed areas across four regions in the United States. PLoS One 14 (6), e0217294.

Li, Q., Liang, X., Xue, X., Wang, K., Wu, L., 2020. Lipidomics provides novel insights into understanding the bee pollen lipids transepithelial transport and metabolism in human intestinal cells. J. Agric. Food Chem. 68, 907–917.

Mahboobi, S., Jafarnejad, S., Eftekhari, M.H., 2019. Royal jelly does not improve markers of glycemia: a systematic review and meta-analysis of randomized clinical trials. Complement. Ther. Med. 44, 235–241.

Marletta, L., Camilli, E., 2019. Tabelle di composizione degli alimenti, Aggiornamento 2019. Website https://www.alimentinutrizione.it/tabelle-di-composizione-degli-alimenti.

Medina, A., Gonz'alez, G., S'aez, J.M., Mateo, R., Jim'enez, M., 2004. Bee pollen, a substrate that stimulates ochratoxin a production by aspergillus ochraceus Wilh. Syst. Appl. Microbiol. 27 (2), 261–267.

Morita, H., Ikeda, T., Kajita, K., Fujioka, K., Mori, I., Okada, H., Uno, Y., Ishizuka, T., 2012. Effect of royal jelly ingestion for six months on healthy volunteers. Nutr. J. 11, 77. https://doi.org/10.1186/1475-2891-11-77.

Mujica, V., Orrego, R., Pérez, J., Romero, P., Ovalle, P., Zúñiga-Hernández, J., Arredondo, M., Leiva, E., 2017. The role of propolis in oxidative stress and lipid metabolism: a randomized controlled trial. Evid. Based Complement. Alternat. Med. 2017. https://doi.org/10.1155/2017/4272940, 4272940. Epub 2017 Apr 30 28539963. PMC5429932.

Münstedt, K., Voss, B., Kullmer, U., Schneider, U., Hübner, J., 2015. Bee pollen and honey for the alleviation of hot flushes and other menopausal symptoms in breast cancer patients. Mol. Clin. Oncol. 3 (4), 869–874.

Oddo, P.L., Bogdanov, S., 2004. Determination of honey botanical origin: problems and issues. Apidologie 35, S2–S3. © INRA/DIB-AGIB/EDP Sciences https://doi.org/10.1051/apido:2004044.

Özcan, M., Arsla, D., Ali, D., 2006. Ceylan effect of inverted saccharose on some properties of honey. Food Chem. 99, 24–29.

Piana, M., Oddo, L., Bentabol, A., Bruneau, E., Bogdanov, S., 2004. Sensory analysis applied to honey: state of the art. Apidologie 35 (Suppl. 1), S26–S37. https://doi.org/10.1051/apido:2004048ff.ffhal-00891892f.

Ranneh, Y., Akim, A.M., Hamid, H.A., Khazaai, H., Fadel, A., Zakaria, Z.A., Albujja, M., Bakar, M.F.A., 2021. Honey and its nutritional and anti-inflammatory value. BMC Complement. Med. Ther. 21 (1), 30. https://doi.org/10.1186/s12906-020-03170-5. 33441127. PMC7807510.

Rossi, M., Marrazzo, P., 2021. The potential of honeybee products for biomaterial applications. Biomimetics 6, 6. https://doi.org/10.3390/biomimetics6010006.

Sabatini, A.G., 2007. Il miele: origine, composizione e proprieta. In: Sabatini, A.G., Botolotti, L., Marcazzan, G.L. (Eds.), Conscere Il Miele. Avenue Media, Bologna-Milano, pp. 3–37.

Samarghandian, S., et al., 2017. Honey and health: a review of recent clinical research. Pharmacognosy Res. 9 (2), 121–127. https://doi.org/10.4103/0974-8490.204647.

Schieber, M., Chandel, N.S., 2014. ROS function in redox signaling and oxidative stress. Curr. Biol. 24. https://doi.org/10.1016/j.cub.2014.03.034.

Schramm, D.D., Karim, M., Schrader, H.R., Holt, R.R., Cardetti, M., Keen, C.L., 2003. Honey with high levels of antioxidants can provide protection to healthy human subjects. J. Agric. Food Chem. 12, 1732–1735.

Sforcin, J.M., 2007. Propolis and the immune system: a review. J. Ethnopharmacol. 113 (1), 1–14.

Silva, B., Carina Biluca, F.C., Gonzaga, L.V., Fett, R., Monguilhott Dalmarco, E., Caon, T., Oliveira Costa, A.C., 2021. In vitro anti-inflammatory properties of honey flavonoids: a review. Food Res. Int. 0963-9969. 141. https://doi.org/10.1016/j.foodres.2020.110086, 110086.

Soares, S., Amaral, J.S., Oliveira, M.B.P., Mafra, I., 2017. A comprehensive review on the main honey authentication issues: production and origin. Compr. Rev. Food Sci. Food Saf. 16, 1072–1100.

Steinberg, D., Kaine, G., Gedalia, I., 1996. Antibacterial effect of propolis and honey on oral bacteria. Am. J. Dent. 9 (6), 236–239.

Toorani, M.Q., 2019. The therapeutic role of honey for treating acute cough in the pediatric population. A systematic review. J. Pediatr. Neonat. Individual Med. 8 (2). https://doi.org/10.7363/080205, e080205.

Tosi, S., Costa, C., Vesco, U., Quaglia, G., Guido, G., 2018. A 3-year survey of Italian honey bee-collected pollen reveals widespread contamination by agricultural pesticides. Sci. Total Environ. 615, 208–218. https://doi.org/10.1016/j.scitotenv.2017.09.226. 28968582.

Ulbricht, C., Conquer, J., Giese, N., Khalsa, K.P.S., Sklar, J., Weissner, W., Woods, J., 2009. An evidence-based systematic review of bee pollen by the natural standard research collaboration. J. Diet. Suppl. 6 (3), 290–312.

Vardi, A., Barzilay, Z., Linder, N., Cohen, H.A., Paret, G., Barzilai, A., 1998. Local application of honey for treatment of neonatal postoperative wound infection. Acta Paediatr. 87, 429–432.

WHO, 2015. Guideline: Sugars Intake for Adults and Children. World Health Organization, Geneva. Available from: https://www.who.int/publications/i/item/9789241549028. (Accessed 6 June 2021).

Wiedenfeld, H., 2011. Plants containing pyrrolizidine alkaloids – toxicity and problems. Food Addit. Contam. Part A Chem. Anal. Control Expo. Risk Assess. 28, 282–292. https://doi.org/10.1080/19440049.2010.541288.

Yaghoobi, N., Al-Waili, N., Ghayour-Mobarhan, M., et al., 2008. Natural honey and cardiovascular risk factors; effects on blood glucose, cholesterol, triacylglycerole, CRP, and body weight compared with sucrose. Sci. World J. 20, 463–469.

Zhang, S., Wu, X., Bian, S., Zhang, Q., Liu, L., Meng, G., Niu, K., 2021. Association between consumption frequency of honey and non-alcoholic fatty liver disease: results from a cross-sectional analysis based on the Tianjin chronic low-grade systemic inflammation and health (TCLSIH) cohort study. Br. J. Nutr. 125 (6), 712–720. https://doi.org/10.1017/S0007114520003190.

Zulhendri, F., Felitti, R., Fearnley, J., Ravalia, M., 2021. The use of propolis in dentistry, oral health, and medicine: a review. J. Oral Biosci. 63 (1), 23–34.

CHAPTER 17

Future prospects of propolis, bee pollen, royal jelly, and bee venom

Dilek Boyacioglu[a,b], Asli Elif Tanugur Samanci[b], and Taylan Samanci[b]
[a]Freelance Consultant for the Food Industry, Lee's Summit, MO, United States
[b]Scientific Bio Solutions, Inc., Istanbul, Turkey

Contents

Introduction

Bee products have been used by humankind for their healthy attributes for many ages and applied as a traditional medicine in many countries. Nowadays, they are gaining more attention from different industry sectors, including pharmaceutical, dietary supplements, food, and cosmetics, based on technological and scientific developments due to the increasing demands of consumers for natural products. A vast amount of research demonstrated their valuable nutrients and bioactive compounds for many years and significantly contributed to the current knowledge, explaining the factors leading to variations in their contents and the significance of postharvest applications, like processing and storage. Many scientific articles illustrated the pharmaceutical value of bee products about their bioactive components for various biological properties. The data generated for years address the great potential of these traditional products for developing natural and effective therapeutic agents, dietary supplements, skincare products, and functional food products. The current research focuses on linking their

Bee Products and Their Applications in the Food and Pharmaceutical Industries
https://doi.org/10.1016/B978-0-323-85400-9.00015-0

biological properties to specific diseases with advanced techniques used in drug discovery research. Nevertheless, due to the variations on their bioactive components, it is difficult to standardize these products, develop specifications for global regulatory standards, and claim health benefits legally. Such differences present a challenge for the producers to comply with regulatory quality standards and define nutritional value for the human diet in developing food supplements.

This chapter discusses the obstacles relevant to their applications in food, dietary supplements, medicine, and cosmetics by focusing on the variation of their contents, needs for standardization of methodologies, and health claim approach at the global level. Current approaches for using propolis, pollen, royal jelly, and bee venom to develop therapeutic applications and many factors affecting their future like safe doses, toxicity, bioavailability of nutrients, allergenicity, and health claim regulations are discussed together with future trends. Especially, the food and dietary supplement industry applications were emphasized by considering further research needs in that area. Outlook discussions were based on the trending research using computational tools for drug discovery to develop therapeutic agents, advanced analytical technology in enhancing the knowledge about their components and developing international standards, as well as the factors such as consumer demands, climate change, international trade, harmonization of standards, and safety.

Challenges in various applications of bee products

The applications of bee products as natural ingredients for new food products and cosmetics, pharmaceutics on human health, and dietary supplements represent many challenges to possess the claimed therapeutic attributes for the producers. Those difficulties differ depending on each bee product's character; however, the significant variation in their bioactive components presents a major challenge in defining the approaches for standardization of products, developing specifications for regulatory standards, and making claims for their health benefits legally on product labels. The composition and content of bioactive components highly depend on botanic and geographic origin, apicultural practices, seasons, and the application of various postharvest technologies. Therefore, it is challenging to define specifications for bee products such as propolis, bee pollen, bee bread, bee venom, and royal jelly. On the other hand, the producers intend to make health claims to explain the therapeutic value of their products to the consumers, while the regulation of such health claims widely differs in each country and might be very restrictive for bee products.

This following section focuses on the most common difficulties by discussing the variations in their bioactive compounds, determining their therapeutic values, safe doses, the bioavailability of nutrients, and regulations for health claims to improve the utilization of propolis, bee pollen, royal jelly, and bee venom, more efficiently and help consumers get health benefits from those natural products.

Variation in bioactive components of bee products

Propolis

Early research showed that propolis has a complex chemical composition with hundreds of substances composed of waxes, resins, balsams, aromatic and essential oils, pollen, and other organic matters (Marcucci, 1995). Bankova (2009) addressed future research to characterize these components in experiments evaluating the biological activity of propolis. Based on scientific data generated in the last two decades, it is clear that the most abundant constituents of propolis are flavonoids, phenylpropanoids, terpenes, stilbenes, lignans, coumarins, and their prenylated derivatives (Alvarenga et al., 2021). There is less research on propolis volatiles as essential oils constituted 10% of the samples against 50% of the resinous substances (Miguel and Antunes, 2011).

The biological activities of propolis are primarily attributed to their significant chemical constituents, including phenolic acids, phenolic acid esters, flavonoids, and terpenoids, such as caffeic acid phenethyl ester (CAPE), artepillin C, caffeic acid, chrysin, galangin, quercetin, apigenin, kaempferol, pinobanksin 5-methyl ether, pinobanksin, pinocembrin, and pinobanksin 3-acetate (Huang et al., 2014). Many scientific articles illustrated the pharmaceutical value of propolis in relation to these polyphenolic structures for various biological properties (Alvarenga et al., 2021; Miguel and Antunes, 2011; Toreti et al., 2013; Zabaiou et al., 2017). However, research on the chemical composition of propolis from diverse geographical origins also demonstrated a great variation in the amount and types of these bioactive compounds (Bankova, 2009; Bankova et al., 2014; Huang et al., 2014; Zabaiou et al., 2017).

Today, it is widely accepted that the chemical composition of propolis depends on its type, local botanic flora, climate zone, and regions (Bankova, 2009, 2005; Bankova et al., 2014), and bee species (Huang et al., 2014), determining its biological activity and hence the expected health benefits. For instance, propolis from temperate regions (Europe, North America, and nontropical regions of Asia), which are called poplar-type propolis, contains mainly free aromatic (phenolic) acids and their esters, flavonoids including flavones, flavanones, flavonols, and dihydroflavonols, chalcones, and dihydrochalcones, terpenoids, and other compounds, representing more than a total of 300 compounds (Ristivojević et al., 2015). Nevertheless, although it is called poplar-type, the phenolic and polyphenolic structures may also vary depending on the harvesting regions of the same country, including the volatile oils (Bankova et al., 2014). Ristivojević et al. (2018) reported that the phenolic profiles of Turkish propolis samples from different botanical origins contained mainly galangin, pinocembrin, chrysin, kaempferol, quercetin, caffeic acid, CAPE, luteolin, and apigenin. However, Ozdal et al. (2019) identified *p*-coumaric acid, ferulic acid, pinobanksin, *t*-cinnamic acid, and pinostrobin in addition to these structures by LC–MS/MS analysis and reported differences in the phenolic profile of Anatolian propolis depending on the collection region in Turkey. Similar to these findings, Ahn et al. (2007) described differences in antioxidant activity of Chinese propolis, a poplar type, collected from various regions of their country. The researchers found large amounts of caffeic acid, ferulic acid, and caffeic acid phenethyl ester (CAFE) in all samples, except for ones harvested from two regions. Park et al. (2002) classified Brazilian propolis into 12 groups depending on their physicochemical characteristics; then a 13th type was added (Freires et al., 2016). Brazilian green and red propolis contain prealkylated derivatives of *p*-coumaric acid and acetophenone, diterpenes, lignans, and flavonoids (Bankova, 2009), while Artepillin C is a characteristic component of Brazilian green propolis (Berretta et al., 2017).

Obviously, due to the natural variation in its major components and relevant bioactivity, the definition of medical or pharmaceutical quality is quite challenging for using propolis as an adjuvant agent. Therefore, the application of propolis in the pharmaceutical and food industries either as an ingredient or a dietary supplement requires more profound knowledge about its chemical composition that directly affects its biological properties.

Furthermore, the extraction method of propolis after being harvested from the beehives determines the amount and type of these phenolic and flavonoid structures and consequently the biological activity. Most propolis phytochemicals are lipophilic, making them readily dissolved in ethanol and methanol solvents. Although water-soluble extracts are considered safe biologically for the cells, they have lesser amounts and kind of these significant bioactive compounds (Kubiliene et al., 2018), rendering a lower antioxidant capacity (Cottica et al., 2011). The differences between ethanolic and water-soluble extracts yield different phytochemicals varying in their biological activity (Dantas Silva et al., 2017). Kubiliene et al. (2018)

recommended the addition of cosolvent polyethylene glycol to improve propolis compounds' solubility and assist in penetration of active substances into cells. Although the ethanolic extract had a 10 times larger content of total phenolic compounds, the use of polyethylene glycol-aqueous extract might help design pharmaceutical products if the ethanolic extract is not an option (Kubiliene et al., 2018). In addition, the pharmacological potential of oil extracted propolis has been investigated for controlling anxiety and the central nervous system (Reis et al., 2014), antitumoral activity (Carvalho et al., 2011) in animal studies, and antifungal activity against *Candida albicans* under in vitro conditions (Finger et al., 2013). Ramanauskienė and Inkėnienė (2011) recommended a mixture of olive oil/96% ethanol extractant to design propolis products for dermatological pharmaceuticals and cosmetic preparations considering its antimicrobial properties. Ethanol extraction is evidently the most popular technique for propolis extraction since it offers a great advantage of removing wax and, most significantly, obtaining very high amounts of bioactive compounds. However, strong residual flavor and the limitations of its application in cosmetics and pharmaceutical products may require the development of nonethanolic extraction techniques (Kubiliene et al., 2015).

Pollen and bee bread

Bee pollen is a valuable food supplement with great potential for medical and nutritional applications due to many claimed activities such as antifungal, antimicrobial, antiviral, antiinflammatory, hepatoprotective, anticancer immunostimulant, and local analgesic (Komosinska-Vassev et al., 2015) due to its content of proteins, amino acids, carbohydrates, lipids, and fatty acids, phenolic compounds, enzymes, and coenzymes as well as vitamins and bio elements (Campos et al., 2008, 2010; De-Melo and de Almeida-Muradian, 2017; Khalifa et al., 2021). Bee bread has a similar composition of pollen as they have a similar origin to plant pollen. However, the composition of bee pollen and, hence, bee bread varies according to botanical and geographical origins (Liolios et al., 2019; Martins et al., 2011; Morgano et al., 2012) as well as many other factors such as plant species in the surrounding area, season (Al-Kahtani et al., 2021; Simanonok et al., 2020), processing methods (Dias et al., 2016; Kanar and Mazı, 2019), and storage conditions (Anjos et al., 2019b; de Arruda et al., 2013b; Estevinho et al., 2019; Fernández et al., 2020). Undoubtedly, such observed variations present a challenge for the producers to comply with regulatory quality standards and define nutritional value for the human diet in developing food supplements.

Royal Jelly

Royal jelly, secreted from the glands in the hypopharynx of nurse bees, is explicitly fed to larvae intended to become queens. It has been used in traditional human medicine for ages and is commercially available as a medication, healthy food, and cosmetic in many countries, especially in China, Japan, and Egypt (Cornara et al., 2017; Xue et al., 2017; Zheng et al., 2011). In general, royal jelly comprises 60%–70% moisture, 7%–18% proteins, 9%–18% sugars, and 3%–8% lipids (Sabatini et al., 2009). Several bioactive substances identified by many researchers include 10-hydroxy-2-decenoic acid (10-HDA), phenolics, methylglyoxal, royal jelly proteins (MRJPs), oligosaccharides, and the antimicrobial jellies and royalisin peptides (Collazo et al., 2021; Ramadan and Al-Ghamdi, 2012). The pharmaceutical properties of royal jelly for humans have been extensively studied. Several claims were made regarding its beneficial effects on the immune system, vascular and glandular function, skin health, heart function, and cholesterol and lipid metabolism due to its claimed antimicrobial, antioxidant, antiaging, immunomodulatory properties (Ahmad et al., 2020; Kanelis et al., 2015). However, the composition of the royal jelly, like all bee products, varies according to many factors such as season, regions, honeybee species, and beekeeping practices (Cornara et al., 2017; Kanelis et al., 2015; Zheng et al., 2011). These variations represent difficulties in setting the limits for specifications in a royal jelly standard harmonized at the international level and using its bioactive compounds in developing drugs. More information is needed on the efficacy, safety, doses, and quality of royal jelly.

Bee venom

Bee venom, secreted from a highly specific venom gland connected to the venom reservoir located in the abdominal cavity of honeybee workers, is rich in many bioactive compounds, such as peptides (melittin, apamin, adolapin, the mast-cell degranulating (MCD) peptide), enzymes (phospholipase A2), biologically active amines (histamine and epinephrine), and non-peptide components. Its bioactive compounds mainly vary according to the different honeybee races and strains, place of origin, years, and season of venom collection (Abd El-Wahed et al., 2019; Kokot et al., 2011).

Safe doses and bioavailability of nutrients

Propolis is generally recognized as a safe bee product derived from animals, of which its phenolic and polyphenolic content is obtained from plants by the bees. Based on reported studies by applying a safety margin for humans, the safe dose of propolis is 1.4 mg/kg body weight/day, or approximately 70 mg/day (Burdock, 1998). Animal and human studies indicate that propolis and its constituents are generally well-tolerated and nontoxic except for administrating extreme quantities (Braakhuis, 2019). However, propolis may still exert adverse reactions as it has been identified clinically as an allergen (Burdock, 1998). Despite propolis being commonly used in cosmetic and medicinal preparations due to its antiseptic, antiinflammatory, and antioxidant properties, there is a concern over hypersensitive skin reactions. The allergy to propolis is more often observed after the external administration into the skin or mucous membranes than after oral administration, occurring as contact dermatitis as observed in beekeepers (Basista-Sołtys, 2013). However, as these products attract producers for use in many products like lip balms, cosmetics, lotions and ointments, shampoos, conditioners, and toothpaste, there is a concern that the incidence of these allergic reactions might be rising (Miguel and Antunes, 2011).

On the other hand, consumers' health consciousness and familiarity with natural remedies may negatively affect their risk perception (Royne et al., 2014), rendering ignorance of any potential side effects. It is crucial to declare that the product might be allergic on product labels due to the presence of bee products. In addition, a safe dose of propolis requires additional research and clinical trials with a larger group of subjects, proper dose–response, and different frequency of propolis administration before recommending it as a therapeutic agent (Farooqui and Farooqui, 2012). Furthermore, as the phenolic compounds of propolis may differ by factors like geographical regions, botanical sources, and type, the differences in the bioactivity will also make defining a correct dosage difficult (Braakhuis, 2019).

The polyphenolic compounds in a food matrix undergo a digestion process in the gut after ingestion. Propolis, rich in phenolics and polyphenols, is the subject of much research to examine how the digestion process affects the stability and function of these compounds in the gut system. The data on bioaccessibility of these bioactive compounds provide additional information for their beneficial effects depending on the chemical state of the compound, its release from the food matrix, possible interactions with other food components, and the presence of suppressors or cofactors (Ozdal et al., 2019). In vitro simulation of GI digestion methods shows digestion's effects on the metabolism and absorption of polyphenols. The test does not thoroughly create in vivo conditions in humans; however, an increase observed in the total antioxidant capacity of samples from the oral phase to the gastric phase is related to the increase in the extractability of phenolics and flavonoids by the application of enzymes and incubation during the intestinal phase (Ozdal et al., 2019; Yesiltas et al., 2014). Many polyphenols can be absorbed in the stomach in their free forms, but not when they form into complexes. On the other hand, the aglycone structure is largely absorbed on an intestinal level. The bioavailability of polyphenols is primarily controlled by intestinal absorption, splanchnic metabolism, and intestinal secretion (Stromsnes et al., 2021). However, there will also be differences in bioavailability and, consequently, the bioactivity of this product, considering the vast amount of variation in the composition of propolis, based on many factors as explained above.

Bee pollen is generally accepted as safe food, except for individuals showing allergic reactions to the bee products. Its recommended dose is about 3–5 teaspoons for adults and 1–2 teaspoons for children. Similarly, one tablespoon of bee bread is a daily recommended dose to regenerate and strengthen the human body (Komosinska-Vassev et al., 2015). Various methods are cited to increase its digestibility and bioavailability because the exine and intine of the pollen wall are highly resistant to chemical, physical, and biological degradation (Wu et al., 2020). Pollen must be thoroughly chewed because its nutrients are used by an organism only on a level of 10%–15% while in raw form (Kieliszek et al., 2018). In addition, grinding, soaking in warm water, or mixing with various foods is recommended to dissolve the outer layer and release pollen nutrients. After the mechanical grinding process, the bioavailability of this product's ingredients for organisms increases by 60%–80% (Komosinska-Vassev et al., 2015). Wu et al. (2019) successfully applied a combination of ultrasonication and high shear technique to disrupt bee pollen walls for releasing nutrients as it can be challenging to digest and absorb valuable nutrients due to the complex wall structure of bee pollen. The dynamic in vitro digestion of protein and crude fat of pollen exposed to this high shearing technology increased to more than 80%, and the amino acids and reducing sugars in all wall-disrupted samples were released almost 1.5 and 2 times faster than these of the unbroken pollen grains (Wu et al., 2020).

Despite a high level of nutrients, bee pollen, like other bee products, is susceptible to the environmental conditions of bee forage areas. Therefore, pollen might be exposed to contaminations with pesticides, heavy metals, and mycotoxin-producing fungi. Another safety concern is about certain plant species that may initially contain hepatotoxic pyrrolizidine alkaloids in high amounts (Végh et al., 2021).

There is no recommended dose for royal jelly; however, some clinical trials reported using various amounts of royal jelly. A metaanalysis evaluated six trials to determine the influence of royal jelly supplementation on blood lipid parameters at doses of 0.35–10 g for 30–180 days (Hadi et al., 2018). According to the reported pooled analysis of six trials, royal jelly reduces total cholesterol blood levels; however, triglycerides, low-density lipoprotein cholesterol, and high-density lipoprotein cholesterol blood concentrations remain the same. Seven selected randomized clinical trials in a metaanalysis study by Mahboobi et al. (2019) summarized the royal jelly supplementation levels between 0.5 and 6.0 g daily in these trials. The results of this evaluation claimed no beneficial effect of royal jelly on the markers of glycemia. However, the authors emphasized the differences in methodology and other variables such as diet and diverse populations and addressed a need for further well-controlled and well-designed research. On the other hand, Münstedt et al. (2009) administered 20 g of royal jelly on 20 volunteers to study its effect on the glucose metabolism of healthy humans and observed lower serum glucose levels after 2 h following royal jelly administration. Another clinical trial conducted with 50 females having type-2 diabetes were given 1 g of royal jelly for 8 weeks to study its influence on glucose metabolism (Pourmoradian et al., 2014).

Royal jelly may also cause allergic reactions, such as contact dermatitis, anaphylaxis, and asthma, due to the presence of major allergens MRJP-1 and MRJP-2 in addition to other adverse effects of eczemas and skin rashes, respiratory stress, bronchospasm, hemorrhagic colitis, and death (Ahmad et al., 2020). Nevertheless, royal jelly, like other bee products, may also be contaminated with pesticides, heavy metals (Matuszewska et al., 2021), bisphenols (Tu et al., 2019), and antibiotic residues resulting from apicultural treatments or agricultural practices, or environmental pollution (Bogdanov, 2006).

The allergic reactions against bee venom could be fatal as it produces a complex cascade of reactions in the human body. Its median lethal dose (LD50) ranges according to its main components, melittin, and phospholipase A2, based on the determination method for predicting the lethal activity of crude bee venom (Salem et al., 2015). Bee venom has been reported to have many therapeutic applications in traditional medicine for treating some diseases, such as arthritis, rheumatism, pain, cancerous tumors, and skin diseases in many countries due to its bioactive components, bee venom exhibits various pharmaceutical properties (Son et al., 2007). However, to obtain official approval as a safe drug, the chemical composition of honeybee venom needs to be consistent to achieve a pharmacological quality (Kokot et al., 2011).

Regulations on health claims for bee products

Increased awareness of the relationship between diet and human health has considerably influenced the food preferences of consumers. This trend has led to the growth of the functional food sector, offering more healthier or disease-preventing food products. The health claims suggesting the presence of such healthy ingredients, or their health benefits are tightly regulated worldwide (Díaz et al., 2020). In the EU, health claims are regulated by the Nutrition and Health Claim Regulation (NHCR), explaining the conditions to declare beneficial characteristics based on the nutritional content of foods by Regulation (EC) No 1924/2006 and EC Regulation No 1169/2011 Annex XIII. The use of a health claim is allowed or refused by the European Commission after consulting the expert opinion of the European Food Safety Authority (EFSA) on the submitted claim. The health claims are evaluated according to publicly available scientific evidence (Martínez and Siani, 2017) list of authorized and nonauthorized health claims is published on the website of the EU Register (European Commission, 2021). There are no authorized health claims on individual bee products like honey, propolis, and royal jelly in the EC countries. The health claim applications for these products were rejected by finding noncompliance with the regulation and insufficient evidence for a scientific assessment of the claimed health effects (European Commission, 2021).

In the US, claims are divided into three categories: health claims, nutrient content claims, and structure/function claims (FDA, 2018). The authorized health claims in food labeling are reviewed by FDA and allowed on food products or dietary supplements to declare that a food or food component may reduce the risk of a disease or help a health-related condition (FDA, 2018). These health claims either to be used on a label or in labeling for a conventional food or dietary supplement are categorized by three pathways: (1) According to the 1990 Nutrition Labeling and Education Act (NLEA), FDA issues regulations authorizing health claims for foods and dietary supplements after reviewing and evaluating the scientific evidence; (2) According to the 1997 Food and Drug Administration Modernization Act (FDAMA), health claims are allowed by an authoritative statement of the National Academy of Sciences (NAS) or a scientific body of the U.S. government with responsibility for public health protection or nutrition research; and (3) the third pathway is about the "Qualified Health Claims" if the agency finds that the evidence supporting the proposed claim is credible and the claim can be qualified to prevent it from misleading consumers (FDA, 2018). This type of claim is advantageous as it presents a procedure to ask FDA to review the scientific evidence when it does not meet the significant scientific agreement standard (SSAS); however, the FDA does not "approve" qualified health claim petitions and only issues a letter showing the specific claim language (FDA, 2018). The structure/function claims in the US have been generally used on the labels of conventional foods and food supplements as defined in the Dietary Supplement Health and Education Act of 1994 (Díaz et al., 2020; Kušar et al., 2021). There are two related types of dietary supplement labeling claims, claims of general well-being and claims related to a nutrient deficiency disease. These types of claims describe the role of a food constituent in body structures and functions. In contrast, general well-being claims describe general well-being arising from consuming a nutrient or other food constituents. Nutrient deficiency disease claims describe a benefit related to a nutrient deficiency disease. Still, such claims are allowed only if they also say how widespread such a disease is in the USA (FDA, 2018). However, FDA does not require conventional food manufacturers to notify FDA about their structure/function claims (e.g., "calcium builds strong bones" or, e.g., "fiber maintains bowel regularity"), and disclaimers are not required for claims on conventional foods (European Commission, 2021). Within this scope, there are currently no authorized claims, including the qualified health claims, for honey or other bee products that can meet the Significant Scientific Agreement (SSA) Standard in the US.

Although bee products have been used since ancient times due to their experienced health benefits, many countries do not allow a health claim for them. Current legislative approaches are, unquestionably, based on scientific evidence, aiming to protect consumers from false nutrition communication (Kušar et al., 2021). On the other hand, the health claims give the food industry a powerful tool to communicate to the consumers for informing about the potential health benefits of their products and push them toward investments for innovation and research to develop healthier formulations or products. In the future, a possible pathway for health claims in bee products would be facilitating the utilization of these products through research and development studies.

Food applications and dietary supplements with bee products

Consumers have become more interested in natural products and ingredients as they have become more health-conscious in the last decade. Bee products consumed either as a supplement in various forms or as food directly also attracted the food industry to utilize them as a natural ingredient in various food formulations and develop food supplements.

Propolis has an enormous application potential as a supplement (Berretta et al., 2017), a valuable ingredient for cosmetics (Barros et al., 2019) and pharmaceutical industries (Santos et al., 2020; Sforcin and Bankova, 2011), and for food producers as it may improve the product's shelf-life and quality due to its antioxidant and antimicrobial properties (Pobiega et al., 2019). Propolis has also been shown to exhibit many health benefits such as enhanced immunity, resistance to viral and bacterial infections, lowered blood pressure, wound healing and skin protection, hepatoprotective and cardioprotective activity, and support for the gut system (Braakhuis, 2019) with the help of its rich polyphenolic content. However, the difficulties of incorporating propolis into the products are primarily associated with its distinct sharp flavor and color that may alter the sensory properties, sticky character, and very low water solubility that can be an obstacle during food processing. Many researchers evaluated various food processing technologies without harming their biological activity to overcome those problems. (Pratami et al., 2019) microencapsulated propolis using a coating mixture of maltodextrin and gum arabic at a 10:1 ratio, resulting in a highly water-soluble product (about 92.48%) about 25% and 33% lower values in antioxidant capacity and total phenolic content, respectively. da Silva et al. (2013) spray dried propolis with gum arabic and octenyl succinic anhydride (OSA) starch as carriers to obtain powder, alcohol-free, water-dispersed, and shelf-stable form of propolis while retaining the antioxidant power. Similarly, Onbas et al. (2016) microencapsulated propolis alcohol extract by Complex Coacervation using alginate and gelatin. At 98.77% efficiency, the encapsulation of ethanolic extract of propolis showed significant cytotoxic effects on cancer cells and nitric oxide production on RAW 264.7 macrophage cells, proposing propolis as a promising natural supplement (Onbas et al., 2016).

These scientific approaches offer a promising future to produce foods with the addition of propolis, thereby providing an opportunity to produce more healthier foodstuff; for example, one food application of microencapsulated propolis was to produce fortified biscuits (Afif et al., 2021). Another study carried out by (Pastor et al., 2010) revealed the incorporation of propolis into an edible film formulation based on hydroxypropyl methylcellulose improved the water vapor permeability and resulted in a more rigid, less flexible and deformable, more opaque and colored films with lower gloss and transparency, in addition to antifungal activity against *Aspergillus niger*.

In the last decade, propolis became very popular due to its health benefits since many scientific studies elucidated the potential of propolis as a therapeutic agent against various diseases, including dental and buccal diseases. Juliano et al. (2007) proposed polymeric film formulations of propolis incorporating alginate, alginate-chitosan, and agar for local delivery into the oral cavity as an alternative therapeutic agent. During the darkness of the COVID-19 pandemic, propolis showed potential in treating patients for blocking coronavirus-induced fibrosis of lungs and stimulating the immune system (Berretta et al., 2020; Elmahallawy et al., 2021; Maruta and He, 2020; Scorza et al., 2020). Although propolis is a safe, natural remedy, adverse effects have also been reported, especially allergic reactions. Therefore, hypersensitivity reactions should be considered when used to prevent or treat COVID-19 (Kurek-Górecka et al., 2020).

Bee pollen is a highly nutritious food packed with proteins, sugars, lipids, fiber, minerals, amino acids, phenolic compounds, and vitamins (de Arruda et al., 2013a; Estevinho et al., 2012; Thakur and Nanda, 2020). Therefore, bee pollen grabbed most of the attention of the food industry not only because of its vital biological activity but also due to its functional properties improving the quality of food incorporated, and its easy integration into formulations for replacing other ingredients. Besides, food processes can also help overcome low microbiological quality (de Arruda et al., 2017; De-Melo et al., 2015, 2016; Hani et al., 2012) in case of insufficient drying to appropriate moisture contents. Yerlikaya (2014) supplemented a fermented milk beverage with bee pollen and observed enhanced probiotic viability and beverage viscosity without any change in sensory properties. Similarly, acidophilus milk and probiotic yogurt enriched with bee pollen (0.6% *w*/w) increased lactic acid production regardless of fat level (Glusac et al., 2015). The use of bee pollen in bakery products is investigated by Krystyjan et al. (2015) in fortified biscuits and by (Conte et al., 2020), who supplemented the gluten-free bread using bee pollen. Conte et al. (2020) demonstrated an improvement of technological properties, decreased rate of staling, and an increase in overall sensory acceptance of bread with the addition of bee pollen at 3%–5% in addition to higher levels of nutrients—protein, sugar, ash, fiber polyphenols, and antioxidant power—at 10% level; however, the taste like control samples could be achieved with 5% of pollen addition (Krystyjan et al., 2015). Some recent studies recommended bee pollen as a natural antioxidant substitute to inhibit the fat oxidation in the black pudding (Anjos et al., 2019a) and refrigerated pork sausages (Almeida et al., 2017), which is attributed to the increased antioxidant potential and higher levels of phenolic compounds. Karabagias et al. (2018) added bee pollen into yogurt and observed an increase in total phenolic content and antioxidant capacity in addition to improvements in taste, odor, appearance, and cohesion. Thakur and Nanda (2019) developed polyphenol-rich vacuum-dried milk powder using bee pollen, showing the potential of pollen addition into the formulation of functional foods.

These studies provide evidence for further exploration of bee pollen as a promising ingredient in other food products. The applications of bee pollen in food products highly depend on the uniformity in the amounts of bioactive compounds and how its incorporation affects the product's sensory properties, quality, and safety.

Outlook for bee products

The composition and content of bioactive components of bee products highly depend on many factors like botanical and geographical origin, apicultural practices, seasons, application of various postharvest technologies. Apparently, due to the natural variation in significant components and relevant bioactivity, producing food, supplement, cosmetics, medical or pharmaceutical quality from these products requires profound knowledge in the applications with additional information on efficacy, safety, dose, bioavailability nutrients, and quality. Moreover, the observed variations, undoubtedly, present a challenge for the producers to comply with regulatory quality standards and to define nutritional value for the human diet in developing healthy and natural products.

Recent research focuses on evaluating the potential of propolis's individual phenolic and flavonoid structures as a dietary supplement or as a pharmaceutical agent since it is an excellent source of these bioactive compounds (Gao et al., 2021). A 34-year data from 1981 to 2014 based on the approvals of FDA showed an increasing interest for molecules from natural products or their derivatizations (Newman and Cragg, 2016). Besides, the interest in developing new drugs and therapeutic alternatives is growing with drug delivery systems such as nanocarriers, polymeric carriers, fibers, or gels in combination with other materials, to overcome the low water solubility and bioavailability of propolis (Mendez-Pfeiffer et al., 2021). For example, folic acid conjugated chitosan nanoparticles synthesized by encapsulating propolis aimed to deliver propolis to the cell target and improve its bioavailability (Tan and Rahayu, 2021). Folic acid, as a ligand, directed nanoparticles to targeted cells with high folic acid receptors such as tumor cells. Tong and Kutty (2021) nanoencapsulated CAPE, as the target phenolic compound in propolis, with a biodegradable polymer, d-α-tocopheryl polyethylene glycol 1000 succinate (TPGS), also known as vitamin E TPGS-the water-soluble form of the vitamin, as a drug delivery system. Similarly, propolis entrapped in sodium caseinate–maltodextrin nanocomplex is suitable for food and pharmaceutical applications, according to Soleimanifard et al. (2021). In addition, a liposomal formulation for the efficient delivery of flavonoid structures of Egyptian propolis components was studied against COVID-19 3CL-protease and S1 spike protein to enhance delivery to the target cells and the improved cellular uptake of encapsulated propolis. Molecular docking has shown a high binding affinity of propolis components to COVID 3-CL protease, and rutin and CAPE had the highest affinity to both targets (Refaat et al., 2021). Guler et al. (2021) investigated the binding kinetics of flavonoid structures of ethanol extract of Anatolian propolis on ACE-II enzyme, a functional SARS-CoV receptor, in a molecular docking study. Based on the results of this study, rutin, myricetin, quercetin, CAPE, and hesperetin were reported to show a better affinity against the ACE-II enzyme and effectively inhibit the molecule. The mechanism of action of Egyptian propolis constituents on human prostate cancer (DU-145), breast adenocarcinoma (MCF-7), and colorectal adenocarcinoma (Caco-2) cell lines was investigated by (Ibrahim and El-Banna, 2021), using an integrated approach of network pharmacology-based analysis, molecular docking and in vitro cytotoxicity testing. The results suggested propolis with its phytoconstituents could be implemented for cancer prevention or treatment, although further extensive in vivo and clinical studies are required to confirm its anticancer potential (Ibrahim and El-Banna, 2021).

Research on propolis and its components using pharmacological tools, such as computational (in silico) methods as an important tool for drug discovery, has been increasing remarkably in recent years. Such methods, including databases, quantitative structure–activity relationships, pharmacophores, homology models and other molecular modeling approaches, machine learning, data mining, network analysis, and data analysis tools, allow understanding the interactions between a small molecule and a protein as well as characterizing the behavior of molecules in the binding site of target proteins, and explaining fundamental biochemical processes (Ekins et al., 2007). Such research enhancing the current understanding of the action mechanisms of polyphenolic molecules of propolis and evaluating suitable carrier systems to deliver these compounds to targeted cells constitutes important steps in developing human medicine. Unquestionably, further studies on tissue and species variation in ADME (Absorption, Distribution, Metabolism, Excretion) of individual bioactive compounds would improve our understanding of developing therapeutic agents using animal models and clinical studies (Gao et al., 2021). In addition, research on the interaction of nanoparticles with the immune system and human cells would provide a better assessment of the suitability of the drug delivery systems for propolis as an alternative for treatment (Mendez-Pfeiffer et al., 2021).

Nowadays, propolis is available worldwide as a supplement either in capsules, tablets, tinctures, sprays or as food with mixtures with honey or other natural ingredients or an ingredient of cosmetics. The current opinion is that standardized preparations of propolis are safe and less toxic than many synthetic medicines. Still, the components of propolis are variable, and it isn't easy to standardize and apply propolis for the formulations of dietary supplements. Robust manufacturing processes, standardized quality controls, and sound design clinical trials are critical in verifying these claims (Toreti et al., 2013). However, the recommendations for propolis-containing products may vary according to the producers or sellers, not necessarily provide information about its chemical composition, botanical source, geographic origin, and extraction methods (Sforcin, 2016). However, it is valuable information to include concentrations of polyphenols or flavonoids, antioxidant capacity, safe doses, and consumption period and frequency on product labels. Besides, Huang et al. (2014) recommends considering the bee varieties and subspecies together with geographical factors and plant species surrounding the beehives in future studies on propolis. It is evident that propolis obtained from different locations and plant sources should be characterized for its composition and related to its biological activity in clinical studies to develop a standard product. Promoting the use of bee products in the concept of "food as medicine" requires more studies to answer questions about the dose and type of propolis prescribed with controlled research to understand better the effects of biochemical pathways underlying the diverse pathogenic mechanisms involved in many diseases (Alvarenga et al., 2021).

Discussions around the need for propolis standardization also require an approach to the analytical tests defining its content for these bioactive components since significant variations have been observed in many propolis products currently marketed in various countries (Kosalec et al., 2005; Tagliacollo and Orsi, 2011). The advanced methods based on high-performance liquid chromatography (HPLC), high-performance thin-layer chromatography (HPTLC), and gas chromatography (GC), as well as structure elucidation techniques such as mass spectrometry (MS) and nuclear magnetic resonance (NMR), helped to identify many compounds, revealing significant differences in propolis samples originating from different regions (Ristivojević et al., 2015).

Another bee product, pollen, is an excellent source of proteins, ω-3 fatty acids, B-complex vitamins, minerals, polyphenols. In recent years the rising demands for healthier and natural foods motivated many researchers to characterize these products and the global market to accelerate their use in various commodities. However, significant variation in its composition as affected by the botanical and geographical origins presents a challenge for its promotion in the market. In addition, such variation in its biologically active components also affects its functional properties when incorporated into a food product. Therefore, investigating the monofloral pollen of varying geographical and plant sources allows the evaluation of their unique composition and safety in the diet (Thakur and Nanda, 2020). However, Campos et al. (2010) emphasize the difficulties in harvesting monofloral pollen and proposes more standardized bee pollen by mixing different pollen types with a constant composition for consistent biological activity and including antioxidant activity vitamin content in a bee pollen standard. The tests of biological properties of pollen may help to determine the pharmacological properties after active substances are identified. The standardized pollen products with optimal pharmacological properties can be further supported with clinical studies to evaluate its potential for human therapy (Campos et al., 2010). However, further applications of bee pollen in food products highly depend on the uniformity in the amounts of bioactive compounds and knowledge on its effects on sensory properties, quality, and safety.

Royal jelly, a traditional bee product, has been used in most Asian countries and Egypt for ages. It is mainly consumed as a functional food, dietary supplement, and medicinal formulation for its claimed health benefits. These include antilipidemic, antioxidant, antiproliferative, antimicrobial, neuroprotective, antiinflammatory, immunomodulatory, antiaging, and estrogenic activities. Future lines of research still require identifying any unknown bioactive compounds and their functions and more evidence for their claimed biological activities by implementing in vivo and clinical experiments and the use of cutting-edge genomics, transcriptomics, proteomics, and metabolomics (Ahmad et al., 2020).

Bee venom, also known as apitoxin, has been used to treat many diseases with its antiinflammatory, antibacterial, antimutagenic, radioprotective, antinociceptive immunity promoting, hepatocyte protective, and anticancer activity (Hossen et al., 2017). Bee venom contains many active substances (peptides, enzymes, and biogenic amines) showing different pharmacological attributes on the human body through different pathways. Research with experimental and clinical studies has shown that bee venom therapy can be used as an alternative medicine to treat various diseases such as arthritis, pain, and cancerous tumors (Son et al., 2007). It is apparent that the active components of bee venom would be increasingly investigated in drug discovery research against many diseases as a complementary therapy for protection against SARS-CoV-2 (Zakar et al., 2020), neurological diseases like Parkinson's disease (Awad et al., 2017), Alzheimer's disease and amyotrophic lateral sclerosis (Wehbe et al., 2019; Zhang et al., 2018), and autoimmune diseases (Hwang et al., 2015). However, further studies are needed to answer the concerns over injection route, dose, frequency of administration, delivery systems, possible side effects, and long-term effects. Therefore, it is necessary to conduct further research to understand the mechanisms related to their pharmacological activities, their allergenicity, and the safe amounts for obtaining the expected health benefits (Pasupuleti et al., 2017). Precautions need to be taken in case of allergens associated with bee products and finding the proper intake dosage.

The future of beekeeping and these products depends on many factors jeopardizing their sustainable supply (García, 2018). Those factors are excessive and indiscriminate use of agricultural pesticides (Fernando dos Santos et al., 2018), climate change (Flores et al., 2019; Vercelli et al., 2021) and, consequently increase in the spread of forest fires (de Faria et al., 2021), and unfortunately, the destruction of their natural environment.

Moreover, bee products are especially vulnerable to food frauds that cost approximately EUR 30 billion every year for the global food industry (European Commission, 2018). Honey, unfortunately, is one of the most frequently adulterated food products in the global trade, putting the future of beekeeping in jeopardy if economically motivated adulteration (EMA), customs fraud, and the violation of international and national trade laws continue (García, 2018). Hence, it is easy to predict that such scams would also affect the application potential of other bee products for the food, pharmaceutical, and cosmetics industries. It is essential to develop specifications and standards defining their characteristics, in addition to the methods. A joint effort of beekeepers, honest traders, scientists, private and public laboratories, and legal authorities (García, 2018) will be promoting a sustainable future for these products at the global level. As an example of such an approach, very recently, Campos et al. (2021) defined the standard methods for researching bee pollen. Their approach included a method for the production and storage of bee pollen to assure the quality of the product, routine methods for the identification of the pollen's floral sources, determination of the contents of water, proteins, carbohydrates, fatty acids, vitamins, alkaloids, phenolic and polyphenolic compounds, and finally, methods for the determination of antioxidant, antiinflammatory, antimicrobial and antimutagenic properties. Such international harmonization efforts for defining the specifications of bee pollen provide a powerful tool to the producers in developing many formulations entering into the global trade and gaining customers' trust.

The current trends concerning consumer behavior address a pathway for developing new products and technologies, particularly in the segments of functional and health-beneficial products (Kieliszek et al., 2018). In addition, the increased environmental awareness of the consumers and the fears for the effects of climate change on future generations are the driving forces for sustainable and environmentally friendly agricultural practices. Accordingly, the health-promoting and saving nature features of bee products comply with these trends and present natural ingredients to the producers to design products able to prevent humans against many diseases and develop all-natural cosmetics and pharmaceuticals. The global apiculture market evaluated at US$8.943 billion for the year 2019 is projected to grow at a CAGR of 3.02%, reaching a market size of US$ 11.015 billion by the year 2026, according to the executive summary of a market report (ResearchandMarkets, 2021). Despite many challenges, these figures also support increased interest in bee products for various food, dietary supplements, pharmaceuticals, and cosmetics applications. From the perspective of developing functional foods and dietary supplements, the allowance of health claims for bee products on their labels would be rendering a new direction for the market, opening a new era in these traditional products used for ages by humankind.

Global trade of bee products primarily is based on honey imports by major countries, like the USA, Germany, Japan, the UK, and other European countries. As the demand for natural and healthy products steadily rises due to the population increase and awareness, many countries cannot meet their honey demand with domestic production. To meet such high demands, the international market exporters may re-export those cheap imports, straight or in blends, as locally produced (García, 2018). However, during the pandemic of COVID-19, the food and medicine supply chains faced disruption through panic buying behavior of the consumers, sharp increases in demands of some products, and short labor, leading to difficulties in the logistics of products globally. Consequently, the manufacturing industries started preferring the local suppliers for raw material procurement rather than the usual contract suppliers and marketing by the local retailers and distributors, making the local supply chain a viable solution for marketing the products (Nagarajan et al., 2021). In addition, increasing consumers' preference toward sustainable agricultural practices emphasizes local food production's importance (Migliore, 2021). Such shifts to the local products may also impact the international trade of bee products and the development of value-added products in food, dietary supplements, and cosmetics.

Conclusion

It is evident that bee products have a great potential for developing natural and effective therapeutic agents, dietary supplements, skin care products, and functional food products. As a natural and nutritious source, the applications of bee products present many challenges due to the great variation in their bioactive components, leading to difficulties in standardizing the products, developing specifications for global regulatory standards, and making health claims legally. Obviously, due to the natural variation in their major components and relevant bioactivity, the definition of medical or pharmaceutical quality is also quite challenging to use bee products as an adjuvant agent. However, many of them are accepted as traditional medicine in some countries. In addition to these differences, many other factors also influence their bioactive compounds and, thereby, biological properties, such as the preparation methods and the bioavailability of nutrients for the cells.

The allergenicity of bee products causing hypersensitivities, skin reactions, and even death presents another challenge for bee products in developing healthy functional foods, dietary supplements, pharmaceuticals, and cosmetics. Although bee products are generally accepted as safe products as used by humankind for ages and recommended doses are available for some of them, there are concerns for the allergic reactions and safe doses of long-term use. Besides, bee products are susceptible to the environmental conditions of bee forage areas. Therefore, precautionary measures should be developed to control the risk of contamination with pesticides, heavy metals, antibiotics, and mycotoxin-producing fungi at every step of production, including beekeeping practices, harvesting, and postharvest processing and storage.

The limitations for declaring health benefits strictly regulated in many countries may also pose a barrier for the attention of the food industry to utilize and invest in these products more efficiently in the segment of dietary supplements or functional food products. Demands for natural products and ingredients increased in parallel to the increase in the health-consciousness level of consumers at the global level in the last decade; hence it is pretty reasonable to expect the same trend going on in future years. Bee products have tremendous potential as a supplement, food, or natural ingredient with reported health benefits. Propolis is a trending dietary supplement and a valuable ingredient for the cosmetics and pharmaceutical industries. Similarly, in addition to the interest from the cosmetics industry, bee pollen is a promising ingredient for food formulations due to its vital biological activity and functional properties, improving the quality of food incorporated, and its easy integration into formulations for replacing other ingredients.

Recent research on propolis and its components using pharmacological tools, such as computational (in silico) methods, as an important tool for drug discovery, is increasing remarkably in recent years in addition to investigating drug delivery systems to overcome the challenges of its low water solubility and bioavailability. Many advanced methods applied for drug discovery from natural molecules enhanced the current understanding of the action mechanisms of polyphenolic molecules of propolis, and the evaluation of suitable carrier systems to deliver these compounds to targeted cells represent very significant steps in developing a human medicine from propolis. Similarly, traditional bee products used to treat various diseases, royal jelly and bee venom, have many bioactive compounds showing interesting pharmacological properties on the human body. It is obvious that bee venom and propolis would be increasingly investigated in drug discovery research against many diseases as complementary therapy agents for treatment and protection against many diseases in coming years. However, there is still a need for further studies explaining diverse molecular mechanisms due to the complexity of their metabolites, considering possible synergistic actions on multitargets in the cells simultaneously. Besides, the interest of the cosmetics industry utilizing these bee products—propolis, bee pollen, royal jelly, and bee venom—will continue to rise in the future as long as the demands of consumers continue for all-natural, chemical-free products.

Furthermore, the future of beekeeping and its products is governed by many factors threatening their sustainable supply, like climate change, environmental conditions, food frauds, food safety issues, agricultural applications, beekeeping practices, and the strength of the local food chain in meeting demands. All the efforts and attention should be given to maintain a sustainable supply of these beneficial insects and support beekeeping practices worldwide, concerning the enormous potential of bee products on human health and nature.

References

Abd El-Wahed, A.A., Khalifa, S.A.M., Sheikh, B.Y., Farag, M.A., Saeed, A., Larik, F.A., Koca-Caliskan, U., AlAjmi, M.F., Hassan, M., Wahabi, H.A., Hegazy, M.-E.F., Algethami, A.F., Büttner, S., El-Seedi, H.R., 2019. Bee venom composition: from chemistry to biological activity. In: Studies in Natural Products Chemistry. Elsevier, pp. 450–484, https://doi.org/10.1016/B978-0-444-64181-6.00013-9.

Afif, M., Pratami, K., Sahlan, M., 2021. Scale-up production simulation of biscuit fortified with propolis as immunomodulator. AIP Conf. Proc. 2344. https://doi.org/10.1063/5.0047171. 060003.

Ahmad, S., Campos, M.G., Fratini, F., Altaye, S.Z., Li, J., 2020. New insights into the biological and pharmaceutical properties of royal jelly. Int. J. Mol. Sci. 21, 382. https://doi.org/10.3390/ijms21020382.

Ahn, M.R., Kumazawa, S., Usui, Y., Nakamura, J., Matsuka, M., Zhu, F., Nakayama, T., 2007. Antioxidant activity and constituents of propolis collected in various areas of China. Food Chem. 101, 1383–1392. https://doi.org/10.1016/j.foodchem.2006.03.045.

Al-Kahtani, S.N., Taha, E.-K.A., Farag, S.A., Taha, R.A., Abdou, E.A., Mahfouz, H.M., Durazzo, A., 2021. Harvest season significantly influences the fatty acid composition of bee pollen. Biology 10, 495. https://doi.org/10.3390/biology10060495.

Almeida, J.D.F., dos Reis, A.S., Heldt, L.F.S., Pereira, D., Bianchin, M., de Moura, C., Plata-Oviedo, M.V., Haminiuk, C.W.I., Ribeiro, I.S., da Luz, C.F.P., Carpes, S.T., 2017. Lyophilized bee pollen extract: a natural antioxidant source to prevent lipid oxidation in refrigerated sausages. LWT Food Sci. Technol. 76, 299–305. https://doi.org/10.1016/j.lwt.2016.06.017.

Alvarenga, L., Cardozo, L.F.M.F., Borges, N.A., Chermut, T.R., Ribeiro, M., Leite, M., Shiels, P.G., Stenvinkel, P., Mafra, D., 2021. To bee or not to bee? The bee extract propolis as a bioactive compound in the burden of lifestyle diseases. Nutrition 83. https://doi.org/10.1016/j.nut.2020.111094, 111094.

Anjos, O., Fernandes, R., Cardoso, S.M., Delgado, T., Farinha, N., Paula, V., Estevinho, L.M., Carpes, S.T., 2019a. Bee pollen as a natural antioxidant source to prevent lipid oxidation in black pudding. LWT Food Sci. Technol. 111, 869–875. https://doi.org/10.1016/j.lwt.2019.05.105.

Anjos, O., Paula, V., Delgado, T., Estevinho, L.M., 2019b. Influence of the storage conditions on the quality of bee pollen. Zemdirbyste-Agriculture 106, 87–94. https://doi.org/10.13080/z-a.2019.106.012.

Awad, K., Abushouk, A.I., Abdel Karim, A.H., Mohammed, M., Negida, A., Shalash, A.S., 2017. Bee venom for the treatment of Parkinson's disease: how far is it possible? Biomed. Pharmacother. 91, 295–302. https://doi.org/10.1016/j.biopha.2017.04.065.

Bankova, V., 2005. Chemical diversity of propolis and the problem of standardization. J. Ethnopharmacol. 100. https://doi.org/10.1016/j.jep.2005.05.004.

Bankova, V., 2009. Chemical diversity of propolis makes it a valuable source of new biologically active compounds. J. ApiProd. ApiMed. Sci. 1, 23–28. https://doi.org/10.3896/IBRA.4.01.2.01.

Bankova, V., Popova, M., Trusheva, B., 2014. Propolis volatile compounds: chemical diversity and biological activity: a review. Chem. Cent. J. 8, 1–8. https://doi.org/10.1186/1752-153X-8-28.

Barros, K.B.N.T., Neto, E.M.R., de França Fonteles, M.M., 2019. Propolis and its cosmetic applications: a technological prospection. J. Young Pharm. 11 (4), 350 (11, 350–352).

Basista-Sołtys, K., 2013. Allergy to propolis in beekeepers—a literature review. Occup. Med. Health Aff. 1, 1000105. https://doi.org/10.4172/2329-6879.1000105.

Berretta, A.A., Arruda, C., Miguel, F.G., Baptista, N., Nascimento, A.P., Marquele-Oliveira, F., Hori, J.I., Barud, H.D.S., Damaso, B., Ramos, C., Ferreira, R., Bastos, J.K., 2017. Functional properties of Brazilian propolis: from chemical composition until the market. In: Viduranga, W., Naofumi, S. (Eds.), Superfood and Functional Food—An Overview of their Processing and Utilization. InTechOpen, pp. 55–96, https://doi.org/10.5772/65932.

Berretta, A.A., Silveira, M.A.D., Cóndor Capcha, J.M., de Jong, D., 2020. Propolis and its potential against SARS-CoV-2 infection mechanisms and COVID-19 disease. Biomed. Pharmacother. 131. https://doi.org/10.1016/j.biopha.2020.110622, 110622.

Bogdanov, S., 2006. Contaminants of bee products. Apidologie. https://doi.org/10.1051/apido:2005043.

Braakhuis, A., 2019. Evidence on the health benefits of supplemental Propolis. Nutrients 11, 2705. https://doi.org/10.3390/nu11112705.

Burdock, G.A., 1998. Review of the biological properties and toxicity of bee propolis (propolis). Food Chem. Toxicol. 36, 347–363. https://doi.org/10.1016/S0278-6915(97)00145-2.

Campos, M.G.R., Bogdanov, S., de Almeida-Muradian, L.B., Szczesna, T., Mancebo, Y., Frigerio, C., Ferreira, F., 2008. Pollen composition and standardisation of analytical methods. J. Apic. Res. 47, 154–161. https://doi.org/10.1080/00218839.2008.11101443.

Campos, M.G.R., Frigerio, C., Lopes, J., Bogdanov, S., 2010. What is the future of bee-pollen? J. ApiProd. ApiMed. Sci. 2, 131–144. https://doi.org/10.3896/IBRA.4.02.4.01.

Campos, M.G., Anjos, O., Chica, M., Campoy, P., Nozkova, J., Almaraz-Abarca, N., Barreto, L.M.R.C., Nordi, J.C., Estevinho, L.M., Pascoal, A., Paula, V.B., Chopina, A., Dias, L.G., Tešić, Ž.L.J., Mosić, M.D., Kostić, A.Ž., Pešić, M.B., Milojković-Opsenica, D.M., Sickel, W., Ankenbrand, M.J., Grimmer, G., Steffan-Dewenter, I., Keller, A., Förster, F., Tananaki, C.H., Liolios, V., Kanelis, D., Rodopoulou, M.-A., Thrasyvoulou, A., Paulo, L., Kast, C., Lucchetti, M.A., Glauser, G., Lokutova, O., de Almeida-Muradian, L.B., Szczęsna, T., Carreck, N.L., 2021. Standard methods for pollen research. J. Apic. Res. 60, 1–109. https://doi.org/10.1080/00218839.2021.1948240.

Carvalho, A.A., Finger, D., MacHado, C.S., Schmidt, E.M., da Costa, P.M., Alves, A.P.N.N., Morais, T.M.F., de Queiroz, M.G.R., Quináia, S.P., da Rosa, M.R., dos Santos, J.M.T., Pessoa, C., de Moraes, M.O., Costa-Lotufo, L.V., Sawaya, A.C.H.F., Eberlin, M.N., Torres, Y.R., 2011. In vivo antitumoural activity and composition of an oil extract of Brazilian propolis. Food Chem. 126, 1239–1245. https://doi.org/10.1016/j.foodchem.2010.12.035.

Collazo, N., Carpena, M., Nuñez-Estevez, B., Otero, P., Simal-Gandara, J., Prieto, M.A., 2021. Health promoting properties of bee royal jelly: food of the Queens. Nutrients 13, 543. https://doi.org/10.3390/nu13020543.

Conte, P., del Caro, A., Urgeghe, P.P., Petretto, G.L., Montanari, L., Piga, A., Fadda, C., 2020. Nutritional and aroma improvement of gluten-free bread: is bee pollen effective? LWT 118. https://doi.org/10.1016/J.LWT.2019.108711, 108711.

Cornara, L., Biagi, M., Xiao, J., Burlando, B., 2017. Therapeutic properties of bioactive compounds from different honeybee products. Front. Pharmacol. 8, 412. https://doi.org/10.3389/fphar.2017.00412.

Cottica, S.M., Sawaya, A.C.H.F., Eberlin, M.N., Franco, S.L., Zeoulae, D.L.M., Visentainera, J.V., 2011. Antioxidant activity and composition of Propolis obtained by different methods of extraction. J. Braz. Chem. Soc. 22, 929–935.

da Silva, F.C., da Fonseca, C.R., de Alencar, S.M., Thomazini, M., Balieiro, J.C.D.C., Pittia, P., Favaro-Trindade, C.S., 2013. Assessment of production efficiency, physicochemical properties and storage stability of spray-dried propolis, a natural food additive, using gum Arabic and OSA starch-based carrier systems. Food Bioprod. Process. 91, 28–36. https://doi.org/10.1016/j.fbp.2012.08.006.

Dantas Silva, R., Machado, B.A.S., Barreto, G.D.A., Costa, S.S., Andrade, L.N., Amaral, R.G., Carvalho, A.A., Padilha, F.F., Barbosa, J.D.V., Umsza-Guez, M.A., 2017. Antioxidant, antimicrobial, antiparasitic, and cytotoxic properties of various Brazilian propolis extracts. PLoS One 12. https://doi.org/10.1371/journal.pone.0172585, e0172585.

de Arruda, V.A.S., Pereira, A.A.S., de Freitas, A.S., Barth, O.M., de Almeida-Muradian, L.B., 2013a. Dried bee pollen: B complex vitamins, physicochemical and botanical composition. J. Food Compos. Anal. 29, 100–105. https://doi.org/10.1016/j.jfca.2012.11.004.

de Arruda, V.A.S., Santos Pereira, A.A., Estevinho, L.M., de Almeida-Muradian, L.B., 2013b. Presence and stability of B complex vitamins in bee pollen using different storage conditions. Food Chem. Toxicol. 51, 143–148. https://doi.org/10.1016/j.fct.2012.09.019.

de Arruda, V.A.S., Vieria dos Santos, A., Figueiredo Sampaio, D., da Silva Araújo, E., de Castro Peixoto, A.L., Estevinho, M.L.F., Bicudo de Almeida-Muradian, L., 2017. Microbiological quality and physicochemical characterization of Brazilian bee pollen. J. Apic. Res. 56, 231–238. https://doi.org/10.1080/00218839.2017.1307715.

de Faria, B.L., Staal, A., Silva, C.A., Martin, P.A., Panday, P.K., Dantas, V.L., Silva, T., 2021. Climate change and deforestation increase the vulnerability of Amazonian forests to post-fire grass invasion. Glob. Ecol. Biogeogr. https://doi.org/10.1111/geb.13388.

De-Melo, A.A.M., de Almeida-Muradian, L.B., 2017. Chemical composition of bee pollen. In: Alvarez-Suarez, J.M. (Ed.), Bee Products-Chemical and Biological Properties. Springer International Publishing, Cham, pp. 221–259, https://doi.org/10.1007/978-3-319-59689-1_11.

De-Melo, A.A.M., Estevinho, M.L.M.F., Almeida-Muradian, L.B., 2015. A diagnosis of the microbiological quality of dehydrated bee-pollen produced in Brazil. Lett. Appl. Microbiol. 61, 477–483. https://doi.org/10.1111/lam.12480.

De-Melo, A.A.M., Estevinho, M.L.M.F., Sattler, J.A.G., Souza, B.R., Freitas, A.D.S., Barth, O.M., Almeida-Muradian, L.B., 2016. Effect of processing conditions on characteristics of dehydrated bee-pollen and correlation between quality parameters. LWT Food Sci. Technol. 65, 808–815. https://doi.org/10.1016/j.lwt.2015.09.014.

Dias, L.G., Tolentino, G., Pascoal, A., Estevinho, L.M., 2016. Effect of processing conditions on the bioactive compounds and biological properties of bee pollen. J. Apic. Res. 55, 357–365. https://doi.org/10.1080/00218839.2016.1248109.

Díaz, L.D., Fernández-Ruiz, V., Cámara, M., 2020. An international regulatory review of food health-related claims in functional food products labeling. J. Funct. Foods 68. https://doi.org/10.1016/j.jff.2020.103896, 103896.

Ekins, S., Mestres, J., Testa, B., 2007. In silico pharmacology for drug discovery: methods for virtual ligand screening and profiling. Br. J. Pharmacol. 152, 9–20. https://doi.org/10.1038/sj.bjp.0707305.

Elmahallawy, E.K., Mohamed, Y., Abdo, W., El-Gohary, F.A., Ahmed Awad Ali, S., Yanai, T., 2021. New insights into potential benefits of bioactive compounds of bee products on COVID-19: a review and assessment of recent research. Front. Mol. Biosci. 7, 513. https://doi.org/10.3389/fmolb.2020.618318.

Estevinho, L.M., Rodrigues, S., Pereira, A.P., Feá, X., 2012. Portuguese bee pollen: palynological study, nutritional and microbiological evaluation. Int. J. Food Sci. Technol. 47, 429–435. https://doi.org/10.1111/j.1365-2621.2011.02859.x.

Estevinho, L.M., Dias, T., Anjos, O., 2019. Influence of the storage conditions (frozen vs. dried) in health-related lipid indexes and antioxidants of bee pollen. Eur. J. Lipid Sci. Technol. 121, 1800393. https://doi.org/10.1002/ejlt.201800393.

European Commission, 2018. The EU Food Fraud Network and the System for Administrative Assistance-Food Fraud. Annual Report 2018. Available from: https://ec.europa.eu/food/sites/food/files/safety/docs/ff_ffn_annual-report_2018.pdf. (Accessed 10 September 2021).

European Commission, 2021. Nutrition and Health Claims. [WWW Document]. Available from: https://ec.europa.eu/food/safety/labelling_nutrition/claims/register/public/?event=search. (Accessed 5 September 2021).

Farooqui, T., Farooqui, A.A., 2012. Beneficial effects of propolis on human health and neurological diseases. Front. Biosci. (Elite Ed.) 4, 779–793.

FDA, 2018. Food Labelling & Nutrition. [WWW Document]. Available from: https://www.fda.gov/food/food-labeling-nutrition. 5 September 2021.

Fernández, L.A., Rodríguez, A., Sánchez, R.M., Pérez, M., Gallez, L.M., Fern Andez, L.A., Rodr Iguez, A., Romina, M.S., Anchez, A., 2020. Long-term microbiological and chemical changes in bee pollen for human consumption: influence of time and storage conditions. J. Apic. Res. 60, 319–325. https://doi.org/10.1080/00218839.2020.1728867.

Fernando dos Santos, C., Otesbelgue, A., Blochtein, B., 2018. The dilemma of agricultural pollination in Brazil: beekeeping growth and insecticide use. PLoS One 13. https://doi.org/10.1371/journal.pone.0200286, e0200286.

Finger, D., Machado, C.S., Torres, Y.R., Quináia, S.P., Thomaz, A.C.G., Gobbo, A.R., Monteiro, M.C., Ferreira, A.G., Sawaya, A.C.H.F., Eberlin, M.N., 2013. Antifungal bioassay-guided fractionation of an oil extract of Propolis. J. Food Qual. 36, 291–301. https://doi.org/10.1111/jfq.12039.

Flores, J.M., Gil-Lebrero, S., Gámiz, V., Rodríguez, M.I., Ortiz, M.A., Quiles, F.J., 2019. Effect of the climate change on honey bee colonies in a temperate Mediterranean zone assessed through remote hive weight monitoring system in conjunction with exhaustive colonies assessment. Sci. Total Environ. 653, 1111–1119. https://doi.org/10.1016/j.scitotenv.2018.11.004.

Freires, I.A., de Alencar, S.M., Rosalen, P.L., 2016. A pharmacological perspective on the use of Brazilian red Propolis and its isolated compounds against human diseases. Eur. J. Med. Chem. https://doi.org/10.1016/j.ejmech.2016.01.033.

Gao, S., Siddiqui, N., Etim, I., Du, T., Zhang, Y., Liang, D., 2021. Developing nutritional component chrysin as a therapeutic agent: bioavailability and pharmacokinetics consideration, and ADME mechanisms. Biomed. Pharmacother. 142. https://doi.org/10.1016/J.BIOPHA.2021.112080, 112080.

García, N.L., 2018. The current situation on the international honey market. Bee World 95, 89–94. https://doi.org/10.1080/0005772X.2018.1483814.

Glusac, J., Stijepic, M., Milanovic, S., Djurdjevic-Milosevic, D., 2015. Physicochemical properties of honeybee pollen enriched acidophilus milk and probiotic yoghurt. Acta Period. Technol. 46, 45–54. https://doi.org/10.2298/APT1546045G.

Guler, H.I., Tatar, G., Yildiz, O., Belduz, A.O., Kolayli, S., 2021. Investigation of potential inhibitor properties of ethanolic propolis extracts against ACE-II receptors for COVID-19 treatment by molecular docking study. Arch. Microbiol. 203, 3557–3564. https://doi.org/10.1007/s00203-021-02351-1.

Hadi, A., Najafgholizadeh, A., Aydenlu, E.S., Shafiei, Z., Pirivand, F., Golpour, S., Pourmasoumi, M., 2018. Royal jelly is an effective and relatively safe alternative approach to blood lipid modulation: a meta-analysis. J. Funct. Foods 41, 202–209. https://doi.org/10.1016/j.jff.2017.12.005.

Hani, B., Dalila, B., Saliha, D., Daoud, H., Mouloud, G., Seddik, K., 2012. Microbiological sanitary aspects of pollen. Adv. Environ. Biol. 6, 1415–1420.

Hossen, M.S., Shapla, U.M., Gan, S.H., Khalil, M.I., 2017. Impact of bee venom enzymes on diseases and immune responses. Molecules. https://doi.org/10.3390/molecules22010025.

Huang, S., Zhang, C.-P., Wang, K., Li, G., Hu, F.-L., 2014. Recent advances in the chemical composition of propolis. Molecules 19, 19610–19632. https://doi.org/10.3390/molecules191219610.

Hwang, D.-S., Kim, S.K., Bae, H., 2015. Therapeutic effects of bee venom on immunological and neurological diseases. Toxins 7, 2413–2421. https://doi.org/10.3390/toxins7072413.

Ibrahim, R.S., El-Banna, A.A., 2021. Network pharmacology-based analysis for unraveling potential cancer-related molecular targets of Egyptian propolis phyto constituents accompanied with molecular docking and in vitro studies. RSC Adv. 11, 11610–11626. https://doi.org/10.1039/d1ra01390d.

Juliano, C., Pala, C.L., Cossu, M., 2007. Preparation and characterisation of polymeric films containing propolis. J. Drug Deliv. Sci. Technol. 17, 177–181. https://doi.org/10.1016/S1773-2247(07)50033-X.

Kanar, Y., Mazı, B.G., 2019. Effect of different drying methods on antioxidant characteristics of bee-pollen. Journal of Food Measurement and Characterization 13, 3376–3386. https://doi.org/10.1007/s11694-019-00283-5.

Kanelis, D., Tananaki, C., Liolios, V., Dimou, M., Goras, G., Rodopoulou, M.A., Karazafiris, E., Thrasyvoulou, A., 2015. A suggestion for royal jelly specifications/Prijedlog globalnih standarda za matičnu mliječ. Arch. Ind. Hyg. Toxicol. 66, 275–284. https://doi.org/10.1515/aiht-2015-66-2651.

Karabagias, I.K., Karabagias, V.K., Gatzias, I., Riganakos, K.A., 2018. Bio-functional properties of bee pollen: the case of "Bee Pollen Yoghurt". Coatings 8, 483. https://doi.org/10.3390/coatings8120423.

Khalifa, S.A.M., Elashal, M.H., Yosri, N., Du, M., Musharraf, S.G., Nahar, L., Sarker, S.D., Guo, Z., Cao, W., Zou, X., Abd El-Wahed, A.A., Xiao, J., Omar, H.A., Hegazy, M.-E.F., El-Seedi, H.R., 2021. Bee pollen: current status and therapeutic potential. Nutrients 13, 1876. https://doi.org/10.3390/nu13061876.

Kieliszek, M., Piwowarek, K., Kot, A.M., Błażejak, S., Chlebowska-Śmigiel, A., Wolska, I., 2018. Pollen and bee bread as new health-oriented products: a review. Trends Food Sci. Technol. 71, 170–180. https://doi.org/10.1016/J.TIFS.2017.10.021.

Kokot, Z.J., Matysiak, J., Urbaniak, B., Dereziński, P., 2011. New CZE-DAD method for honeybee venom analysis and standardization of the product. Anal. Bioanal. Chem. 399, 2487–2494. https://doi.org/10.1007/s00216-010-4627-2.

Komosinska-Vassev, K., Olczyk, P., Kafmierczak, J., Mencner, L., Olczyk, K., 2015. Bee pollen: chemical composition and therapeutic application. Evid. Based Complement. Alternat. Med. https://doi.org/10.1155/2015/297425.

Kosalec, I., Pepeljnjak, S., Bakmaz, M., Vladimir-Knežević, S., 2005. Flavonoid analysis and antimicrobial activity of commercially available propolis products. Acta Pharma. 55, 423–430.

Krystyjan, M., Gumul, D., Ziobro, R., Korus, A., 2015. The fortification of biscuits with bee pollen and its effect on physicochemical and antioxidant properties in biscuits. LWT Food Sci. Technol. 63, 640–646. https://doi.org/10.1016/j.lwt.2015.03.075.

Kubiliene, L., Laugaliene, V., Pavilonis, A., Maruska, A., Majiene, D., Barcauskaite, K., Kubilius, R., Kasparaviciene, G., Savickas, A., 2015. Alternative preparation of propolis extracts: comparison of their composition and biological activities. BMC Complement. Altern. Med. 15, 1–7. https://doi.org/10.1186/s12906-015-0677-5.

Kubiliene, L., Jekabsone, A., Zilius, M., Trumbeckaite, S., Simanaviciute, D., Gerbutaviciene, R., Majiene, D., 2018. Comparison of aqueous, polyethylene glycol-aqueous and ethanolic propolis extracts: antioxidant and mitochondria modulating properties. BMC Complement. Altern. Med. 18, 1–10. https://doi.org/10.1186/s12906-018-2234-5.

Kurek-Górecka, A., Górecki, M., Rzepecka-Stojko, A., Balwierz, R., Stojko, J., 2020. Bee products in dermatology and skin care. Molecules 25, 556. https://doi.org/10.3390/molecules25030556.

Kušar, A., Žmitek, K., Lähteenmäki, L., Raats, M.M., Pravst, I., 2021. Comparison of requirements for using health claims on foods in the European Union, the USA, Canada, and Australia/New Zealand. Compr. Rev. Food Sci. Food Saf. 20, 1307–1332. https://doi.org/10.1111/1541-4337.12716.

Liolios, V., Tananaki, C., Papaioannou, A., Kanelis, D., Rodopoulou, M.-A., Argena, N., 2019. Mineral content in monofloral bee pollen: investigation of the effect of the botanical and geographical origin. J. Food Meas. Charact. 13, 1674–1682. https://doi.org/10.1007/s11694-019-00084-w.

Mahboobi, S., Jafarnejad, S., Eftekhari, M.H., 2019. Royal jelly does not improve markers of glycemia: a systematic review and meta-analysis of randomized clinical trials. Complement. Ther. Med. 44, 235–241. https://doi.org/10.1016/j.ctim.2019.04.017.

Marcucci, M., 1995. Propolis: chemical composition, biological properties and therapeutic activity. Apidologie 26 (2), 83–99 (26, 83–89).

Martínez, S.V., Siani, A., 2017. Health claims made on food in the EU: the edge between scientific knowledge and regulatory requirements. Trends Food Sci. Technol. 69, 315–323. https://doi.org/10.1016/J.TIFS.2017.01.005.

Martins, M., Morgano, M.A., Baggio, S., 2011. Physicochemical composition of bee pollen from eleven Brazilian states. J. Apicult. Sci. 107, 107–116.

Maruta, H., He, H., 2020. PAK1-blockers: potential therapeutics against COVID-19. Med. Drug Discov. 6. https://doi.org/10.1016/j.medidd.2020.100039, 100039.

Matuszewska, E., Klupczynska, A., Maciołek, K., Kokot, Z.J., Matysiak, J., 2021. Multielemental analysis of bee pollen, propolis, and royal jelly collected in West-Central Poland. Molecules 26, 2415. https://doi.org/10.3390/molecules26092415.

Mendez-Pfeiffer, P., Juarez, J., Hernandez, J., Taboada, P., Virués, C., Valencia, D., Velazquez, C., 2021. J. Drug Deliv. Sci. Technol. 65, 1773–2247. Available from: https://doi.org/10.1016/j.jddst.2021.102762.

Migliore, G., 2021. Sustainable food consumption practices: insights into consumers' experiences. Sustainability 13. https://doi.org/10.3390/su13115979.

Miguel, M., Antunes, M., 2011. Is propolis safe as an alternative medicine? J. Pharm. Bioallied Sci. 3, 479. https://doi.org/10.4103/0975-7406.90101.

Morgano, M.A., Martins, M.C.T., Rabonato, L.C., Milani, R.F., Yotsuyanagi, K., Rodriguez-Amaya, D.B., 2012. A comprehensive investigation of the mineral composition of brazilian bee pollen: geographic and seasonal variations and contribution to human diet. J. Braz. Chem. Soc. 23, 727–736. https://doi.org/10.1590/S0103-50532012000400019.

Münstedt, K., Bargello, M., Hauenschild, A., 2009. Royal Jelly Reduces the serum glucose levels in healthy subjects. J. Med. Food 12, 1170–1172. https://doi.org/10.1089/jmf.2008.0289.

Nagarajan, L., Mahalingam, S.K., Kandasamy, J., Gurusamy, S., 2021. Supply chain adoption during COVID-19 lockdown: global scenario. In: Sakthivel, A.R., Kandasamy, J.D.J.P. (Eds.), Managing Supply Chain Risk and Disruptions: Post COVID-19. Springer, Cham, pp. 57–64, https://doi.org/10.1007/978-3-030-72575-4_5.

Newman, D.J., Cragg, G.M., 2016. Natural products as sources of new drugs from 1981 to 2014. J. Nat. Prod. 79, 629–661. https://doi.org/10.1021/acs.jnatprod.5b01055.

Onbas, R., Kazan, A., Nalbantsoy, A., Yesil-Celiktas, O., 2016. Cytotoxic and nitric oxide inhibition activities of propolis extract along with microencapsulation by complex coacervation. Plant Foods Hum. Nutr. 71, 286–293. https://doi.org/10.1007/s11130-016-0558-1.

Ozdal, T., Ceylan, F.D., Eroglu, N., Kaplan, M., Olgun, E.O., Capanoglu, E., 2019. Investigation of antioxidant capacity, bioaccessibility and LC-MS/MS phenolic profile of Turkish propolis. Food Res. Int. 122, 528–536. https://doi.org/10.1016/j.foodres.2019.05.028.

Park, Y.K., Alencar, S.M., Aguiar, C.L., 2002. Botanical origin and chemical composition of Brazilian propolis. J. Agric. Food Chem. 50, 2502–2506. https://doi.org/10.1021/jf011432b.

Pastor, C., Sánchez-González, L., Cháfer, M., Chiralt, A., González-Martínez, C., 2010. Physical and antifungal properties of hydroxypropylmethylcellulose based films containing propolis as affected by moisture content. Carbohydr. Polym. 82, 1174–1183. https://doi.org/10.1016/j.carbpol.2010.06.051.

Pasupuleti, V.R., Sammugam, L., Ramesh, N., Gan, S.H., 2017. Honey, propolis, and royal jelly: a comprehensive review of their biological actions and health benefits. Oxidative Med. Cell. Longev. 2017. https://doi.org/10.1155/2017/1259510.

Pobiega, K., Kraśniewska, K., Gniewosz, M., 2019. Application of propolis in antimicrobial and antioxidative protection of food quality – a review. Trends Food Sci. Technol. https://doi.org/10.1016/j.tifs.2018.11.007.

Pourmoradian, S., Mahdavi, R., Mobasseri, M., Faramarzi, E., Mobasseri, M., 2014. Effects of royal jelly supplementation on glycemic control and oxidative stress factors in type 2 diabetic female: a randomized clinical trial. Chin. J. Integr. Med. 20, 347–352. https://doi.org/10.1007/s11655-014-1804-8.

Pratami, D.K., Mun'im, A., Yohda, M., Hermansyah, H., Gozan, M., Putri, Y.R.P., Sahlan, M., 2019. Total phenolic content and antioxidant activity of spray-dried microcapsules propolis from Tetragonula species. AIP Conf. Proc. 2085 (1), 020040/1–020040/10. https://doi.org/10.1063/1.5095018.

Ramadan, M.F., Al-Ghamdi, A., 2012. Bioactive compounds and health-promoting properties of royal jelly: a review. J. Funct. Foods 4, 39–52. https://doi.org/10.1016/j.jff.2011.12.007.

Ramanauskienė, K., Inkėnienė, A.M., 2011. Propolis oil extract: quality analysis and evaluation of its antimicrobial activity. Nat. Prod. Res. 25, 1463–1468. https://doi.org/10.1080/14786419.2010.529440.

Refaat, H., Mady, F.M., Sarhan, H.A., Rateb, H.S., Alaaeldin, E., 2021. Optimization and evaluation of propolis liposomes as a promising therapeutic approach for COVID-19. Int. J. Pharm. 592. https://doi.org/10.1016/j.ijpharm.2020.120028, 120028.

Reis, J.S.S., Oliveira, G.B., Monteiro, M.C., Machado, C.S., Torres, Y.R., Prediger, R.D., Maia, C.S.F., 2014. Antidepressant- and anxiolytic-like activities of an oil extract of propolis in rats. Phytomedicine 21, 1466–1472. https://doi.org/10.1016/j.phymed.2014.06.001.

ResearchandMarkets, 2021. Apiculture Market - Forecasts from 2021 to 2026. Available from: https://www.researchandmarkets.com/reports/5317989/apiculture-market-forecasts-from-2021-to-2026?w=5&utm_source=CI&utm_medium=PressRelease&utm_code=k4wmqn. (Accessed 22 September 2021).

Ristivojević, P., Trifković, J., Andrić, F., Milojković-Opsenica, D., 2015. Poplar-type propolis: chemical composition, botanical origin and biological activity. Nat. Prod. Commun. 10, 1869–1876.

Ristivojević, P., Dimkić, I., Guzelmeric, E., Trifković, J., Knežević, M., Berić, T., Yesilada, E., Milojković-Opsenica, D., Stanković, S., 2018. Profiling of Turkish propolis subtypes: comparative evaluation of their phytochemical compositions, antioxidant and antimicrobial activities. LWT 95, 367–379. https://doi.org/10.1016/j.lwt.2018.04.063.

Royne, M.B., Fox, A.K., Deitz, G.D., Gibson, T., 2014. The effects of health consciousness and familiarity with DTCA on perceptions of dietary supplements. J. Consum. Aff. 48 (3), 515–534. https://doi.org/10.1111/joca.12051.

Sabatini, A.G., Marcazzan, G.L., Fiorenza Caboni, M., Bogdanov, S., Bicudo De Almeida-Muradian, L., 2009. Quality and standardisation of Royal Jelly. J. ApiProd. ApiMed. Sci. 1, 1–6. https://doi.org/10.3896/IBRA.4.1.01.04.

Salem, A.Z., Wael, R.H., Abir, A.E., Basma, I.H., 2015. The antibacterial activity of bee venom on selected pathogenic bacteria. Egypt Acad. J. Biol. Sci. 7, 69–75.

Santos, L.M., Fonseca, M.S., Sokolonski, A.R., Deegan, K.R., Araújo, R.P., Umsza-Guez, M.A., Barbosa, J.D., Portela, R.D., Machado, B.A., 2020. Propolis: types, composition, biological activities, and veterinary product patent prospecting. J. Sci. Food Agric. 100, 1369–1382. https://doi.org/10.1002/jsfa.10024.

Scorza, C.A., Gonçalves, V.C., Scorza, F.A., Fiorini, A.C., de Almeida, A.C.G., Fonseca, M.C.M., Finsterer, J., 2020. Propolis and coronavirus disease 2019 (COVID-19): lessons from nature. Complement. Ther. Clin. Pract. https://doi.org/10.1016/j.ctcp.2020.101227.

Sforcin, J.M., 2016. Biological properties and therapeutic applications of Propolis. Phytother. Res. 30, 894–905. https://doi.org/10.1002/ptr.5605.

Sforcin, J.M., Bankova, V., 2011. Propolis: is there a potential for the development of new drugs? J. Ethnopharmacol. 133. https://doi.org/10.1016/j.jep.2010.10.032.

Simanonok, M.P., Otto, C.R.V., Smart, M.D., 2020. Do the quality and quantity of honey bee-collected pollen vary across an agricultural land-use gradient? Environ. Entomol. 49, 189–196. https://doi.org/10.1093/ee/nvz139.

Soleimanifard, M., Feizy, J., Maestrelli, F., 2021. Nanoencapsulation of propolis extract by sodium caseinate-maltodextrin complexes. Food Bioprod. Process. 128, 177–185. https://doi.org/10.1016/J.FBP.2021.05.005.

Son, D., Lee, J., Lee, Y., Song, H., Lee, C., Hong, J., 2007. Therapeutic application of anti-arthritis, pain-releasing, and anti-cancer effects of bee venom and its constituent compounds. Pharmacol. Ther. 115, 246–270. https://doi.org/10.1016/j.pharmthera.2007.04.004.

Stromsnes, K., Lagzdina, R., Olaso-Gonzalez, G., Gimeno-Mallench, L., Gambini, J., 2021. Biomedicines pharmacological properties of polyphenols: bioavailability, mechanisms of action, and biological effects in in vitro studies, animal models, and humans. Biomedicine 9, 1074. https://doi.org/10.3390/biomedicines9081074.

Tagliacollo, V.A., Orsi, R.D.O., 2011. Quality of propolis commercialized in the informal market. Cienc. Tecnol. Aliment. 31, 752–757. https://doi.org/10.1590/S0101-20612011000300031.

Tan, M.I., Rahayu, A.K., 2021. Synthesis of chitosan-folic acid nanoparticles as a drug delivery system for propolis compounds. In: Azura, A. (Ed.), Multifaceted Protocols in Biotechnology. vol. 2. Springer International Publishing, Cham, pp. 145–159, https://doi.org/10.1007/978-3-030-75579-9_10.

Thakur, M., Nanda, V., 2019. Process optimization of polyphenol-rich milk powder using bee pollen based on physicochemical and functional properties. J. Food Process Eng. 42. https://doi.org/10.1111/jfpe.13148, e13148.

Thakur, M., Nanda, V., 2020. Composition and functionality of bee pollen: a review. Trends Food Sci. Technol. 98, 82–106. https://doi.org/10.1016/j.tifs.2020.02.001.

Tong, K.Y., Kutty, R.V., 2021. Physical characterization of propolis encapsulated vitamin E TPGS as nanomedicine. Mater. Today Proc. https://doi.org/10.1016/j.matpr.2021.02.320.

Toreti, V.C., Sato, H.H., Pastore, G.M., Park, Y.K., 2013. Recent progress of propolis for its biological and chemical compositions and its botanical origin. Evid. Based Complement. Alternat. Med. 2013. https://doi.org/10.1155/2013/697390, 697390.

Tu, X., Wu, S., Liu, W., Gao, Z., Huang, S., Chen, W., 2019. Sugaring-out assisted liquid-liquid extraction combined with high-performance liquid chromatography-fluorescence detection for the determination of bisphenol a and bisphenol B in royal jelly. Food Anal. Methods 12, 705–711. https://doi.org/10.1007/s12161-018-1398-4.

Végh, R., Csóka, M., Sörös, C., Sipos, L., 2021. Food safety hazards of bee pollen – a review. Trends Food Sci. Technol. 114, 490–509. https://doi.org/10.1016/j.tifs.2021.06.016.

Vercelli, M., Novelli, S., Ferrazzi, P., Lentini, G., Ferracini, C., 2021. A qualitative analysis of beekeepers' perceptions and farm management adaptations to the impact of climate change on honey bees. Insects 12, 228. https://doi.org/10.3390/insects12030228.

Wehbe, R., Frangieh, J., Rima, M., El Obeid, D., Sabatier, J.-M., Fajloun, Z., 2019. Bee venom: overview of main compounds and bioactivities for therapeutic interests. Molecules 24, 2997. https://doi.org/10.3390/molecules24162997.

Wu, W., Wang, K., Qiao, J., Dong, J., Li, Z., Zhang, H., 2019. Improving nutrient release of wall-disrupted bee pollen with a combination of ultrasonication and high shear technique. J. Sci. Food Agric. 99, 564–575. https://doi.org/10.1002/jsfa.9216.

Wu, W., Qiao, J., Xiao, X., Kong, L., Dong, J., Zhang, H., 2020. In vitro and in vivo digestion comparison of bee pollen with or without wall-disruption. J. Sci. Food Agric. 101, 2744–2755. https://doi.org/10.1002/jsfa.10902.

Xue, X., Wu, L., Wang, K., 2017. Chemical composition of royal jelly. In: Alvarez-Suarez, J.M. (Ed.), Bee Products—Chemical and Biological Properties. Springer International Publishing, Cham, pp. 181–190, https://doi.org/10.1007/978-3-319-59689-1_8.

Yerlikaya, O., 2014. Effect of bee pollen supplement on antimicrobial, chemical, rheological, sensorial properties and probiotic viability of fermented milk beverages. Mljekarstvo 64, 268–279. https://doi.org/10.15567/mljekarstvo.2014.0406.

Yesiltas, B., Capanoglu, E., Firatligil-Durmus, E., Sunay, A.E., Samanci, T., Boyacioglu, D., Sunay, A.E., 2014. Investigating the in-vitro bioaccessibility of propolis and pollen using a simulated gastrointestinal digestion system. J. Apic. Res. 53, 101–108. https://doi.org/10.3896/IBRA.1.53.1.10.

Zabaiou, N., Fouache, A., Trousson, A., Baron, S., Zellagui, A., Lahouel, M., Lobaccaro, J.M.A., 2017. Biological properties of propolis extracts: something new from an ancient product. Chem. Phys. Lipids. https://doi.org/10.1016/j.chemphyslip.2017.04.005.

Zakar, T., Lavie, C.J., Kasozi, K.I., Niedbała, G., El-Saber Batiha, G., Welburn, S.C., Alqarni, M., Zirintunda, G., Ssempijja, F., Musinguzi, S.P., Usman, I.M., Matama, K., Hetta, H.F., Mbiydzenyuy, N.E., Beshbishy, A.M., 2020. Bee venom-a potential complementary medicine candidate for SARS-CoV-2 infections. Front. Public Health 8. https://doi.org/10.3389/fpubh.2020.594458, 594458.

Zhang, S., Liu, Y., Ye, Y., Wang, X.-R., Lin, L.-T., Xiao, L.-Y., Zhou, P., Shi, G.-X., Liu, C.-Z., 2018. Bee venom therapy: potential mechanisms and therapeutic applications. Toxicon 148, 64–73. https://doi.org/10.1016/j.toxicon.2018.04.012.

Zheng, H.-Q., Hu, F.-L., Dietemann, V., 2011. Changes in composition of royal jelly harvested at different times: consequences for quality standards. Apidologie 42, 39–47. https://doi.org/10.1051/apido/2010033.

Index

Note: Page numbers followed by *f* indicate figures and *t* indicate tables.

E

F

G

H

I

L

FSC
www.fsc.org
MIX
Papier | Fördert
gute Waldnutzung
FSC® C083411

Kolibri 360 GmbH
Ferdinand-Jühlke-Straße 7
99095 Erfurt, Deutschland
produktsicherheit@kolibri360.de